Inspired by

S. S. CHERN

A Memorial Volume in Honor of
A Great Mathematician

NANKAI TRACTS IN MATHEMATICS

Series Editors: Yiming Long and Weiping Zhang
Nankai Institute of Mathematics

Published

Nankai Tracts in Mathematics – Vol. 11

Inspired by
S. S. CHERN

A Memorial Volume in Honor of
A Great Mathematician

Editor

Phillip A. Griffiths

Institute for Advanced Study, Princeton, USA

W♭ **World Scientific**

NEW JERSEY · LONDON · SINGAPORE · BEIJING · SHANGHAI · HONG KONG · TAIPEI · CHENNAI

Published by

World Scientific Publishing Co. Pte. Ltd.

5 Toh Tuck Link, Singapore 596224

USA office: 27 Warren Street, Suite 401-402, Hackensack, NJ 07601

UK office: 57 Shelton Street, Covent Garden, London WC2H 9HE

British Library Cataloguing-in-Publication Data
A catalogue record for this book is available from the British Library.

Nankai Tracts in Mathematics — Vol. 11
INSPIRED BY S S CHERN
A Memorial Volume in Honor of A Great Mathematician

ISBN 13 978-981-270-061-2
ISBN 10 981-270-061-7
ISBN 13 978-981-270-062-9 (pbk)
ISBN 10 981-270-062-5 (pbk)

Printed by Mainland Press Pte Ltd

Preface

This is a volume of papers in honor and celebration of the life and contributions to mathematics of Shiing-Shen Chern. The works included cover a very wide spectrum of mathematical areas, all of which have some relation to geometry. As such they represent the breadth and lasting influence of Chern's mathematical career, both in his published works and his personal interactions with colleagues. The contributions that follow are from established leaders in the field as well as from younger mathematicians working in areas deeply influenced by Chern. The latter exemplify Chern's lifelong interest in and support of the emerging generation of mathematicians. Also included are a number of works by mathematicians in China, reflecting Chern's return to his homeland and his profound influence on the development of mathematics in that country.

One of the characteristics of Chern's mathematical life was that in addition to producing pioneering works that opened up whole new areas of mathematics, he revisited several classical topics in geometry that had attracted his attention as a young mathematician. Web geometry was the subject of Chern's thesis, written under Blaschke in Hamburg, Germany. He returned to the subject in the late 1970's, and there has been a resurgence of interest in the field among younger mathematicians. Three of the contributions to the present volume are in this area. The paper on Finsler geometry reflects similarly a field to which Chern returned in his later years.

The several papers on the relation between PDEs and geometry exemplify Chern's long interest in this subject. Through his own work and that of his students, especially S. T. Yau, this has become one of the major areas of contemporary mathematics. As noted in the accompanying essay, Chern had a lifelong interest in complex geometry, and this is also reflected in papers in this volume. Finally, although Chern did not work directly in the fields of algebraic geometry and arithmetic algebraic geometry, his research profoundly influenced these fields; the ubiquitous role played by Chern classes and secondary differential invariants, and the geometric consequences of differential constraints, are addressed in some of the contributions herein.

I would like to express my personal appreciation to all who have contributed to this volume in memory of S. S. Chern. Hongjun Li deserves special mention for her excellent assistance in the preparation of this collection.

Phillip Griffiths
September, 2006

In Appreciation of S. S. Chern

S. S. Chern is widely considered to be the most influential geometer of his era. His mathematical achievements, centering on geometry and spanning almost seven decades, has shaped whole areas of modern mathematics. More than any other scholar he has defined the subject of global differential geometry, a central pillar of contemporary mathematics. The depth and originality of his contributions are manifest in the many basic concepts of modern mathematics to which his name is attached: Chern classes, the Chern–Weil map, the Chern connection, the Bott-Chern forms, Chern–Moser invariants, and the Chern–Simons invariants. His selection in 2004 as the first recipient of the Shaw Prize in Mathematics recognized his singular contributions to and influence on the mathematical life of our time.

Chern's potential for mathematics became apparent when he was still a student. He completed his undergraduate work at Nankai University and in the 1930s was sent to Europe, where he studied with Wilhelm Blaschke in Hamburg and Elie Cartan in Paris. Under their influence he wrote, in effect, two theses: one on web geometry and one on the differential invariants of a third-order ordinary differential equation. Both works were published and remain of interest today.

Chern then returned his homeland to teach at Tsinghua University in Beijing, and during the war he — and the university — moved to Kunming, in southwestern China. He took the opportunity to leave Kunming, and by a circuitous route through Africa he managed to arrive in the United States. There, at the invitation of Oswald Veblen and André Weil, Chern began a highly productive stay at the Institute for Advanced Study in Princeton. During this period he gave the first intrinsic proof of the general Gauss–Bonnet formula. From this proof evolved many fundamental concepts in topology, such as transgression in sphere bundles, introduced from a differential geometric perspective. In addition, he began the monumental work that introduced Chern classes and, as a byproduct, initiated the subject of Hermitian differential geometry. This work clarified the relationship between differential geometry and topology, and it opened fertile new areas for other mathematicians, remaining of central importance to this day.

During a brief return to China after the war he completed his work on the Chern classes. Then he moved to the University of Chicago, where together with Weil and others he helped establish one of the nation's leading mathematics departments. It was during this period that his work arrived at the center of mathematical life. Through his growing productivity and influence on colleagues, Chern led the field of differential geometry and demonstrated how it interacted with essentially all aspects of geometry, including topology, algebraic geometry, integral geometry, geometric function theory, exterior differential systems, global analysis, and partial differential equations.

As a personal observation, I might offer a thumbnail sketch of the "Chern phenomenon" — the way in which he has brought his influence to bear on geometry time and again. His contributions typically begin when he engages a specific problem; then, through his geometric insight and computational mastery, he brings it to resolution; eventually, and often, his work then opens up productive new areas for other mathematicians to develop. This pattern, which continues to this day, is an extension of the classic tradition of Cartan, supplemented by a deep and far-reaching global perspective.

Two examples may serve to illustrate the continuing influence of Chern's mathematical work in the scientific community. A specific example of great current interest derives from the Chern–Simons invariants, which inform both theoretical physics and three-dimensional topology. A more general example is Chern's recognition of the special role of a complex structure in differential geometry. Examples of this role are evident throughout his work, including the introduction of the Chern classes of holomorphic vector bundles via curvature forms, the study of minimal surfaces and harmonic maps using the conformal structure, Chern's geometrization of complex function theory, and the geometry of CR-structures. The differential geometric properties of complex algebraic varieties are of central importance in modern theoretical physics and number theory because they reflect the pervasive role of the complex structures.

Toward the end of the 1950s Chern moved to the University of California at Berkeley, where he became a professor of mathematics and, in 1980, founding director of the Mathematical Sciences Research Institute (MSRI), one of the world's leading mathematics institutes. He continued his professional life in Berkeley until the late 1990s, when he returned to his undergraduate institution of the University of Nankai, where he had established a mathematics institute. There he remained mathematically active, among other things initiating a revitalization of the subject of Finsler geometry, until his passing in November 2004.

At Berkeley, as in all his positions of leadership, Chern's influence as a practitioner of mathematics was matched by his warmth and skill as a teacher and leader. He was always an attentive and generous mentor to younger people, and he liked nothing better than to gather people of all ages and perspectives for long talks about mathematics. He has always taken special pleasure in advancing the careers of his students and colleagues. He has always been quick to recognize important aspects of the work of his colleagues and to bring them to the attention of the wider mathematical community. Chern's former students now populate the mathematics departments of major universities across the US, and his influence has always been widely felt in China.

CONTENTS

Chern, 91-years-old, giving a talk to undergraduates in Nankai on 24th November 2002.

Chapter 1

IN MEMORY OF PROFESSOR S. S. CHERN

C. N. Yang

Tsinghua University, Beijing and
The Chinese University of Hong Kong, China

The year was 1930 and I was just turning eight. It was autumn when Mr Shiing-Shen Chern (陈省身) entered a Ph.D program in Mathematical Science at Tsinghua University. Incidentally, my father, Mr Yang Ko-Chuen (杨克纯), was a professor in the mathematics department and my family were living on the campus. I remember seeing Mr Chern on a few occasions. He often came by our home while I was in primary four. Many years later, Mr Chern and I carved out our separate careers in science. But as fate would have it, both of us finally came full circle and "reunited" once again in the academic world. Today, I would like to dedicate this piece to the memory of Mr Chern. Looking back, it suddenly struck me that all those years when we were running the marathon of our dreams, even though the routes we took were very different, we were in fact moving towards the same goal.

After graduation, Mr Chern went to Europe to further his studies. He returned home in 1937 and became a mathematics Professor at Southwest Associated University in Kunming. I was an undergraduate physics student in 1938–1942. Mr Chern was a young professor and was popular with the students. I attended his classes on differential geometry. He was a very serious teacher and his lessons were very well-structured and presented. I especially enjoyed his lessons on the geometry of curves and surfaces. But what I could not forget were the little things. Like the time I tried to prove the existence of conformal relations between any 2 dimensional geometry and the flat plane. Though I was able to convert metric tensor to $A^2 d\mu^2 + B^2 d\omega^2$, I was unable to prove that A and B could be made equal. After numerous attempts, I was getting really frustrated. It was Mr Chern who gave me the missing piece of the puzzle — complex variables. The problem was solved instantly. I will never forget that feeling of sudden illumination.

In 1943, Mr Chern was offered a research position at the Institute for Advanced Study in Princeton. During his two-year stint there, he successfully led research on differential geometry to a whole new level. And it swiftly became one of the major components of basic mathematics in the twentieth century.

Mr Chern became a professor at the University of Chicago in 1949. His next stop was Berkeley in the early 1960s. During the sixties and seventies, the two of us met up quite often in Chicago, Princeton and Berkeley. I was also a regular stay-in guest at his house in the city of El Cerrito which was just located north of Berkeley. Needless to say, I knew his family very well. He and his wife, Shih-ning Cheng (郑士宁), had a son Paul (陈伯龙) and a daughter May (陈璞). Mr Chern and I talked about practically everything under the sun — friends, relatives, family matters or politics, you name it. Even though I knew Mr Chern was widely regarded as one of the greatest mathematician of the twentieth century, the two of us hardly ever discussed our work.

Mathematics and physics used to share a close relationship aeons ago. But all of this has changed since the mid-nineteenth century. As mathematicians and physicists gradually move towards cutting-edge R&D initiatives, the gap between the two has widened and they have eventually become independent of each other. In "The Revolution in Mathematics" which was published in the *American Mathematical Monthly 68*, the former chairman of the mathematics department of the University of Chicago, Professor Marshall Stone (1903–1989) wrote:

> "... since 1900 in our conception of mathematics or in our view concerning it, the one which truly involves a revolution in ideas is the discovery that mathematics is entirely independent of the physical world."

The article was first published in 1961. During the same period, Mr Chern was already regarded as the father of modern differential geometry. He was the first mathematician who succeeded in bridging local geometry and topology. But as Professor Stone suggested, it was a time when everybody believed that modern mathematics had nothing to do with "the physical world".

But as it has turned out, what Professor Stone said could not be further from the truth. Exterior differential forms or fiber bundles is very much intertwined with gauge field theory of theoretical physics. It was during the early seventies that I began to see the intricate link between the two disciplines. I had related my experience of this understanding in an article in 1980:

"In 1975, impressed with the fact that gauge fields are connections on fiber bundles, I drove to the house of Shiing-Shen Chern in El Cerrito near Berkeley. (I had taken courses with him in the early 1940's when he was a young professor and I an undergraduate student at the National Southwest Associated University in Kunming, China. That was before fiber bundles had become important in differential geometry and before Chern had made history with his contributions to the generalized Gauss–Bonnet theorem and the Chern classes.) We had much to talk about: friends, relatives, China. When our conversation turned to fiber bundles, I told him that I had finally learned from Jim Simons the beauty of fiber-bundle theory and the profound Chern–Weil theorem. I said I found it amazing that gauge fields are exactly connections on fiber bundles, which the mathematicians developed *without reference to the physical world.* I added, "this is both thrilling and puzzling, since you mathematicians dreamed up these concepts out of nowhere." He immediately protested, "No, no. These concepts were not dreamed up. They were natural and real.""

Why would the creator of the universe use "natural and real" concepts of abstract mathematics to create the physical world? I suppose this question will never be answered. But I penned a poem <In praise of the Chern class> in the seventies:

天衣岂无缝　匠心剪接成

浑然归一体　广邃妙绝伦

造化爱几何　四力纤维能

千古寸心事　欧高黎嘉陈

As a matter of fact, the line "creator loves geometry" had already been stated by Einstein. After the 1970s this idea was gradually accepted widely by theoretical physicists.

Mr Chern once said that he would not be held responsible for what was said in my poem. It is true I did not consult him when I wrote the poem. But I do believe not only that there is a link between physics and geometry, that differential geometry holds the key to theoretical physics, but also that beyond the geometrical concepts that Mr Chern and I had helped to introduce into physics, there are more abstract, more intrinsic,

Fig. 1. Chinese artist Fan Zeng's impression of S. S. Chern and C. N. Yang.

but "natural and real" mathematical concepts, that are favored by the creator of the universe.

The renowned Chinese artist Fan Zeng is a good friend of both Mr Chern and me. He had presented a huge painting to Nankai University in the summer of 2004, with a poem (Figure 1):

纷繁造化赋玄黄

宇宙浑茫即大荒

递变时空皆有数

迁流物类总成场

天衣剪掇丛无缝

太极平衡律是纲

巨擘从来诗作魄

真情妙语铸文章

Mr Fan's poetry and painting reminded me of a passage in an article written by Mr Chern in 1987 about our long relationship:

"Fifty odd years, from Lianda to Nankai, the creator was kind to us."

Translated from *Mathematics and Mathematical People, A Festscrift in Honor of Mr. S. S. Chern*, ed. by Yau *et al.*

Translated by Tan Hwee Chiang

Chapter 2

TWISTED K-THEORY AND COHOMOLOGY

Michael Atiyah

School of Mathematics, University of Edinburgh,
James Clerk Maxwell Building, King's Buildings, Edinburgh EH9 3JZ, UK

Graeme Segal

All Souls College, Oxford OX1 4AL, UK

1. Introduction

In a previous paper [AS] we introduced the twisted K-theory of a space X, with the twisting corresponding to a 3-dimensional cohomology class of X. If the class is of finite order then the theory, particularly when tensored with the rationals $\mathbb{Q}$, is very similar to ordinary K-theory. For elements of infinite order, however, there are substantial differences. In particular the relations with cohomology theory are very different, and it is these that are the subject of this paper.

There are three different ways in which ordinary K-theory and cohomology are related:

(1) by the Atiyah–Hirzebruch spectral sequence,
(2) by the Chern character,
(3) by the Chern classes.

In the first and third of these integral cohomology is involved, while in the second only rational cohomology enters.

In the twisted case we shall examine the counterparts of all three of these relations, finding essential new features in all of them. We shall also briefly examine *operations* in twisted K-theory, the analogue of the well-known λ-ring operations in ordinary K-theory, because these have some bearing on the cohomology questions.

Twisted K-theory arises because K-theory admits natural automorphisms given by tensoring with line-bundles. We review this briefly in
Section 2, and then use the basic formulae in Sections 3 and 4 to compute the relevant differentials of the appropriate spectral sequences. The
first key result is the explicit identification of the differential d_3 of the
Atiyah–Hirzebruch spectral sequence already forecast in [AS] (cf. also [R]).
In Section 5 we enter new territory by showing that *the higher differentials $d_5, d_7, \ldots$ are non-zero even over the rationals or reals*, and given by
Massey products. (In the Appendix we give explicit examples, pointed out
to us by Elmer Rees, of spaces for which the higher differentials do not
vanish.) In Section 6 we introduce (over the reals) the appropriate twisted
cohomology. A key feature here is that the theory is not integer graded but
(like K-theory) is filtered (and only mod 2 graded). In terms of differential
forms such a theory has already appeared, at least implicitly, in the physics
literature. This twisted cohomology has its own "Atiyah–Hirzebruch spectral sequence". Comparing this with twisted K-theory we show in Section
7 the existence of a *Chern character* in the twisted context.

The last three sections are then devoted to Chern classes for twisted
K-theory. First we show that the ring of characteristic classes (in ordinary
real cohomology) is the subring of the algebra of Chern classes invariant
under tensoring by line-bundles (Proposition (8.8)). We then examine how
these might be lifted to twisted characteristic classes.

In Section 9 we discuss a very geometric set of characteristic classes
studied by Koschorke [K], and show how they fit into the twisted theory.

Finally in Section 10 we discuss internal operations in twisted K-theory.

These last sections identify a number of purely algebraic problems which
arise, and, though it would be nice to know the answers, we have not
felt that there was sufficient motivation or application as yet to justify
embarking on the possibly lengthy algebraic investigation.

In a differential-geometric context there is a very rich theory relating
curvature to differential forms representing the (real) Chern classes. These
are of course of special interest to both geometers and physicists, and they
can lead to more precise local formulae. It is natural to ask for a similar
theory for twisted K-theory, but we have given only a brief sketch of it in
Section 7. Fuller treatments of this subject can be found in [BCMMS] and
in unpublished work of Bunke. (Cf. also [TX].) There are in fact some
significant technical difficulties, at least from our approach. Although the
space of Fredholm operators is an elegant model for K-theory from the point
of view of topology, it is too large or too crude for differential geometry.

This point arises clearly in Connes's non-commutative geometry, where C^*-algebras are too crude to provide explicit cocycle formulae, and the theory has to be refined to incorporate some differential calculus. It seems to us that Connes's theory should provide the right framework to deal with curvature and Chern forms.

2. The Action of the Automorphism Group

In our previous paper [AS] we explained briefly how automorphisms of a cohomology theory lead to twisted versions of that theory. In particular, for K-theory, we studied the automorphisms given by tensoring with line bundles, and the twisting that this determines. In this section we shall examine the cohomological implications of these automorphisms.

For a compact space X we have the group $\text{Pic}(X)$ formed by (classes of) line bundles with the tensor product as multiplication. The first Chern class gives an isomorphism

$$c_1 : \text{Pic}(X) \longrightarrow H^2(X; \mathbb{Z}) \tag{2.1}$$

Moreover $\text{Pic}(X)$ is naturally a subgroup of the multiplicative group of the ring $K(X)$, and hence it acts by group homomorphisms on $K(X)$. This action is functorial in X, and so it can equally well be described in terms of classifying spaces.

The classifying space of K-theory can be taken to be $\text{Fred}(\mathcal{H})$, the space of Fredholm operators on Hilbert space $\mathcal{H}$, while the classifying space of Pic can be taken to be $PU(\mathcal{H})$, the projective unitary group of $\mathcal{H}$. The group structure of $PU(\mathcal{H})$ gives the group structure of $\text{Pic}(X)$.

Proposition 2.1. *The action of* $\text{Pic}(X)$ *on* $K(X)$ *is induced by the conjugation action of* $PU(\mathcal{H})$ *on* $\text{Fred}(\mathcal{H})$.

Proof. If L is a line bundle on X, and an element ξ of $K(X)$ is represented by a family $F = \{F_x\}$ of Fredholm operators in a fixed Hilbert space $\mathcal{H}$, then $L \otimes \xi \in K(X)$ is represented by the family of Fredholm operators $1 \otimes F = \{1 \otimes F_x\}$ in the Hilbert bundle $L \otimes \mathcal{H}$ on X. This bundle can be trivialized by an isomorphism $u : L \otimes \mathcal{H} \to X \times \mathcal{H}$, which exists and is unique up to homotopy because $U(\mathcal{H})$ is contractible. So the class $L \otimes \xi$ can be represented by the family $u \circ (1 \otimes F) \circ u^{-1}$ in the fixed space $\mathcal{H}$. This proves the proposition, for we can regard u as a map $u : X \to PU(\mathcal{H})$, and the line bundle on X pulled back from $PU(\mathcal{H})$ by u is precisely L. $\square$

We will now consider the effect on cohomology of the map

$$PU(\mathcal{H}) \times \mathrm{Fred}(\mathcal{H}) \longrightarrow \mathrm{Fred}(\mathcal{H}) \qquad (2.2)$$

or equivalently the effect on cohomology of the multiplication

$$\mathrm{Pic}(X) \times K(X) \longrightarrow K(X). \qquad (2.3)$$

In other words we need to describe the Chern classes of a vector bundle $L \otimes E$ in terms of the Chern classes of the vector bundle E and the line bundle L. These are given by universal polynomials, and since the classifying spaces in (2.2) have no torsion we lose nothing by working with rational cohomology.

As usual we define the Chern classes $c_n(E)$ of an N-dimensional bundle E in terms of variables x_j by

$$\sum_{n=0}^{N} c_n = \prod_{j=1}^{N} (1 + x_j),$$

and the power sums s_n by

$$s_n = \sum_{j=1}^{N} x_j^n.$$

In particular, $s_0(E) = \sum x_j^0$ is the *dimension* of E. As a function on $\mathrm{Fred}(\mathcal{H})$ it becomes the *index* map, labelling the connected components of $\mathrm{Fred}(\mathcal{H})$.

The Chern character is given by

$$\mathrm{ch}\, E = \sum_{n=0}^{\infty} \frac{s_n}{n!},$$

and it extends to a ring homomorphism

$$\mathrm{ch} : K^*(X) \longrightarrow H^*(X, \mathbb{Q})$$

which becomes an isomorphism after tensoring with $\mathbb{Q}$.

If L is a line-bundle with $c_1(L) = u$, and E is a vector bundle, then the Chern character of $L \otimes E$ is given by

$$\mathrm{ch}(L \otimes E) = e^u\, \mathrm{ch} E.$$

This implies the same formula for any element E in $K(X)$. If s_n and $s_n(u)$ denote the power sums for E and $L \otimes E$ then

$$\sum \frac{s_n(u)}{n!} = \left(1 + u + \frac{u^2}{2!} + \dots \right) \sum \frac{s_n}{n!} \qquad (2.4)$$

so that

$$s_n(u) = s_n + n s_{n-1} u + \binom{n}{2} s_{n-2} u^2 + \cdots + s_0 u^n. \qquad (2.5)$$

This formula can be interpreted as giving the homomorphism in cohomology induced by the map (2.2).

In the next section we shall deduce various consequences from the formula (2.5).

3. The Universal Fibration

The group action (2.2) of $PU(\mathcal{H})$ on $\mathrm{Fred}(\mathcal{H})$ gives rise to a "universal bundle" with fibre $\mathrm{Fred}(\mathcal{H})$ over the classifying space $BPU(\mathcal{H})$. This classifies $PU(\mathcal{H})$-bundles over a space X as homotopy classes of maps $X \to BPU(\mathcal{H})$. Note that a $PU(\mathcal{H})$-bundle determines a $P(\mathcal{H})$-bundle and conversely; in [AS] we worked primarily with $P(\mathcal{H})$-bundles.

Since $PU(\mathcal{H}) = U(\mathcal{H})/U(1)$, it is an Eilenberg–MacLane space $K(\mathbb{Z}, 2)$, and so its classifying space $BPU(\mathcal{H})$ is a $K(\mathbb{Z}, 3)$. Thus, as observed in [AS] Prop.2.1, isomorphism classes of $PU(\mathcal{H})$-bundles over X are classified by elements of $H^3(X, \mathbb{Z})$.

In the notation of [AS] our universal bundle over $BPU(\mathcal{H})$ with fibre $\mathrm{Fred}(\mathcal{H})$ is written as $\mathrm{Fred}(P)$, where P is the universal $P(\mathcal{H})$ — or $PU(\mathcal{H})$ — bundle over $BPU(\mathcal{H})$. We now want to study the rational cohomology of this fibration, using the formula (2.5). Note that because the index is preserved by the group action the connected components $\mathrm{Fred}_k(P)$ of $\mathrm{Fred}(P)$ also correspond to the index.

Now, over the rationals, $K(\mathbb{Z}, 3)$ is indistinguishable from a 3-sphere. To have an explicit map $S^3 \to BPU(\mathcal{H})$, note that, for any group G, we have an inclusion of its suspension ΣG into BG (corresponding to G-bundles with just one transition function, when ΣG is regarded as the union of two cones on G). Hence we have inclusions

$$\Sigma PU(\mathcal{H}) \to BPU(\mathcal{H}) \qquad (3.1)$$

$$S^3 = \Sigma(\mathbb{C}P^1) \to \Sigma(\mathbb{C}P^\infty) \simeq \Sigma(PU(\mathcal{H})). \qquad (3.2)$$

The composite map

$$S^3 \longrightarrow BPU(\mathcal{H}) \qquad (3.3)$$

induces an isomorphism on $H^3(\ ; \mathbb{Z})$, while all higher cohomology groups of $BPU(\mathcal{H}) \simeq K(\mathbb{Z}, 3)$ are finite.

Consequently, if Y_k denotes the restriction of the bundle $\mathrm{Fred}_k(P)$ to S^3, our task is to examine the Leray spectral sequence (over $\mathbb{Q}$) of Y_k. Since the base is S^3 the only non-zero differential of the spectral sequence is d_3, and the spectral sequence reduces to the Wang exact sequence which is valid for any fibration over a sphere S^m, with total space Y. This is just the exact sequence of the pair (Y, F_0), where F_0 is the fibre over the base-point $0 \in S^m$.

$$\cdots \longrightarrow H^{q-1}(F_0) \xrightarrow{\delta} H^q(Y, F_0) \longrightarrow H^q(Y) \longrightarrow H^q(F_0) \longrightarrow \cdots \quad (3.4)$$

If $\infty \in S^m$ is the antipodal base-point, and F_∞ is the fibre over it, then we have a suspension isomorphism

$$\sigma : H^{q-m}(F_\infty) \longrightarrow H^q(Y, F_0).$$

Substituting this in (3.4) we get the Wang exact sequence

$$\cdots \longrightarrow H^{q-1}(F_0) \xrightarrow{d} H^{q-m}(F_\infty) \longrightarrow H^q(Y) \longrightarrow H^q(F_0) \longrightarrow \cdots, \quad (3.5)$$

where (identifying $H^m(S^m, H^{q-m}(F_\infty))$ with $H^{q-m}(F_\infty)$)

$$d = \sigma^{-1}\delta \quad (3.6)$$

is the differential d_m of the Leray spectral sequence

$$d_m : E_2^{0,q-1} \longrightarrow E_2^{m,q-m}.$$

Because the base of the fibration is simply connected we can identify $H^*(F_\infty)$ canonically with $H^*(F_0)$, and then the homomorphism $d :$ $H^*(F_0) \to H^*(F_0)$ is a *derivation*. We can explicitly calculate d from the homomorphism ϕ^* induced by the gluing map

$$\phi : S^{m-1} \times F_\infty \longrightarrow S^{m-1} \times F_0 \quad (3.7)$$

which (homotopically) defines the fibration $Y \to X$. Using (3.6) one finds the formula

$$\phi^*(1 \otimes x) = (1 \otimes x) + (s \otimes dx), \quad (3.8)$$

where $x \in H^{q-1}(F_0)$ and 1 and s are the generators of $H^0(S^{m-1})$ and $H^{m-1}(S^{m-1})$ respectively. (Notice that d is a derivation $\iff \phi^*$ is a ring homomorphism.)

We now take $m = 3$ and $F_0 = \mathrm{Fred}_k(\mathcal{H})$, with the fibration over S^3 induced by the inclusion of S^3 in $BPU(\mathcal{H})$ as in (3.3). The map ϕ is then the restriction of the multiplication map (2.2) to

$$S^2 = \mathbb{C}P^1 \subset \mathbb{C}P^\infty \simeq PU(\mathcal{H}).$$

Cohomologically this means that, in formula (2.5), we drop all powers of u except the linear term. In particular this implies that only for $n = 1$ does the index $k = s_0$ of the component $\mathrm{Fred}_k(\mathcal{H})$ enter:

$$s_1(u) = s_1 + ku.$$

Putting all this together, we see that we have established:

Proposition (3.9). *In the spectral sequence for the universal fibration over S^3, with fibre $\mathrm{Fred}_k(\mathcal{H})$, the differential d_3 is given on the cohomology generators s_n by the formulae:*

$$d_3 s_n = n s_{n-1} w \text{ for } n \geq 2$$
$$d_3 s_1 = s_0 w = kw,$$

where w is the generator of $H^3(S^3; \mathbb{Z})$.

The coefficients in (3.9) can be taken to be the integers. Over the rationals, as noted before, S^3 is equivalent to $BPU(\mathcal{H})$ and so the formula in (3.9) will hold over $\mathbb{Q}$ for all fibrations. If the fibration over X is induced by a class $\eta \in H^3(X; \mathbb{Z})$, then the spectral sequence for the induced fibration over X, has differential d_3 given by

$$d_3 s_n = n s_{n-1} \eta \qquad n \geq 2$$
$$d_3 s_1 = k\,\eta\,.$$

In particular, the spectral sequence yields the exact sequence

$$H^2(\mathrm{Fred}_k) \xrightarrow{d_3} H^3(X) \xrightarrow{\pi^*} H^3(\mathrm{Fred}_k(P))\,,$$

where π is the fibre projection. Hence $\pi^* d_3 = 0$ and so $\pi^* \eta = 0$ if $k \neq 0$. But if the bundle possesses a section then π^* is injective, so if $\eta \neq 0$ rationally — i.e. if $\eta \in H^3(X; \mathbb{Z})$ has infinite order — we must have $k = 0$. This shows that, as forecast in [AS] Section 3 Remark (iv)], *for twistings of infinite order there are no sections with non-zero index.* In other words, the index map $K_P(X) \to \mathbb{Z}$ is zero for $[P]$ of infinite order.

In subsequent sections we shall pursue further the cohomological implications of the formula in (3.9), but since our main concern is to relate twisted K-theory to cohomology we next turn to this.

4. The Atiyah–Hirzebruch Spectral Sequence

For ordinary K-theory there is the Atiyah–Hirzebruch spectral sequence [AH] which relates it to cohomology. As already briefly explained in [AS]

Prop (4.1), there is an analogous spectral sequence in the twisted case. This has the same E_2 term

$$E_2^{p,q} = H^p(X, K^q(\text{point}))$$

as in the untwisted case, but converges to the graded group $E_\infty^{p,q}$ associated to a filtration on $K_P^*(X)$. Because $K^q(\text{point}) = 0$ when q is odd the even differentials $d_2, d_4, \ldots$ are zero. In the untwisted case the first non-zero differential d_3 is (see [AH]) the Steenrod operation

$$Sq_{\mathbb{Z}}^3 : H^p(X; \mathbb{Z}) \longrightarrow H^{p+3}(X; \mathbb{Z}).$$

Our first task is to find d_3 in the twisted case. We shall show that

$$d_3(x) = Sq_{\mathbb{Z}}^3(x) - \eta x \tag{4.1}$$

where $\eta = \eta[P]$ is the class of P.

By a standard argument in homotopy theory, the only universal operations on $H^p(\ ; \mathbb{Z})$, raising dimension by 3, defined for spaces with a given class $\eta \in H^3$, are given by elements of

$$H^{p+3}(K(\mathbb{Z}, p) \times K(\mathbb{Z}, 3); \mathbb{Z}).$$

This group is isomorphic to

$$H^{p+3}(K(\mathbb{Z}, p)) \oplus H^{p+3}(K(\mathbb{Z}, 3)) \oplus \mathbb{Z},$$

where the third summand is generated by the product of the generators of $H^p(K(\mathbb{Z}, p))$ and $H^3(K(\mathbb{Z}, 3))$. We deduce that the operation d_3 must be of the form

$$d_3(x) = Sq_{\mathbb{Z}}^3(x) + b\eta x \tag{4.2}$$

for some $b \in \mathbb{Z}$, for the formula must agree with the untwisted case when $\eta = 0$, and the operation must vanish when $x = 0$. It remains to compute the integer b. For this it is sufficient to consider the spectral sequence for the twisted K-theory of S^3, with the generator η of $H^3(S^3; \mathbb{Z})$ as the class of the twisting. We just have to compute

$$d_3 : H^0(S^3) \longrightarrow H^3(S^3) \tag{4.3}$$

and show that $d_3(1) = -\eta$. This proceeds analogously to the calculation of the d_3 of Section 3.

The spectral sequence for a space X is obtained by applying the functor K_P^* to the filtered space X, filtered by its skeletons. When $X = S^3$ the filtration has just a single step $X^0 = X^1 = X^2 \subset X^3 = X$, where X^0 is the basepoint, and so the spectral sequence reduces to the long exact sequence for the pair (X, X^0). The differential (4.3) is easily seen to be the boundary map $K^0_P(X^0) \to K^1_P(X, X^0)$. Equivalently, it is the boundary map $K^0_P(D_0) \to K^1_P(S^3, D_0)$, when S^3 is written as the union of two hemispheres D_0 and D_∞ intersecting in the equatorial sphere S^2. This map fits into the commutative square

$$
\begin{array}{ccc}
K^0_P(D_0) & \to & K^1_P(S^3, D_0) \\
\downarrow & & \downarrow \\
K^0_P(S^2) & \to & K^1_P(D_\infty, S^2).
\end{array}
$$

The differential of (4.3) is the passage from top left to bottom right in this diagram, when the groups are identified with $K^0(D_0) \cong \mathbb{Z}$ and $K^1(D_\infty, S^2) \cong \mathbb{Z}$ by trivializing P over D_0 and D_∞ respectively. Now $K^0(S^2) \cong \mathbb{Z} \oplus \mathbb{Z}$, with generators 1 and $[L]$, where L is the standard line bundle whose first Chern class is the generator of $H^2(S^2)$. The lower horizontal map takes 1 to 0 and $[L]$ to 1 when the trivialization from D_∞ is used for both groups. The left-hand vertical map takes 1 to 1 if we use the D_0-trivialization on S^2, but takes 1 to $[L^{-1}] = 2 - [L]$ when P is trivialised from D_∞. So, finally, the differential takes 1 to -1, as we claimed. We have thus established the result anticipated in [AS].

Proposition (4.6). *In the Atiyah–Hirzebruch spectral sequence for the functor K_P the differential d_3 is given by*

$$
d_3(x) = Sq^3_{\mathbb{Z}}(x) - \eta x,
$$

where η is the class of P in $H^3(X; \mathbb{Z})$.

5. The Higher Differentials

For ordinary K-theory all the differentials of the Atiyah–Hirzebruch spectral sequence are of finite order and so disappear on tensoring with the rationals, showing that the graded group associated to a filtration of $K^* \otimes \mathbb{Q}$ is isomorphic to $H^* \otimes \mathbb{Q}$. If the twisting class $\eta(P)$ is of finite order then the same holds for K^*_P, but if $\eta(P)$ is of infinite order then already, as shown in Section 3, d_3 can be of infinite order. At first sight one might be tempted to think that this might be all that happens, and might expect

$$
E_4 \otimes \mathbb{Q} \cong E_\infty \otimes \mathbb{Q}. \tag{5.1}
$$

However, as pointed out to us by Michael Hopkins, this is **not** the case
in general, although there are important cases where it does hold. The
purpose of this section is to explain precisely what does happen to the
spectral sequence when we tensor with $\mathbb{Q}$.

As explained in Section 4, general homotopy theory implies that, in
our category of spaces having a preferred class η in H^3, the only *primary*
cohomology operations over $\mathbb{Q}$ are given by products with η. Since η is
odd-dimensional, $\eta^2 = 0$ and so we only get multiplication by η, as in the
d_3 for K_P. But now let us consider possible *secondary* operations, i.e. ones
defined not on E_2, but on E_4. To get E_4 we start from $E_2 = E_3 = H^*(X, \mathbb{Q})$
and form first the d_3-cycles

$$Z_4 = \{x \in H^*(X) \text{ with } \eta x = 0\}$$

and then the boundaries

$$B_4 = \{y \in H^*(X) \text{ of the form } y = \eta x\}$$

giving

$$E_4 = Z_4/B_4 \,.$$

Now on $E_4 = E_5$ there is a new universal operation that can be defined,
namely the *Massey triple product*. Let us rapidly review the theory of such
products. We need to introduce cochains, and although it is technically
possible [Su] to do this over $\mathbb{Q}$, it is simpler to assume our spaces are manifolds and to work over $\mathbb{R}$, where we can use differential forms as cochains.
In fact little is lost by this assumption, and for any application in physics
the restriction is quite natural.

In general, let x, y, z be closed differential forms on X of degrees p, q, r
with

$$[x][y] = 0 \,, \quad [y][z] = 0 \,,$$

where [ ] denotes the de Rham cohomology class. Thus

$$xy = du \,, \quad yz = dv \,.$$

Then the $(p + q + r - 1)$ - form

$$w = uz + (-1)^{p-1}xv$$

satisfies

$$dw = duz - xdv = xyz - xyz = 0 \,.$$

The cohomology class

$$[w] = \{x, y, z\}$$

is called the *Massey triple product*. Because of the choices of u and v , it is ambiguous up to the addition of arbitrary multiples of x and z.

There is a whole hierarchy of higher Massey products, and they play a basic role in rational homotopy theory [Su]. Essentially they give all the additional information (beyond the cohomology ring structure) that is needed to determine the rational homotopy type. With these general explanations out of the way, let us return to our twisted K-theory over $\mathbb{R}$. Let η denote a fixed closed 3-form on X representing the class of P. Then a class in E_4 in the spectral sequence may be represented by a closed p-form x with $[\eta][x] = 0$. Since $[\eta]^2 = 0$, we can form the Massey triple product $\{\eta, \eta, x\}$. Since we actually have $\eta^2 = 0$ as a form, we can take

$$\{\eta, \eta, x\} = v\eta \quad \text{where} \quad \eta x = dv\,, \tag{5.2}$$

and the Massey product is less ambiguous than in the general case. If we change the choice of v by adding v_0 with $dv_0 = 0$, then ηv_0 belongs to the boundaries B_4 and so its class in $E_4 = E_5$ is well-defined. Thus we have exhibited an operation

$$\delta_5 : E_4^p \longrightarrow E_4^{p+5}\,.$$

In Section 7 we shall show that $-\delta_5$ is the differential d_5 of the spectral seqence for twisted K-theory.

If $\{\eta, \eta, x\} = 0$ we can repeat the process, defining a Massey product

$$\{\eta, \eta, \eta, x\} = \eta w \quad \text{where} \quad \{\eta, \eta, x\} \equiv dw (\text{mod multiples of } \eta)\,,$$

giving an operation

$$\delta_7 = E_6^p \longrightarrow E_6^{p+7}$$

which we shall see in Section 7 is minus the differential d_7 of the spectral sequence.

Continuing in this way, we find arbitrarily many potentially non-vanishing differentials. This means that, in general, $K_P^*(X)$ will have much smaller rank than $H^*(X)$. A more precise formulation will emerge in the subsequent sections. However, one fact can already be noted. If, for a given manifold X, all Massey products vanish, then the spectral sequence stops at E_4, so that (5.1) holds and

$$\operatorname{gr} K_P^*(X) \cong E_4\,. \tag{5.3}$$

An important class of manifolds for which this holds are *compact Kähler manifolds*, where a deep theorem [DGMS] asserts that all Massey products vanish. On the other hand we shall give examples in the Appendix of manifolds for which the Massey products of arbitrarily high order do not vanish.

In ordinary K-theory the collapsing of the Atiyah–Hirzebruch spectral sequence over $\mathbb{Q}$ (or $\mathbb{R}$) is intimately tied up with the existence of the *Chern character*, which gives a ring isomorphism

$$\mathrm{ch} : K^*(X) \otimes \mathbb{Q} \longrightarrow H^*(X, \mathbb{Q}).$$

As we have seen, the twisted K-theory given by a twisting element η of infinite order is far from being isomorphic over $\mathbb{Q}$ to cohomology: typically it is considerably smaller. So the question arises: is there an appropriate twisted cohomology theory with a corresponding twisted Chern character? In the next section we shall describe such a theory.

6. Twisted Cohomology

Given a manifold X and a closed 3-form η we shall introduce a "twisted" cohomology theory (over $\mathbb{R}$). On the de Rham complex $\Omega(X)$ of differential forms on X we define an operator D_η (or just D)

$$D = d - \eta. \tag{6.1}$$

Note that

$$D^2 = (d - \eta)(d - \eta) = d^2 - d\eta - \eta d + \eta^2 = 0,$$

since η has odd degree. However, D is not homogeneous, though it preserves the grading mod 2. Thus the cohomology groups of D can be defined and they are graded mod 2. We denote these groups by

$$H_\eta^0(X) \ , \ H_\eta^1(X).$$

If Y is a submanifold of X we can of course define relative groups $H_\eta^*(X, Y)$ as the D-cohomology of the relative de Rham complex $\Omega(X, Y)$ of forms which vanish on Y; and we shall have the usual long exact sequence.

Remarks.

(i) It is not clear how one could define twisted cohomology with *integer* coefficients. Even on the level of cohomology we need not have $\eta^2 = 0$ integrally (though, as we saw in Proposition 4.6, we do have $(Sq_{\mathbb{Z}}^3 - \eta)^2 = 0$).

(ii) The groups $H^*_\eta(X)$ depend on the closed form η and not just on its cohomology class. If η and η' are cohomologous — say $\eta' - \eta = d\zeta$ — then multiplication by the even-dimensional form e^ζ induces an isomorphism $H^*_\eta(X) \to H^*_{\eta'}(X)$, as $D_{\eta'} \circ e^\zeta = e^\zeta \circ D_\eta$. Because we have $(d\alpha)\beta = D_\eta(\alpha\beta)$ if $D_\eta\beta = 0$ the isomorphism of twisted cohomology evidently does not change if ζ is altered by the addition of an exact 2-form. One consequence of this — taking ζ to be closed — is that the group $H^2(X;\mathbb{R})$ acts on the twisted cohomology $H^*_\eta(X)$. We shall see in the next section that this action corresponds by the Chern character to the natural action of the Picard group $H^2(X;\mathbb{Z})$ on $K^*_P(X)$ which was described in [AS]. Another consequence is that if we are given not just one closed 3-form η but a family $\{\eta_a\}_{a \in A}$ of 3-forms together with a choice of a 2-form ζ_{ab} for each $a, b \in A$ such that $\eta_b - \eta_a = d\zeta_{ab}$, and if

$$\zeta_{ab} - \zeta_{ac} + \zeta_{bc} = 0$$

for each a, b, c, then the groups $H^*_{\eta_a}(X)$ are *canonically* isomorphic. We shall sometimes write them as $H^*_{\mathcal{F}}(X)$, where $\mathcal{F}$ denotes the family $\{\{\eta_a\}, \{\zeta_{ab}\}\}$. In fact the forms ζ_{ab} need to be given only up to the addition of exact forms. We shall refer to such a family $\mathcal{F}$ as a *coherent family* of closed 3-forms.

(iii) We shall often encounter the situation where we have an open covering $\mathcal{U} = \{U_a\}$ of X and a closed 2-form ω_{ab} defined in each intersection $U_{ab} = U_a \cap U_b$ such that

$$\omega_{ab} - \omega_{ac} + \omega_{bc} = 0$$

in the triple intersection U_{abc}. By choosing a partition of unity $\{\lambda_a\}$ subordinate to the covering $\mathcal{U}$ we can then define a global closed 3-form η whose value in U_a is dB_a, where $B_a = \sum_b \lambda_b.\omega_{ab}$. (Indeed, in string theory it is usually the "B-field" represented locally by the B_a which is given, and ω_{ab} is defined as $B_a - B_b$.) The previous remark shows that the mod 2 graded de Rham complex $\Omega(X)$ of X with the twisted differential D_η is isomorphic to the complex $\Omega_\omega(X)$ in which a form is a collection $\alpha_a \in \Omega(U_a)$ such that $\alpha_b = e^{\omega_{ab}}\alpha_a$ for all a, b, with the usual exterior differential. The map takes the collection α_a to the globally defined form which is $e^{B_a}\alpha_a$ in U_a. (Furthermore, if the covering $\mathcal{U}$ and the forms ω_{ab} are given, then varying the partition of unity gives rise to a coherent family $\mathcal{F}$ of 3-forms in the sense of the previous remark.)

Although the twisted differential D does not preserve the grading of the de Rham complex, it does preserve the filtration whose pth stage is the sum of the forms of all degrees $\geq p$. The filtration therefore gives us a spectral sequence converging to the twisted cohomology. The E_1 term is clearly just the differential forms with the usual d (since η raises filtration by 3), and so

$$E_2^p = H^*(X, \mathbb{R})$$

is the usual de Rham cohomology. To compute the higher differentials we proceed as follows. Let

$$x = x_p + x_{p+2} + x_{p+4} + \dots$$

be an (inhomogenous) form with $dx_p = 0$, so that x_p represents a class $[x_p] \in H^p$. Then

$$Dx = (d - \eta)x = dx_p + (dx_{p+2} - \eta x_p) + (dx_{p+4} - \eta x_{p+2}) + \dots$$

i.e.

$$y = Dx = y_{p+1} + y_{p+3} + y_{p+5} \dots$$

where

$$y_{p+1} = dx_p = 0$$
$$y_{p+3} = dx_{p+2} - \eta x_p$$
$$y_{p+5} = dx_{p+4} - \eta x_{p+2}$$

$$\dots\dots\dots\dots\dots\dots\dots\dots .$$

The class $[y_{p+3}] = -[\eta][x_p]$ represents $d_3 x$, so that the differential d_3 of the spectral sequence is just multiplication by $-\eta$. Hence E_4 is the same as the E_4 term of the Atiyah–Hirzebruch spectral sequence (over $\mathbb{R}$) that we found in Section 4.

Proceeding further, if $[\eta x_p] = 0$ so that $\eta x_p = dv$, we can choose $x_{p+2} = v$ and make $y_{p+3} = 0$. Then $y_{p+5} = dx_{p+4} - \eta v$ represents $d_5 x_p$. In other words

$$d_5[x_p] = -\{\eta, \eta, x_p\}$$

is given by the Massey triple product. Continuing this way we see that

$$d_7[x_p] = -\{\eta, \eta, \eta, u\}$$

and so on. Thus we have proved

Proposition 6.1. *The iterated Massey products with η give (up to sign) all the higher differentials of the spectral sequence for the twisted cohomology.*

The similarity of this spectral sequence with that of twisted K-theory described in Section 5 is clear. In the next section we shall bring them together via the *twisted Chern character* which we shall define.

Let us first, however, point out a simple corollary of the existence of the filtration spectral sequence which makes working with twisted cohomology a lot more flexible. Obviously we can define twisted cohomology $H_\eta^*(A)$ whenever we have a differential graded algebra A which is graded-commutative, and a chosen closed element η in degree 3. The spectral sequence tells us that if $\phi : A \to A'$ is a homomorphism of differential graded algebras which induces an isomorphism of cohomology in the usual sense, then it induces an isomorphism

$$H_\eta^*(A) \to H_{\phi(\eta)}^*(A')$$

for every closed $\eta \in A^3$. Furthermore, if instead of η we have a closed 3-form η' in A', then we can define a coherent family $\phi^*(\eta')$ in A such that

$$H_{\phi^*(\eta')}^*(A) \cong H_{\eta'}^*(A').$$

The family is indexed by the set of pairs (η, ξ') in $A^3 \oplus (A')^2$ such that $d\eta = 0$ and $\phi(\eta) + d\xi' = \eta'$: this gives us a coherent family because the mapping-cone[1] of the equivalence $A \to A'$ is acyclic.

An important application of this remark is the following. Any topological space X has a "singular de Rham complex" $\Omega_{\text{sing}}(X)$, in which a p-form is a family of p-forms $\alpha_\sigma \in \Omega^p(\Delta^m)$, one for each singular simplex $\sigma : \Delta^m \to X$, which are compatible under the face and degeneracy maps. It is well-known [Su] that the cohomology of $\Omega_{\text{sing}}(X)$ — which is clearly a commutative differential graded algebra — is the singular cohomology $H^*(X; \mathbb{R})$. Now if X is a smooth manifold there are natural homomorphisms of differential graded algebras

$$\Omega(X) \to \Omega_{\text{sm-sing}}(X) \leftarrow \Omega_{\text{sing}}(X)$$

inducing isomorphisms of cohomology, where $\Omega_{\text{sm-sing}}(X)$ is defined like $\Omega_{\text{sing}}(X)$ but using only *smooth* singular simplexes. Coherent families of 3-forms in these algebras correspond one-to-one, and the same twisted cohomology is obtained from any of them. In other words, we need not hesitate to speak of the twisted cohomology of spaces which are not manifolds.

---

[1]The mapping-cone of $A \to A'$ is the total complex of the double complex $\ldots \to 0 \to A \to A' \to 0 \to \ldots$.

We end this section with a technical result needed in the next section. Let us first notice that if X is a space for which $H^{\text{odd}}(X; \mathbb{R}) = 0$ then the spectral sequence shows that $H^{\text{odd}}_\eta(X) = 0$ for any η; and the same principle holds for relative groups. We shall apply this to the situation studied in Section 3, where we have a fibration $Y \to S^3$ over the 3-sphere, with fibres F_0 and F_∞ over the poles 0 and ∞ in S^3. We are interested in the twisted cohomology $H_\eta(Y)$, where $\eta \in \Omega^3(Y)$ is pulled back from a form on S^3 which generates $H^3(S^3)$, but vanishes near the base-point 0. Thus η vanishes in the neighbourhood of F_0, and so there is a restriction map $H^{\text{ev}}_\eta(Y) \to H^{\text{ev}}(F_0)$.

Proposition 6.2. *For the fibration $Y \to S^3$ with fibre F_0, if $H^{\text{odd}}(F_0; \mathbb{R}) = 0$ then*

$$H^{\text{ev}}_\eta(Y) \to H^{\text{ev}}(F_0)$$

is injective.

Proof. As explained in Section 3, we have $H^i(Y, F_0) \cong H^{i-3}(F_\infty)$, so the hypothesis implies that the relative group $H^{\text{ev}}_\eta(Y, F_0)$ vanishes. $\square$

7. The Chern Character

For each projective bundle P on X we should like to have a natural Chern character map

$$\text{ch} : K^*_P(X) \to H^*_\eta(X)$$

which induces an isomorphism $K^*_P(X) \otimes \mathbb{R} \to H^*_\eta(X)$. Here η is a 3-form on X representing the class of the projective bundle P. In fact it is unreasonable to try to assign a specific 3-form to the bundle P without choosing any additional structure,[2] and we shall instead define a coherent family of forms in the sense explained in Remark (ii) of the previous section.

In this paper we are using the traditional methods of algebraic topology — though we have compromised a little by defining twisted cohomology in terms of the de Rham complex — and in that spirit we shall define the Chern character as a universal form on the representing space for twisted K-theory. Before doing so, however, it seems worthwhile to sketch briefly how things would look from the point of view of Chern–Weil theory, for that is more geometrical, and more relevant in applications of the theory.

---

[2] What is needed is a choice of a *string connection* in the sense of [S3], or of some version of a *gerbe connection* [BCMMS].

To treat the Chern–Weil theory fully would require a careful discussion of smooth Hilbert bundles and families of unbounded operators in them, and we shall not embark on that.

The Chern character of a smooth vector bundle E on X is represented by the inhomogeneous differential form $\operatorname{tr}(e^{F_E})$, where $F_E \in \Omega^2(X; \operatorname{End}(E))$ is the curvature of an arbitrary connection in E. (We recall that if

$$d_E : \Omega^i(X; E) \to \Omega^{i+1}(X; E)$$

is the covariant exterior derivative operator associated to a connection in E then d_E^2 is multiplication by F_E: a little cryptically, we write $d_E^2 = F_E$.)

It is perhaps slightly less well known that there is a generalization — due to Quillen [Q] — of this description of the Chern character to the case of an element of $K(X)$ represented by a family of unbounded self-adjoint degree 1 Fredholm operators $\{T_x\}$ in a mod 2 graded bundle $\mathcal{H}$ of Hilbert spaces on X, at least when $e^{-T_x^2}$ is of trace class for all $x \in X$. We simply form the *superconnection* $d_{\mathcal{H}} + iT$, an operator on $\Omega(X; \mathcal{H})$, and define its curvature as

$$\mathcal{F} = (d_{\mathcal{H}} + iT)^2 .$$

Then the inhomogeneous differential form[3] $\operatorname{str}(e^{\mathcal{F}})$ represents the Chern character. (We are assuming here that the bundle $\mathcal{H}$ is smooth in the sense that one can speak of differential forms on X with values in $\mathcal{H}$, that we have a connection in $\mathcal{H}$ giving an operation $d_{\mathcal{H}}$ of covariant differentiation on such forms, and that the family $\{T_x\}$ is appropriately smooth. For more details, see [BGV].)

Quillen's description of the Chern character is well adapted to the definition of twisted K-theory $K_P(X)$ by means of families of Fredholm operators in a projective Hilbert bundle P. It is simplest to explain this in terms of local trivializations. Let us choose local trivializations of P over the sets of an open covering $\mathcal{U} = \{U_a\}$ as above, and represent P and its connection by an actual Hilbert bundle $\mathcal{H}_a$ with a unitary connection $d_{\mathcal{H}_a}$ in U_a. Then over $U_{ab} = U_a \cap U_b$ we have

$$\mathcal{H}_b \cong \mathcal{H}_a \otimes L_{ab} ,$$

---

[3] The *supertrace* of an operator T in a mod 2 graded vector space $\mathcal{H} = \mathcal{H}^0 \oplus \mathcal{H}^1$ is defined by

$$\operatorname{str} T = \operatorname{tr} T|\mathcal{H}^0 - \operatorname{tr} T|\mathcal{H}^1 .$$

where L_{ab} is a line bundle with a unitary connection. Let the curvature of L_{ab} be the closed 2-form ω_{ab} on U_{ab}. Then the curvature forms $\mathcal{F}_a$ defined in each U_a are related by

$$\mathcal{F}_b = \mathcal{F}_a + \omega_{ab}$$

in U_{ab}, and so the character forms $\mathrm{str}(e^{\mathcal{F}_a})$ constitute an element of the twisted de Rham complex $\Omega_\omega(X)$ which — as we have pointed out in Remark (iii) in Section 6 — calculates the twisted cohomology $H_\eta^*(X)$. If the twisting is given by a B-field $\{B_a\}$ then the Chern character is given by the globally defined form $\mathrm{str}(e^{\mathcal{F}_a+B_a})$. This is the Chern–Weil definition of the twisted Chern character.

Now let us give a less problematic definition of the Chern character as a twisted cohomology class of the universal fibration $\mathrm{Fred}(P)$ of Section 3, with the twisting defined by the generator of $H^3(K(\mathbb{Z},3);\mathbb{Z})$. The first point to notice is that, by the remark at the end on Section 6, we can replace the universal fibration by its restriction Y to the sphere S^3 contained in the base-space $K(\mathbb{Z},3)$, for the two spaces are rationally homotopy equivalent, and so the restriction map of their singular de Rham complexes induces an isomorphism of both ordinary and twisted cohomology. As in Section 3 we shall denote the fibres of the universal bundle over the north and south poles of S^3 by F_∞ and F_0. We shall prove that there is a *unique* twisted class of Y which restricts to the standard Chern character on the fibre F_0. (We shall explain the significance of the uniqueness presently.) Since we are concentrating on the case when the twisting class η is of infinite order we may (as explained in Section 3) restrict ourselves to the index zero component of $\mathrm{Fred}(P)$. Let s_n be a closed form on F_0 representing the cohomology generator denoted by the same symbol in Section 3, and let η be a 3-form on Y pulled back from a generating 3-form on S^3 whose support does not contain either pole of S^3. Our task is to construct a sequence of closed forms S_n on Y with the two properties

(i) $S_n|F_0 = s_n$,

and

(ii) $dS_n = n\eta S_{n-1}$.

We shall do this by induction on n. Suppose that S_m has been found for $m < n$. Because the bundle Y can be trivialized in the complement of the single fibre F_∞ we can pull back s_n to a closed form on this complement, and then extend it further to a not-necessarily-closed form S_n' on all of

Y. (This extension is especially easy using the singular de Rham complex, as we can extend by zero; if we were working with honest smooth forms we should have to multiply the pull-back by a suitable real-valued function equal to 1 outside a neighbourhood of F_∞ and vanishing in a smaller neighbourhood of F_∞.) Then $dS'_n = 0$ away from F_∞, and so it defines a class in $H^{2n+1}(Y, F_0)$. By the arguments in Section 3 (leading up to Proposition (3.9)) this class can also be represented by the form $n\eta S_{n-1}$, which also vanishes near F_0. Hence on the level of forms we have

$$dS'_n - n\eta S_{n-1} = d\theta_n , \qquad (7.1)$$

where θ_n is form on Y with support away from F_0. So we define $S_n = S'_n - \theta_n$.

Note that the formulae (i) and (ii) still hold for $n = 1$, where $S_0 = 0$ in the zero-component.

Thus the even form on Y

$$\mathrm{ch} = \sum_{n=1}^{\infty} \frac{S_n}{n!} \qquad (7.2)$$

satisfies $(d-\eta)\mathrm{ch} = 0$, showing that it is a cocyle for the twisted cohomology of the universal fibration.

The construction of the form in (7.2) provides a class in $H_\eta(Y)$ which lifts the usual Chern character in $H^*(F_0)$. It defines a twisted Chern character in the sense that when we have a manifold X with a projective bundle P we can choose a smooth map $f : X \to K(\mathbb{Z}, 3)$ which pulls back P from the universal $\mathbb{C}P^\infty$-bundle $\hat{P}$ on $K(\mathbb{Z}, 3)$. Then $K_P(X)$ is defined as the set of homotopy classes of lifts F of f to Y, and for each such lift F we define $\mathrm{ch}\, F = F^*(\mathrm{ch})$. This is an element of the twisted cohomology $H^*_{\eta_f}(X)$, where $\eta_f = f^*(\eta)$. The definition of the twisting form η_f depends on the choice of the classifying map f, but as f varies we obtain a coherent family $\mathcal{F}_P$ of forms in the sense of the preceding section. For though the space of maps f pulling back a bundle equivalent to P is not contractible, the space of pull-back diagrams

$$\begin{array}{ccc} P & \longrightarrow & \hat{P} \\ \downarrow & & \downarrow \\ X & \longrightarrow & K(\mathbb{Z}, 3) , \end{array}$$

i.e. of *pairs* (f, α), where α is an isomorphism $\alpha : P \to f^*\hat{P}$, *is* contractible, and we index $\mathcal{F}_P$ by this set of pairs. Then there is a path, unique up to homotopy, from one choice of f to any other — say f' —, and hence, by the usual argument of de Rham cohomology, a choice of $\zeta_{ff'}$, unique up to an exact form, such that

$$d\zeta_{ff'} = \eta_f - \eta_{f'}.$$

The uniqueness of the twisted class of Y which lifts the Chern character on F_0 is important.[4] The classical Chern character defines a ring homomorphism from K-theory to cohomology, but if we wanted just an additive homomorphism we could equally well use $\sum \lambda_k \mathrm{ch}_k$ for any sequence of rational numbers $\{\lambda_k\}$. The position is different with twisted K-theory. A class $\xi \in H^*_\eta(Y)$ defines an additive transformation from twisted K to twisted cohomology if and only if it is *primitive* (i.e. maps to $\xi \otimes 1 + 1 \otimes \xi$) under the fibrewise Whitney sum map

$$Y \times_B Y \to Y,$$

where $B = K(\mathbb{Z}, 3)$. Now Proposition 6.2 applies to $Y \times_B Y$, and it follows that a twisted class ξ defines an additive transformation if and only if its restriction to the fibre is primitive, and hence of the form $\sum \lambda_k \mathrm{ch}_k$. But only if all the λ_k are equal does this class lift to a twisted class. Thus the twisted Chern character is singled out by additivity alone, without mentioning any multiplicativity property. On the other hand, the argument just given also proves that our twisted character is indeed multiplicative in the natural sense with respect to the pairings

$$K_P(X) \otimes K_{P'}(X) \to K_{P \otimes P'}(X)$$

and

$$H_\eta(X) \otimes H_{\eta'}(X) \to H_{\eta + \eta'}(X).$$

We have now defined the twisted character on the even group $K_P(X)$. But we can extend the definition to a homomorphism

$$\mathrm{ch} : K^*_P(X) \longrightarrow H^*_{\mathcal{F}_P}(X)$$

of mod 2 graded theories simply by replacing X by $X \times S^1$ in the usual way.

The remaining formal properties of the twisted Chern character also follow from those of the untwisted case, as we shall now explain.

Because the twisted Chern character is functorial it is compatible with the filtrations which are used to construct the Atiyah–Hirzebruch spectral sequences, and so it induces a homomorphism of the associated spectral sequences of Sections 4 and 6. But when the K-theory spectral sequence is

---

[4] We are indebted to U. Bunke for a helpful discussion of this point.

tensored with $\mathbb{R}$ the E_2 terms coincide, and the homomorphism there is an isomorphism. Hence the whole spectral sequence map is an isomorphism, and we conclude that

$$\mathrm{ch} : K_P^*(X) \otimes \mathbb{R} \to H_{\mathcal{F}_P}^*(X)$$

is an isomorphism. This argument also shows that the differentials of the spectral sequences must agree. Since the higher differentials for twisted cohomology were shown in Section 6 to be given precisely by the iterated Massey products, the same follows for the twisted K-theory spectral sequence.

To summarize, we have established

Proposition (7.4). *There is a functorial twisted Chern character*

$$\mathrm{ch} : K_P^*(X) \longrightarrow H_{\mathcal{F}_P}^*(X)$$

which becomes an isomorphism after tensoring with $\mathbb{R}$.

Proposition (7.5). *In the twisted Atiyah–Hirzebruch spectral sequence over $\mathbb{R}$ the higher differentials $d_5, d_7, \ldots$ are given by the iterated Massey products*

$$d_5(x) = -\{\eta, \eta, x\}$$
$$d_7(x) = -\{\eta, \eta, \eta, x\}$$

$$\cdots$$

8. Chern Classes

Ordinary (untwisted) K-theory is related to cohomology in three different ways: via the Atiyah–Hirzebruch spectral sequence, via the Chern character, and via Chern classes. These are all connected in well-known ways. In particular the Chern character can be expressed in terms of the Chern classes, and the converse holds over the rationals. However, in the twisted case (for a twisting element of infinite order) there are more substantial differences, and one cannot pass so easily from one to another. For example, the twisted Chern character, not being graded, does not lead to a sequence of rational twisted Chern classes.

The basic result about characteristic classes in the untwisted case is that any cohomology class naturally associated to an element $\xi \in K(X)$ is a polynomial in its Chern classes $c_n(\xi)$. For a twisted class $\xi \in K_P(X)$ we can look for characteristic classes in either the ordinary cohomology or the

twisted cohomology of X. Although the latter question might seem more
logical, we shall begin with the former, which is easier and more tractable.
Our results are summarized in Proposition 8.8 below.

For any generalized cohomology theory characteristic classes correspond
to cohomology classes of the classifying space. For twisted K-theory the
appropriate classifying space is the universal fibration studied in Section 3,
with fibre $\mathrm{Fred}(\mathcal{H})$ and base the Eilenberg–MacLane space $K(\mathbb{Z}, 3)$.

The cohomology classes of the total space define characteristic classes for
our twisted K-theory. For a twisting by an element η of infinite order only
the index zero component of $\mathrm{Fred}(\mathcal{H})$ is relevant, and so we shall restrict
ourselves to this case.

We now consider only real or rational coefficients, so that, as in Section
3, we can replace the Eilenberg–MacLane space by the 3-sphere S^3. The key
result of Section 3 was Proposition (3.9), determining the only non-trivial
differential d_3 for the Leray spectral sequence of the fibration. As noted
there, the Leray spectral sequence reduces to the Wang exact sequence
(3.5), and hence gives the exact sequence:

$$0 \to H^{\mathrm{ev}}(Y) \xrightarrow{j} H^{\mathrm{ev}}(F) \xrightarrow{d} H^{\mathrm{ev}}(F) \to H^{\mathrm{odd}}(Y) \to 0, \qquad (8.1)$$

where we have lumped together the even and odd degrees. Here $F = \mathrm{Fred}_0(\mathcal{H})$ and Y is the total space.

The homomorphism d is the derivation given by (3.9) with $k = 0$:

$$ds_n = n\, s_{n-1} \qquad n \geq 2$$
$$ds_1 = 0 \qquad\qquad\qquad\qquad\qquad (8.2)$$

and (8.1) shows that

$$H^{\mathrm{ev}}(Y) = J, \qquad\qquad\qquad (8.3)$$

where $J \subset Q[s_1, s_2, \dots]$ is the kernel of d in (8.2). The sequence (8.1)
also determines $H^{\mathrm{odd}}(Y)$ as the cokernel of d, but, as the following lemma
shows, this is zero except for H^3.

Lemma 8.4. *The derivation d given by (8.2) is surjective in degree > 0.*

Proof.   For simplicity of notation it is convenient to introduce new variables $x_n (n \geq 0)$ by

$$x_n = \mathrm{ch}_n = \frac{s_n}{n!} \qquad n \geq 1,$$

and to put $x_0 = 0$. Then d acting on a polynomial $f(x_1, \dots)$ becomes

$$df = \sum_{n \geq 1} \frac{\partial f}{\partial x_n} x_{n-1} .$$

It is sufficient to consider monomials of degree $n > 0$, and we assume, as an inductive hypothesis, that the Lemma is true for all smaller positive values of n. If f has degree n and is not in the image of d, then we can write $f = x_k g$ with k minimal but $k < n$ (since $x_k = dx_{k+1}$). Then $0 < \deg(g) < n$, so by the inductive hypothesis $g = dh$ for some h. Then

$$f = x_k g = x_k dh = d(x_k h) - x_{k-1} h$$

so that $x_{k-1} h$ is also not in the image of d. But, by the minimality of k, this contradicts our assumption, proving the lemma. $\qquad\square$

Thus $H^q(Y) = 0$ for odd $q \neq 3$, while $H^3(Y) \cong H^3(S^3)$ with the generator pulled back from the base. Thus, apart from the twisting class η itself, *there are no odd (rational) characteristic classes for twisted K-theory.*

The even characteristic classes are then given by the subring J. We shall describe some aspects of this ring, but it does not seem easy to give its full structure in terms of generators and relations.

One consequence of (8.1) and (8.4) is a formula for the Poincaré series of J. Let $A = \underset{n}{\oplus} A_n = \mathbb{Q}[x_1, x_2, \dots]$ be the graded polynomial algebra, where degree $x_n = n$, and let

$$J = \oplus J_n$$

be the grading of J. (Note that d in (8.1) just shifts the grading by 1.) Then (8.1) and (8.4) lead to the short exact sequence

$$0 \to J_n \to A_n \overset{d}{\to} A_{n-1} \to 0$$

for $n \geq 2$, while

$$J_0 = A_0 \qquad J_1 = A_1 .$$

Put $a_n = \dim A_n$, $j_n = \dim J_n$. Then

$$j_n = a_n - a_{n-1} \qquad (n \geq 2)$$
$$j_1 = a_1 = 1$$
$$j_0 = a_0 = 1 .$$

Defining the Poincaré series $J(t)$, $A(t)$ by

$$J(t) = \sum_{n \geq 0} j_n t^n \qquad A(t) = \sum_{n \geq 0} a_n t^n ,$$

we deduce that

$$\sum_2^\infty j_n t^n = \sum_2^\infty a_n t^n - \sum_2^\infty a_{n-1} t^n \,,$$

or

$$J(t) - (1+t) = A(t) - (1+t) - t[A(t) - 1]$$
$$J(t) = A(t) - tA(t) + t$$
$$= (1-t)A(t) + t \,.$$

But, since A is the polynomial algebra,

$$A(t) = \frac{1}{(1-t)(1-t^2)\dots} \,,$$

and we have proved

Proposition 8.1. *The Poincaré series of J is given by*

$$J(t) = \frac{1}{(1-t^2)(1-t^3)\dots} + t. \tag{8.5}$$

The first few explicit values of j_n are

$$j_0 = 1, \ j_1 = 1, \ j_2 = 1, \ j_3 = 1, \ j_4 = 2, \ j_5 = 2,$$
$$j_6 = 4, \ j_7 = 4, \ j_8 = 7 \,.$$

This shows that J is not just a polynomial algebra (on generators of degrees ≥ 2), as one might have expected from comparison with the finite-dimensional groups $U(N) \to PU(N)$. The extra term t in (8.5) arises from the class $x_1 = s_1$, but J does not even contain a polynomial subalgebra on generators of degrees ≥ 2, for x_1^2 and x_1^3 are the only elements in degrees 2 and 3, and they are algebraically dependent.

Although $J \subset A$ is explicitly defined by the formulae (8.2) there is a more fundamental way of describing it that was already implicit in Section 2, where we studied the action of tensoring by line-bundles. We continue with our present notation, where we have replaced the power sums s_n by the components x_n of the Chern character. In terms of a variable u of degree 2, introduce variables $y_1, y_2, \dots$ related to $x_1, x_2, \dots$ by

$$\sum y_n = \left(\sum x_n \right) \left(\sum \frac{u^k}{k!} \right) . \tag{8.6}$$

Thus, explicitly

$$y_1 = x_1$$
$$y_2 = x_2 + ux_1$$
$$y_3 = x_3 + ux_2 + \frac{u^2}{2}x_1$$

. . . .

These equations are, of course, just a rewriting of equations 2.4 and 2.5.

Given a polynomial $f(x_1, x_2 \ldots)$ in A, we define it to be *invariant* if

$$f(y) \equiv f(x) \tag{8.7}$$

identically in u. Now equation (8.6) is invertible (giving x in terms of y), and it can be written symbolically as

$$Y = X \, e^u$$

is invertible (giving x_k in terms of $y_1, \ldots, y_k$). It defines an algebraic action of the additive group $\mathbb{C}$ (on which u is the coordinate function) on A by algebra automorphisms. This explains why we used the word "invariant". (To keep track of the grading: if we think of the grading of A as corresponding to an action of $\mathbb{C}^\times$ on it, then the action of $\mathbb{C}$ on A is $\mathbb{C}^\times$-equivariant when $\mathbb{C}^\times$ acts by multiplication on $\mathbb{C}$.)

As noted in Section 3, our homomorphism $d : A \to A$ is given by just using the linear term in u in (8.6). In other words it is the action of the Lie algebra of the 1-parameter group. But Lie algebra invariance implies Lie group invariance. Hence the apparently much stronger identity (8.7), given by using all higher powers of u, is actually equivalent to the simpler one using the linear terms. Thus *our subring $J \subset A$ is just the invariant subring*.

A topological explanation of this fact is implicit in the formula (3.2) of Section 3. A class in $H^{\mathrm{ev}}(F)$ which is in J extends by definition to $H^{\mathrm{ev}}(Y)$, where Y is fibred over S^3. Because of the rational homotopy equivalence (3.3) it then also extends over $BPU(H)$, and in particular over the subspace $\Sigma \mathbb{C}P^\infty$ of (3.2). But this passage from $\mathbb{C}P^1$ to $\mathbb{C}P^\infty$ is just the extension from the term linear in u to all powers of u, recovering the algebraic argument above.

We summarize all this in:

Proposition (8.8). *Each polynomial in the universal Chern classes which is invariant under the tensor product operation (8.6) extends uniquely to a* *rational characteristic class of twisted K-theory, and (apart from the twisting class) all such characteristic classes are obtained in this way.*

Furthermore, all of the characteristic classes are annihilated by multiplication by the twisting class η.

Having identified the ring J of characteristic classes of twisted K-theory in *ordinary* cohomology, we can now ask for characteristic classes in *twisted* cohomology. As explained in Section 6 we get the twisted (real) cohomology as a mod 2 graded theory by starting with the de Rham complex and using the total differential $D = d - \eta$ of (6.1). We then get a spectral sequence, whose E_2-term is the usual real cohomology and whose first differential d_3 is multiplication by the class of $-\eta$. The E_∞ is then the associated graded group of our twisted cohomology.

We apply this to our universal space Y. In this section we have shown that

$$H^{\mathrm{ev}}(Y) = J \qquad H^{\mathrm{odd}}(Y) = H^3(Y) = \mathbb{R},$$

and the only non-trivial differential d_3 gives an isomorphism

$$H^0(Y) \to H^3(Y).$$

Thus the E_4-term of our spectral sequence is just the part J^+ of J of positive degree:

$$E_4 \cong J^+. \tag{8.9}$$

Since there are no more terms of odd degree, all remaining differentials are zero, and so

$$J^+ \cong E_\infty \cong \mathrm{gr}\, H_\eta(Y),$$

where H_η is the universal twisted cohomology.

To get characteristic twisted classes for twisted K-theory we therefore have to choose liftings of J^+ back into the filtered ring $H_\eta(Y)$. For example, for the first class $x_1 = s_1 \in J^2$ the *Chern character*, as constructed in Section 7, provides a natural lift. In general, a lift can be constructed by the following simple but not very illuminating formula.

Define a new derivation δ of the polynomial algebra $A = \mathbb{C}[x_1, x_2, \ldots]$ by

$$\delta(x_k) = k x_{k+1}$$

for $k \geq 1$. The commutator $\Delta = [d, \delta]$ is then also a derivation: it multiplies each monomial in the x_k by its degree, when each x_k is given degree 1 rather than k. (This new grading of A is preserved by both d and δ, and is hence inherited by J.) As d and δ commute with Δ we have $[d, \exp(\lambda\delta)] = \lambda\Delta\exp(\lambda\delta)$ for any scalar λ, i.e.

$$(d - \lambda\Delta) \circ \exp(\lambda\delta) = \exp(\lambda\delta) \circ d.$$

If f is an invariant element of A with $\Delta(f) = mf$ for some $m > 0$, we accordingly have

$$(d - 1)\exp(\lambda\delta)(f) = 0$$

when $\lambda = 1/m$. This shows that $\exp(\lambda\delta)(f)$ defines a class in the twisted cohomology of the universal fibration which lifts f. When $f = s_1 = x_1$ it is precisely the Chern character.

To find more natural lifts for classes in J^+ one procedure would be to first construct *internal operations* in twisted K-theory and then take their Chern characters. This is one motivation for our study of such internal operators in Section 10. But, before we proceed down that route, we will in the next section say something about an interesting subset of elements of J defined geometrically — the Koschorke classes.

9. Koschorke Classes

The classical treatment of Chern classes has several aspects, all of importance. One can introduce them as cohomology classes or perhaps differential forms of the classifying space. Dually one can define them more geometrically by explicit cycles on the classifying space — the famous "Schubert cycles" on the Grassmannian manifolds of subspaces of a vector space. These Schubert cycles are defined relative to some fixed (partial) flag in the ambient vector space. If we work with Fredholm operators in a Hilbert space we would need flags formed by subspaces of finite codimension. But for a twisting of infinite order such flags do not exist for reasons explained in [AS]. However, even though we cannot use the direct analogue of the Schubert cycles, there is an alternative and very natural class of geometric cycles that we can use. These are the cycles studied by Koschorke [K].

In $\mathrm{Fred}(\mathcal{H})$ we have the locally closed submanifolds $F_{p,q}$ consisting of Fredholm operators T with

$$\dim \mathrm{Ker}\, T = p \quad \dim\ \mathrm{Coker}\, T = q.$$

Of course $k = \mathrm{index}\, T = p - q$ is constant on the components $\mathrm{Fred}_k(\mathcal{H})$, so that for each k we have a simple sequence of these submanifolds.

The submanifold $F_{p,q}$, or rather its closure $\bar{F}_{p,q}$, defines an integral cohomology class $k_{p,q}$ of dimension $2pq$ called the *Koschorke class* .

Clearly the submanifolds $F_{p,q}$ are invariant under the conjugation action of $PU(\mathcal{H})$ on $\mathrm{Fred}(\mathcal{H})$, and they therefore define a submanifold of the total space of the associated fibration over $BPU(\mathcal{H})$, and so over any pull-back to a space X by a twisting class η. This exhibits a natural lift of the Koschorke classes in the fibre $\mathrm{Fred}(\mathcal{H})$ to the total space, and hence to characteristic classes in the twisted cohomology $H_\eta(X)$. We shall refer to these also as *Koschorke classes.*

There is an explicit formula expressing the $k_{p,q}$ in terms of the classical Chern classes $c_1, c_2, \ldots$ as Hankel determinants. The proof is given in [K]. In particular, since we are interested in the case where the index k is zero, we have $p = q$, and $k_{p,p}$ has dimension $2p^2$. The first few Hankel determinants are:

$$h_{1,1} = c_1$$

$$h_{2,2} = \begin{vmatrix} c_2 & c_3 \\ c_1 & c_2 \end{vmatrix} = c_2^2 - c_3 c_1$$

$$h_{3,3} = \begin{vmatrix} c_3 & c_4 & c_5 \\ c_2 & c_3 & c_4 \\ c_1 & c_2 & c_3 \end{vmatrix} .$$

In general $h_{p,q}$ is the determinant of the $q \times q$-matrix whose $(i,j)^{\mathrm{th}}$ entry is c_{i-j+p}. (Here we interpret c_0 as 1, and c_k as 0 when $k < 0$.)

To see where these formulae come from, let us suppose that the formal series

$$c(t) = 1 + c_1 t + c_2 t^2 + c_3 t^3 + \ldots$$

is the expansion of a rational function

$$c(t) = \frac{a(t)}{b(t)} = \frac{1 + a_1 t + \ldots + a_p t^p}{1 + b_1 t + \ldots + b_q t^q} .$$

This means that each c_n is expressed as a polynomial of degree $2n$ in the variables $a_1, \ldots, a_p; b_1, \ldots, b_q$, when a_k and b_k are given weight $2k$. The resulting map of polynomial rings

$$\mathbb{Q}[c_1, c_2, c_3, \ldots] \to \mathbb{Q}[a_1, \ldots, a_p] \otimes \mathbb{Q}[b_1, \ldots, b_q] \qquad (9.1)$$

can be identified with the map of rational cohomology induced by the map

$$BU_p \times BU_q \to \mathrm{Fred}_{p-q}(\mathcal{H})$$

which classifies the formal difference of bundles $E_p \otimes 1 - 1 \otimes E_q$, where E_p and E_q are the universal p- and q-dimensional bundles on BU_p and BU_q, and $a_1, \ldots, a_p$ and $b_1, \ldots, b_q$ are the universal Chern classes in $H^*(BU_p)$ and $H^*(BU_q)$. We can identify BU_n for any n with the Grassmannian manifold of n-dimensional subspaces of $\mathcal{H}$, and then we have an obvious fibration

$$F_{p,q} \to BU_p \times BU_q \,, \tag{9.2}$$

whose fibre is the general linear group of Hilbert space, which is contractible.

It will also be helpful to introduce the embedding of polynomial rings

$$\mathbb{Q}[a_1, \ldots, a_p] \otimes \mathbb{Q}[b_1, \ldots, b_q] \to \mathbb{Q}[x_1, \ldots, x_p] \otimes \mathbb{Q}[y_1, \ldots, y_q] \tag{9.3}$$

defined by the factorizations

$$a(t) = (1 + x_1 t) \ldots (1 + x_p t) \,,$$
$$b(t) = (1 + y_1 t) \ldots (1 + y_q t) \,.$$

The polynomial $R(a, b)$ of degree $2pq$ in the a_i and b_j that maps to $\prod(x_i - y_j)$ is called the *resultant* of the polynomials $a(t)$ and $b(t)$. It is well-known in classical algebra because its vanishing is the condition on the coefficients of the polynomials $a(t)$ and $b(t)$ for them to have a common factor.

The following proposition gives us the basic facts about Koschorke classes and Hankel determinants.

Proposition 9.4.

  (i) *The homomorphism 9.1 is injective in degrees* $< 2(p+1)(q+1)$.
 (ii) *The Hankel determinant* $h_{p,q}(c)$ *maps to the resultant* $R(a, b)$.
(iii) *The determinant* $h_{p',q'}(c)$ *maps to zero if* $p' > p$ *and* $q' > q$.
(iv) *The Koschorke class* $k_{p,q}$ *is given, up to sign, by the Hankel determinant* $h_{p,q}(c)$.

Proof. We begin with (iii). The relation $c(t)b(t) = a(t)$ expresses the polynomial $a(t)$ as a complex linear combination of the $q + 1$ formal power series

$$c(t), tc(t), t^2 c(t), \ldots, t^q c(t).$$

In other words, the row vector

$$(1 \quad a_1 \quad a_2 \quad \ldots \quad a_p \quad 0 \quad 0 \quad 0 \quad \ldots)$$

is a linear combination of the $q + 1$ rows

$$
\begin{matrix}
(1 & c_1 & c_2 & c_3 & \cdots & c_p & \cdots) \\
(0 & 1 & c_1 & c_2 & \cdots & c_{p-1} & \cdots) \\
(0 & 0 & 1 & c_1 & \cdots & c_{p-2} & \cdots) \\
& & & \cdots & & & \\
(0 & 0 & 0 & 0 & \cdots & c_{p-q} & \cdots).
\end{matrix}
$$

The Hankel determinant $h_{p',q+1}$ is the determinant of the $(q+1) \times (q+1)$-submatrix extracted from these rows, beginning with the entry $c_{p'}$ in the top left-hand corner. But because $a_n = 0$ if $n > p$ the rows of the submatrix are linearly dependent if $p' > p$, i.e. $h_{p',q+1}(c) = 0$ if $p' > p$. The result (iii) follows, as when $q' > q + 1$ the determinant $h_{p',q'}$ is that of a $q' \times q'$-submatrix of an array with q' rows which are a fortiori linearly dependent.

To prove (i), consider the map (9.1) with p and q replaced by $p + k$ and $q + k$. Let I_k be its kernel. Clearly I_k is zero in degrees $\leq 2n$ if $p + k \geq n$. So if a polynomial $f(c)$ of degree $2n$ belongs to I_0 we can find $k \leq n - p$ such that $f(c) \in I_{k-1}$ but $f(c) \notin I_k$. Then $f(c)$ gives rise to a non-zero polynomial of degree $2n$ in $a_1, \ldots, a_{p+k}; b_1, \ldots, b_{q+k}$ which vanishes if the polynomials $a(t)$ and $b(t)$ have a common factor. It is therefore divisible by $R(a, b)$, and so $n \geq (p + k)(q + k) \geq (p + 1)(q + 1)$, as we want.

To prove (ii), we consider the determinant $h_{p,q}(c)$ as a function of the coefficients of the rational function $a(t)/b(t)$. If the polynomials $a(t)$ and $b(t)$ have a common factor, then the proof of (iii) just given, but with q replaced by $q - 1$, tells us that $h_{p,q}(c)$ will vanish. It must therefore be divisible by the resultant polynomial $R(a, b)$. As it has the same degree $2pq$ as $R(a, b)$, it must be a rational multiple of it. But if we specialize by setting $a_i = 0$ and $b_j = 0$ except for a_p, then both $R(a, b)$ and $h_{p,q}$ become $(a_p)^q$.

Finally, to prove (iv), it is enough by (i) to show that $k_{p,q}$ and $h_{p,q}(c)$ have the same restriction to $F_{p,q}$. But let us recall that the cohomology class represented by $\bar{F}_{p,q}$ becomes, when restricted to $F_{p,q}$, the Euler class of the normal bundle to $F_{p,q}$. This normal bundle is $(E_p \otimes 1)^* \otimes (1 \otimes E_q)$, and its Euler class is $\prod(y_j - x_i)$. $\qquad\square$

The geometry shows that the Koschorke classes extend from the fibre $\mathrm{Fred}(\mathcal{H})$ to the total space Y of the universal bundle, and hence that they are *invariant* in the sense of Section 8. But we can also see the invariance from Proposition 9.4. The maps (9.1) and (9.3) are equivariant with respect to the action of the 1-parameter group of section 8, which acts on the variables x_i and y_j of (9.3) by $x_i \mapsto x_i + u$ and $y_j \mapsto y_j + u$. The resultant $\prod(x_i - y_j)$ is clearly invariant under this action, and so, therefore, is the class $h_{p,q}(c)$. It is worth remarking that the action of the 1-parameter group on the total Chern class $c(t)$ of a class of virtual dimension 0 is $c(t) \mapsto c(\tilde{t})$, where $\tilde{t} = t/(1 - ut)$.

It is clearly attractive to have such nicely defined geometric invariant classes, but unfortunately they are far from generating all invariant classes. They are too thin on the ground, as we see by noticing that their dimensions $2p^2$, i.e. 2, 8, 18, ... rise rapidly. In fact they are invariant not only under the 1-parameter group described in (8.6) but even under an action of the infinite dimensional group of formal reparametrizations

$$t \mapsto \tilde{t} = t + u_1 t^2 + u_2 t^3 + u_3 t^4 + \cdots,$$

of the line, in which the previous group sits as the subgroup of Möbius transformations

$$t \mapsto t/(1 - ut).$$

This is proved in a computational way in [Se2] Prop. 6.4, and unfortunately we do not know any conceptual proof. One complicating thing is that the larger reparametrization group does not act on the cohomology of $\mathrm{Fred}(\mathcal{H})$ by ring automorphisms, and so it does not change the Chern classes simply by $c(t) \mapsto c(\tilde{t})$, but by a twisted version of that.

It is not clear what special role the Koschorke classes play, but we believe that they deserve further study.

10. Operations in Twisted K-Theory

Ordinary K-theory has operations arising from representation theory. Thus we have the exterior powers λ^k, the symmetric powers s^k, and the Adams operations ψ^k. In [A1] the ring of operations is described in terms of the decomposition of the tensor powers of a bundle under the symmetric groups. If we define

$$R_* = \sum_k \mathrm{Hom}_{\mathbb{Z}}(R(S_k), \mathbb{Z})$$

where $R(G)$ is the character ring of a group G, then there is a natural homomorphism

$$R_* \to \mathrm{Op}(K)$$

which is an embedding. If we think of $\mathrm{Op}(K)$ in terms of self-maps of the classifying space $\mathrm{Fred}(\mathcal{H})$ of K then it acquires a topology from the skeleton filtration, and it is not hard to see that R_* has dense image in $\mathrm{Op}(K)$, so that $\mathrm{Op}(K)$ can be viewed as the completion $\hat{R}_*$.

In [A1] it is shown that R_* is a polynomial ring on the λ^k, or on the s^k. Over $\mathbb{Q}$, the ψ^k are also polynomial generators.

There is actually another natural set of operations not included in R_*. These involve the operation $*$ given by taking the dual E^* of a vector bundle E. It is natural to denote $*$ by ψ^{-1}, as can be seen by computing the Chern character of ψ^k applied to the universal bundle ξ:
if

$$\mathrm{ch}\,\xi = \sum_n \mathrm{ch}_n\,\xi$$

then

$$\mathrm{ch}\,\psi^k\xi = \sum_n k^n \mathrm{ch}_n\,\xi$$

while

$$\mathrm{ch}\,\xi^* = \sum_n (-1)^n \mathrm{ch}_n\,\xi\,.$$

The usual composition rules

$$\psi^k o \psi^l = \psi^{kl} = \psi^l o \psi^k$$

for positive integers k, l then extend also to negative integers when we define (for $k > 0$)

$$\psi^{-k} = \psi^k o \psi^{-1}\,,$$

i.e.

$$\psi^{-k}(E) = \psi^k(E^*) = [\psi^k(E)]^*\,.$$

Note. We must distinguish carefully between the *composition* $\psi^k o \psi^l$ and the *product* $\psi^k \psi^l$. Thus

$$\psi^k o \psi^l(E) = \psi^k(\psi^l E) = \psi^{kl} E$$

is quite different from the product $\psi^k(E)\psi^l(E)$.

The operation ψ^{-1} automatically belongs to the completed ring $\hat{R}_*$, and in fact

$$\psi^{-1}(x) = -\left(\sum \gamma^p(x)\right)^{-1} \sum p\gamma^p(x)$$

in terms of the operations γ^p defined below. Thus if we allow ourselves to work within the completion $\hat{R}_*$ there is nothing new to be gained from the ψ^k for negative k, but if we want to avoid completion then the negative exponents do give something new. This will become important shortly when we pass to the twisted K-theory.

The treatment in [A1] is in terms of complexes of vector bundles, but it extends naturally to Fredholm complexes [S1]. Here a single Fredholm operator

$$T : \mathcal{H} \to \mathcal{H}$$

is generalized to a complex of Hilbert spaces

$$0 \to \mathcal{H}_0 \xrightarrow{T_0} \mathcal{H}_1 \xrightarrow{T_1} \mathcal{H}_2 \cdots \to \mathcal{H}_n \to 0$$

whose homology is finite-dimensional. We can still take tensor products and decompose under the symmetric group.

Now we are ready to consider the twisted case. A twisted family of Fredholm operators over a base space X can be tensored with itself n times to give a twisted family of Fredholm complexes over X, and then decomposed under the symmetric group S_n. The important point to note is that the twisting element ζ we get is now n times the original twisting element η. This follows from Proposition (2.1)(vii) in [AS]. Thus it would seem that (for η of infinite order) we always get operations that shift the twisting, and so are not "internal". This is where the duality operation $*$ or ψ^{-1} comes to our rescue, as we shall now explain.

Consider any monomial μ in the ψ^k:

$$\mu = \psi^{k_1} \psi^{k_2} \ldots \psi^{k_l}$$

where the total degree $\sum k_i = 1$. Of course this requires that some of the k_i are *negative*. We can apply this to our Fredholm family with twisting element η and get a new Fredholm family with twisting element

$$\left(\sum k_i \right) \eta = \eta \,,$$

as follows by using Proposition (2.1)(viii) of [AS]. Thus the monomial μ does define an *internal* operation which preserves the twisting.

It seems reasonable to expect that in this way we get (over $\mathbb{Q}$) a dense set of all internal operations. However, it is not obvious how to establish this. The basic reason is that the filtration that is normally used in K-theory is generated by the Grothendieck γ^k-operations which are defined in terms of $\lambda_t = \sum \lambda^i t^i$ by

$$\gamma_t = \sum \gamma^k t^k = \lambda_{t/1-t} \,.$$

This means that γ^n is a linear combination of λ^k for $k \le n$, and so in the twisted case it mixes up the twistings. We cannot therefore use the γ^n to give a filtration on our internal operations for twisted K-theory. This might of course suggest that our whole question is unnatural and that one should consider only the K-theory got by summing over all twistings.

It also seems clear that we should not need all the monomials μ of total degree 1 — they would be over-abundant. One might, for example, hope to restrict oneself to those μ with at most one negative exponent, say k_l: these are determined by the arbitrary string of positive exponents $k_1, k_2, \ldots k_{l-1}$. We would also include the case $\mu = \psi^1$ (the identity), where there is no negative exponent.

Since we are only interested in elements of augmentation or index 0, a monomial μ will raise the filtration from 1 to l.

We now revert to the question raised in Section 8 of finding natural lifts of the invariant cohomology classes into twisted cohomology classes. We now have the twisted classes given by the μ and taking the leading term (i.e. dimension $2l$) of their Chern characters we must end up with invariant classes. Thus our internal operations, followed by the Chern character, give in principle some natural lifts. In other words we have a method to construct characteristic classes which end up in the twisted cohomology. It remains a purely algebraic problem to find some natural basis.

Note that we could use monomials in the λ^k

$$\lambda^{k_1} \lambda^{k_2} \ldots \lambda^{k_l},$$

with $\sum k_i = 1$, in exactly the same way, if we define λ^k for negative k by

$$\lambda^{-k}(\xi) = \lambda^k(\xi^*) = [\lambda^k(\xi)]^*.$$

These are more likely to produce some form of integral basis, but they are much more difficult to handle algebraically because their Chern characters are complicated.

Finally we make a remark about the Koschorke classes. Since these are defined by complex submanifolds $F_{p,q}$ of $\mathrm{Fred}(H)$, their closures $\bar{F}_{p,q}$ should define complex analytic subspaces and hence, by using a resolution of the ideal-sheaf, an element of $K(\mathrm{Fred}(\mathcal{H}))$. But this requires some care, since we are in infinite dimensions, and the approximations are more difficult to handle in K-theory because it is only filtered and not graded. However, assuming these difficulties are overcome, we shall end up with classes $\tilde{k}_{p,q}$ in the ordinary (untwisted) K-theory of the base. These provide "lifts" for the Koschorke classes in the Atiyah–Hirzebruch spectral sequence.

11. Appendix

In this appendix we shall give simple examples of spaces with non-zero Massey products of the kind occurring in Section 6 as higher differentials in the Atiyah–Hirzebruch spectral sequence of twisted K-theory. We are indebted to Elmer Rees for these examples. We shall also discuss some general issues arising from them.

We begin by introducing a 3-manifold Y which has a non-zero Massey triple product $\{x, x, y\}$ with $x, y \in H^1(Y)$. We will then take the product $Y \times \mathbb{C}P^2$ with the complex projective plane. This will have the Massey triple product $\{xt, xt, y\}$ non-zero, where t generates $H^2(\mathbb{C}P^2)$.

To construct Y we just take the $U(1)$-bundle over a 2-torus $T = S^1 \times S^1$ with Chern class 1. If x, y are closed 1-forms on T, coming from the two factors, and if z is the standard connection 1-form on Y (normalized so that its integral over the $U(1)$-fibre is 1), then $dz = xy$ gives the (normalized) curvature 2-form. The cohomology of Y has the following generators.

$$\left.\begin{array}{ll} 1 & \text{in } H^0 \\ x, y & \text{in } H^1 \\ xz, yz & \text{in } H^2 \\ xyz & \text{in } H^3 \end{array}\right\} . \tag{A1}$$

We see that the Massey triple product

$$\{x, x, y\} = xz \in H^2$$

is non-zero. Moreover, it is not in the image of multiplication by x (note that z is **not** a closed form). In fact multiplication by x annihilates all of H^1.

Now consider $M = Y \times \mathbb{C}P^2$. The generators in $(A1)$ get multiplied by $1, t, t^2$ to give the generators of $H^*(M)$. We see that

$$\{xt, xt, y\} = xzt^2 \in H^6 . \tag{A2}$$

Moreover this class is not in the image of multiplication by xt from H^3 to H^6. In fact xt annihilates all three generators xt, yt, xyz of H^3.

Thus on M, for the twisting class xt, the differential

$$d_5 : E_4 \longrightarrow E_4$$

of the Atiyah–Hirzebruch spectral sequence is non-zero on the element $y \in E_4$.

A little computation shows in fact that the ranks of $E_2, E_4, E_6 = E_\infty$ are 18, 14, 10 respectively. Thus the rank of $K_P^*(M)$ is 10 (where $[xt]$ is the class of P). This can also be checked directly by computing the cohomology of the operator $d + xt$ on differential forms as in Section 6.

Note that Y is just the Eilenberg–MacLane space $K(\pi, 1)$ for the Heisenberg group π generated by three elements a, b, c with c central and equal to the commutator of a, b. The group π is a central extension of $\mathbb{Z}^2$ by $\mathbb{Z}$. All the computations above are just computations in the cohomology of this discrete group.

We can generalize this example to a tower of examples, each a $U(1)$-bundle over its predecessor with Chern class the previous Massey product. The first step beyond Y gives a 4-manifold Z with Chern class xz, having a connection 1-form with $xz = du$. The Massey product

$$\{x, x, x, y\} = xu$$

is then non-zero. The construction provides a sequence of examples in which the successive Massey products are, at each stage, non-zero. Note that all these examples arise from discrete groups which are successive central extensions by $\mathbb{Z}$, so that again all calculations are just calculations in the cohomology of discrete groups.

A more direct construction of the $(n + 1)$-dimensional manifold Y_n in this tower is as a bundle over the circle whose fibre is the n-dimensional torus $T^n = \mathbb{R}^n / \mathbb{Z}^n$, and whose monodromy is the integral unipotent matrix $1 + N$, where N is the Jordan block with 1s in the superdiagonal. Then $H^1(Y_n) = \mathbb{Z}^2$, generated by x coming from the base circle, and y which restricts to the generator of $H^1(T^n)$ invariant under the monodromy. In Y_n the $(n + 1)$-fold Massey product $\{x, \ldots, x, y\} \in H^n(Y_n)$ is non-zero.

To get to Massey products involving a class in H^3 we again multiply by a complex projective space of the relevant dimension. Thus for Y_3 we use $Y_3 \times \mathbb{C}P^3$ and find

$$\{xt, xt, xt, y\} = xut^3 \in H^8,$$

and verify that the element

$$d_7(y) = xut^3$$

in the E_6-term of the spectral sequence is non-zero.

Clearly, in all these examples, we could replace the finite dimensional $\mathbb{C}P^n$ by $\mathbb{C}P^\infty$ and we could interpret our results in terms of equivariant calculations for the group $G = U(1)$. The class xt that we have been using is just the equivariant class generating $H^3_G(G)$ that was studied in Section 6 of [AS] (note that G acts on G by conjugation, and so trivially when $G = U(1)$). We observed there that this is the class of a natural $P(\mathcal{H})$-bundle over $G = U(1)$, where $\mathcal{H}$ is a graded Hilbert space, and the grading shifts by 1 on flowing round the circle. Our example therefore deals with this equivariant bundle P on S^1 lifted from the first factor of T^2 to Y. The example is therefore not as *ad hoc* as it might seem.

Instead of treating t as an equivariant cohomology class we can also treat it simply as a real or complex variable. Then the operator

$$D_t = d + xt$$

can be viewed as a covariant derivative for a connection, depending on a parameter t. Regarding t as constant, we have $D_t^2 = 0$, so that the connection is flat. Its cohomology is then just the standard cohomology of a flat connection. When we regard t as a variable the cohomology of D_t becomes naturally a module over the polynomial ring $\mathbb{C}[t]$.

Since $S^1 \times \mathbb{C}P^\infty$ is the universal space for pairs of integer cohomology classes $x \in H^1$, $t \in H^2$, the interpretation of twisted cohomology in terms of the cohomology of flat connections applies whenever our twisting class η decomposes as xt. More generally, the same applies whenever η is a linear combination $\eta = \sum x_i t_i$. In particular, this occurs in the following situation related to the twisted equivariant K-theory $K_{G,P}(G)$ in Section 6 of [AS], where G is a compact Lie group acting on itself by conjugation. The case we have just been examining is when $G = U(1)$, so let us now restrict to the case where G is simply-connected. In [A2] the ordinary K-theory of G was usefully studied via the natural map

$$\pi : G/T \times T \longrightarrow G \,,$$

where T is a maximal torus, and $\pi(gT, u) = gug^{-1}$. Clearly this map is G-equivariant for the conjugation action on G, so it induces a map in G-equivariant K-theory or G-equivariant cohomology. Since

$$H^*_G(G/T) = H^*_T(\text{point}) = H^*(BT) \,,$$

we get a homomorphism

$$H^*(B_T) \otimes H^*(T) \xleftarrow{\pi^*} H^*_G(G) \,.$$

The twisting class

$$[\eta] \in H^3_G(G)$$

that we introduced in [AS] then necessarily gives a decomposable element

$$\pi^*[\eta] = \sum x_i t_i\,,$$

where $t_i \in H^1(T)$ and $x_i \in H^2(BT)$. (It is easy to see there can be no term in $H^3(T)$). Thus, the twisted equivariant K-theory of $G/T \times T$ (twisted by $\pi^*\eta$) can be understood in terms of families of flat connections, parametrized by $(t_1, \ldots, t_n)$.

A crucial point, exploited in [A2], is that π preserves orientation and dimension. (Its degree is the order of the Weyl group of G). This implies, purely formally, that π^* on real cohomology is injective and canonically split. The same argument applies to twisted (real) cohomology, and to twisted K-theory (tensored with $\mathbb{R}$). It also extends to the equivariant case. Thus we can view the twisted equivariant K-theory of G (modulo torsion) as a subspace of the corresponding group for $G/T \times T$. The corresponding cohomology is therefore interpretable in terms of families of flat connections.

The groups $K_{G,P}(G)$ are the subject of the work by Freed, Hopkins, and Teleman [FHT], referred to in [AS], and are shown by them to be related to the Verlinde algebra. This in turn is connected to conformal field theory and to Chern-Simons theory. There were speculations in [A3] that Chern-Simons theory for G might be "abelianized", i.e. reduced to its maximal term T. This programme has been carried out in detail by Yoshida [Y]. There might be some connection between this programme and the process described above replacing G by $G/T \times T$.

References

[A1] M. F. Atiyah, Power operations in K-theory, *Quarterly J. Math., Oxford* (2), **17** (1966), 165–193.

[A2] M. F. Atiyah, On the K-theory of compact Lie groups, *Topology* **4** (1965), 95–99.

[A3] M. F. Atiyah, *The Geometry and Physics of Knots*, Lincei Lectures, Cambridge Univ. Press, 1990.

[AH] M. F. Atiyah and F. Hirzebruch, Vector bundles and homogeneous spaces, *Proc. Symp. Pure Mathematics* **3**, Amer. Math. Soc. 1961.

[AS] M. F. Atiyah and G. B. Segal, Twisted K-theory, *Ukrainian Math. Bull.* **1** (2004).

[BCMMS] P. Bouwknegt, A. L. Carey, V. Mathai, M. K. Murray, and D. Stevenson, Twisted K-theory and K-theory of bundle gerbes, *Comm. Math. Phys.* **228** (2002), 17–45.

[BGV] N. Berline, E. Getzler, and M. Vergne, *Heat Kernels and Dirac Operators*, Springer, Berlin, 1992.

[DGMS] P. Deligne, P. Griffiths, J. Morgan, and D. Sullivan, Real homotopy theory of Kähler manifolds, *Invent. Math.*, **29** (1975), 245–274.

[FHT] D. Freed, M. Hopkins, and C. Teleman, Twisted equivariant K-theory with complex coefficients, math.AT/0206257.

[K] U. Koschorke, Infinite dimensional K-theory and characteristic classes of Fredholm bundle maps, *Proc. Symp. Pure Math.* **15** Amer. Math. Soc. 1970, 95–133.

[Q] D. G. Quillen, Superconnections and the Chern character, *Topology* **24** (1985), 89–95.

[R] J. Rosenberg, Continuous-trace algebras from the bundle-theoretic point of view, *J. Austral. Math. Soc.* **A47** (1989), 368–381.

[S1] G. B. Segal, Fredholm complexes, *Quarterly J. Math., Oxford* **21** (1970), 385–402.

[S2] G. B. Segal, Unitary representations of some infinite dimensional groups, *Comm. Math. Phys.* **80** (1981), 301–342.

[S3] G. B. Segal, Topological structures in string theory, *Phil. Trans. Roy. Soc. London* **A359** (2001), 1389–1398.

[Su] D. Sullivan, Infinitesimal computations in topology, *Inst. Hautes Études Sci. Publ. Math.* **47** (1977), 269–331.

[TX] J.-L. Tu and P. Xu, Chern character for twisted K-theory of orbifolds, math.KT/0505267.

[Y] T. Yoshida, An abelianization of SU(2) WZW model, *Annals of Math.* (2005) (to appear).

Chapter 3

YANGIAN AND ITS APPLICATIONS

Cheng-Ming Bai, Mo-Lin Ge

Theoretical Physics Division, Chern Institute of Mathematics,
Nankai University, Tianjin 300071, P. R. China

Kang Xue

Department of Physics, Northeast Normal University,
Changchun 130024, P. R. China

In this paper, the Yangian relations are tremendously simplified for Yangians associated to $SU(2)$, $SU(3)$, $SO(5)$ and $SO(6)$ based on RTT relations that much benefit the realization of Yangian in physics. The physical meaning and some applications of Yangian have been shown.

1. Introduction

Yangian was presented by Drinfel'd ([1–3]) twenty years ago. It receives more attention for the following reasons. It is related to the rational solution of Yang–Baxter equation and the RTT relation. It is a simple extension of Lie algebras and the representation theory of the Yangian associated to $SU(2)$ has been given. Some physical models, say, two component nonlinear Schrodinger equation, Haldane–Shastry model and 1-dimensional Hubbard chain do have Yangian symmetry. Yangian may be viewed as the consequence of a "bi-spin" system. How to understand the physical meaning of Yangian is an interesting topic. In this paper, there is nothing with mathematics. Rather, we try to use the language of quantum mechanics and Lie algebraic knowledge to show the effects of Yangian.

2. Yangian and RTT Relations

Let $\mathcal{G}$ be a complex simple Lie algebra. The Yangian algebra $Y(\mathcal{G})$ associated to $\mathcal{G}$ was given as follows ([1–3]). For a given set of Lie algebraic generators I_μ of $\mathcal{G}$ the new generators J_ν were introduced to satisfy

$$[I_\lambda, I_\mu] = C_{\lambda\mu\nu} I_\nu, \quad C_{\lambda\mu\nu} \text{ are structural constants;} \qquad (2.0.1)$$

$$[I_\lambda, J_\mu] = C_{\lambda\mu\nu} J_\nu; \qquad (2.0.2)$$

and, for $\mathcal{G} \neq SU(2)$:

$$[J_\lambda, [J_\mu, I_\nu]] - [I_\lambda, [J_\mu, J_\nu]] = a_{\lambda\mu\nu\alpha\beta\gamma} \{I_\alpha, I_\beta, I_\gamma\}, \qquad (2.0.3)$$

where

$$a_{\lambda\mu\nu\alpha\beta\gamma} = \frac{1}{4!} C_{\lambda\alpha\sigma} C_{\mu\beta\tau} C_{\nu\gamma\rho} C_{\sigma\tau\rho}, \qquad (2.0.4)$$

$$\{x_1, x_2, x_3\} = \sum_{i \neq j \neq k} x_i x_j x_k, \quad \text{(symmetric summation);} \qquad (2.0.5)$$

or for $\mathcal{G} = SU(2)$:

$$[[J_\lambda, J_\mu], [I_\sigma, J_\tau]] + [[J_\sigma, J_\tau], [I_\lambda, J_\mu]]$$
$$= (a_{\lambda\mu\nu\alpha\beta\gamma} C_{\sigma\tau\nu} + a_{\sigma\tau\nu\alpha\beta\gamma} C_{\lambda\mu\nu}) \{I_\alpha, I_\beta, J_\gamma\}. \qquad (2.0.6)$$

When $C_{\lambda\mu\nu} = i\varepsilon_{\lambda\mu\nu}(\lambda, \mu, \nu = 1, 2, 3)$, equation (2.0.3) is identically satisfied from the Jacobi identities. Besides the commutation relations there are co-products as follows:

$$\Delta(I_\lambda) = I_\lambda \otimes 1 + 1 \otimes I_\lambda; \qquad (2.0.7)$$

$$\Delta(J_\lambda) = J_\lambda \otimes 1 + 1 \otimes J_\lambda + \frac{1}{2} C_{\lambda\mu\nu} I_\mu \otimes I_\nu. \qquad (2.0.8)$$

Further, the Yangian can be derived through RTT relations where R is a rational solution of Yang–Baxter equation (YBE) ([1–12]).

After lengthy calculations, we found the independent relations for $Y(SU(2))$, $Y(SU(3))$, $Y(SO(5))$ and $Y(SO(6))$ by expanding the RTT relations and also checked through equations (2.0.1)–(2.0.3) and (2.0.6) by substituting the structural constants ([13–17]), where RTT relation (Faddeev, Reshetikhin, Takhtajan — RFT [18]) satisfies

$$\check{R}(u - v)(T(u) \otimes 1)(1 \otimes T(v)) = (1 \otimes T(v))(T(u) \otimes 1)\check{R}(u - v). \qquad (2.0.9)$$

2.1. $Y(SU(2))$

Let P_{12} be the permutation. Setting

$$\check{R}_{12}(u) = PR_{12}(u) = uP_{12} + I; \tag{2.1.1}$$

$$T(u) = I + \sum_{n=1}^{\infty} u^{-n} \begin{bmatrix} T_{11}^{(n)} & T_{12}^{(n)} \\ T_{21}^{(n)} & T_{22}^{(n)} \end{bmatrix}$$

$$= I + \sum_{n=1}^{\infty} u^{-n} \begin{bmatrix} \frac{1}{2}(T_0^{(n)} + T_3^{(n)}), & T_+^{(n)} \\ T_-^{(n)}, & \frac{1}{2}(T_0^{(n)} - T_3^{(n)}) \end{bmatrix}, \tag{2.1.2}$$

and substituting the $T(u)$ into RTT relation it turns out that only

$$I_\pm = T_\pm^{(1)}, I_3 = \frac{1}{2}T_3^{(1)}; \tag{2.1.3}$$

$$J_\pm = T_\pm^{(2)}, J_3 = \frac{1}{2}T_3^{(2)} \tag{2.1.4}$$

are independent ones. The quantum determinant

$$\det T(u) = T_{11}(u)T_{22}(u-1) - T_{12}(u)T_{21}(u-1) = C_0 + \sum_{n=1}^{\infty} u^{-n}C_n \tag{2.1.5}$$

gives

$$C_0 = 1, \quad C_1 = T_0^{(1)} = tr T^{(1)}, \tag{2.1.6}$$

$$C_2 = T_0^{(2)} - \mathbf{I}^2 + T_0^{(1)}(1 + \frac{1}{2}T_0^{(1)}), \ldots, . \tag{2.1.7}$$

The independent commutation relations of $Y(SU(2))$ are:

$$[I_\lambda, I_\mu] = i\epsilon_{\lambda\mu\nu}I_\nu \quad (\lambda, \mu, \nu = 1, 2, 3); \tag{2.1.8}$$

$$[I_\lambda, J_\mu] = i\epsilon_{\lambda\mu\nu}J_\nu; \tag{2.1.9}$$

and $(A_\pm = A_1 \pm iA_2)$

$$[J_3, [J_+, J_-]] = (J_-J_+ - I_-J_+)I_3 \tag{2.1.10}$$

that can be checked to generate all of relations of equations (2.0.1), (2.0.2) and (2.0.6) with the help of Jacobi identities.

The co-product is given through (RFT) as

$$\Delta T_{ab} = \sum_c T_{ac} \otimes T_{cb} \,. \tag{2.1.11}$$

The simplest realization of $Y(SU(2))$ is

$$\mathbf{I} = \sum_{i=1}^{N} \mathbf{I}_i \quad (i : \text{lattice indices}) \,, \tag{2.1.12}$$

$$\mathbf{J} = \sum_{i=1}^{N} \mu_i \mathbf{I}_i + \sum_{i<j}^{N} W_{ij} \mathbf{I}_i \times \mathbf{I}_j \,, \tag{2.1.13}$$

where

$$W_{ij} = \begin{cases} 1 & i < j \\ 0 & i = j \quad \text{(for any representation of } SU(2)) \\ -1 & i > j \end{cases} \tag{2.1.14}$$

or

$$W_{jk} = i \cot \frac{(j-k)\pi}{N} \quad \left(\text{only for spin } \frac{1}{2}, \text{ Haldane–Shastry model [19–21]}\right), \tag{2.1.15}$$

and μ_i arbitrary constants. Noting that μ_i plays important role for the representation theory of $Y(SU(2))$ given by Chari and Pressley ([22–24]).

The big difference between representations of Lie algebra and Yangian is in that in Yangian there appear free parameters μ_i depending on models.

Another example for single particle is finite W-algebra ([25–26]). Denoting by $\mathbf{L}$ and $\mathbf{B}$ angular momentum and Lorentz boost, respectively, as well as D the dilatation operator, the set of $\mathbf{L}$ and $\mathbf{J}$ satisfies $Y(SU(2))$ where ([13], [25])

$$\mathbf{I} = \mathbf{L} \,, \tag{2.1.16}$$

$$\mathbf{J} = \mathbf{I} \times \mathbf{B} - i(D-1)\mathbf{B} \,, \tag{2.1.17}$$

and

$$[J_\alpha, J_\beta] = i\epsilon_{\alpha\beta\gamma}(2\mathbf{I}^2 - c_2' - 4)\mathbf{I}_\gamma, \quad c_2' \text{ casimir of } SO(4,2) \,. \tag{2.1.18}$$

There are the following models whose Hamiltonians do commute with $Y(SU(2))$.

- Two component nonlinear Schrodinger equation (Murakami and Wadati [27])

$$i\psi_t = -\psi_{xx} + 2c|\psi|^2\psi\,, \tag{2.1.19}$$

$$\mathbf{I} = \int dx\psi_\alpha^+(x)\left(\frac{\sigma}{2}\right)_{\alpha\beta}\psi_\beta(x); \tag{2.1.20}$$

$$\mathbf{J} = -i\int dx\psi_\alpha^+(x)\left(\frac{\sigma}{2}\right)_{\alpha\beta}\psi_\beta(x)$$

$$-\frac{ic}{2}\int dxdy\varepsilon(y-x)\left(\frac{\sigma}{2}\right)_{\beta\lambda}\psi_\beta^+(x)\psi_\alpha^+(y)\psi_\alpha(x)\psi_\lambda(y)\,. \tag{2.1.21}$$

- One-dimensional Hubbard model (for $N \to \infty$, [28])

$$H = -\sum_{i=1}^{N}(a_i^+a_{i+1} + a_{i+1}^+a_i + b_i^+b_{i+1} + b_{i+1}^+b_i)$$

$$-U\sum_{i=1}^{N}\left(a_i^+a_i - \frac{1}{2}\right)\left(a_i^+a_i - \frac{1}{2}\right); \tag{2.1.22}$$

$$J_\pm = J_1 \pm iJ_2\,,$$

$$J_+ = \sum_{i,j}\theta_{i,j}a_i^+b_j - U\sum_{i\neq j}\varepsilon_{i,j}I_i^+I_j^3\,,$$

$$J_- = \sum_{i,j}\theta_{i,j}b_i^+a_j + U\sum_{i\neq j}\varepsilon_{i,j}I_i^-I_j^3\,,$$

$$J_3 = \frac{1}{2}\left[\sum_{i,j}\theta_{i,j}(a_i^+a_j - b_i^+b_j) + U\sum_{i<j}\varepsilon_{i,j}I_i^+I_j^-\right], \tag{2.1.23}$$

where

$$\theta_{i,j} = \delta_{i,j-1} - \delta_{i,j+1}, \quad \varepsilon_{i,j} = \begin{cases} 1 & i < j\,, \\ 0 & i = j\,, \\ -1 & i > j\,. \end{cases} \tag{2.1.24}$$

Essler, Korepin and Schoutens found the complete solutions ([29–30]) and excitation spectrum ([31]) of 1-D Hubbard model chain.

- Haldane–Shastry model ([19–21]) whose Hamiltonian is given by a family. The first member is

$$H_2 = \sum_{i,j}'\left(\frac{Z_iZ_j}{Z_{ij}Z_{ji}}\right)(P_{ij} - 1)\,, \tag{2.1.25}$$

where and henceforth the $'$ stands for $i \neq j$ in the summation and $P_{ij} = 2(\mathbf{S}_i \cdot \mathbf{S}_j + \frac{1}{4})$, $Z_k = \exp^{i\pi \frac{k}{N}}$, $Z_{ij} = Z_i - Z_j$. The next reads

$$H_3 = \sum_{i,j,k}' \left(\frac{Z_i Z_j Z_k}{Z_{ij} Z_{jk} Z_{ki}} \right) (P_{ijk} - 1), \tag{2.1.26}$$

and

$$H_4 = \sum_{i,j,k,l}' \left(\frac{Z_i Z_j Z_k Z_l}{Z_{ij} Z_{jk} Z_{kl} Z_{li}} \right) (P_{ijkl} - 1) + H_4', \tag{2.1.27}$$

$$H_4' = -\frac{1}{3} H_2 - 2 \sum_{i,j}' \left(\frac{Z_i Z_j}{Z_{ij} Z_{ji}} \right)^2 (P_{ij} - 1), \tag{2.1.28}$$

where

$$P_{ijk} = P_{ij} P_{jk} + P_{jk} P_{ki} + P_{ki} P_{ij},$$
$$P_{ijkl} = P_{ij} P_{jk} P_{kl} + (\text{cyclic for } i, j, k \text{ and } l). \tag{2.1.29}$$

The eigenvalues of H_2 and H_3 have been solved in Ref. [21] and numerical calculations were made for H_4. The H_2 and H_3 were shown to be obtained in terms of quantum determinant ([32]).

- Hydrogen atom (with and without monopole, [33])

$$H = \frac{\pi^2}{2\mu} + \frac{1}{2\mu} \frac{q^2}{r^2} - \frac{\kappa}{r}, \quad \pi = p - zeA \tag{2.1.30}$$

where μ is mass, $q = zeg$, $\kappa = ze^2$ and g being monopole charge.
- Super Yang–Mills Theory ($N = 4$): $Y(SO(6))$ ([34])

$$H = 2 \sum_{\alpha} \sum_j h(j) P_{\alpha\alpha+1}^j, \quad h(j) = \sum_{k=1}^{j} \frac{1}{k}, \quad h(0) = 1, \tag{2.1.31}$$

where P^j is projector for the weight j of $SU(2)$ and α stands for "lattice" index.

2.2. $Y(SU(3))$

For the Yangian associated to $SU(3)$, there are the following independent relations

$$[I_\lambda, I_\mu] = i f_{\lambda\mu\nu} I_\nu, \quad [I_\lambda, J_\mu] = i f_{\lambda\mu\nu} J_\nu \quad (\lambda, \mu, \nu = 1, \ldots, 8). \tag{2.2.1}$$

Define

$$I_{\pm}^{(1)} = I_1 \pm iI_2, \quad U_{\pm}^{(1)} = I_6 \pm iI_7, \quad V_{\pm}^{(1)} = I_4 \mp iI_5, \quad \frac{\sqrt{3}}{2}I_8^{(1)} = I_8 \quad (2.2.2)$$

and J_μ represents the corresponding operator for $I_{\pm}^{(2)}, U_{\pm}^{(2)}, V_{\pm}^{(2)}$ and $I_8^{(2)}, I_3^{(2)}$. After lengthy calculation one finds that based on RTT relation there is only one independent relation for $Y(SU(3))$ additional to equation (2.2.1):

$$[I_8^{(2)}, I_3^{(2)}] = \frac{1}{3!}(\{I_+^{(1)}, U_+^{(1)}, V_+^{(1)}\} - \{I_-^{(1)}, U_-^{(1)}, V_-^{(1)}\}) \quad (2.2.3)$$

where $\{\cdots\}$ stands for the symmetric summation. The conclusion can be verified through both the Drinfel'd formula $(C_{\lambda\mu\nu} = if_{\lambda\mu\nu})$ and RTT relations with replacing P_{12} in $SU(2)$ by

$$P_{12} = \frac{1}{3}I + \frac{1}{2}\sum_\mu \lambda_\mu \lambda_\mu, \quad (2.2.4)$$

where λ_μ are the Gell–Mann matrices. Setting

$$T(u) = \sum_{n=0}^{\infty} u^{-n}T^{(n)}, \quad (2.2.5)$$

$$T^{(n)} = \begin{bmatrix} \frac{1}{3}T_0^{(n)} + T_3^{(n)} + \frac{1}{\sqrt{3}}T_8^{(n)} & T_1^{(n)} - iT_2^{(n)} & T_4^{(n)} - iT_5^{(n)} \\ T_1^{(n)} + iT_2^{(n)} & \frac{1}{3}T_0^{(n)} - T_3^{(n)} + \frac{1}{\sqrt{3}}T_8^{(n)} & T_6^{(n)} - iT_7^{(n)} \\ T_4^{(n)} + iT_5^{(n)} & T_6^{(n)} + iT_7^{(n)} & \frac{1}{3}T_0^{(n)} - \frac{2}{\sqrt{3}}T_8^{(n)} \end{bmatrix},$$

$$(2.2.6)$$

and substituting them into RTT relation we find equations (2.2.1)–(2.2.3) are independent relations together with the co-product, for example,

$$\Delta I_{\pm}^{(2)} = I_{\pm}^{(2)} \otimes 1 + 1 \otimes I_{\pm}^{(2)} \pm 2(I_3^{(1)} \otimes I_{\pm}^{(1)} - I_{\pm}^{(1)} \otimes I_3^{(1)})$$

$$+ \frac{1}{2}(V_{\mp}^{(1)} \otimes U_{\mp}^{(1)} - U_{\mp}^{(1)} \otimes V_{\mp}^{(1)}) \quad (2.2.7)$$

and others.

The quantum determinant of $T(u)$ which is 3 by 3 matrix for the fundamental representation of $gl(3)$ takes the form

$$\tilde{\det}_3 T(u) = T_{11}(u)\{T_{22}(u-1)T_{33}(u-2) - T_{23}(u)T_{32}(u-2)\}$$
$$- T_{12}(u)\{T_{21}(u-1)T_{33}(u-2) - T_{23}(u-1)T_{31}(u-2)\}$$
$$+ T_{13}(u)\{T_{21}(u-1)T_{32}(u-2) - T_{22}(u-1)T_{31}(u-2)\}$$
$$= \sum_p (-1)^P T_{1p_1}(u)T_{2p_2}(u-1)T_{3p_3}(u-2) \qquad (2.2.8)$$

where p stands for all the possible arrangements of (p_1, p_2, p_3). In comparison with the quantum determinant

$$\tilde{\det}_2 T(u) = \sum_{k,l,m=0}^{\infty} \frac{(l-m-1)!}{(m-1)!l!} u^{-(m+l+k)}(T_{11}^{(k)}T_{22}^{(m)} - T_{12}^{(k)}T_{21}^{(m)}), \qquad (2.2.9)$$

now we have

$$\tilde{\det}_3 T(u) = \sum_{k,l,m,p,q=0}^{\infty} \frac{(l+m-1)!}{(m-1)!l!} \frac{2^q(p+q-1)!}{(p-1)!q!} u^{-(m+l+k+p+q)}$$

$$\{T_{11}^{(k)}(T_{22}^{(m)}T_{33}^{(p)} - T_{23}^{(m)}T_{32}^{(p)}) - T_{12}^{(k)}(T_{21}^{(m)}T_{33}^{(p)} - T_{23}^{(m)}T_{31}^{(p)})$$

$$+ T_{13}^{(k)}(T_{21}^{(m)}T_{32}^{(p)} - T_{22}^{(m)}T_{31}^{(p)})\}$$

$$= \sum_{n=0}^{\infty} u^{-n} C_n, \qquad (2.2.10)$$

i.e.,

$$C_0 = 1, C_1 = T_0^{(1)}, C_2 = T_0^{(2)} + T_0^{(1)} + 2(T_0^{(1)})^2 - \mathbf{I}^2, \qquad (2.2.11)$$

$$\mathbf{I}^2 = \sum_{\lambda=1}^{\infty} \mathbf{I}_\lambda^2. \qquad (2.2.12)$$

When we constrain $\tilde{\det}T(u) = 1$ it leads to $Y(SU(2))$ and $Y(SU(3))$ that are formed by the set $\{I_\lambda, J_\lambda\}$, $\lambda = 1, 2, 3$ and $\lambda = 1, 2, \dots, 8$ for $SU(2)$ and $SU(3)$, respectively.

An example of realization of $Y(SU(3))$ is the generalization of Haldane–Shastry model ([19–21]) for the fundamental representation of generators of $SU(3)$:

$$I_\mu = \sum_i F_i^\mu, \qquad (2.2.13)$$

$$J_\mu = \sum_i \mu_i F_i^\mu + \lambda f_{\mu\lambda\nu} \sum_{i\neq j} W_{ij} F_i^\nu F_j^\lambda , \qquad (2.2.14)$$

where W_{ij} satisfies the same relation as in Haldane–Shastry model given in section 2.1 and F^μ are the Gell–Mann matrices.

2.3. $Y(SO(5))$ and $Y(SO(6))$

For $SO(N)$ it holds

$$[L_{ij}, L_{kl}] = iC_{ij,kl}^{st} L_{st} , \qquad (2.3.1)$$

where

$$C_{ij,kl}^{st} = \delta_{ik}\delta_{js}\delta_{lt} - \delta_{il}\delta_{js}\delta_{kt} - \delta_{jk}\delta_{is}\delta_{lt} + \delta_{jl}\delta_{is}\delta_{kt} . \qquad (2.3.2)$$

The rational solutions of YBE for $SO(N)$ were firstly given by Zamolodchikov's ([35]). They are also re-derived by taking the rational limit of the trigonometric R-Matrix:

$$\check{R}(u) = f(u)\left[u^2 P + u\left(A - I - \frac{3}{2}P\right)\xi + \frac{3}{2}I\xi^2\right], \qquad (2.3.3)$$

where u stands for spectral parameter and ξ the other free parameter ([36–37]). The elements of $\check{R}(u)$ are ($a, b, c, d = -2, -1, 0, 1, 2$)

$$[\check{R}(u)]_{cd}^{ab} = u^2\delta_{ab}\delta_{bc} + u(\delta_{a-b}\delta_{c-d} - \delta_{ac}\delta_{bd} - \frac{3}{2}\delta_{ad}\delta_{bc})\xi + \frac{3}{2}\delta_{ac}\delta_{bd}\xi^2 . \qquad (2.3.4)$$

For $SO(5)$, we introduce

$$T^{(1)} = \xi \begin{bmatrix} E_3 - \frac{3}{2} & U_+ & E_+ & V_+ & 0 \\ U_- & F_3 - \frac{3}{2} & F_+ & 0 & -V_+ \\ E_- & F_- & -\frac{3}{2} & -F_+ & -E_+ \\ V_- & 0 & -F_- & -F_3 - \frac{3}{2} & -U_+ \\ 0 & -V_- & -E_- & -U_- & -E_3 - \frac{3}{2} \end{bmatrix} , \qquad (2.3.5)$$

where

$$\begin{aligned}
E_3 &= E_{22} - E_{-2,-2}, & F_3 &= E_{11} - E_{-1-1}, & U_+ &= E_{21} - E_{-1-2}, \\
V_+ &= E_{2-1} - E_{1-2}, & E_+ &= E_{20} - E_{0,-2}, & F_+ &= E_{10} - E_{0-1}, \\
U_- &= E_{12} - E_{-2-1}, & V_- &= E_{-12} - E_{-2} & E_- &= E_{02} - E_{-20}, \\
F_- &= E_{01} - E_{-10;}
\end{aligned}$$

$$(2.3.6)$$

$$T_{ab}^{(2)} = \frac{3}{2}\xi^2 E_{ab}^{(2)} \quad (a,b = -2,-1,0,1,2)\,. \tag{2.3.7}$$

Substituting $T^{(n)}$ (only $n = 1, 2$ are needed to be considered) into RTT relation, there appears 35 relations for J_μ besides the Jacobi identities. However, a lengthy computation shows that besides

$$\begin{aligned}[I_\alpha, I_\beta] &= C_{\alpha\beta}^\gamma I_\gamma \\ [I_\alpha, J_\beta] &= C_{\alpha\beta}^\gamma J_\gamma\end{aligned} \qquad (\alpha = i, j)\,, \tag{2.3.8}$$

there is only one independent relation

$$[E_3^{(2)}, F_3^{(2)}] = \frac{1}{4!}(\{U_-, E_+, F_-\} - \{U_+, E_-, F_+\}$$

$$- \{V_+, E_-, F_-\} + \{V_-, E_+, F_+\})\,, \tag{2.3.9}$$

where again { } stands for the symmetric summation.

A realization of $Y(SO(5))$ is given as follows. Set

$$I_{ab}(x) = \frac{1}{2}\psi_\alpha^+(x)(I^{ab})_{\alpha\beta}\psi_\beta(x) \quad (a,b = -2,-1,0,1,2)\,, \tag{2.3.10}$$

$$\{\psi_\alpha^+(x), \psi_\beta(y)\}_+ = \delta(x - y)\delta_{\alpha\beta}\,. \tag{2.3.11}$$

Then

$$I_{ab} = \sum_x I_{ab}(x)\,, \tag{2.3.12}$$

$$J_{ab} = \sum_{x,y,c \neq a,b} \epsilon(x - y)I_{ac}(x)I_{cb}(y) \tag{2.3.13}$$

satisfies the commuting relations for $Y(SO(5))$. The following Hamiltonian of ladder model not only commutes with I_{ab}, i.e., it possesses $SO(5)$ symmetry, but also commutes with J_{ab}.

$$H = H_1 + \sum_x H_2(x) + \sum_x H_3(x); \tag{2.3.14}$$

$$H_1 = 2t_1 \sum_{\langle x,y \rangle} [c_\sigma^+(x)c_\sigma(y) + d_\sigma^+(x)d_\sigma(y) + H.C.]; \tag{2.3.15}$$

$$H_2(x) = U\left(n_{c\uparrow} - \frac{1}{2}\right)\left(n_{c\downarrow} - \frac{1}{2}\right) + (c \to d) + V(n_c - 1)(n_d - 1) + J\mathbf{S}_c \cdot \mathbf{S}_d$$

$$= \frac{J}{4}\sum_{a<b} I_{ab}^2 + \left(\frac{1}{8}J + \frac{1}{2}U\right)(\psi_\alpha^+\psi_\alpha - 2); \tag{2.3.16}$$

$$H_3(x) = -2t_3(c_\sigma^+(x)d_\sigma(x) + H.C.)\,. \tag{2.3.17}$$

Because locally $SO(6) \simeq SU(4)$ we introduce (15 generators)

$$T_{ab}^{(1)} = I_{ab}, \quad T_{ab}^{(2)} = I_{ab}^{(2)} (a, b = 1, 2, \ldots, 6) \tag{2.3.18}$$

and the $\check{R}(u)$-matrix reads

$$\check{R}(u) = f(u)[u^2 P + u\xi(A - 2P - I) + 2\xi^2 I]. \tag{2.3.19}$$

The RTT relation gives $4 + 4 + 441 + 315 + 225$ more relations. After careful calculations one finds ([15-16]) that there are the following independent relations for J_{ab} themselves:

$$[I_{12}^{(2)}, I_{34}^{(2)}] = \frac{i}{24}(\{I_{23}, I_{16}, I_{46}\} + \{I_{23}, I_{15}, I_{45}\} + \{I_{14}, I_{25}, I_{35}\}$$

$$+ \{I_{14}, I_{26}, I_{36}\} - \{I_{13}, I_{26}, I_{46}\} - \{I_{13}, I_{25}, I_{45}\}$$

$$- \{I_{24}, I_{15}, I_{35}\} - \{I_{24}, I_{16}, I_{36}\}); \tag{2.3.20}$$

$$[I_{12}^{(2)}, I_{56}^{(2)}] = \frac{i}{24}(\{I_{15}, I_{23}, I_{36}\} + \{I_{15}, I_{24}, I_{46}\} + \{I_{26}, I_{13}, I_{35}\}$$

$$+ \{I_{26}, I_{14}, I_{45}\} - \{I_{25}, I_{13}, I_{36}\} - \{I_{25}, I_{14}, I_{46}\}$$

$$- \{I_{16}, I_{23}, I_{35}\} - \{I_{16}, I_{24}, I_{45}\}); \tag{2.3.21}$$

$$[I_{34}^{(2)}, I_{56}^{(2)}] = \frac{i}{24}(\{I_{45}^{(1)}, I_{13}^{(1)}, I_{16}^{(1)}\} + \{I_{45}^{(1)}, I_{23}^{(1)}, I_{26}^{(1)}\} + \{I_{36}^{(1)}, I_{14}^{(1)}, I_{16}^{(1)}\}$$

$$+ \{I_{36}^{(1)}, I_{24}^{(1)}, I_{26}^{(1)}\} - \{I_{35}^{(1)}, I_{14}^{(1)}, I_{16}^{(1)}\} - \{I_{35}^{(1)}, I_{24}^{(1)}, I_{26}^{(1)}\}$$

$$- \{I_{46}^{(1)}, I_{13}^{(1)}, I_{16}^{(1)}\} - \{I_{46}^{(1)}, I_{23}^{(1)}, I_{26}^{(1)}\}). \tag{2.3.22}$$

3. Applications of Yangian

The first example was given by Belavin ([38]) in deriving the spectrum of nonlinear σ model. Here we only show briefly some interpretations of Yangian through the particular realizations of Yangian.

3.1. *Yangian and Hydrogen atom*

In this subsection, we re-derive the correct spectrum of Hydrogen atom (H-A) by applying the representation theory of $Y(SU(2))$.

First we recall some facts about H-A. Since Pauli, people are familiar with the $SO(4)$ symmetry for H-A, i.e. there are two conserved vectors-angular momentum $\mathbf{L}$ and Pauli–Lunge–Lenz vector $\mathbf{A} = \frac{1}{2}(\mathbf{L} \times \mathbf{P} - \mathbf{P} \times \mathbf{L}) + \frac{\kappa \mathbf{r}}{r}$ where $\mathbf{P}$ the momenta and κ the charge. Both of them commute with the

Hamiltonian $H_0 = \frac{1}{2}\mathbf{P}^2 - \frac{\kappa}{r}$. For the bound state, putting $\mathbf{B} = \frac{1}{\sqrt{-2H_0}}\mathbf{A}$, then $\mathbf{L}$ and $\mathbf{B}$ form the relations:

$$[L_\lambda, L_\mu] = i\varepsilon_{\lambda\mu\nu}L_\nu$$

$$[L_\lambda, B_\mu] = i\varepsilon_{\lambda\mu\nu}B_\nu \qquad (\lambda, \mu, \nu = 1, 2, 3) \qquad (3.1.1)$$

$$[B_\lambda, B\mu] = i\varepsilon_{\lambda\mu\nu}L_\nu$$

that lead to $SO(4)$ symmetry by introducing

$$\mathbf{I}_1 = \frac{1}{2}(\mathbf{L} + \mathbf{B}), \mathbf{I}_2 = \frac{1}{2}(\mathbf{L} - \mathbf{B}) \qquad (3.1.2)$$

satisfying

$$[I_{i\lambda}, I_{j\mu}] = i\varepsilon_{\lambda\mu\nu}\delta_{ij}I_{i\nu} \qquad (i = 1, 2). \qquad (3.1.3)$$

For H-A without monopole

$$\mathbf{L} \cdot \mathbf{B} = \mathbf{B} \cdot \mathbf{L} = 0, \qquad (3.1.4)$$

hence, one has

$$\mathbf{I}_1^2 = \mathbf{I}_2^2 = -\frac{1}{4}\left(\frac{1}{2}H_0^{-1}\kappa^2 + 1\right). \qquad (3.1.5)$$

If we denote by $k(k+1)$ the eigenvalues of $\mathbf{I}_1^2 = \mathbf{I}_2^2$, $(k = 0, \frac{1}{2}, 1, \dots)$ from equation (3.1.5) it follows immediately the spectrum of H-A:

$$E_n = -\frac{\kappa^2}{2n^2} \qquad (n = 2k + 1). \qquad (3.1.6)$$

All the above historical derivations are based on the properties of vector space. However, actually the H-A is described in terms of "spinors" $\mathbf{I}_1$ and $\mathbf{I}_2$. The vector $\mathbf{L}$ obeys

$$\mathbf{L} = \mathbf{I}_1 + \mathbf{I}_2, \qquad (3.1.7)$$

and eigenvalues of $\mathbf{L}^2 = l(l+1)$ are $l = 2k, 2k-1, \dots, 0$. In general, l can be written as

$$l = 2k - p, \qquad p = 0, 1, \dots, 2k. \qquad (3.1.8)$$

Therefore, essentially H-A takes quantum tensor space formed by $\mathbf{I}_1$ and $\mathbf{I}_2$.

On the other hand, we consider the representations of $Y(SU(2))$ in the two spin systems. Then we introduce $\mathbf{I}, \mathbf{J}$ as (see equations (2.1.12)–(2.1.14))

$$\mathbf{I} = \mathbf{I}_1 + \mathbf{I}_2; \qquad (3.1.9)$$

$$\mathbf{J} = \frac{h}{4}(a\mathbf{I}_1 + b\mathbf{I}_2 + i\mathbf{I}_1 \times \mathbf{I}_2) \qquad (3.1.10)$$

where a and b are arbitrary constants. They act on the tensor space $V_1 \otimes V_2$, which V_1 and V_2 are the irreducible representations of $SU(2)$ with the highest weight j_1, j_2 respectively. Suppose $j_1 \geq j_2$, then as representations of $SU(2)$, we have

$$j_1 \otimes j_2 = j_1+j_2 \oplus j_1+j_2-1 \oplus \cdots \oplus j_1-j_2 = \sum_{p=0}^{2j_2} j_1+j_2-p. \quad (3.1.11)$$

Since $[\mathbf{J}^2, \mathbf{I}^2] = 0$ and $[\mathbf{J}^2, I_z] = 0$, we know that $\mathbf{J}^2$ acts on the irreducible subrepresentation of weight $j_1 + j_2 - p$ by constant, that is,

$$\mathbf{J}^2 \alpha_{j_1+j_2-p,m} = J_{j_1,j_2}^2(p)\alpha_{j_1+j_2-p,m}\,, \qquad (3.1.12)$$

where $\alpha_{j_1+j_2-p,m}$ is the weight vector of the subrepresentation with the highest weight $j_1 + j_2 - p$, and

$$J_{j_1,j_2}^2(p) = \frac{h^2}{16}\{j_1(j_1+1)a^2 + j_2(j_2+1)b^2 + [2(j_1-p)(j_2-p) - p(p+1)]ab\}$$

$$- \frac{h^2}{4}\bigg\{(j_1+j_2)j_1j_2 + \frac{p}{4}(2j_1+2j_2+1-p)$$

$$\times [4j_1j_2 + 2 - p(2j_1+2j_2+1-p)]\bigg\}. \qquad (3.1.13)$$

It was pointed out in Ref. [13] that H-A possesses Yangian symmetry. In fact, for H-A the set formed by

$$\mathbf{I} = \mathbf{L} = \mathbf{I}_1 + \mathbf{I}_2\,, \qquad (3.1.14)$$

and

$$\mathbf{J} = \frac{ih}{4}\mathbf{L} \times \mathbf{B} \qquad (3.1.15)$$

satisfies $Y(SU(2))$ and $[H_0, \mathbf{I}] = [H_0, \mathbf{J}] = 0$, i.e. $[H_0, Y(SU(2))] = 0$, where $\mathbf{B} = \frac{1}{\sqrt{-2H_0}}\mathbf{A}$ and $\mathbf{A}$ is Pauli–Lunge–Lenz vector. In comparison to equation (3.1.10) on account of equation (3.1.2), we have

$$a = -1, \qquad b = 1\,. \qquad (3.1.16)$$

For $j_1 = j_2 = k$ and $b = -a = 1$, equation (3.1.13) is simplified to

$$J_k^2 = \left(+\frac{h^2}{16}\right)(2k-p)(2k-p+1)[-4k(k+1) + (2k-p)(2k-p+1) + 1]\,. \qquad (3.1.17)$$

We emphasize that equation (3.1.17) is completely the consequence of representation of $Y(SU(2))$ for $b = -a = 1$ and $j_1 = j_2 = k$ ($\mathbf{I}_1^2 = j_1(j_1 + 1)$) and $\mathbf{J}^2$ only works in tensor space.

On account of equation (3.1.15) it is easy to find

$$\mathbf{J}^2 = \left(-\frac{h^2}{16}\right)\mathbf{L}^2(\mathbf{B}^2 - 1). \tag{3.1.18}$$

Noting that

$$\mathbf{L}^2 = \mathbf{I}_1^2 + \mathbf{I}_2^2 + 2\mathbf{I}_1 \cdot \mathbf{I}_2 \tag{3.1.19}$$

whose eigenvalue is $l(l + 1)$, whereas

$$\mathbf{B}^2 = \mathbf{I}_1^2 + \mathbf{I}_2^2 - 2\mathbf{I}_1 \cdot \mathbf{I}_2 \tag{3.1.20}$$

whose eigenvalue is $4k(k + 1) - l(l + 1)$ for H-A without monopole which leads to $\mathbf{I}_1^2 = \mathbf{I}_2^2 = k(k + 1)$. Of course, since $\mathbf{L} = \mathbf{I}_1 + \mathbf{I}_2$, the eigenvalues of l take $(2k + 1)$ values: $l = 2k - p$, $(p = 0, 1, \ldots, 2k)$. Substituting equation (3.1.19) and equation (3.1.20) into equation (3.1.18) it follows the eigenvalue of $\mathbf{J}^2$:

$$J^2 = \left(\frac{h^2}{16}\right)l(l + 1)[-4k(k + 1) + l(l + 1) + 1]. \tag{3.1.21}$$

On the other hand the direct calculation of $\mathbf{J}^2$ in terms of $\mathbf{L}$ and $\mathbf{B}$ gives

$$\mathbf{J}^2 = \frac{h^2}{16}\mathbf{L}^2\left(\mathbf{L}^2 + \frac{1}{2}H_0^{-1}\kappa^2 + 2\right) \tag{3.1.22}$$

whose eigenvalue

$$J^2 = \frac{h^2}{16}l(l + 1)\left[l(l + 1) + \frac{\kappa^2}{2E} + 2\right]. \tag{3.1.23}$$

Identifying equation (3.1.21) with equation (3.1.23) it yields

$$E = -\frac{\kappa^2}{2(2k + 1)^2} = -\frac{\kappa^2}{2n^2} \quad \left(k = 0, \frac{1}{2}, 1, \ldots\right). \tag{3.1.24}$$

Substituting $l = 2k - p$ into equation (3.1.23) the eigenvalues of $\mathbf{J}^2$ on the basis of direct calculation for H-A are given by

$$J^2 = \frac{h^2}{16}(2k - p)(2k - p + 1)[(2k - p)(2k - p + 1) + 1 - 4k(k + 1)] \tag{3.1.25}$$

that is exactly the same as equation (3.1.17) given by Yangian representation independently.

Therefore based on the tensor space of $\mathbf{I}_1$ and $\mathbf{I}_2$ i.e. $SO(4)$ space we have shown that the "current" $\mathbf{J}$ provides the correct spectrum of H-A in terms of the representation of Yangian. The deep reason for this is H-A is related to RTT relation ([14]) which is also the origin of Yangian.

3.2. Reduction of $Y(SU(2))$

As was pointed out in Section 2.1, the simplest realization of $Y(SU(2))$ is made of two-spin system with $\mathbf{S_1}$ and $\mathbf{S_2}$ (any dimensional representations of $SU(2)$):

$$\mathbf{J}' = \frac{1}{\mu+\nu}\mathbf{J} = \frac{1}{\mu+\nu}(\mu\mathbf{S_1}\times\mathbf{1} + \nu\mathbf{1}\times\mathbf{S_2} + \lambda\mathbf{S_1}\times\mathbf{S_2}). \tag{3.2.1}$$

For $S_1 = S_2 = 1/2$, when

$$\mu\nu = \lambda^2, \tag{3.2.2}$$

we prove that after the following similar transformation

$$\mathbf{Y} = A\mathbf{J}'A^{-1}, \qquad A = \begin{bmatrix} 1 & 0 & 0 & 0 \\ 0 & \nu & i\lambda & 0 \\ 0 & i\lambda & \nu & 0 \\ 0 & 0 & 0 & 1 \end{bmatrix}, \tag{3.2.3}$$

the Yangian reduces to $SO(4)$: $(\rho = \nu + i\lambda = \sqrt{\nu^2 + \lambda^2}e^{i\theta})$

$$Y_1 = \begin{bmatrix} M_1 & 0 \\ 0 & L_1 \end{bmatrix}, \quad M_1 = \frac{1}{2}\begin{bmatrix} 0 & \rho \\ \rho^{-1} & 0 \end{bmatrix}, \quad L_1 = \frac{1}{2}\begin{bmatrix} 0 & \rho^{-1} \\ \rho & 0 \end{bmatrix},$$

$$Y_2 = \begin{bmatrix} M_2 & 0 \\ 0 & L_2 \end{bmatrix}, \quad M_2 = \frac{1}{2}\begin{bmatrix} 0 & -i\rho \\ i\rho^{-1} & 0 \end{bmatrix}, \quad L_2 = \frac{1}{2}\begin{bmatrix} 0 & -i\rho^{-1} \\ i\rho & 0 \end{bmatrix},$$

$$Y_3 = \begin{bmatrix} \frac{1}{2}\sigma_3 & 0 \\ 0 & \frac{1}{2}\sigma_3 \end{bmatrix}, \quad M_3 = \frac{1}{2}\sigma_3. \tag{3.2.4}$$

and

$$\mathbf{Y}^2 = \frac{1}{2}\left(\frac{1}{2}+1\right) = \frac{3}{4}. \tag{3.2.5}$$

Namely, under $\mu\nu = \lambda^2$, the $\mathbf{Y}$ reduces to $SO(4)$ by $M_\pm = M_1 \pm iM_2$, $M_+ = \rho\sigma_+$, $M_- = \rho^{-1}\sigma_-$. The scaled $M_\pm$ and M_3 still satisfy the $SU(2)$ relations:

$$[M_3, M_\pm] = \pm M_\pm, \qquad [M_+, M_-] = 2M_3, \tag{3.2.6}$$

and there are the similar relations for $\mathbf{L}$.

It should be emphasized that here the new "spin" $\mathbf{M}$ (and $\mathbf{L}$) is the consequence of two spin($\frac{1}{2}$) interaction. As usual for two 2-dimensional representations of $SU(2)$ (Lie algebra)

$$\underline{2} \bigotimes \underline{2} = \underline{3} \quad \text{(spin triplet)} \bigoplus \underline{1} \quad \text{(singlet)}. \tag{3.2.7}$$

However, here we meet a different decomposition:

$$\underline{2} \bigotimes \underline{2} = \underline{2}(\mathbf{M}) \bigoplus \underline{2}(\mathbf{L}). \tag{3.2.8}$$

Furthermore, we also find that the condition (3.2.2) holds if and only if the $\mathbf{J}^2$ acts on the whole space $\frac{1}{2} \bigotimes \frac{1}{2}$ by a constant, that is $J^2_{j_1,j_2}(p)$ given in equation (3.1.13) does not depend on p, where $j_1 = j_2 = \frac{1}{2}$ and $p = 0, 1$. Comparing the realizations of $Y(SU(2))$ given by equation (3.1.10) and equation (3.2.1), we know that when $ab = -\frac{1}{4}$, the Yangian given by

$$\mathbf{J} = \frac{h}{4} \left(a\mathbf{I}_1 + b\mathbf{I}_2 + \frac{\sqrt{-1}}{2}\mathbf{I}_1 \times \mathbf{I}_2 \right) \tag{3.2.9}$$

for $j_1 = j_2 = \frac{1}{2}$ reduces to $SO(4)$.

For a general case, if $Y(SU(2))$ can be reduced to $SO(4)$, it is natural to consider that action of $\mathbf{J}^2$ on the whole tensor space should be a constant. Let $\mathbf{J}$ be given in equation (3.2.9). Then we rewrite equation (3.1.13) for it as

$$J^2_{j_1,j_2}(p) = \frac{h^2}{16}[j_1(j_1+1)a^2 + j_2(j_2+1)b^2 + 2j_1j_2ab - (j_1+j_2)j_1j_2]$$

$$- \frac{h^2}{64}p(2j_1 + 2j_2 - p + 1)[4ab + 4j_1j_2 + 2 - p(2j_1 + 2j_2 + 1 - p)]. \tag{3.2.10}$$

Hence the term involving p is

$$- \frac{h^2}{64}p(2j_1 + 2j_2 - p + 1)[4ab + 4j_1j_2 + 2 - p(2j_1 + 2j_2 + 1 - p)]. \tag{3.2.11}$$

Since $p = 0, 1, \ldots, 2\min(j_1, j_2)$, equation (3.2.11) does not depend on p if and only if there are exactly two choices of the value of p, that is, $p = 0, 1$. Under this sense $j_1 = l$ for any l and $j_2 = \frac{1}{2}$. Hence by equation (3.2.11), we know that if

$$ab = -\frac{1}{4}, \tag{3.2.12}$$

then the action of $\mathbf{J}^2$ is given by

$$J^2_{l,\frac{1}{2}}(p) = \frac{h^2}{16}\left[l(l+1)a^2 + \frac{3}{4}b^2 + lab - \frac{1}{2}l\left(l + \frac{1}{2}\right) \right] \tag{3.2.13}$$

which does not depend on p.

Introduce the component of **J** as

$$J_{\pm} = \frac{h}{4}\left[aI_{1,\pm} + bI_{2,\pm} \pm \frac{1}{2}(I_{1,z}I_{2,\pm} - I_{1,\pm}I_{2,z})\right]; \tag{3.2.14}$$

$$J_3 = \frac{h}{4}[aI_{1,z} + bI_{2,z} + (I_{1,+}I_{2,-} - I_{1,-}I_{2,+})]. \tag{3.2.15}$$

Let $V_{\pm}$ be the subspace spanned by the vectors

$$J_{\pm}^t \alpha_{l+\frac{1}{2}, \mp l \mp \frac{1}{2}}, t = 0, 1, \ldots, 2l, \tag{3.2.16}$$

where $\alpha_{l+\frac{1}{2}, l+\frac{1}{2}}$ is the highest weight vector ($\alpha_{l+\frac{1}{2}, -l-\frac{1}{2}}$ is the lowest weight vector) of the irreducible subrepresentation of $SU(2)$ with the highest weight $l + \frac{1}{2}$ according to the decomposition (3.1.11) for $j_1 = l, j_2 = \frac{1}{2}$,

$$l \bigotimes \frac{1}{2} = l + \frac{1}{2} \bigoplus l - \frac{1}{2}. \tag{3.2.17}$$

The two vector spaces $V_{\pm}$ are stable under the action of $J_{\pm}$ and J_3. Moreover, the action of **J** on both V_+ and V_- is as the same as the action $\frac{h}{4}(a+b)\mathbf{I}$ (of the representation of $SU(2)$) with weight l. Hence we have a different decomposition (see Figure 1):

$$\underline{2l+1} \bigotimes \underline{2} = \underline{2l+1} \bigoplus \underline{2l+1}. \tag{3.2.18}$$

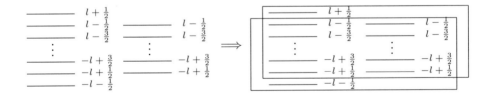

Fig. 1.   Reduction of $Y(SU(2))$.

The above study can be generalized to $SU(3)$'s fundamental representation

$$J_{\lambda} = uI_1^{\lambda} + vI_2^{\lambda} + \lambda f_{\lambda\mu\nu} \sum_{i<j} F_{1i}^{\mu} F_{2j}^{\nu}, \tag{3.2.19}$$

$$[F_{i\mu}, F_{j\nu}] = if_{\mu\nu\lambda}F_{i\lambda}\delta_{ij} \quad (\lambda, \mu, \nu = 1, 2, \ldots, 8). \tag{3.2.20}$$

Under the condition

$$uv = \lambda^2, \qquad v + i\lambda = \rho, \tag{3.2.21}$$

and the similar transformation

$$Y_\mu = AJ_\mu A^{-1}/(u+v), \quad A = \begin{bmatrix} 1 & 0 & 0 & 0 & 0 & 0 & 0 & 0 & 0 \\ 0 & \nu & 0 & i\lambda & 0 & 0 & 0 & 0 & 0 \\ 0 & 0 & \nu & 0 & 0 & 0 & i\lambda & 0 & 0 \\ 0 & i\lambda & 0 & \nu & 0 & 0 & 0 & 0 & 0 \\ 0 & 0 & 0 & 0 & 1 & 0 & 0 & 0 & 0 \\ 0 & 0 & 0 & 0 & 0 & \nu & 0 & i\lambda & 0 \\ 0 & 0 & i\lambda & 0 & 0 & 0 & \nu & 0 & 0 \\ 0 & 0 & 0 & 0 & 0 & i\lambda & 0 & \nu & 0 \\ 0 & 0 & 0 & 0 & 0 & 0 & 0 & 0 & 1 \end{bmatrix},$$

$$\tag{3.2.22}$$

the Yangian then reduces to

$$Y(I_-) = \begin{bmatrix} \rho^{-1}I_- & 0 & 0 \\ 0 & \rho I_- & 0 \\ 0 & 0 & I_- \end{bmatrix}, \quad Y(I_+) = \begin{bmatrix} \rho I_+ & 0 & 0 \\ 0 & \rho^{-1}I_- & 0 \\ 0 & 0 & I_3 \end{bmatrix},$$

$$Y(I_8) = \frac{\sqrt{3}}{3}\begin{bmatrix} \lambda_3 & 0 & 0 \\ 0 & \lambda_3 & 0 \\ 0 & 0 & \lambda_3 \end{bmatrix}, \quad Y(I_3) = \frac{1}{2}\begin{bmatrix} \lambda_3 & 0 & 0 \\ 0 & \lambda_3 & 0 \\ 0 & 0 & \lambda_3 \end{bmatrix},$$

$$Y(U_+) = \begin{bmatrix} U_+ & 0 & 0 \\ 0 & \rho U_+ & 0 \\ 0 & 0 & \rho^{-1}U_+ \end{bmatrix}, \quad Y(U_-) = \begin{bmatrix} U_- & 0 & 0 \\ 0 & \rho^{-1}U_- & 0 \\ 0 & 0 & \rho U_- \end{bmatrix},$$

$$Y(V_+) = \begin{bmatrix} \rho^{-1}V_- & 0 & 0 \\ 0 & V_- & 0 \\ 0 & 0 & \rho V_- \end{bmatrix}, \quad Y(V_-) = \begin{bmatrix} \rho V_- & 0 & 0 \\ 0 & V_- & 0 \\ 0 & 0 & \rho^{-1}V_- \end{bmatrix}.$$

$$\tag{3.2.23}$$

The usual decomposition through the Clebsch-Gordan coefficients for the representations of Lie algebra $SU(3)$ is $\underline{3} \otimes \underline{3} = \underline{6} \oplus \underline{3}$. However, here we have

$$\underline{3} \otimes \underline{3} = \underline{3} \oplus \underline{3} \oplus \underline{3}, \tag{3.2.24}$$

and

$$\sum_{\lambda=1}^{8} Y_\lambda^2 = \frac{1}{u+v} \sum_{\lambda=1}^{\infty} J_\lambda^2 = \frac{1}{3}.$$ (3.2.25)

It is easy to check that the rescaling factor ρ does not change the commutation relations for $SU(3)$ formed by $I_\pm$, $U_\pm$, $V_\pm$, I_3 and I_8. In general, we guess for the fundamental representation of $SU(n)$ we shall meet

$$\underline{n} \bigotimes \underline{n} = \underline{n} \bigoplus \underline{n} \bigoplus \underline{n} \bigoplus \cdots \bigoplus \underline{n} \quad (n \text{ times}).$$ (3.2.26)

Next we consider Yang–Mills gauge field for reduced $Y(SU(2))$. For a tensor wave function ($x \equiv \{x_1, x_2, x_3, x_0\}$),

$$\Psi(x) = \|\psi_{ij}(x)\| \quad (i, j = 1, 2, 3, 4).$$ (3.2.27)

An isospin transformation yields

$$\Psi'(x) = U(x)\Psi(x), \quad U(x) = 1 - i\theta^a J_a,$$ (3.2.28)

where

$$J^a = uS_a \otimes \mathbf{1} + v\mathbf{1} \otimes S_a + 2\lambda\epsilon_{abc}S^b \otimes S^c,$$ (3.2.29)

or

$$[J_a]_{\gamma\delta}^{\alpha\beta} = u(S^a)_{\alpha\gamma}\delta_{\beta\delta} + v(S^a)_{\beta\delta}\delta_{\alpha\gamma} + i\alpha\varepsilon_{abc}(S^b)_{\alpha\gamma}(S^c)_{\beta\delta}.$$ (3.2.30)

Define

$$D_\mu = \partial_\mu + gA_\mu,$$ (3.2.31)

i.e.,

$$[D_\mu\psi]_{\alpha\beta} = \partial_\mu\psi_{\alpha\beta} + gA_\mu^a[Y_a]_{\gamma\delta}^{\alpha\beta}\psi_{\gamma\delta}(x), \quad A_\mu = A_\mu^a J_a.$$ (3.2.32)

The gauge-covariant derivative should preserve

$$\delta(D_\mu\psi) = 0,$$ (3.2.33)

i.e.,

$$(-i\partial_\mu\theta^a(x) + g\delta A_\mu^a)[Y_a]_{\gamma\delta}^{\alpha\beta} - ig\theta^a(x)A_\mu^b[J_b, J_a]_{\gamma\delta}^{\alpha\beta} = 0.$$ (3.2.34)

When $uv = \lambda^2$ and by rescaling

$$Y_a = (u+v)J_a,$$ (3.2.35)

we have

$$\delta A_\mu^a = \epsilon_{abc}\theta^b(x)A_\mu^c(x) + \frac{i}{g}\partial_\mu\theta^a(x)\,, \qquad (3.2.36)$$

and

$$F_{\mu\nu} = \frac{1}{g}[D_\mu, D_\nu] = F_{\mu\nu}^a Y_a\,, \qquad (3.2.37)$$

$$F_{\mu\nu}^a = \partial_\mu A_\gamma^a - \partial_\nu A_\mu^a + ig\epsilon_{abc}A_\mu^b A_\gamma^c\,. \qquad (3.2.38)$$

Here the tensor isospace has been separated to two irrelevant spaces, i.e.,
$\Psi - \begin{bmatrix} \Psi_1 & 0 \\ 0 & \Psi_2 \end{bmatrix}$ where Ψ_1 and Ψ_2 are 2×2 wavefunction.

3.3. $\mathbf{J}^2$ *as a new quantum number*

Because $[\mathbf{I}^2, \mathbf{J}^2] = 0$, $[\mathbf{I}^2, I_z] = 0, [\mathbf{J}^2, I_z] = 0$, but $[\mathbf{J}^2, J_z] \neq 0$, we can take $\{\mathbf{I}^2, I_z, \mathbf{J}^2\}$ as a conserved set.

First we consider the case $\mathbf{S}_1 \otimes \mathbf{S}_2 \otimes \mathbf{S}_3$, where $S_1 = l_1, S_2 = l_2, S_3 = l_3$. We shall show that instead of 6-j coefficients and Young diagrams, $\mathbf{J}^2$ can be viewed as a "collective" quantum number that describes the "history" besides $\mathbf{S}^2$ ($\mathbf{S} = \mathbf{S}_1 + \mathbf{S}_2 + \mathbf{S}_3$) and S_z.

We introduce $\mathbf{J}$ as

$$\mathbf{J} = \sum_{i=1}^3 u_i\mathbf{S}_i + \frac{\sqrt{-1}}{2}h\sum_{i<j}^3 (\mathbf{S}_i \times \mathbf{S}_j)\,. \qquad (3.3.1)$$

Then we know that

$$\mathbf{J}^2 = [u_1^2 l_1(l_1 + 1) + u_2^2 l_2(l_2 + 1) + u_3^2 l_3(l_3 + 1)]$$

$$- \frac{1}{4}h^2[l_1(l_1 + 1)l_3(l_3 + 1) + l_1(l_1 + 1)l_2(l_2 + 1) + l_2(l_2 + 1)l_3(l_3 + 1)]$$

$$+ \left[2u_1u_2 - \frac{1}{2}h^2 l_3(l_3 + 1)\right]\mathbf{S}_1 \cdot \mathbf{S}_2 + \left[2u_2u_3 - \frac{1}{2}h^2 l_1(l_1 + 1)\right]\mathbf{S}_2 \cdot \mathbf{S}_3$$

$$+ \left[2u_1u_3 + \frac{1}{2}h^2 l_2(l_2 + 1)\right]\mathbf{S}_1 \cdot \mathbf{S}_3 + ih(u_1 - u_2 + u_3)\mathbf{S}_1 \cdot (\mathbf{S}_2 \times \mathbf{S}_3)$$

$$+ \frac{1}{4}h^2[(\mathbf{S}_1 \cdot \mathbf{S}_2 + \mathbf{S}_2 \cdot \mathbf{S}_3 + \mathbf{S}_1 \cdot \mathbf{S}_3)^2 + (\mathbf{S}_1 \cdot \mathbf{S}_2 + \mathbf{S}_2 \cdot \mathbf{S}_3 + \mathbf{S}_1 \cdot \mathbf{S}_3)]$$

$$- \frac{1}{2}h^2[(\mathbf{S}_1 \cdot \mathbf{S}_2)(\mathbf{S}_2 \cdot \mathbf{S}_3) + (\mathbf{S}_2 \cdot \mathbf{S}_3)(\mathbf{S}_1 \cdot \mathbf{S}_2)]\,. \qquad (3.3.2)$$

Here we give several examples with the explicit action of $\mathbf{J}^2$.

(1) $S_1 = S_2 = S_3 = \frac{1}{2}$.

As representations of Lie algebra $SU(2)$, we have

$$\left(\frac{1}{2} \otimes \frac{1}{2}\right) \otimes \frac{1}{2} = (1 \oplus 0) \otimes \frac{1}{2} = \frac{3}{2} \oplus \frac{1}{2} \oplus \frac{1'}{2}. \tag{3.3.3}$$

Noting that $|\frac{1}{2}\rangle$ and $|\frac{1'}{2}\rangle$ are degenerate regarding the total spin $\frac{1}{2}$. The usual Lie algebraic base can be easily written as

$$\phi_{\frac{3}{2},\frac{3}{2}} = |\uparrow\uparrow\uparrow\rangle,$$

$$\phi_{\frac{3}{2},\frac{1}{2}} = \frac{1}{\sqrt{3}}(|\uparrow\uparrow\downarrow\rangle + |\uparrow\downarrow\uparrow\rangle + |\downarrow\uparrow\uparrow\rangle),$$

$$\phi_{\frac{3}{2},-\frac{1}{2}} = \frac{1}{\sqrt{3}}(|\uparrow\downarrow\downarrow\rangle + |\downarrow\uparrow\downarrow\rangle + |\downarrow\downarrow\uparrow\rangle),$$

$$\phi_{\frac{3}{2},-\frac{3}{2}} = |\downarrow\downarrow\downarrow\rangle, \tag{3.3.4}$$

and the two degeneracy states with respect to $\mathbf{S}^2$ and S_z are given by:

$$\phi'_{\frac{1}{2},\frac{1}{2}} = \frac{1}{\sqrt{6}}(|\downarrow\uparrow\uparrow\rangle + |\uparrow\downarrow\uparrow\rangle - 2|\uparrow\uparrow\downarrow\rangle),$$

$$\phi'_{\frac{1}{2},-\frac{1}{2}} = \frac{1}{\sqrt{6}}(|\uparrow\downarrow\downarrow\rangle + |\downarrow\uparrow\downarrow\rangle - 2|\downarrow\downarrow\uparrow\rangle),$$

$$\phi_{\frac{1}{2},\frac{1}{2}} = \frac{1}{\sqrt{2}}(|\downarrow\uparrow\uparrow\rangle - \uparrow\downarrow\uparrow\rangle),$$

$$\phi_{\frac{1}{2},-\frac{1}{2}} = \frac{1}{\sqrt{2}}(|\uparrow\downarrow\downarrow\rangle - |\downarrow\uparrow\downarrow\rangle). \tag{3.3.5}$$

By equation (3.3.2), we know that

$$\mathbf{J}^2 \phi_{\frac{3}{2},m} = \left[\frac{3}{4}(u_1^2 + u_2^2 + u_3^2) + \frac{1}{2}(u_1 u_2 + u_2 u_3 + u_1 u_3) - h^2\right] \phi_{\frac{3}{2},m};$$

$$\mathbf{J}^2 \phi'_{\frac{1}{2},m} = \left[\frac{3}{4}(u_1^2 + u_2^2 + u_3^2) + \frac{1}{2}u_1 u_2 - u_2 u_3 - u_1 u_3 - \frac{7}{4}h^2\right] \phi'_{\frac{1}{2},m}$$

$$- \frac{\sqrt{3}}{2}(u_1 - u_2 + h)(u_3 + h)\phi_{\frac{1}{2},m};$$

$$\mathbf{J}^2 \phi_{\frac{1}{2},m} = -\frac{\sqrt{3}}{2}(u_1 - u_2 - h)(u_3 - h)\phi'_{\frac{1}{2},m}$$

$$+ \left[\frac{3}{4}(u_1 - u_2)^2 + \frac{3}{4}u_3^2 - \frac{3}{4}h^2\right]\phi_{\frac{1}{2},m}. \tag{3.3.6}$$

In order to make the matrix of $\mathbf{J}^2$ be symmetric (then it surely can be diagonalized), one should put

$$u_2 = u_1 + u_3 . \tag{3.3.7}$$

The eigenvalues of $\mathbf{J}^2$ are given by

$$\lambda_{\frac{3}{2}} = 2u_1^2 + 2u_3^2 + 3u_1 u_3 - h^2,$$

$$\lambda_{\frac{1}{2}}^{\pm} = u_1^2 + u_3^2 - \frac{5}{4}h^2 \pm \frac{1}{2}[(2u_1^2 - u_3^2 - h^2)^2 + 3(u_3^2 - h^2)^2]^{\frac{1}{2}} . \tag{3.3.8}$$

The eigenstates of $\mathbf{J}^2$ are the rotation of $\phi'_{\frac{1}{2},m}$ and $\phi_{\frac{1}{2},m}$:

$$\begin{pmatrix} \alpha_{\frac{1}{2},m}^+ \\ \alpha_{\frac{1}{2},m}^- \end{pmatrix} = \begin{pmatrix} \cos\frac{\varphi}{2} & -\sin\frac{\varphi}{2} \\ \sin\frac{\varphi}{2} & \cos\frac{\varphi}{2} \end{pmatrix} \begin{pmatrix} \phi'_{\frac{1}{2},m} \\ \phi_{\frac{1}{2},m} \end{pmatrix}, \quad \mathbf{J}^2 \alpha_{\frac{1}{2}}^{\pm} = \lambda_{\frac{1}{2}}^{\pm} \alpha_{\frac{1}{2},m}^{\pm}, \tag{3.3.9}$$

where

$$\sin\varphi = \sqrt{3}(u_3^2 - h^2)/\omega, \quad \omega^2 = (2u_1^2 - u_3^2 - h^2)^2 + 3(u_3^2 - h^2)^2 . \tag{3.3.10}$$

It is worth noting that the conclusion is independent of the order, say, $(\frac{1}{2} \otimes \frac{1}{2}) \otimes \frac{1}{2}, \frac{1}{2} \otimes (\frac{1}{2} \otimes \frac{1}{2})$ and the other way. The difference is only in the value of φ.

(2) $S_1 = S_2 = \frac{1}{2}, S_3 = l$.

The above example can be generalized to $\mathbf{S}_1 \otimes \mathbf{S}_2 \otimes \mathbf{L}$ where $S_1 = S_2 = \frac{1}{2}$ and $\mathbf{L}^2 = l(l+1)$. As representations of Lie algebra $SU(2)$, we have

$$\left(\frac{1}{2} \otimes \frac{1}{2}\right) \otimes l = \left(1 \oplus 0\right) \otimes l = l+1 \oplus \quad l \quad \oplus l-1$$
$$l \tag{3.3.11}$$

There are no degeneracy for $l \pm 1$, but two l states can be distinguished in terms of $\mathbf{J}^2$

$$\mathbf{J}^2 \Phi_{l+1,m} = \left\{ \frac{3}{4}(u_1^2 + u_2^2) + l(l+1)u_3^2 + \frac{1}{2}u_1 u_2 + l(u_2 u_3 + u_1 u_3) \right.$$

$$\left. - h^2 \left[l(l+1) + \frac{1}{4} \right] \right\} \Phi_{l+1,m} ,$$

$$\mathbf{J}^2 \Phi_{l-1,m} = \left\{ \frac{3}{4}(u_1^2 + u_2^2) + l(l+1)u_3^2 + \frac{1}{2}u_1 u_2 - (l+1)u_1 u_3 - (l+1)u_2 u_3 \right.$$

$$\left. - h^2 \left[l(l+1) + \frac{1}{4} \right] \right\} \Phi_{l-1,m} ,$$

$$\mathbf{J}^2\Phi^1_{l,m} = \left\{ \frac{3}{4}(u_1^2 + u_2^2) + l(l+1)u_3^2 + \frac{1}{2}u_1 u_2 - u_2 u_3 - u_1 u_3 \right.$$

$$\left. - 2h^2\left[l(l+1)\frac{1}{8} \right] \right\}\Phi^1_{l,m} - \sqrt{l(l+1)}(u_1 - u_2 + h)(u_3 + h)\Phi^2_{l,m} \,,$$

$$\mathbf{J}^2\Phi^2_{l,m} = -\sqrt{l(l+1)}(u_1 - u_2 - h)(u_3 - h)\Phi^1_{l,m}$$

$$+ \left[\frac{3}{4}(u_1 - u_2)^2 + l(l+1)u_3^2 - \frac{3}{4} \right]\Phi^2_{l,m} \,. \tag{3.3.12}$$

Again in order to guarantee the symmetric form of the matrix we put

$$u_2 = u_1 + u_3 \,, \tag{3.3.13}$$

then the eigenvalues and eigenstates of $\mathbf{J}^2$ are given by

$$\lambda_l^\pm = u_1^2 + \left[l(l+1) + \frac{1}{4} \right]u_3^2 - h^2\left[l(l+1) + \frac{1}{2} \right] \pm \frac{1}{2}\sqrt{P} \,, \tag{3.3.14}$$

$$\begin{pmatrix} \alpha^+_{l,m} \\ \alpha^-_{l,m} \end{pmatrix} = \begin{pmatrix} \cos\frac{\varphi}{2} & -\sin\frac{\varphi}{2} \\ \sin\frac{\varphi}{2} & \cos\frac{\varphi}{2} \end{pmatrix}\begin{pmatrix} \Phi^1_{l,m} \\ \Phi^2_{l,m} \end{pmatrix} \,, \tag{3.3.15}$$

where

$$\omega^2 = P = \left[2u_1^2 - u_3^2 - h^2\left(2l(l+1) - \frac{1}{2} \right) \right]^2 + 4l(l+1)(u_3^2 - h^2)^2 \,, \tag{3.3.16}$$

$$\sin\varphi = \frac{2\sqrt{l(l+1)}}{\omega}(u_3^2 - h^2) \,. \tag{3.3.17}$$

The above conclusion is independent of the order of the decomposition (3.3.11), too.

(3) $S_1 = S_2 = 1, S_3 = 2$.

As representations of Lie algebra $SU(2)$, we have

$$\left(1 \otimes 1\right) \otimes 2 = \left(2 \oplus 1 \oplus 0\right) \otimes 2 = 4 \oplus 3 \oplus 2 \oplus 1 \oplus 0$$
$$3 \oplus 2 \oplus 1$$
$$2 \tag{3.3.18}$$

There are no degeneracy for 4 and 0 states, but the two 3 states, three 2 states, two 1 states can be distinguished in terms of $\mathbf{J}^2$.

$$\mathbf{J}^2\Phi^1_{3,m} = (2u_1^2 + 2u_2^2 + 2u_3^2 + 2u_1u_2 - 38h^2)\Phi^1_{3,m}$$
$$+ 2\sqrt{2}(u_2 - u_1 - 2h)(u_3 + 2h)\Phi^2_{3,m}$$

$$\mathbf{J}^2\Phi^2_{3,m} = 2\sqrt{2}(u_2 - u_1 - 2h)(2h - u_3)\Phi^1_{3,m}$$
$$+ (2u_1^2 + 2u_2^2 + 6u_3^2 - 2u_1u_2 + 2u_2u_3 + 2u_1u_3 - 10h^2)\Phi^2_{3,m}$$

$$\mathbf{J}^2\Phi^1_{2,m} = (2u_1^2 + 2u_2^2 + 6u_3^2 + 2u_1u_2 - 3u_1u_3 - 3u_2u_3 - 32h^2)\Phi^1_{2,m}$$
$$+ \sqrt{7}(u_2 - u_1 - 2h)(u_3 + 2h)\Phi^2_{2,m}$$

$$\mathbf{J}^2\Phi^2_{2,m} = \sqrt{7}(u_2 - u_1 - 2h)(2h - u_3)\Phi^1_{2,m}$$
$$+ (2u_1^2 + 2u_2^2 + 6u_3^2 - 2u_1u_2 - u_2u_3 - u_1u_3 - 16h^2)\Phi^2_{2,m}$$
$$+ 4(u_2 - u_1 - h)(u_3 + h)\Phi^3_{2,m}$$

$$\mathbf{J}^2\Phi^3_{2,m} = 4(u_1 - u_2 - h)(-u_3 + h)\Phi^2_{2,m}$$
$$+ (2u_1^2 + 2u_2^2 + 6u_3^2 - 4u_1u_2 - 2h^2)\Phi^3_{2,m}$$

$$\mathbf{J}^2\Phi^1_{1,m} = (2u_1^2 + 2u_2^2 + 6u_3^2 + 2u_1u_2 - 5u_1u_3 - 5u_2u_3 - 18h^2)\Phi^1_{1,m}$$
$$+ \sqrt{3}(u_2 - u_1 - 2h)(u_3 + 2h)\Phi^2_{1,m}$$

$$\mathbf{J}^2\Phi^2_{1,m} = \sqrt{3}(u_1 - u_2 - 2h)(2h - u_3)\Phi^1_{1,m}$$
$$+ (2u_1^2 + 2u_2^2 + 6u_3^2 - 2u_1u_2 - 3u_2u_3 - 3u_1u_3 - 10h^2)\Phi^2_{2,m}\,.$$
$$(3.3.19)$$

The above matrix is symmetric if we set

$$u_2 = u_1 + u_3\,. \qquad (3.3.20)$$

Then $\mathbf{J}^2$ acts on the 3 states, 2 states and 1 states respectively by

$$\mathbf{J}^2\begin{pmatrix}\Phi^1_{3,m}\\\Phi^2_{3,m}\end{pmatrix} = \begin{pmatrix}6u_1^2 + 4u_3^2 + 6u_1u_3 - 38h^2 & 2\sqrt{2}(u_3^2 - 2h^2)\\2\sqrt{2}(u_3^2 - 2h^2) & 2u_1^2 + 10u_3^2 + 4u_1u_3 - 10h^2\end{pmatrix}\begin{pmatrix}\Phi^1_{3,m}\\\Phi^2_{3,m}\end{pmatrix};$$

$$\mathbf{J}^2\begin{pmatrix}\Phi^1_{2,m}\\\Phi^2_{2,m}\\\Phi^3_{2,m}\end{pmatrix} = \begin{pmatrix}6u_1^2 + 5u_3^2 - 32h^2 & \sqrt{7}(u_3^2 - h^2) & 0\\\sqrt{7}(u_3^2 - 4h^2) & 2u_1^2 + 7u_3^2 - 16h^2 & 4(u_3^2 - h^2)\\0 & 4(u_3^2 - h^2) & 8u_3^2 - 2h^2\end{pmatrix}\begin{pmatrix}\Phi^1_{2,m}\\\Phi^2_{2,m}\\\Phi^3_{2,m}\end{pmatrix};$$

$$\mathbf{J}^2\begin{pmatrix}\Phi^1_{1,m}\\\Phi^2_{1,m}\end{pmatrix} = \begin{pmatrix}6u_1^2 + 3u_3^2 - 2u_1u_3 - 18h^2 & \sqrt{3}(u_3^2 - 4h^2)\\\sqrt{3}(u_3^2 - 4h^2) & 2u_1^2 + 5u_3^2 - 4u_1u_3 - 10h^2\end{pmatrix}\begin{pmatrix}\Phi^1_{1,m}\\\Phi^2_{1,m}\end{pmatrix}.$$
$$(3.3.21)$$

We would like to point out again that the action of $\mathbf{J}^2$ is independent of the order of the decomposition (3.3.18).

Next we consider the case $\mathbf{S}_1 \otimes \mathbf{S}_2 \otimes \mathbf{S}_3 \otimes \mathbf{S}_4$, where $S_1 = S_2 = S_3 = S_4 = \frac{1}{2}$. As representations of Lie algebra $SU(2)$, we have

$$\left(\frac{1}{2} \otimes \frac{1}{2}\right) \otimes \left(\frac{1}{2} \otimes \frac{1}{2}\right) = (1 \oplus 0) \otimes (1 \oplus 0) = 2 \bigoplus \quad 1 \quad \bigoplus 0$$

$$1(1 \otimes 0)$$
$$1(0 \otimes 1)$$
$$0$$

$$(3.3.22)$$

The Lie algebraic basis can be written as (without renormalization)

$$\phi_{2,2} = |\uparrow\uparrow\uparrow\uparrow\rangle,$$

$$\phi_{2,1} = |\uparrow\uparrow\uparrow\downarrow\rangle + |\uparrow\uparrow\downarrow\uparrow\rangle + |\uparrow\downarrow\uparrow\uparrow\rangle + |\downarrow\uparrow\uparrow\uparrow\rangle,$$

$$\phi_{2,0} = |\uparrow\uparrow\downarrow\downarrow\rangle + |\uparrow\downarrow\uparrow\downarrow\rangle + |\uparrow\downarrow\downarrow\uparrow\rangle + |\downarrow\uparrow\downarrow\uparrow\rangle + |\downarrow\downarrow\uparrow\uparrow\rangle + |\downarrow\uparrow\uparrow\downarrow\rangle,$$

$$\phi_{2,-1} = |\uparrow\downarrow\downarrow\downarrow\rangle + |\downarrow\uparrow\downarrow\downarrow\rangle + |\downarrow\downarrow\uparrow\downarrow\rangle + |\downarrow\downarrow\downarrow\uparrow\rangle,$$

$$\phi_{2,-2} = |\downarrow\downarrow\downarrow\downarrow\rangle;$$

$$\phi_{1,1}^1 = |\uparrow\uparrow\uparrow\downarrow\rangle + |\uparrow\uparrow\downarrow\uparrow\rangle - |\uparrow\downarrow\uparrow\uparrow\rangle - \downarrow\uparrow\uparrow\uparrow\rangle,$$

$$\phi_{1,0}^1 = 2(|\uparrow\uparrow\downarrow\downarrow\rangle - \downarrow\downarrow\uparrow\uparrow\rangle)),$$

$$\phi_{1,-1}^1 = |\uparrow\downarrow\downarrow\downarrow\rangle + |\downarrow\uparrow\downarrow\downarrow\rangle - |\downarrow\downarrow\uparrow\downarrow\rangle - \downarrow\downarrow\downarrow\uparrow\rangle;$$

$$\phi_{1,1}^2 = |\uparrow\downarrow\uparrow\uparrow\rangle - \downarrow\uparrow\uparrow\uparrow\rangle,$$

$$\phi_{1,0}^2 = |\uparrow\downarrow\uparrow\downarrow\rangle + |\uparrow\downarrow\downarrow\uparrow\rangle - |\downarrow\uparrow\uparrow\downarrow\rangle - |\downarrow\uparrow\downarrow\uparrow\rangle,$$

$$\phi_{1,-1}^2 = |\uparrow\downarrow\downarrow\downarrow\rangle + |\downarrow\uparrow\downarrow\downarrow\rangle;$$

$$\phi_{1,1}^3 = |\uparrow\uparrow\uparrow\downarrow\rangle - \uparrow\uparrow\downarrow\uparrow\rangle,$$

$$\phi_{1,0}^3 = |\uparrow\downarrow\uparrow\downarrow\rangle + |\downarrow\uparrow\uparrow\downarrow\rangle - |\downarrow\uparrow\downarrow\uparrow\rangle - \uparrow\downarrow\downarrow\uparrow\rangle,$$

$$\phi_{1,-1}^3 = |\downarrow\downarrow\uparrow\downarrow\rangle - |\downarrow\downarrow\downarrow\uparrow\rangle;$$

$$\phi_{0,0}^1 = 2(|\uparrow\uparrow\downarrow\downarrow\rangle + |\downarrow\downarrow\uparrow\uparrow\rangle) - (|\uparrow\downarrow\uparrow\downarrow\rangle + |\uparrow\downarrow\downarrow\uparrow\rangle + |\downarrow\uparrow\downarrow\uparrow\rangle + +|\downarrow\uparrow\uparrow\downarrow\rangle));$$

$$\phi_{0,0}^2 = |\uparrow\downarrow\uparrow\downarrow\rangle + |\downarrow\uparrow\downarrow\uparrow\rangle - |\uparrow\downarrow\downarrow\uparrow\rangle - |\downarrow\uparrow\uparrow\downarrow\rangle. \quad (3.3.23)$$

We introduce $\mathbf{J}$ (for simplicity, we set $h = 1$) as

$$\mathbf{J} = \sum_{i=1}^{4} u_i \mathbf{S}_i + \frac{\sqrt{-1}}{2} \sum_{i<j}^{4} (\mathbf{S}_i \times \mathbf{S}_j). \quad (3.3.24)$$

Then the action of $\mathbf{J}^2$ turns out that

$$\mathbf{J}^2 \phi_{2,m} = \left[\frac{3}{8}(u_1 + u_2 + u_3 + u_4)^2 + \frac{1}{8}(u_1 + u_2 - u_3 - u_4)^2 + \frac{1}{4}(u_1 - u_2)^2 \right.$$
$$\left. + \frac{1}{4}(u_3 - u_4)^2 - \frac{5}{2} \right] \phi_{2,m};$$

$$\mathbf{J}^2 \phi_{1,m}^1 = \left[\frac{1}{2}(u_1^2 + u_2^2 + u_3^2 + u_4^2) + \frac{1}{4}(u_1 + u_2 - u_3 - u_4)^2 - \frac{9}{2} \right] \phi_{1,m}^1$$
$$- (u_1 - u_2 + 1)(u_3 + u_4 + 2)\phi_{1,m}^2 + (u_3 - u_4 + 1)(u_1 - u_2 - 2)\phi_{1,m}^3;$$

$$\mathbf{J}^2 \phi_{1,m}^2 = -\frac{1}{2}(u_1 - u_2 - 1)(u_3 + u_4 - 2)\phi_{1,m}^1$$
$$+ \left[\frac{1}{2}(u_3 + u_4)^2 + \frac{1}{4}(u_3 - u_4)^2 + \frac{3}{4}(u_1 - u_2)^2 - 1 \right] \phi_{1,m}^2$$
$$+ \frac{1}{2}(u_3 - u_4 + 1)(u_1 - u_2 - 1)\phi_{1,m}^3;$$

$$\mathbf{J}^2 \phi_{1,m}^3 = \frac{1}{2}(u_1 + u_2 + 2)(u_3 - u_4 - 1)\phi_{1,m}^1 + \frac{1}{2}(u_3 - u_4 - 1)(u_1 - u_2 + 1)\phi_{1,m}^2$$
$$+ \left[\frac{1}{2}(u_1 + u_2)^2 + \frac{1}{4}(u_1 - u_2)^2 + \frac{3}{4}(u_3 - u_4)^2 - 1 \right] \phi_{1,m}^3;$$

$$\mathbf{J}^2 \phi_{0,0}^1 = \left\{ \frac{1}{2}(u_1 + u_2 - u_3 - u_4 - 2)(u_1 + u_2 - u_3 - u_4 + 2) \right.$$
$$\left. + \frac{1}{4}[(u_1 - u_2)^2 + (u_3 - u_4)^2 - 2] \right\} \phi_{0,0}^1$$
$$- \frac{3}{2}(u_1 - u_2 + 1)(u_3 - u_4 + 1)\phi_{0,0}^2;$$

$$\mathbf{J}^2 \phi_{0,0}^2 = -\frac{1}{2}(u_1 - u_2 - 1)(u_3 - u_4 - 1)\phi_{0,0}^1$$
$$+ \frac{3}{4}[(u_1 - u_2)^2 + (u_3 - u_4)^2 - 2]\phi_{0,0}^2. \tag{3.3.25}$$

As the same as the 3-spin systems, the action of $\mathbf{J}^2$ for the 4-spin systems is also independent of the order of the decomposition (3.3.22). On the space spanned by the two 0 states $\phi_{0,0}^1$, $\phi_{0,0}^2$, the eigenvalue of $\mathbf{J}^2$ is given by

$$\lambda_0^\pm = \frac{1}{2}\left[(u_1 - u_2)^2 + (u_3 - u_4)^2 + \frac{1}{2}(u_1 + u_2 - u_3 - u_4)^2 - 4 \right] \pm \frac{1}{2}P^{\frac{1}{2}}, \tag{3.3.26}$$

where

$$P = \frac{1}{4}[(u_1 - u_2)^2 + (u_3 - u_4)^2 - (u_1 + u_2 - u_3 - u_4)^2 + 2]^2$$

$$+ 3[(u_1 - u_2)^2 - 1)][(u_3 - u_4)^2 - 1]. \tag{3.3.27}$$

The corresponding eigenstates are

$$\alpha_{0,0}^{\pm} = 2a_{\pm}(|\uparrow\uparrow\downarrow\downarrow\rangle + |\downarrow\downarrow\uparrow\uparrow\rangle) + (b_{\pm} - a_{\pm})(|\uparrow\downarrow\uparrow\downarrow\rangle + |\downarrow\uparrow\downarrow\uparrow\rangle)$$

$$- (a_{\pm} + b_{\pm})(|\uparrow\downarrow\downarrow\uparrow\rangle + |\downarrow\uparrow\uparrow\downarrow\rangle), \tag{3.3.28}$$

where

$$a_{\pm} = (2u_1u_2 + 2u_3u_4 - u_2u_3 - u_1u_4 - u_1u_3 - u_2u_4 - 1) \pm P^{\frac{1}{2}},$$

$$b_+ = b_- = -3(u_1 - u_2 + 1)(u_3 - u_4 + 1). \tag{3.3.29}$$

The action of $\mathbf{J}^2$ acts on the space spanned by the 1 states $\phi_{1,m}^1, \phi_{1,m}^2, \phi_{1,m}^3$ can be given by solving the eigenvalues and eigenstates of 3×3 matrices.

As an example, we give the explicit solutions in the case $u_1 = u_2 = u_3 = u_4 = 0$. First we have $\mathbf{J}^2\phi_{2,m} = -\frac{5}{2}\phi_{2,m}$. The eigenstates of $\mathbf{J}^2$ on the space spanned by $\phi_{1,1}^1, \phi_{1,1}^2, \phi_{1,1}^3$ are given by

$$\alpha_{1,1}^1 = -|\uparrow\uparrow\uparrow\downarrow\rangle + |\uparrow\uparrow\downarrow\uparrow\rangle + |\uparrow\downarrow\uparrow\uparrow\rangle - |\downarrow\uparrow\uparrow\uparrow\rangle;$$

$$\alpha_{1,1}^2 = 3|\uparrow\uparrow\uparrow\downarrow\rangle + |\uparrow\uparrow\downarrow\uparrow\rangle - |\uparrow\downarrow\uparrow\uparrow\rangle - 3|\downarrow\uparrow\uparrow\uparrow\rangle;$$

$$\alpha_{1,1}^3 = |\uparrow\uparrow\uparrow\downarrow\rangle - 3|\uparrow\uparrow\downarrow\uparrow\rangle + 3|\uparrow\downarrow\uparrow\uparrow\rangle - |\downarrow\uparrow\uparrow\uparrow\rangle, \tag{3.3.30}$$

which correspond to the eigenvalues $-\frac{1}{2}, -\frac{11}{2}, -1$ respectively. And the eigenstates of $\mathbf{J}^2$ on the space spanned by $\phi_{0,0}^1, \phi_{0,0}^2$ are given by

$$\alpha_{0,0}^+ = 2(|\uparrow\uparrow\downarrow\downarrow\rangle + |\downarrow\downarrow\uparrow\uparrow\rangle) - 4(|\uparrow\downarrow\uparrow\downarrow\rangle + |\downarrow\uparrow\downarrow\uparrow\rangle) + 2(|\uparrow\downarrow\downarrow\uparrow\rangle + |\downarrow\uparrow\uparrow\downarrow\rangle);$$

$$\alpha_{0,0}^- = -6(|\uparrow\uparrow\downarrow\downarrow\rangle + |\downarrow\downarrow\uparrow\uparrow\rangle) + 6(|\uparrow\downarrow\downarrow\uparrow\rangle + |\downarrow\uparrow\uparrow\downarrow\rangle), \tag{3.3.31}$$

which correspond to the eigenvalues $-1, -3$ respectively.

The action of $\mathbf{J}^2$ on the spin systems $\mathbf{S}_1 \otimes \cdots \otimes \mathbf{S}_n$ with $n \geq 5$ is very complicated. At the end of this subsection, we give an explicit action of $\mathbf{J}^2$ on $S_1 \otimes \cdots \otimes S_5$ with $S_i = \frac{1}{2}, i = 1, \ldots, 5$, where ($h = 1, u_i = 0$)

$$\mathbf{J} = \frac{\sqrt{-1}}{2} \sum_{i<j}^{5} (\mathbf{S}_i \times \mathbf{S}_j). \tag{3.3.32}$$

There are one $\frac{5}{2}$ state, four $\frac{5}{2}$ states and five $\frac{1}{2}$ states given in the following decomposition of Lie algebra representations of $SU(2)$: (the action of $\mathbf{J}^2$ is independent of the order)

$$\frac{1}{2} \otimes \frac{1}{2} \otimes \frac{1}{2} \otimes \frac{1}{2} \otimes \frac{1}{2} = \left[\left(\frac{1}{2} \otimes \frac{1}{2} \right) \otimes \left(\frac{1}{2} \otimes \frac{1}{2} \right) \right] \otimes \frac{1}{2}$$

$$= \left(2 \bigoplus 1^1 \bigoplus 1^2 \bigoplus 1^3 \bigoplus 0^1 \bigoplus 0^2 \right) \otimes \frac{1}{2}$$

$$=
\begin{array}{cc}
\frac{5}{2} & \frac{3}{2}\,0 \\
 & \frac{3}{2}\,1 \quad \frac{1}{2}\,1 \\
 & \frac{3}{2}\,2 \quad \frac{1}{2}\,2 \\
 & \frac{3}{2}\,3 \quad \frac{1}{2}\,3 \\
 & \frac{1}{2}\,4 \\
 & \frac{1}{2}\,5\, .
\end{array}$$

$$(3.3.33)$$

The action of $\mathbf{J}^2$ is given by

$$\mathbf{J}^2 \phi_{\frac{5}{2},m} = -5\phi_{\frac{5}{2},m};$$

$$\mathbf{J}^2
\begin{pmatrix}
\phi^0_{\frac{3}{2},m} \\
\phi^1_{\frac{3}{2},m} \\
\phi^2_{\frac{3}{2},m} \\
\phi^3_{\frac{3}{2},m}
\end{pmatrix}
=
\begin{pmatrix}
-\frac{17}{4} & -6 & -3 & -3 \\
-2 & -\frac{21}{4} & -2 & -2 \\
-\frac{1}{2} & -1 & -\frac{7}{4} & -\frac{1}{2} \\
-\frac{1}{2} & -1 & -\frac{1}{2} & -\frac{7}{4}
\end{pmatrix}
\begin{pmatrix}
\phi^0_{\frac{3}{2},m} \\
\phi^1_{\frac{3}{2},m} \\
\phi^2_{\frac{3}{2},m} \\
\phi^3_{\frac{3}{2},m}
\end{pmatrix};$$

$$\mathbf{J}^2
\begin{pmatrix}
\phi^1_{\frac{1}{2},m} \\
\phi^2_{\frac{1}{2},m} \\
\phi^3_{\frac{1}{2},m} \\
\phi^4_{\frac{1}{2},m} \\
\phi^5_{\frac{1}{2},m}
\end{pmatrix}
=
\frac{1}{2}
\begin{pmatrix}
-\frac{112}{9} & -4 & -4 & -4 & 0 \\
-2 & -5 & -1 & -\frac{5}{9} & -3 \\
-2 & -1 & -5 & -\frac{13}{9} & -3 \\
-2 & -1 & -1 & -\frac{37}{9} & -3 \\
0 & -1 & -1 & -1 & -3
\end{pmatrix}
\begin{pmatrix}
\phi^1_{\frac{1}{2},m} \\
\phi^2_{\frac{1}{2},m} \\
\phi^3_{\frac{1}{2},m} \\
\phi^4_{\frac{1}{2},m} \\
\phi^5_{\frac{1}{2},m}
\end{pmatrix} .$$

$$(3.3.34)$$

3.4. *The Hamiltonians commuting with* $\mathbf{J}^2$

First we consider the case $\mathbf{S}_1 \otimes \mathbf{S}_2 \otimes \mathbf{S}_3$, where $S_1 = S_2 = S_3 = \frac{1}{2}$. Set

$$H = t_{12}\mathbf{S}_1 \cdot \mathbf{S}_2 + t_{23}\mathbf{S}_2 \cdot \mathbf{S}_3 + t_{13}\mathbf{S}_1 \cdot \mathbf{S}_3 . \qquad (3.4.1)$$

Since H is symmetric, in order to let $\mathbf{J}^2$ be commute with H, $\mathbf{J}^2$ is needed to be symmetric, too. That is, $u_2 = u_1 + u_3$ for the $\mathbf{J}$ given in equation (3.3.1). We also let $h = 1$ for simplicity.

Then by a direct calculation, besides the case $t_{12} = t_{13} = t_{23}$ which H is nothing but the identity action, $[H, \mathbf{J}^2] = 0$ if and only if t_{12}, t_{13}, t_{23} satisfies

$$4Tt_{12} + (T^2 - 2T - 3)t_{13} + (-T^2 - 2T + 3)t_{23} = 0, \tag{3.4.2}$$

where

$$T[2u_1u_2 - u_1u_3 - u_2u_3 - 1 \pm \sqrt{(2u_1^2 - u_3^2 - 1)^2 + 3(u_3^2 - 1)^2}]$$
$$= 3(u_1 - u_2 + 1)(u_3 + 1). \tag{3.4.3}$$

Notice that $2u_1u_2 - u_1u_3 - u_2u_3 - 1 \pm \sqrt{(2u_1^2 - u_3^2 - 1)^2 + 3(u_3^2 - 1)^2} = 0$ if and only $u_3^2 = 1$, and in this case, $t_{13} = t_{23}$.

If $u_1 = u_2 = u_3 = 0$, then $[H, \mathbf{J}^2] = 0$ if and only if $t_{12} = t_{23}$. That is,

$$H = t_{12}(\mathbf{S}_1 \cdot \mathbf{S}_2 + \mathbf{S}_2 \cdot \mathbf{S}_3) + t_{13}\mathbf{S}_1 \cdot \mathbf{S}_3. \tag{3.4.4}$$

The actions of H and $\mathbf{J}^2$ on the space spanned by the two $\frac{1}{2}$ states are given respectively by

$$H\alpha'_{\frac{1}{2},m} = \left(\frac{1}{4}t_{13} - t_{12}\right)\alpha'_{\frac{1}{2},m}, \quad H\alpha_{\frac{1}{2},m} = -\frac{3}{4}t_{13}\alpha_{\frac{1}{2},m}; \tag{3.4.5}$$

$$\mathbf{J}^2\alpha'_{\frac{1}{2},m} = -\frac{1}{4}\alpha'_{\frac{1}{2},m}, \quad \mathbf{J}^2\alpha_{\frac{1}{2},m} = -\frac{9}{4}\alpha_{\frac{1}{2},m}. \tag{3.4.6}$$

For the case $\frac{1}{2} \otimes \frac{1}{2} \otimes l$, we consider the spin structure of rare gas

$$H = -a\mathbf{L} \cdot \mathbf{S}_1 - b\mathbf{S}_1 \cdot \mathbf{S}_2, \quad \left(\lambda = \frac{b}{a}\right). \tag{3.4.7}$$

It describes the interaction of spin $\mathbf{S}_1$ of an electron exited from l-shell and the left hole $\mathbf{S}_2$.

$$H\Phi_{l+1,m} = -\frac{1}{2}\left(al + \frac{1}{2}b\right)\Phi_{l+1,m},$$

$$H\Phi_{l-1,m} = \frac{1}{2}\left[(l+1)a - \frac{1}{2}b\right]\Phi_{l-1,m},$$

$$H\begin{bmatrix} \Phi_{l,m}^1 \\ \Phi_{l,m}^2 \end{bmatrix} = \frac{1}{2}\begin{bmatrix} (a - \frac{1}{2}b) & a\sqrt{l(l+1)} \\ a\sqrt{l(l+1)} & \frac{3}{2}b \end{bmatrix}\begin{bmatrix} \Phi_{l,m}^1 \\ \Phi_{l,m}^2 \end{bmatrix}. \tag{3.4.8}$$

The eigenstates of H associated to l, m are

$$
\begin{pmatrix} \alpha_{l,m}^+ \\ \alpha_{l,m}^- \end{pmatrix} = \begin{pmatrix} \cos\frac{\varphi}{2} & -\sin\frac{\varphi}{2} \\ \sin\frac{\varphi}{2} & \cos\frac{\varphi}{2} \end{pmatrix} \begin{pmatrix} \Phi_{l,m}^1 \\ \Phi_{l,m}^2 \end{pmatrix}, \tag{3.4.9}
$$

where

$$
\sin\varphi = \frac{\sqrt{l(l+1)}}{\omega}, \quad \omega^2 = (\frac{1}{2} - \lambda)^2 + l(l+1), \lambda = \frac{b}{a}. \tag{3.4.10}
$$

The eigenvalues are

$$
\lambda_{l+1} = -\frac{1}{2}\left(la + \frac{b}{2}\right), \quad \lambda_{l-1} = \frac{1}{2}\left[(l+1)a - \frac{b}{2}\right];
$$

$$
\lambda_l^\pm = \frac{1}{4}(a+b) \pm \frac{1}{2}\left[l(l+1)a^2 + \left(\frac{a}{2} - b\right)^2\right]^{\frac{1}{2}}. \tag{3.4.11}
$$

The rotation should be made in such a way that

$$
[H, \mathbf{J}^2] = 0 \tag{3.4.12}
$$

which is satisfied if the matrix $\mathbf{J}^2$ is symmetric, i.e.,

$$
\gamma = \frac{\{2u_1^2 - 2h^2[l(l+1) + \frac{1}{4}]\}}{(u_3^2 - h^2)} = 2(1 - \lambda). \tag{3.4.13}
$$

Therefore, the parameter γ in $Y(SU(2))$ determines the rotation angle φ. It is reasonable to think that the appearance of "rotation" of degenerate states is closely related to the "quantum number" of $\mathbf{J}^2$.

Next we consider the case $\frac{1}{2} \otimes \frac{1}{2} \otimes \frac{1}{2} \otimes \frac{1}{2}$. Let

$$
H = \sum_{i<j} a_{ij} \mathbf{S}_i \cdot \mathbf{S}_j. \tag{3.4.14}
$$

Let $\mathbf{J}$ be given in equation (3.3.24) with $u_1 = u_2 = u_3 = u_4 = 0$. Then $[H, \mathbf{J}^2] = 0$ if and only if

$$
H = a_{12}(\mathbf{S}_1 \cdot \mathbf{S}_2 + \mathbf{S}_3 \cdot \mathbf{S}_4) + a_{13}(\mathbf{S}_1 \cdot \mathbf{S}_3 + \mathbf{S}_2 \cdot \mathbf{S}_4) + \frac{1}{3}(a_{12} + 2a_{13})\mathbf{S}_1 \cdot \mathbf{S}_4
$$

$$
+ \frac{1}{3}(5a_{12} - 2a_{13})\mathbf{S}_2 \cdot \mathbf{S}_3. \tag{3.4.15}
$$

The eigenstates of $\mathbf{J}^2$ given by equations (3.3.30)–(3.3.31) are also eigenstates of H:

$$
H\begin{pmatrix} \phi_{2,m} \\ \alpha_{1,m}^1 \\ \alpha_{1,m}^2 \\ \alpha_{1,m}^3 \\ \alpha_{0,0}^+ \\ \alpha_{0,0}^- \end{pmatrix} = \begin{pmatrix} a_{12}+\frac{1}{2}a_{13} & 0 & 0 & 0 & 0 & 0 \\ 0 & -\frac{1}{2}a_{13} & 0 & 0 & 0 & 0 \\ 0 & 0 & \frac{a_{12}-4a_{13}}{6} & 0 & 0 & 0 \\ 0 & 0 & 0 & -\frac{4}{3}a_{12}+\frac{5}{6}a_{13} & 0 & 0 \\ 0 & 0 & 0 & 0 & -2a_{12}+\frac{1}{2}a_{13} & 0 \\ 0 & 0 & 0 & 0 & 0 & -\frac{3}{2}a_{13} \end{pmatrix}
$$

$$
\times \begin{pmatrix} \phi_{2,m} \\ \alpha_{1,m}^1 \\ \alpha_{1,m}^2 \\ \alpha_{1,m}^3 \\ \alpha_{0,0}^+ \\ \alpha_{0,0}^- \end{pmatrix}. \tag{3.4.16}
$$

3.5. J as shift operators

Some compositions of Yangian operators $J_\pm, J_3$ and $\mathbf{J}^2$ can be used to be the shift operators of certain operations where the usual Lie algebra $SU(2)$ cannot do them, like the transitions between the eigenstates with the same weights.

First we consider again the spin structure of rare gas with Hamiltonian (3.4.7) which is the case $\mathbf{S}_1 \otimes \mathbf{S}_2 \otimes \mathbf{S}_3$, where $S_1 = S_2 = \frac{1}{2}, S_3 = l$. The transition between $\alpha_{l,m}^+$ and $\alpha_{l,m}^-$ can be made by J_3. Because there are two independent parameters u_1 and u_3 in $\mathbf{J}$, one can choose a suitable relation between u_3 and $\lambda = \frac{b}{a}$ such that

$$
J_3\alpha^+ \sim \alpha^-, \tag{3.5.1}
$$

i.e., the transition between two degenerate states in Lie-algebra is made through J_3 operator, because of

$$
[\mathbf{J}^2, J_3] \neq 0. \tag{3.5.2}
$$

As an example, we let $l = 1$ and $h = 1$. Then $J_3\alpha_{1,m}^+$ is in the subspace spanned by $\alpha_{1,m}^+$ and $\alpha_{1,m}^-$ if and only if

$$\left[\frac{1}{4} - \frac{1}{2}\lambda + \frac{1}{4}\sqrt{4\lambda^2 - 4\lambda + 9}\right]\left[\sqrt{u_3^2(1-\lambda) + \lambda + \frac{5}{4}} - \frac{u_3}{2} - 2\right] - \frac{u_3}{2} - \frac{1}{2} = 0. \tag{3.5.3}$$

For such cases, we can suppose

$$J_3\alpha_{1,m}^+ = k^+\alpha_{1,m}^+ + k^-\alpha_{1,m}^-. \tag{3.5.4}$$

Thus we have

$$(J_3 - k^-\mathrm{Id})\alpha_{1,m}^+ \longrightarrow \alpha_{1,m}^-, \tag{3.5.5}$$

where Id is identity transformation. Similarly, $J_3\alpha_{1,m}^-$ is in the subspace spanned by $\alpha_{1,m}^+$ and $\alpha_{1,m}^-$ if and only if

$$\left[\frac{1}{4} - \frac{1}{2}\lambda - \frac{1}{4}\sqrt{4\lambda^2 - 4\lambda + 9}\right]\left[\sqrt{u_3^2(1-\lambda) + \lambda + \frac{5}{4}} - \frac{u_3}{2} - 2\right] - \frac{u_3}{2} - \frac{1}{2} = 0. \tag{3.5.6}$$

Next we consider the case $\mathbf{S}_1 \otimes \mathbf{S}_2 \otimes \cdots \otimes \mathbf{S}_n$, where $S_1 = S_2 = \cdots = S_n = \frac{1}{2}$. It is known that the usual Lie algebra operators of $SU(2)$ cannot shift the eigenstates with the same weight. On the other hand, set

$$\mathbf{J} = \sum_{i=1}^n u_i\mathbf{S}_i + \frac{\sqrt{-1}}{2}\sum_{i<j}^n (\mathbf{S}_i \times \mathbf{S}_j). \tag{3.5.7}$$

We have already known that $\mathbf{J}^2$ as a new quantum number can indicate clearly these eigenstates with the same weight. Suppose there are t eigenstates $\alpha_{l_1,m}^1, \ldots, \alpha_{l_t,m}^t$ with the action of I_3 by multiplying m. By choosing the spectral parameters $u_1, \ldots, u_n$, we can have

$$\mathbf{J}^2\alpha_{l_i,m}^i = d^i\alpha_{l_i,m}^i, \quad 1 = 1, \ldots, t, \tag{3.5.8}$$

where $d^i \neq 0$ and $d^i \neq d^j$ for $i \neq j$. Suppose

$$J_3\alpha_{l_i,m}^i = \sum_{j=1}^t c^{ij}\alpha_{l_i,m}^j, i = 1, \ldots, t. \tag{3.5.9}$$

Therefore there is the following shift operator from $\alpha_{l_i,m}^i$ to $\alpha_{l_k,m}^k$ given by

$$\prod_{\substack{j\neq k}}^t \left(1 - \frac{1}{d^j}\mathbf{J}^2\right)J_3\alpha_{l_i,m}^i = c^{ik}\prod_{\substack{j\neq k}}^t \left(1 - \frac{d^i}{d^j}\right)\alpha_{l_k,m}^k. \tag{3.5.10}$$

With the composition of such an action and the usual Lie algebra, we can obtain the shift operators from an eigenstate to any other eigenstate. We also can find the shift operators between the eigenstates with weights l and $l \pm 1$ directly by

$$J'_\pm = \prod_{j \neq k} \left(1 - \frac{1}{d^j}\mathbf{J}^2\right) J_\pm . \tag{3.5.11}$$

As an example, we consider the case $n = 3$ with $u_1 = u_2 = u_3 = 0$ and $h = 1$. Then by equation (3.3.8), we have

$$\mathbf{J}^2 \alpha_{\frac{3}{2},m} = -\alpha_{\frac{3}{2},m}, \quad \mathbf{J}^2 \alpha^+_{\frac{1}{2},m} = -\frac{1}{4}\alpha^+_{\frac{1}{2},m}, \quad \mathbf{J}^2 \alpha^-_{\frac{1}{2},m} = -\frac{9}{4}\alpha^-_{\frac{1}{2},m} . \tag{3.5.12}$$

The corresponding eigenstates are

$$\alpha_{\frac{3}{2},\frac{1}{2}} = \frac{1}{\sqrt{3}}(|\downarrow\uparrow\uparrow\rangle + |\uparrow\downarrow\uparrow\rangle + |\uparrow\uparrow\downarrow\rangle),$$

$$\alpha^+_{\frac{1}{2},\frac{1}{2}} = \frac{1}{\sqrt{6}}(|\downarrow\uparrow\uparrow\rangle - 2|\uparrow\downarrow\uparrow\rangle + |\uparrow\uparrow\downarrow\rangle),$$

$$\alpha^-_{\frac{1}{2},\frac{1}{2}} = \frac{1}{\sqrt{2}}(-|\downarrow\uparrow\uparrow\rangle + |\uparrow\uparrow\downarrow\rangle), \tag{3.5.13}$$

respectively. For the shift operators between $\alpha^+_{\frac{1}{2},m}$ and $\alpha^-_{\frac{1}{2},m}$, we have

$$J_3 \alpha^+_{\frac{1}{2},m} \longrightarrow \alpha^-_{\frac{1}{2},m}, \quad (1 + \mathbf{J}^2) J_3 \alpha^-_{\frac{1}{2},m} \longrightarrow \alpha^+_{\frac{1}{2},m} . \tag{3.5.14}$$

3.6. *Yangian and Bell basis*

Yangian $\mathbf{J}$ also can be used to construct the shift operators for the Bell basis.

First we consider the case $\frac{1}{2} \otimes \frac{1}{2}$. The Bell basis is given by

$$\Phi_1 = \frac{1}{\sqrt{2}}(|\uparrow\uparrow\rangle - |\downarrow\downarrow\rangle);$$

$$\Phi_2 = \frac{1}{\sqrt{2}}(|\uparrow\uparrow\rangle + |\downarrow\downarrow\rangle);$$

$$\Phi_3 = \frac{1}{\sqrt{2}}(|\uparrow\downarrow\rangle + |\downarrow\uparrow\rangle);$$

$$\Phi_4 = \frac{1}{\sqrt{2}}(|\uparrow\downarrow\rangle - |\downarrow\uparrow\rangle). \tag{3.6.1}$$

We introduce $\mathbf{J}$ as

$$\mathbf{J} = a\mathbf{S}_1 + b\mathbf{S}_2 + \frac{\sqrt{-1}}{2}\mathbf{S}_1 \otimes \mathbf{S}_2 . \tag{3.6.2}$$

Set

$$J_1 = \frac{1}{2}(J_+ + J_-), \quad J_2 = \frac{1}{2\sqrt{-1}}(J_+ - J_-) \,. \tag{3.6.3}$$

Then we have

$$J_1 \Phi_4 = \frac{1}{2}(-a + b + 1)\Phi_1 \,;$$

$$J_2 \Phi_4 = \frac{1}{2\sqrt{-1}}(-a + b + 1)\Phi_2 \,;$$

$$J_3 \Phi_4 = \frac{1}{2}(a - b - 1)\Phi_3 \,. \tag{3.6.4}$$

Let $b - a = 1$ and

$$J_1' = J_1, \quad J_2' = \sqrt{-1}J_2, \quad J_3' = -J_3 \,, \tag{3.6.5}$$

then

$$J_1' \Phi_4 = \Phi_1, \quad J_2' \Phi_4 = \Phi_2, \quad J_3' \Phi_4 = \Phi_3 \,. \tag{3.6.6}$$

Next we consider the 4-spin system $\frac{1}{2} \otimes \frac{1}{2} \otimes \frac{1}{2} \otimes \frac{1}{2}$. We give the transition from the state

$$\phi_{0,0}^1 = 2(|\uparrow\uparrow\downarrow\downarrow\rangle + |\downarrow\downarrow\uparrow\uparrow\rangle) - (|\uparrow\downarrow\uparrow\downarrow\rangle + |\uparrow\downarrow\downarrow\uparrow\rangle + |\downarrow\uparrow\downarrow\uparrow\rangle + +|\downarrow\uparrow\uparrow\downarrow\rangle) \tag{3.6.7}$$

to any GHZ state.

Introduce **J** as be given in equation (3.3.24) and J_1, J_2 are defined by equation (3.6.3). Then for every GHZ state, we have

$$\frac{1}{\sqrt{2}}(|\uparrow\uparrow\uparrow\uparrow\rangle + |\downarrow\downarrow\downarrow\downarrow\rangle) \longleftarrow \left(J_1^2 + \frac{1}{2}J_3^2\right)\phi_{0,0}^1, \quad \text{when } b = a + 1, c = a + 1, d = a + 2;$$

$$\frac{1}{\sqrt{2}}(|\uparrow\uparrow\uparrow\uparrow\rangle - |\downarrow\downarrow\downarrow\downarrow\rangle) \longleftarrow J_2 J_1 \phi_{0,0}^1, \quad \text{when } b = a + 1, c = -a - 1, d = -a, \left(a \neq 0, \frac{1}{2}\right);$$

$$\frac{1}{\sqrt{2}}(|\uparrow\uparrow\uparrow\downarrow\rangle + |\downarrow\downarrow\downarrow\uparrow\rangle) \longleftarrow J_3 J_2 \phi_{0,0}^1, \quad \text{when } b = a + 1, c = 2a + \frac{1}{2}, d = -\frac{3}{2}, \left(a \neq -\frac{1}{2}, -\frac{3}{2}\right);$$

$$\frac{1}{\sqrt{2}}(|\uparrow\uparrow\uparrow\downarrow\rangle - |\downarrow\downarrow\downarrow\uparrow\rangle) \longleftarrow J_3 J_1 \phi_{0,0}^1, \quad \text{when } b = a + 1, c = 2a + \frac{1}{2}, d = -\frac{3}{2}, \left(a \neq -\frac{1}{2}, -\frac{3}{2}\right);$$

$$\frac{1}{\sqrt{2}}(|\uparrow\uparrow\downarrow\uparrow\rangle + |\downarrow\downarrow\uparrow\downarrow\rangle) \longleftarrow J_2 J_3 \phi_{0,0}^1, \quad \text{when } a = 1 + \frac{\sqrt{5}}{2}, b = 1 - \frac{\sqrt{5}}{2}, c = \frac{1}{2}, d = -\frac{1}{2};$$

$$\frac{1}{\sqrt{2}}(|\uparrow\uparrow\downarrow\uparrow\rangle - |\downarrow\downarrow\uparrow\downarrow\rangle) \longleftarrow J_1 J_3 \phi_{0,0}^1, \quad \text{when } a = 1 + \frac{\sqrt{5}}{2}, b = 1 - \frac{\sqrt{5}}{2}, c = \frac{1}{2}, d = -\frac{1}{2};$$

$$\frac{1}{\sqrt{2}}(|\uparrow\downarrow\uparrow\uparrow\rangle + |\downarrow\uparrow\downarrow\downarrow\rangle) \longleftarrow J_3 J_2 \phi_{0,0}^1,$$

$$\text{when } a = 2c + \frac{5}{2}, b = \frac{1}{2}, d = 1 + c, \left(c \neq -\frac{1}{2}, -\frac{3}{2}\right);$$

$$\frac{1}{\sqrt{2}}(|\uparrow\downarrow\uparrow\uparrow\rangle - |\downarrow\uparrow\downarrow\downarrow\rangle) \longleftarrow J_3 J_1 \phi_{0,0}^1,$$

$$\text{when } a = 2c + \frac{5}{2}, b = \frac{1}{2}, d = 1 + c, \left(c \neq -\frac{1}{2}, -\frac{3}{2}\right);$$

$$\frac{1}{\sqrt{2}}(|\downarrow\uparrow\uparrow\uparrow\rangle + |\uparrow\downarrow\downarrow\downarrow\rangle) \longleftarrow J_2 J_3 \phi_{0,0}^1,$$

$$\text{when } a = \frac{1}{2}, b = -\frac{1}{2}, c = -1 + \frac{\sqrt{-3}}{2}, d = -1 - \frac{\sqrt{-3}}{2};$$

$$\frac{1}{\sqrt{2}}(|\downarrow\uparrow\uparrow\uparrow\rangle - |\uparrow\downarrow\downarrow\downarrow\rangle) \longleftarrow J_1 J_3 \phi_{0,0}^1,$$

$$\text{when } a = \frac{1}{2}, b = -\frac{1}{2}, c = -1 + \frac{\sqrt{-3}}{2}, d = -1 - \frac{\sqrt{-3}}{2};$$

$$\frac{1}{\sqrt{2}}(|\uparrow\downarrow\uparrow\downarrow\rangle + |\downarrow\uparrow\downarrow\uparrow\rangle) \longleftarrow J_3^2 \phi_{0,0}^1, \quad \text{when } b = a - 1, c = a + 1, d = a - 4;$$

$$\frac{1}{\sqrt{2}}(|\uparrow\downarrow\uparrow\downarrow\rangle - |\downarrow\uparrow\downarrow\uparrow\rangle) \longleftarrow J_3 \phi_{0,0}^1, \quad \text{when } c = a - 1, d = b - 1, (a - b \neq -1);$$

$$\frac{1}{\sqrt{2}}(|\uparrow\downarrow\downarrow\uparrow\rangle + |\downarrow\uparrow\downarrow\uparrow\rangle) \longleftarrow J_3^2 \phi_{0,0}^1, \quad \text{when } b = a - 2, c = a - 3, d = a - 1;$$

$$\frac{1}{\sqrt{2}}(|\uparrow\downarrow\uparrow\downarrow\rangle - |\downarrow\uparrow\uparrow\downarrow\rangle) \longleftarrow J_3 \phi_{0,0}^1, \quad \text{when } c = b - 2, d = a, (a - b \neq -1);$$

$$\frac{1}{\sqrt{2}}(|\uparrow\uparrow\downarrow\downarrow\rangle + |\downarrow\uparrow\downarrow\uparrow\rangle) \longleftarrow J_3^2 \phi_{0,0}^1, \quad \text{when } b = a + 1, c = a + 5, d = a;$$

$$\frac{1}{\sqrt{2}}(|\uparrow\uparrow\downarrow\downarrow\rangle - |\downarrow\uparrow\downarrow\uparrow\rangle) \longleftarrow J_3 \phi_{0,0}^1, \quad \text{when } b = a + 1, d = c + 1, (a - c \neq 1).$$

3.7. Illustrative examples: NMR of Breit–Rabi Hamiltonian and Yangian

The Breit–Rabi Hamiltonian is given by

$$H = \mathbf{K} \cdot \mathbf{S} + \mu \mathbf{B} \cdot \mathbf{S}, \tag{3.7.1}$$

where $S = \frac{1}{2}$ and $B = \mathbf{B}(t)$ is magnetic field.

The Hamiltonian can easily be diagonalized for any background angular momentum (or spin) $\mathbf{K}$. The $\mathbf{S}$ stands for spin of electron and for simplicity

$\mathbf{K} = \mathbf{S_1}(S_1 = 1/2)$ is an average background spin contributed by other source, say, control spin. Denoting by

$$H = H_0 + H_1(t), \quad H_0 = \alpha \mathbf{S_1} \cdot \mathbf{S_2}, \quad H_1(t) = \mu \mathbf{B}(t) \cdot \mathbf{S_2}. \qquad (3.7.2)$$

Let us work in the interaction picture:

$$H_I = \mu \mathbf{B}(t) \cdot (e^{i\alpha \mathbf{S_1} \cdot \mathbf{S_2}} \mathbf{S_2} e^{-i\alpha \mathbf{S_1} \cdot \mathbf{S_2}}) = \mu \mathbf{B}(t) \cdot \mathbf{J}, \qquad (3.7.3)$$

$$\mathbf{J} = \mu_1 \mathbf{S_1} + \mu_2 \mathbf{S_2} + 2\lambda(\mathbf{S_1} \times \mathbf{S_2}), \qquad (3.7.4)$$

where $\mu_1 = \frac{1}{2}(1 - \cos\alpha)$, $\mu_2 = \frac{1}{2}(1 + \cos\alpha)$, $\lambda = \frac{1}{2}\sin\alpha$. Obviously, here we have $\mu_1\mu_2 = \lambda^2$. It is not surprising that the $Y(SU(2))$ reduces to $SO(4)$ here because the transformation is fully Lie-algebraic operation. This is an exercise in quantum mechanics.

For generalization we regard μ_1 and μ_2 as independent parameters, i.e., drop the relation $\mu_1\mu_2 = \lambda^2$. Looking at

$$\mathbf{J} = \mu_1 \mathbf{S_1} + \mu_2 \mathbf{S_2} - \frac{1}{2}(\mu_1 + \mu_2)(\mathbf{S_1} + \mathbf{S_2}) + \gamma(\mathbf{S_1} + \mathbf{S_2}) + 2\lambda \mathbf{S_1} \times \mathbf{S_2}. \quad (3.7.5)$$

When $\gamma = \frac{1}{2}$, $\mu_2 - \mu_1 = \cos\alpha$ and $\lambda = \frac{1}{2}\sin\alpha$, it reduces to the form in the interacting picture. Putting

$$\mathbf{S_1} + \mathbf{S_2} = \mathbf{S}, \quad 2\lambda = -\frac{h}{2}(h \text{ is not Plank constant}). \qquad (3.7.6)$$

In accordance with the convention we have

$$\mathbf{J} = \gamma \mathbf{S} + \sum_{i=1}^{2} \mu_i \mathbf{S_i} + \frac{h}{2} \mathbf{S_1} \times \mathbf{S_2} - \frac{1}{2}(\mu_1 + \mu_2)\mathbf{S} = \gamma \mathbf{S} + \mathbf{Y}. \qquad (3.7.7)$$

Since $\mathbf{J} \to \xi\mathbf{S} + \mathbf{J}$ still satisfies Yangian relations, it is natural to appear the term $\gamma\mathbf{S}$. The interacting Hamiltonian then reads

$$H_I(t) = -\gamma\mathbf{B}(t) \cdot \mathbf{S} - \mathbf{B}(t) \cdot \mathbf{Y}. \qquad (3.7.8)$$

When $\mu_i = 0$, $h = 0$, it is the usual NMR for spin $1/2$. To solve the equation, we use

$$i\frac{\partial \Psi(t)}{\partial t} = H_I(t)\Psi(t), \quad |\Psi(t)\rangle = \sum_{\alpha = \pm, 3; 0} a_\alpha(t)|\chi_\alpha\rangle, \qquad (3.7.9)$$

where $\{\chi_\pm, \chi_3\}$ is the spin triplet and χ_0 singlet. Setting

$$B_\pm(t) = B_1(t) \pm iB_2(t) = B_1 e^{\mp i\omega_0 t}, \quad \text{and} \quad B_3 = \text{const}. \qquad (3.7.10)$$

and rescaling by

$$a_\pm(t) = e^{\pm i\omega_0 t} b_\pm(t), \qquad (3.7.11)$$

we get

$$i\frac{db_\pm(t)}{dt} = -\gamma\left\{\frac{1}{\sqrt{2}}B_1 a_3(t) \mp (\omega_0\gamma^{-1} - B_3)b_\pm(t)\right\} \pm \frac{1}{2\sqrt{2}}\mu_- B_1 a_0(t)\,,$$

$$i\frac{da_3(t)}{dt} = -\frac{\gamma B_1}{\sqrt{2}}\{b_+(t) + b_-(t)\} - \frac{1}{2}\mu_- B_3 a_0(t)\,,$$

$$i\frac{da_0(t)}{dt} = -\frac{1}{2}\mu_+\left\{\frac{1}{\sqrt{2}}B_1[b_-(t) - b_+(t)]\right\} + B_3 a_3(t)\,, \qquad (3.7.12)$$

where $\mu_\pm = (\mu_1 - \mu_2 \pm i\frac{h}{2})$, i.e.,

$$|\Phi(t)\rangle = \begin{bmatrix} b_1(t) \\ a_3(t) \\ b_-(t) \\ a_0(t) \end{bmatrix}, \mathcal{H}_I = \begin{bmatrix} \omega_0 - \gamma B_3 & -\gamma B_1\frac{1}{\sqrt{2}} & 0 & \frac{1}{2\sqrt{2}}\mu_- B_1 \\ -\gamma B_1\frac{1}{\sqrt{2}} & 0 & -\gamma B_1\frac{1}{\sqrt{2}} & -\frac{1}{2}\mu_- B_3 \\ 0 & -\gamma B_1\frac{1}{\sqrt{2}} & -(\omega_0 - \gamma B_3) & -\frac{1}{2\sqrt{2}}\mu_- B_1 \\ \frac{1}{2\sqrt{2}}\mu_+ B_1 & -\frac{1}{2}\mu_+ B_3 & -\frac{1}{2\sqrt{2}}\mu_+ B_1 & 0 \end{bmatrix},$$

$$(3.7.13)$$

$$i\frac{d|\Phi(t)\rangle}{dt} = H_I|\Phi(t)\rangle\,. \qquad (3.7.14)$$

Noting that $\mathcal{H}_I$ is independent of time, we get

$$|\Phi(t)\rangle = e^{-iEt}|\Phi(t)\rangle\,. \qquad (3.7.15)$$

Then

$$\det|H_I - E| = 0 \qquad (3.7.16)$$

leads to

$$E^4 - \left[(\omega_1 - \gamma B_3)^2 + \gamma^2 B_1^2 + \frac{1}{4}\mu_+\mu_-(B_1^2 + B_3^2)\right]E^2$$

$$+ \frac{1}{4}\mu_+\mu_-[B_3^2(\omega_0 - \gamma B_3)^2 - 2\gamma B_3 B_1^2(\omega_0 - \gamma B_3) + \gamma^2 B_1^4] = 0\,. \quad (3.7.17)$$

There is a transition between the spin singlet and triplet in the NMR process, i.e., the Yangian transfers the quantum information through the evolution. The simplest case is $B_1 = 0$, then the eigenvalues are

$$E = \pm(\omega_0 - \gamma B_3), E = \pm\omega = \pm\frac{B_3}{2}\sqrt{(\mu_1 - \mu_2)^2 + \frac{h^2}{4}}\,. \qquad (3.7.18)$$

It turns out that there is a vibration between $s = 0$ and $s = 1$.

$$\langle s^2 \rangle = 0 \quad \text{at} \quad t = \frac{\pi}{2\omega} \quad (\text{total spin} = 0), \qquad (3.7.19)$$

$$\langle s^2 \rangle = 2 \quad \text{at} \quad t = \frac{\pi}{\omega} \quad (\text{total spin} = 1). \qquad (3.7.20)$$

Under adiabatic approximation it can be proved that it appears Berry's phase. Obviously, only spin vector can make the stereo angle. The role of spin singlet here is a witness that shares energy of spin $= 1$ state.

Actually, if

$$B_{\pm}(t) = B_0 \sin\theta e^{\mp i\omega_0 t}, \quad B_3 = B_0 \cos\theta, \qquad (3.7.21)$$

and

$$|\chi_{11}\rangle = |\uparrow\uparrow\rangle, \quad |\chi_{1-1}\rangle = |\downarrow\downarrow\rangle, \quad |\chi_{10}\rangle = \frac{1}{\sqrt{2}}(|\uparrow\downarrow\rangle + |\downarrow\uparrow\rangle),$$

$$|\chi_{00}\rangle = \frac{1}{\sqrt{2}}(|\uparrow\downarrow\rangle - |\downarrow\uparrow\rangle), \qquad (3.7.22)$$

then let us consider the eigenvalues of

$$H = \alpha \mathbf{S}_1 \cdot \mathbf{S}_2 - \gamma B_0 S_3 - g B_0 J_3, \qquad (3.7.23)$$

under adiabatic approximation which are

$$E_{\pm} = \frac{1}{2}\left(-\frac{\alpha}{2} \pm \sqrt{\alpha^2 + g^2 B_0^2 \mu_+ \mu_-} \right), \qquad (3.7.24)$$

and

$$f_1^{(\pm)} = [2(\alpha^2 + g^2 B_0^2 \mu_+ \mu_-)]^{-1/2}[(\alpha^2 + g^2 B_0^2 \mu_+ \mu_-)^{1/2} \pm \alpha]^{1/2}, \quad (3.7.25)$$

$$f_2^{(\pm)} = [2(\alpha^2 + g^2 B_0^2 \mu_+ \mu_-)]^{-1/2}\left[\frac{\mu_+}{\mu_-}(\alpha^2 + g^2 B_0^2 \mu_+ \mu_-)^{1/2} \mp \alpha \right]^{1/2}.$$

$$(3.7.26)$$

We obtain the eigenstates of H besides $|\chi_{1i}\rangle$ $(i = 1, 2)$

$$|\chi_{\pm}\rangle = f_1^{(\pm)}|\chi_{10}\rangle + f_2^{(\pm)}|\chi_{00}\rangle, \qquad (3.7.27)$$

where

$$|\chi_{11}(t)\rangle = \cos^2\frac{\theta}{2}|\chi_{11}\rangle + \frac{1}{\sqrt{2}}\sin\theta e^{-i\omega_0 t}|\chi_{10}\rangle + \sin^2\frac{\theta}{2}e^{-2i\omega_0 t}|\chi_{1-1}\rangle \,,$$

$$|\chi_{1-1}(t)\rangle = \sin^2\frac{\theta}{2}e^{2i\omega_0 t}|\chi_{11}\rangle - \frac{1}{\sqrt{2}}\sin\theta e^{i\omega_0 t}|\chi_{10}\rangle + \cos^2\frac{\theta}{2}|\chi_{1-1}\rangle \,,$$

$$|\chi_{\pm}(t)\rangle = \frac{1}{\sqrt{2}}f_1^{(\pm)}\{-\sin\theta e^{i\omega_0 t}|\chi_{11}\rangle + \sqrt{2}\cos\theta|\chi_{10}\rangle + \sin\theta e^{-i\omega_0 t}|\chi_{1-1}\rangle\}$$

$$+ f_2^{(\pm)}|\chi_{00}\rangle \,. \tag{3.7.28}$$

We then obtain

$$\langle\chi_{11}(t)|\frac{\partial}{\partial t}|\chi_{11}(t)\rangle = -i\omega_0(1-\cos\theta) \,,$$

$$\langle\chi_{1-1}(t)|\frac{\partial}{\partial t}|\chi_{11}(t)\rangle = i\omega_0(1-\cos\theta) \,,$$

$$\langle\chi_{\pm}(t)|\frac{\partial}{\partial t}|\chi_{\pm}(t)\rangle = 0 \,. \tag{3.7.29}$$

The Berry's phase is then

$$\gamma_{1\pm 1} = \pm\Omega, \quad \Omega = 2\pi(1-\cos\theta) \,, \tag{3.7.30}$$

whereas $\gamma_{10} = \gamma_{00} = 0$. The Yangian changes the eigenstates of H, but preserves the Berry's phase.

3.8. *Transition between S-wave and P-wave superconductivity*

We set for a pair of electrons:

$$S: \quad \text{spin singlet}, \quad L = 0 \,; \tag{3.8.1}$$

$$P: \quad \text{spin triplet}, \quad L = 1 \,. \tag{3.8.2}$$

Due to Balian–Werthamer ([39]), we have

$$\triangle(\mathbf{k}) = -\frac{1}{2}\sum_{\mathbf{k}'}V(\mathbf{k},\mathbf{k}')\frac{\triangle(\mathbf{k}')}{E(\mathbf{k}')}\tanh\frac{\beta}{2}E(\mathbf{k}') \,, \tag{3.8.3}$$

$$E(\mathbf{k}) = (\epsilon^2(k) + |\triangle(\mathbf{k})|^2)^{\frac{1}{2}} \,. \tag{3.8.4}$$

Therefore, still by Balian–Werthamer ([39]), we know

$$\triangle(\mathbf{k}) = \triangle(k)\left(\frac{4\pi}{3}\right)^{\frac{1}{2}} \left[\begin{array}{cc} \sqrt{2}Y_{1,1}(\hat{\mathbf{k}}) & Y_{1,0}(\hat{\mathbf{k}}) \\ Y_{1,0}(\hat{\mathbf{k}}) & \sqrt{2}Y_{1,-1}(\hat{\mathbf{k}}) \end{array} \right]^{*}$$

$$= (-\sqrt{6})\triangle(k)\left(\frac{4\pi}{3}\right)^{\frac{1}{2}}\Phi_{0,0}(\hat{\mathbf{k}}), \qquad (3.8.5)$$

$$\Phi_{0,0}(\hat{\mathbf{k}}) = \frac{1}{\sqrt{3}}\{Y_{1,-1}(\hat{\mathbf{k}})\chi_{11} - Y_{1,0}(\hat{\mathbf{k}})\chi_{10} + Y_{1,1}(\hat{\mathbf{k}})\chi_{1-1}\} = \frac{1}{\sqrt{8}}\left[\begin{array}{cc} \hat{\mathbf{k}}_{-} & -\hat{\mathbf{k}}_{z} \\ -\hat{\mathbf{k}}_{z} & -\hat{\mathbf{k}}_{+} \end{array} \right],$$

$$(3.8.6)$$

where χ_{11}, χ_{10} and χ_{1-1} stand for spin triplet:

$$\Phi_{0,0} \equiv \Phi_{J=0,m=0}. \qquad (3.8.7)$$

The wave function of SC is

$$\phi_{0,0} = \frac{1}{\sqrt{2}}\left[\begin{array}{cc} 0 & Y_{0,0} \\ -Y_{0,0} & 0 \end{array} \right]. \qquad (3.8.8)$$

Introducing

$$I_{\mu} = \sum_{i=1}^{2} S_{\mu}(i); \qquad (\mu = 1,2,3), \qquad (3.8.9)$$

$$J_{\mu} = \sum_{i=1}^{2} \lambda_{i}S_{\mu}(i) - \frac{ihv}{4}\epsilon_{\mu\lambda\nu}(S^{\lambda}(1)S^{\nu}(2) - S^{\lambda}(2)S^{\nu}(1)), \qquad (3.8.10)$$

and noting that $J_{\mu} \rightarrow J_{\mu} + fI_{\mu}$ does not change the Yangian relations, we choose for simplicity $f = -\frac{1}{2}(\lambda_{1} + \lambda_{2})$. Then we obtain for $G = \hat{\mathbf{k}} \cdot (\mathbf{J} + f\mathbf{I})$

$$G\phi_{0,0} = \hat{\mathbf{k}} \cdot (\mathbf{J} + f\mathbf{I})\phi_{0,0} = \frac{\sqrt{3}}{2}\left(\lambda_{2} - \lambda_{1} + \frac{hv}{2}\right)\Phi_{0,0}, \qquad (3.8.11)$$

$$G\Phi_{0,0} = \hat{\mathbf{k}} \cdot (\mathbf{J} + f\mathbf{I})\Phi_{0,0} = \frac{1}{2\sqrt{3}}\left(\lambda_{2} - \lambda_{1} - \frac{hv}{2}\right)\phi_{0,0}. \qquad (3.8.12)$$

The transition directionally depends on the parameters in $Y(SU(2))$. For instance,

$$SC \rightarrow PC: \ G\phi_{0,0} = \frac{\sqrt{3}}{2}\Phi_{0,0}, \ G\Phi_{0,0} = 0, \ \text{if} \ \lambda_{1} - \lambda_{2} = -\frac{hv}{2}, \quad (3.8.13)$$

and

$$PC \to SC: \ G\phi_{0,0} = 0, \ G\Phi_{0,0} = -\frac{hv}{2\sqrt{3}}\phi_{0,0}, \ \text{if} \ \lambda_1 - \lambda_2 = \frac{hv}{2}. \quad (3.8.14)$$

We call the type of the transition "directional transition" ([40]). The controlled parameters are in the Yangian operation. They represent the interaction coming from other controlled spin.

We have got used to apply electromagnetic field A_μ to make transitions between l and $l \pm 1$ states. Now there is Yangian formed by two spins that plays the role changing angular momentum states.

3.9. $Y(SU(3))$-directional transitions

Setting

$$F_\mu = \frac{1}{2}\lambda_\mu, \quad [F_\lambda, F_\mu] = if_{\lambda\mu\nu}F_\nu, \quad (3.9.1)$$

$$I_\mu = \sum_i F_i^\nu, \quad (3.9.2)$$

$$J_\mu = \sum_i \mu_i F_i^\mu - ihf_{\mu\nu\lambda}\sum_{i \neq j} W_{ij}F_i^\nu F_j^\lambda, \quad (W_{ij} = -W_{ji}), \quad (3.9.3)$$

$$[F_i^\lambda, F_j^\mu] = if_{\lambda\mu\nu}\delta_{ij}F_i^\nu, \quad (3.9.4)$$

where $\{F_\mu\}$ is the fundamental representation of $SU(3)$ and $(i, j, k = 1, 2, \ldots, 8)$

$$\triangle_{ijk} = W_{ij}W_{jk} + W_{jk}W_{ki} + W_{ki}W_{ij} = -1. \quad (3.9.5)$$

(Here, no summation over repeated indices, $i \neq j \neq k$.) The reason that such a condition works only for 3-dimensional representation of $SU(3)$ is similar to Haldane's (long-ranged) realization of $Y(SU(2))([19])$. In $SU(2)$ long-ranged form, the property of Pauli matrices leads to $(\sigma^\pm)^2 = 0$. Instead, for $SU(3)$ the conditions of J_μ satisfying $Y(SU(3))$ read

$$\sum_{i \neq j}(1 - w_{ij}^2)(I_j^+ V_i^+ U_i^+ - U_i^- V_i^- I_j^- + I_i^+ V_j^+ U_i^+ - U_i^- V_j^- I_i^-$$

$$+ I_j^+ V_j^+ U_i^+ - U_i^- V_j^- I_j^-) = 0, \quad (3.9.6)$$

and

$$\sum_i(I_i^+ V_i^+ U_i^+ - U_i^- V_i^- I_i^-) = 0, \quad (3.9.7)$$

that are satisfied for Gell–Mann matrices.

The simplest realization of $Y(SU(3))$ is then

$$W_{ij} = \begin{cases} 1 & i > j \\ 0 & i = j \\ -1 & i < j \end{cases} \quad (W_{ij} = -W_{ji}) . \tag{3.9.8}$$

Recalling $(I_8 = \frac{\sqrt{3}}{2} Y)$

$$I^+ = \begin{bmatrix} 0 & 1 & 0 \\ 0 & 0 & 0 \\ 0 & 0 & 0 \end{bmatrix}, \ U^+ = \begin{bmatrix} 0 & 0 & 0 \\ 0 & 0 & 1 \\ 0 & 0 & 0 \end{bmatrix}, \ V^+ = \begin{bmatrix} 0 & 0 & 0 \\ 0 & 0 & 0 \\ 1 & 0 & 0 \end{bmatrix},$$

$$I^3 = \begin{bmatrix} 1 & 0 & 0 \\ 0 & -1 & 0 \\ 0 & 0 & 0 \end{bmatrix}, \ Y = \frac{1}{3} \begin{bmatrix} 1 & 0 & 0 \\ 0 & 1 & 0 \\ 0 & 0 & -2 \end{bmatrix} . \tag{3.9.9}$$

We find

$$J_\mu = \{\bar{I}_\pm, \bar{U}_\pm, \bar{V}_\pm, \bar{I}_3, \bar{I}_8\} ,$$

$$\bar{I}_\pm = \sum_i \mu_i I_i^\pm \mp 2h \sum_{i \neq j} W_{ij} \left(I_i^\pm I_j^3 + \frac{1}{2} U_i^\mp V_j^\mp \right) ,$$

$$\bar{U}_\pm = \sum_i \mu_i U_i^\pm \pm h \sum_{i \neq j} W_{ij} \left[U_i^\pm \left(I_j^3 - \frac{3}{2} Y_j \right) + I_i^\mp V_j^\mp \right] ,$$

$$\bar{V}_\pm = \sum_i \mu_i V_i^\pm \pm h \sum_{i \neq j} W_{ij} \left[V_i^\pm \left(I_j^3 + \frac{3}{2} Y_j \right) + U_i^\mp I_j^\mp \right] ,$$

$$\bar{I}_3 = \sum_i \mu_i I_i^3 + h \sum_{i \neq j} W_{ij} \left[I_i^+ I_j^- - \frac{1}{2} (U_i^+ U_j^- - V_i^+ V_j^-) \right] ,$$

$$\bar{I}_8 = \sum_i \mu_i Y_i + h \sum_{i \neq j} W_{ij} (U_i^+ U_j^- - V_j^+ V_j^-) , \tag{3.9.10}$$

where μ_i and h (not Planck constant) are arbitrary parameters. Notice again that the simplest choice of W_{ij} is given by equation (3.9.8).

When $i = 1, 2$, $Y(SU(2))$ makes transition between spin singlet and triplet. Now $Y(SU(3))$ transits $SU(3)$ singlet and Octet. For instance, setting

$$| \pi^- \rangle = |d\bar{u}\rangle, \quad |\pi^0\rangle = \frac{1}{\sqrt{2}} (|u\bar{u}\rangle - |d\bar{d}\rangle), \quad |K^-\rangle = |d\bar{u}\rangle, \quad |K^0\rangle = |d\bar{s}\rangle,$$

$$| \eta^0 \rangle = \frac{1}{\sqrt{(6)}} (-|u\bar{u}\rangle - |d\bar{d}\rangle + 2|s\bar{s}\rangle), \quad |\eta^{0'}\rangle = \frac{1}{\sqrt{(3)}} (|u\bar{u}\rangle + |d\bar{d}\rangle + |s\bar{s}\rangle) .$$

$$\tag{3.9.11}$$

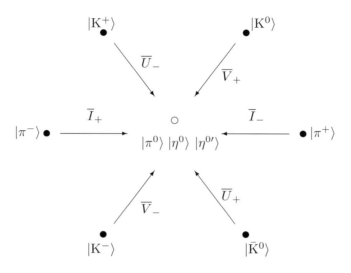

Fig. 2.   Representation of $SU(3)$.

Special interest is the following. When

$$\mu_1 - \mu_2 = -3h, \quad f = -\frac{1}{2}(\mu_1 - \mu_2), \qquad (3.9.12)$$

by acting the Yangian operators on the Octet of $SU(3)$, we obtain (see Figure 2)

$$\overline{I}_-|\pi^+\rangle = \frac{1}{\sqrt{6}}(\mu_1 - \mu_2)|\eta^0\rangle + \frac{1}{\sqrt{2}}(\mu_1 + \mu_2)|\pi^0\rangle - \frac{1}{\sqrt{3}}(\mu_1 - \mu_2 + 3h)|\eta^{0'}\rangle,$$

$$\overline{U}_+|\overline{K}^0\rangle = \frac{1}{\sqrt{6}}(\mu_1 + 2\mu_2)|\eta^0\rangle + \frac{1}{\sqrt{2}}\mu_1|\pi^0\rangle - \frac{1}{\sqrt{3}}(\mu_1 - \mu_2 + 3h)|\eta^{0'}\rangle,$$

$$\overline{U}_-|K^0\rangle = \frac{1}{\sqrt{6}}(2\mu_1 + \mu_2)|\eta^0\rangle + \frac{1}{\sqrt{2}}\mu_2|\pi^0\rangle + \frac{1}{\sqrt{3}}(\mu_1 - \mu_2 + 3h)|\eta^{0'}\rangle,$$

$$\overline{V}_+|K^+\rangle = \frac{1}{\sqrt{6}}(2\mu_1 + \mu_2)|\eta^0\rangle - \frac{1}{\sqrt{2}}\mu_2|\pi^0\rangle + \frac{1}{\sqrt{3}}(\mu_1 - \mu_2 + 3h)|\eta^{0'}\rangle,$$

$$\overline{V}_-|K^-\rangle = -\frac{1}{\sqrt{6}}(\mu_1 + 2\mu_2)|\eta^0\rangle + \frac{1}{\sqrt{2}}\mu_1|\pi^0\rangle + \frac{1}{\sqrt{3}}(\mu_1 - \mu_2 + 3h)|\eta^{0'}\rangle,$$

$$\overline{I}_3|\pi^0\rangle = -\frac{1}{2\sqrt{3}}(\mu_1 - \mu_2)|\eta^0\rangle + \frac{1}{\sqrt{6}}(\mu_1 - \mu_2 + 3h)|\eta^{0'}\rangle \,,$$

$$\overline{I}_8|\eta^0\rangle = -\frac{1}{3}(\mu_1 - \mu_2)|\eta^0\rangle - \frac{\sqrt{2}}{3}(\mu_1 - \mu_2 + 3h)|\eta^{0'}\rangle \,, \qquad (3.9.13)$$

i.e.,

$$(\overline{I}_\pm + fI_\pm)|\eta^{0'}\rangle = \pm 2\sqrt{3}h|\pi^\pm\rangle, \quad (\overline{U}_+ + fU_+)|\eta^{0'}\rangle = -2\sqrt{3}h|K^0\rangle \,,$$

$$(\overline{U}_- + fU_-)|\eta^{0'}\rangle = 2\sqrt{3}h|\overline{K}^0\rangle, \quad (\overline{V}_\pm + fV_\pm)|\eta^{0'}\rangle = -2\sqrt{3}h|K^\mp\rangle \,,$$

$$(\overline{I}_3 + fI_3)|\eta^{0'}\rangle = -\sqrt{6}h|\pi^0\rangle, \quad (\overline{I}_8 + fI_8)|\eta^{0'}\rangle = 2\sqrt{2}h|\eta^0\rangle, \qquad (3.9.14)$$

and

$$(\overline{I}_\pm + fI_\pm)|\pi^\mp\rangle = \pm\sqrt{\frac{3}{2}}h|\eta^0\rangle \,,$$

$$(\overline{U}_+ + fU_+)|K^0\rangle = -\frac{\sqrt{3}}{2\sqrt{2}}h(\sqrt{3}|\pi^0\rangle - |\eta^0\rangle) \,,$$

$$(\overline{U}_- + fU_-)|K^0\rangle = \frac{\sqrt{3}}{2\sqrt{2}}h(\sqrt{3}|\pi^0\rangle - |\eta^0\rangle) \,,$$

$$(\overline{V}_\pm + fV_\pm)|K^\pm\rangle = -\frac{\sqrt{3}}{2\sqrt{2}}h(\sqrt{3}|\pi^0\rangle + |\eta^0\rangle) \,,$$

$$(\overline{I}_3 + fI_3)|\pi^0\rangle = \sqrt{\frac{3}{2}}h|\eta^0\rangle, \quad (\overline{I}_8 + fI_8)|\eta^0\rangle = \sqrt{3}h|\eta^0\rangle \,. \qquad (3.9.15)$$

The Yangian operators play the role to transit the Octet states to the singlet state of $SU(3)$.

Whereas, if

$$\mu_1 - \mu_2 = 3h, \quad f = -\frac{1}{2}(\mu_1 + \mu_2) \,, \qquad (3.9.16)$$

with the notations

$$(\overline{A}^{(2)} + fA^{(1)})|\eta^{0'}\rangle = 0, \quad A = I_\alpha, \quad (\alpha = \pm, 3, 8), \quad U_\pm, \quad V_\pm \,, \qquad (3.9.17)$$

we have

$$(\bar{I}_\pm + fI_\pm)|\pi^\mp\rangle = \mp\sqrt{\frac{3}{2}}h|\eta^0\rangle \pm 2\sqrt{3}h|\eta^{0'}\rangle ,$$

$$(\bar{U}_+ + fU_+)|\overline{K}^0\rangle = \frac{\sqrt{3}}{2\sqrt{2}}h(\sqrt{3}|\pi^0\rangle - |\eta^0\rangle) - 2\sqrt{3}h|\eta^{0'}\rangle ,$$

$$(\bar{U}_- + fU_-)|K^0\rangle = -\frac{\sqrt{3}}{2\sqrt{2}}h(\sqrt{3}|\pi^0\rangle - |\eta^0\rangle) + 2\sqrt{3}h|\eta^{0'}\rangle ,$$

$$(\bar{V}_\pm + fV_\pm)|K^\pm\rangle = \frac{\sqrt{3}}{2\sqrt{2}}h(\sqrt{3}|\pi^0\rangle + |\eta^0\rangle) + 2\sqrt{3}h|\eta^{0'}\rangle ,$$

$$(\bar{I}_3 + fI_3)|\pi^0\rangle = -\frac{\sqrt{3}}{2}h|\eta^0\rangle + \sqrt{6}h|\eta^{0'}\rangle ,$$

$$(\bar{I}_8 + fI_8)|\eta^0\rangle = h|\eta^0\rangle - 2\sqrt{2}h|\eta^{0'}\rangle . \tag{3.9.18}$$

Obviously, in this case the Yangian operators make the transition from the Octet to a "combined" singlet state of $SU(3)$.

3.10. *Happer degeneracy*

In the experiment for ^{87}Rb molecular there appears new degeneracy ([41]) at the special $\pm B_0$ (magnetic field), i.e., the Zeeman effect disappears at $\pm B_0$. The model Hamiltonian reads ([42]) (x is scaled magnetic field)

$$H = \mathbf{K} \cdot \mathbf{S} + x(k + \frac{1}{2})S_z , \tag{3.10.1}$$

where $\mathbf{K}$ is angular momentum and $\mathbf{K}^2 = K(K+1)$. It only occurs for spin $S = 1$. It turns out that when $x = \pm 1$ there appears the curious degeneracy, that is, there is a set of eigenstates corresponding to

$$E = -\frac{1}{2} . \tag{3.10.2}$$

The conserved set is $\{\mathbf{K}^2, G_z = K_z + S_z\}$. For $\mathbf{G} = \mathbf{K} + \mathbf{S}$ we have $G = k \pm 1, k$. The eigenstates are specified in terms of three families: T, B and D. Only D-set possesses the degeneracy.

Happer gives, for example, the eigenstates for $x = \pm 1$ ([42]):

$$\begin{aligned} x = +1 \qquad &H\alpha_{Dm} = (-\tfrac{1}{2})\alpha_{Dm} \\ x = -1 \qquad &H\beta_{Dm} = (-\tfrac{1}{2})\beta_{Dm} , \end{aligned} \tag{3.10.3}$$

and shows that

$$\alpha_{Dm} = \left[2\left(K + \frac{1}{2}\right)\left(K + m + \frac{1}{2}\right)\right]^{-\frac{1}{2}}\left\{-\left[\frac{(K - m + 1)(K + m + 1)}{2}\right]^{\frac{1}{2}}\alpha_1\right.$$

$$\left. + \left[(K + m)(K + m + 1)\right]^{\frac{1}{2}}\alpha_2 + \left[\frac{(K - m)(K + m)}{2}\right]^{\frac{1}{2}}\alpha_3\right\};$$

$$(3.10.4)$$

$$\beta_{Dm} = \left[2\left(K + \frac{1}{2}\right)\left(K - m + \frac{1}{2}\right)\right]^{-\frac{1}{2}}\left\{\left[\frac{(K - m)(K + m)}{2}\right]^{\frac{1}{2}}\alpha_1\right.$$

$$\left. + \left[(K - m)(K - m + 1)\right]^{\frac{1}{2}}\alpha_2 - \left[\frac{(K - m + 1)(K + m + 1)}{2}\right]^{\frac{1}{2}}\alpha_3\right\},$$

$$(3.10.5)$$

where $\alpha_1 = e_1 \otimes e_{m-1}$, $\alpha_2 = e_0 \otimes e_m$ and $\alpha_3 = e_{-1} \otimes e_{m+1}$.

It is natural to ask what is the transition operator between α_{Dm} and β_{Dm}? The answer is Yangian operator. In fact, introducing

$$J_\pm = aS_+ + bK_- \pm (s_\pm K_z - s_z K_\pm), \qquad (3.10.6)$$

we find that by choosing $a = -\frac{k+1}{2}, b = 0$, we have

$$\beta_{Dm} \xrightarrow{J_+} \lambda_1(m)\alpha_{Dm+1} \text{ and } \alpha_{Dm} \xrightarrow{J_-} \lambda_2(m)\beta_{Dm-1}; \qquad (3.10.7)$$

and by choosing $a = \frac{k}{2}, b = 0$, we have

$$\beta_{Dm} \xrightarrow{J_-} \lambda'_1(m)\alpha_{Dm-1} \text{ and } \alpha_{Dm} \xrightarrow{J_+} \lambda'_2(m)\beta_{Dm+1}. \qquad (3.10.8)$$

The Yangian makes the transition between the states with B and $-B$, which here is only for $S = 1$. The reason is that for $S = 1$ there are two independent coefficients in the combination of α_1, α_2 and α_3 and there are two free parameters in $\mathbf{J}$. Hence the number of equations are equal to those of free parameters (a and b), so we can find a solution. The numerical computation shows that only $S = 1$ gives rise to the new degeneracy ([42]) that prefers the Yangian operation ([43]).

3.11. *New degeneracy of extended Breit–Rabi Hamiltonian*

As was shown in the Happer's model ($H = \mathbf{K} \cdot \mathbf{S} + x(k + \frac{1}{2})S_3$) there appeared new degeneracy for $S = 1$. It has been pointed out that the above degeneracy with respect to Zeeman effect cannot appear for spin$=\frac{1}{2}$. Actually, in this case it yields for $S = \frac{1}{2}$ ([42]),

$$E = -\frac{1}{4} - \omega_m S_3 , \qquad (3.11.1)$$

where

$$\omega_m^2 = \left[(1 + x^2) \left(k + \frac{1}{2} \right) + 2xm \right] \left(k + \frac{1}{2} \right) . \qquad (3.11.2)$$

Therefore if the Happer's type of degeneracy can occurs, there should be $\omega_m = 0$ that means

$$x_0 = -\frac{m}{k} \pm i \sqrt{1 - \frac{m^2}{k^2}} \quad \left(k = K + \frac{1}{2} \right) , \qquad (3.11.3)$$

i.e., the magnetic field should be complex.

However, the situation will be completely different, if a third spin is involved. For simplicity we assume $S_1 = S_2 = S_3 = \frac{1}{2}$ in the Hamiltonian:

$$H = -(a\mathbf{S}_2 + b\mathbf{S}_3) \cdot \mathbf{S}_1 + x\sqrt{ab} S_1^z, \lambda = b/a , \qquad (3.11.4)$$

then besides two non-degenerate states, there appears the degenerate family:

$$H\alpha_{D,\pm\frac{1}{2}}^{\pm} = -\left(\frac{a+b}{4} \right) \alpha_{D,\pm\frac{1}{2}}^{\pm}, \quad \text{for} \quad x = \pm 1 , \qquad (3.11.5)$$

where

$$\alpha_{D,+\frac{1}{2}}^{\pm} = -\sqrt{2}\lambda |\uparrow\uparrow\downarrow\rangle \pm \sqrt{\lambda} |\uparrow\downarrow\uparrow\rangle + (1 \pm \sqrt{\lambda}) |\downarrow\uparrow\uparrow\rangle ; \qquad (3.11.6)$$

$$\alpha_{D,-\frac{1}{2}}^{\pm} = -\sqrt{2}\lambda |\downarrow\downarrow\uparrow\rangle \mp \sqrt{\lambda} |\downarrow\uparrow\downarrow\rangle + (1 \mp \sqrt{\lambda}) |\uparrow\downarrow\downarrow\rangle . \qquad (3.11.7)$$

The expecting value of S_1^z are

$$\langle \alpha_{D,\pm\frac{1}{2}}^{+} | S_1^z | \alpha_{D,\pm\frac{1}{2}}^{+} \rangle \sim \sqrt{\lambda} \quad (x = 1) ; \qquad (3.11.8)$$

$$\langle \alpha_{D,\pm\frac{1}{2}}^{-} | S_1^z | \alpha_{D,\pm\frac{1}{2}}^{-} \rangle \sim -\sqrt{\lambda} \quad (x = -1) . \qquad (3.11.9)$$

namely, at the special magnetic field ($x = \pm 1$) the observed $\langle S_1^z \rangle$ still opposite to each other for $x = \pm 1$, but without the usual Zeeman split.

The reason of the appearance of the new degeneracy is obvious. The two spins $\mathbf{S}_2$ and $\mathbf{S}_3$ here play the role of $S = 1$ in comparison with Happer model.

4. Remarks

Although there has been certain progress of Yangian's application in physics, there are still open questions:

(1) How can the Yangian representations help to solve physical models, in particular, in strong correlation models?

(2) Direct evidences of Yangian in the real physics.

(3) What is the geometric meaning of Yangian?

Acknowledgement

We thank Professors F. Dyson and W. Happer for encouragement and enlighten discussions. This work is in part supported by NSF of China.

References

1. V. Drinfel'd, Sov. Math. Dokl. **32**(1985) 32.
2. V. Drinfel'd, Quantum group (PICM, Berkeley, 1986) 269.
3. V. Drinfel'd, Sov. Math. Dokl. **36** (1985) 212.
4. L. D. Faddeev, Sov. Sci. Rev. **C1** (1980) 107.
5. L. D. Faddeev, Les Houches, Session **39**, 1982.
6. L. D. Faddeev, Proc. of Les Houches Summer School, Session **LXIV** (1998) 149.
7. C. N. Yang, Phys. Rev. Lett. **19** (1967) 1312.
8. R. Baxter, Exactly Solved Methods in Statistical Mechanics, Academic, London, 1982.
9. M. Jimbo (ed.), Yang–Baxter Equations in Integrable Systems, World Scientific, Singapore, 1990.
10. P. P. Kulish, I. K. Sklyanin, Lecture Notes in Physics, **151**(1982) 1.
11. V. E. Korepin, N. M. Bogoliubov, A. G. Izergin, Quantum inverse scattering method and correlation function, (Cambridge Univ. Press, Cambridge, 1993).
12. E. K. Sklyanin, *Quantum Inverse Scattering Methods*, Selected Topics, in M. L. Ge (ed.) Quantum Groups and Quantum Integrable Systems, World Scientific, Singapore, 63-88 (1991).
13. M. L. Ge, K. Xue and Y. M. Cho, Phys. Lett. A **260**(1999) 484.
14. M. L. Ge, K. Xue and Y. M. Cho, Phys. Lett. A **249**(1998) 358.
15. H. B. Zhang, M. L. Ge, K. Xue, J. Phys. A **33** (2000) L345.
16. H. B. Zhang, M. L. Ge, K. Xue, J. Phys. A **34** (2001) 919.
17. H. B. Zhang, M. L. Ge, K. Xue, J. Phys. A **35** (2001) L7.
18. L. D. Faddeev, N. Yu. Reshetikhin, L. A. Takhtajan, Algebraic Analysis, **1** (1988) 178 (in Russian).
19. F. D. M. Haldane, Phys. Rev. Lett. **60** (1988) 635.
20. S. Shastry, Phys. Rev. Lett. **60** (1988) 639.

21. F. D. M. Haldane, Physics of the ideal semion gas: spinions and quantum symmetries of the integrable Haldnade-Shastry spin-chain, Proc. 16th Taniguch Symp. on Condensed Matter (Springer, Berlin, 1994).

22. V. Chari, A. Pressley, L'Enseignement mathematique **36** (1990) 267.

23. V. Chari, A. Pressley, J. Reine Angew. Math. **417** (1991) 87.

24. V. Chari, A. Pressley, A Guide to Quantum Groups (Cambridge Univ. Press, Cambridge, 1994).

25. E. Ragoucy, P. Sorba, Comm. Math. Phys. **203** (1999) 551.

26. M. Mintchev, E. Ragoucy, P. Sorba, Ph. Zaugg, J. Phys. A **32** (1999) 5885.

27. S. Murakami, M. Wadati, J. Phys. Soc. Japan **62** (1993) 4203.

28. D. B. Uglov, V. E. Korepin, Phys. Lett. A **190** (1994) 238.

29. F. Essler, V. E. Korepin, K. Schoutens, Nucl. Phys. B **372** (1992) 559.

30. F. Essler, V. E. Korepin, K. Schoutens, Nucl. Phys. B **384** (1992) 431.

31. F. Essler, V. E. Korepin, Phys. Rev. Lett. **72** (1994)908.

32. Z. F. Wang, M. L. Ge, K, Xue, J. Phys. A **30** (1997) 5023.

33. C. M. Bai, M. L. Ge and K. Xue, J. Stat. Phys. **102** (2001) 545.

34. L. Dolan, C. R. Nappi, E. Witten, arXiv: hep-th/0401243 (2004).

35. A. B. Zamolodchikov, Al. B. Zamolodchikov, Ann. Phys. **120** (1979) 253.

36. Y. Cheng, M. L. Ge, K. Xue, Comm. Math. Phys. **136** (1991) 195.

37. M. L. Ge, K. Xue, Phys. Lett. A **120** (1991) 266.

38. A. Belavin, Phys. Lett. B **283** (1992) 67.

39. R. Balian, N. R. Werthamer, Phys. Rev. **131** (1963) 1553.

40. C. M. Bai, M. L. Ge and K. Xue, Directional Transitions in spin systems and representations of $Y(sl(2))$, Nankai preprint, APCTP-98-026.

41. C. J. Erickson, D. Levron, W. Happer, S. Kadlecek, B. Chann, L. W. Anderson, T. G. Walker, Phys. Rev. Lett. **85** (2000) 4237.

42. W. Happer, Degeneracies of the Hamiltonian $x(K+1/2)S_z + \mathbf{K} \cdot \mathbf{S}$, preprint, Princeton University, November, 2000.

43. C. M. Bai, M. L. Ge, K. Xue, Inter. J. Mod. Phys. B **16** (2002) 1876.

Chapter 4

GEODESICALLY REVERSIBLE FINSLER 2-SPHERES OF CONSTANT CURVATURE*

Robert L. Bryant

*Duke University Mathematics Department,
P.O. Box 90320, Durham, NC 27708-0320, USA
bryant@math.duke.edu
http://www.math.duke.edu/~ bryant*

September 20, 2004

Dedicated to the memory of S.-S. Chern

A Finsler space (M, Σ) is said to be *geodesically reversible* if each oriented geodesic can be reparametrized as a geodesic with the reverse orientation. A reversible Finsler space is geodesically reversible, but the converse need not be true.

In this note, building on recent work of LeBrun and Mason [18], it is shown that a geodesically reversible Finsler metric of constant flag curvature on the 2-sphere is necessarily projectively flat.

As a corollary, using a previous result of the author [5], it is shown that a reversible Finsler metric of constant flag curvature on the 2-sphere is necessarily a Riemannian metric of constant Gauss curvature, thus settling a long-standing problem in Finsler geometry.

Keywords: Finsler metrics, flag curvature, projective structures.

1991 *Mathematics Subject Classification*. 53B40, 53C60.

1. Introduction

The purpose of this note is to settle a long-standing problem in Finsler geometry: Whether there exists a *reversible* Finsler metric on the 2-sphere with constant flag curvature that is not Riemannian. By making use of

*Thanks to Duke University for its support via a research grant and to the NSF for its support via DMS-0103884.

some old results and a fundamental new result of LeBrun and Mason, I show that such Finsler structures do not exist.

First, I prove something related: Any geodesically reversible Finsler metric on the 2-sphere with constant flag curvature must be projectively flat. Since the projectively flat Finsler metrics with constant flag curvature on S^2 were classified some years ago [5], the above result then reduces to examining the Finsler structures provided by this classification.

In a famous 1988 paper [1], Akbar-Zadeh showed that a (not necessarily reversible) Finsler structure on a compact surface with constant negative flag curvature was necessarily Riemannian or with zero flag curvature was necessarily a translation-invariant Finsler structure on the standard 2-torus $\mathbb{R}^2/\mathbb{Z}^2$. This naturally raised the question about what happens in the case of constant positive flag curvature.

This problem was made more interesting by the discovery of non-reversible Finsler metrics on the 2-sphere with constant positive flag curvature in [4]. (It should be pointed out that, in 1974, Katok [16] had already constructed non-reversible Finsler metrics on the 2-sphere (further analyzed in a paper of Ziller [24]) that later were shown by Shen [23] to have constant flag curvature, although, this constancy was not known at the time of [4].)

In the interests of brevity, no attempt has been made to give an exposition of the basics of Finsler geometry. There are many sources for this background material however, among them [2], [9, 10, 11], and [20].

For background more specifically suited for studying the case of constant flag curvature, including its proper formulation in higher dimensions, see [3], [13, 14, 15], and [21, 22, 23].

The corresponding question about (geodesically) reversible Finsler metrics of constant positive flag curvature on the n-sphere for $n > 2$ remains open at this writing, since an essential component of the proof for $n = 2$ that is due to LeBrun and Mason has not yet been generalized to higher dimensions.

Contents

2. Structure Equations

In this section, Cartan's structure equations for a Finsler surface will be recalled.

2.1. *Cartan's coframing*

Let M be a surface and let $\Sigma \subset TM$ be a smooth Finsler structure. I.e., Σ is a smooth hypersurface in M such that the basepoint projection $\pi : \Sigma \to M$ is a surjective submersion and such that each fiber

$$(2.1) \qquad \pi^{-1}(x) = \Sigma_x = \Sigma \cap T_x M$$

is a smooth, strictly convex closed curve in $T_x M$ whose convex hull contains the origin 0_x in its interior.

Remark 1 (Reversibility). Note that there is no assumption that $\Sigma = -\Sigma$. In other words, a Finsler structure need not be 'reversible' (some sources call this property 'symmetry'). This assumption is not needed for the development of the local theory.

Remark 2 (Σ versus F). One can think of Σ as the unit vectors of a 'Finsler metric', i.e., a function $F : TM \to \mathbb{R}$ that restricts to each tangent space $T_x M$ to be a not-necessarily-symmetric but strictly convex Banach norm on $T_x M$.

In fact, in most sources, the function F is referred to as the Finsler structure and Σ is called the *tangent indicatrix*. Since I will have little need to refer to F, I will take Σ as the defining structure in this article.

2.1.1. Σ-length of oriented curves

A smooth[1] curve $\gamma : (a, b) \to M$ will be said to be a Σ-*curve* (or *unit speed curve*) if $\gamma'(t)$ lies in Σ for all $t \in (a, b)$. The velocity curve $\gamma' : (a, b) \to \Sigma$ is known as the *tangential lift* of γ.

Any smooth, immersed curve $\gamma : (a, b) \to M$ has an orientation-preserving reparametrization $h : (u, v) \to (a, b)$ such that $\gamma \circ h$ is a Σ-curve. This reparametrization is unique up to translation in the domain of h. Thus, one can unambiguously define the Σ-length of an oriented subcurve $\gamma : (\alpha, \beta) \to M$ to be $h^{-1}(\beta) - h^{-1}(\alpha)$, when $a < \alpha < \beta < b$. The 1-form dh^{-1} on (a, b) is known as the element of Σ-length and is denoted ds_γ.

2.1.2. Cartan's coframing

The fundamental result about the geometry of Finsler surfaces is due to Cartan [8]:

Theorem 1 (Canonical coframing). *Let* $\Sigma \subset TM$ *be a Finsler structure on the oriented surface M with basepoint projection $\pi : \Sigma \to M$. There exists a unique coframing $\omega = (\omega_1, \omega_2, \omega_3)$ on Σ with the properties:*

(1) $\omega_1 \wedge \omega_2$ *is a positive multiple of any π-pullback of a positive 2-form on M,*
(2) *The lift γ' of any Σ-curve γ satisfies $(\gamma')^*\omega_2 = 0$ and $(\gamma')^*\omega_1 = ds_\gamma$,*
(3) $d\omega_1 \wedge \omega_2 = 0$,
(4) $\omega_2 \wedge d\omega_2 = \omega_1 \wedge d\omega_1$, *and*
(5) $d\omega_1 = \omega_3 \wedge \omega_2$ *and* $\omega_3 \wedge d\omega_2 = 0$.

Moreover, there exist unique functions I, J, and K on Σ so that

$$
\begin{aligned}
d\omega_1 &= -\omega_2 \wedge \omega_3, \\
(2.2) \qquad d\omega_2 &= -\omega_3 \wedge (\omega_1 - I\,\omega_2), \\
d\omega_3 &= -(K\,\omega_1 - J\,\omega_3) \wedge \omega_2.
\end{aligned}
$$

Remark 3 (The invariants I, J, and K). The 1-form ω_1 is *Hilbert's invariant integral*. Cartan [8] provides the following interpretations of the invariants: The function I vanishes if and only if Σ is the unit circle bundle of a Riemannian metric on M, in which case, J vanishes as well and the

---

[1]One can get by with weaker differentiability assumptions, but the issues that need such weaker assumptions are not relevant to the discussions in this article.

function K is simply the π-pullback of the Gauss curvature of the underlying metric on M.

The function J vanishes iff Σ is what is called *Landsberg* [2].

The function K is the Finsler-Gauss curvature (also known as the *flag curvature*).

Remark 4 (Bianchi identities). Taking the exterior derivatives of the structure equations (2.2) yields the formulae

$$(2.3) \qquad \begin{pmatrix} \mathrm{d}I \\ \mathrm{d}J \\ \mathrm{d}K \end{pmatrix} = \begin{pmatrix} J & I_2 & I_3 \\ -K_3 - KI & J_2 & J_3 \\ K_1 & K_2 & K_3 \end{pmatrix} \begin{pmatrix} \omega_1 \\ \omega_2 \\ \omega_3 \end{pmatrix}$$

for some functions I_2, I_3, J_2, J_3, K_1, K_2, and K_3 on Σ. These are the *Bianchi identities* of the coframing ω.

Remark 5 (The effect of orientations). If one reverses the orientation of M, then the canonical coframing ω on Σ is replaced by $(\omega_1, -\omega_2, -\omega_3)$.

In fact, Cartan's actual statement of Theorem 1 does not assume that M is oriented and concludes that there is a canonical coframing on Σ up to the sign ambiguity given here. The present version of the statement is a trivial rearrangement of Cartan's that is more easily applied in the situations encountered in this note.

2.1.3. *Reconstruction of M and its Finsler structure*

The information contained in the 3-manifold Σ and its coframing $\omega = (\omega_1, \omega_2, \omega_3)$ is sufficient to recover M, its orientation, and the embedding of Σ into TM, a fact that is implicit in Cartan's analysis:

Proposition 1 (Isometries and automorphisms). *For any orientation-preserving Finsler isometry* $\phi : M \to M$, *its derivative* $\phi' : TM \to TM$ *induces a diffeomorphism* $\phi' : \Sigma \to \Sigma$ *that preserves the coframing* $\omega = (\omega_1, \omega_2, \omega_3)$.

Conversely, any diffeomorphism $\psi : \Sigma \to \Sigma$ *that preserves* ω *is of the form* $\psi = \phi'$ *for a unique orientation-preserving Finsler isometry* $\phi : M \to M$.

Proof. The first statement follows directly from the uniqueness aspect of Theorem 1. I will sketch how the converse goes.

The integral curves of the system $\omega_1 = \omega_2 = 0$ on Σ are closed and the codimension 2 foliation they define has trivial holonomy, so M can be identified with the leaf space of this system and carries a unique smooth structure for which the leaf projection $\pi : \Sigma \to M$ is a smooth submersion.

Because of the connectedness of the π-fibers, there is a unique orientation on M such that a positive 2-form pulls back under π to be a positive multiple of $\omega_1 \wedge \omega_2$. Thus, M, its smooth structure, and its orientation can be recovered from the coframing.

The inclusion $\iota : \Sigma \to TM$ is then seen to be simply given by $\iota(u) = \pi'\big(X_1(u)\big) \in T_{\pi(u)}M$. Thus, even the Finsler structure on M can be recovered from Σ and the coframing.

The desired result now follows by noting that any $\psi : \Sigma \to \Sigma$ that preserves ω will necessarily preserve the integral curves of the system $\omega_1 = \omega_2 = 0$ and hence induce a map $\phi : M \to M$ that is π-intertwined with ψ. The verification that ϕ is an orientation-preserving Finsler isometry is easy and can be left to the reader. $\square$

Corollary 1 (Finsler isometries). *If M is connected, then the group $\mathrm{Iso}(M, \Sigma)$ of orientation-preserving Finsler isometries of M acts freely on Σ and carries a unique Lie group structure for which the evaluation map $\mathrm{Iso}(M, \Sigma) \times \Sigma \to \Sigma$ is a smooth Lie group action. In particular, $\mathrm{Iso}(M, \Sigma)$ has dimension at most 3.*

Proof. The assumption that M be connected implies that Σ is connected. Thus, in light of Proposition 1, the Corollary follows immediately from Kobayashi's results [17] about automorphisms of coframings on connected manifolds. $\square$

Corollary 2 (Orientation-reversing isometries). *Any diffeomorphism $\psi : \Sigma \to \Sigma$ that satisfies $\psi^*(\omega) = (\omega_1, -\omega_2, -\omega_3)$ is of the form $\psi = \phi'$ for a unique orientation-reversing Finsler isometry $\phi : M \to M$.*

2.2. Geodesics

A Σ-curve γ is a *geodesic* of the Finsler structure if and only if its tangential lift satisfies $(\gamma')^*\omega_3 = 0$. (Of course, by definition, $(\gamma')^*\omega_2 = 0$.)

Note that, *by definition*, a geodesic has 'unit speed', i.e., is a Σ-curve. Strictly speaking, these should be referred to as 'Σ-geodesics' but, since, in this article, there will usually be no ambiguity about which Finsler structure is under discussion, the shorter term 'geodesics' will be adequate.

Let X_1, X_2, and X_3 be the vector fields on Σ that are dual to the coframing $(\omega_1, \omega_2, \omega_3)$. Then the flow of X_1 is the geodesic flow on Σ: A

Σ-curve $\gamma : (a, b) \to M$ is a geodesic if and only if $\gamma' : (a, b) \to \Sigma$ is an integral curve of X_1.

Consequently, for each $u \in \Sigma$, there is a unique maximal open interval $I_u \subset \mathbb{R}$ containing 0 for which there exists a geodesic $\gamma_u : I_u \to M$ satisfying $\gamma'_u(0) = u$. The Finsler structure Σ is said to be *geodesically complete* if $I_u = \mathbb{R}$ for all $u \in \Sigma$, i.e., if the flow of X_1 is complete.

Definition 1 (Closed geodesics). A geodesic $\gamma : \mathbb{R} \to M$ is said to be *closed* if it is periodic with some period $L > 0$, i.e., if $\gamma(t + L) = \gamma(t)$. If $L > 0$ is the minimal such period, then γ is said to be of *prime length L*.

2.2.1. *Local minimizing properties and stability*

Just as in the Riemannian case, a geodesic $\gamma : (a, b) \to M$ minimizes the oriented distance from $\gamma(\alpha)$ to $\gamma(\beta)$ when $a < \alpha < \beta < b$ and α and β are sufficiently close.

Regarding the second variation, the function K plays the same role in the Jacobi equation for Finsler geodesics as the Gauss curvature does in the Jacobi equation for Riemannian geodesics. See [8].

2.3. *Simplifications when $K \equiv 1$*

The Finsler structures of interest in this article are the ones that satisfy $K \equiv 1$. In this case, the structure equations simplify to

$$(2.4) \qquad \begin{aligned} \mathrm{d}\omega_1 &= -\omega_2 \wedge \omega_3, \\ \mathrm{d}\omega_2 &= -\omega_3 \wedge \left(\omega_1 - I\,\omega_2\right), \\ \mathrm{d}\omega_3 &= -\left(\omega_1 - J\,\omega_3\right) \wedge \omega_2, \end{aligned}$$

and the Bianchi identities become

$$(2.5) \qquad \begin{pmatrix} \mathrm{d}I \\ \mathrm{d}J \end{pmatrix} = \begin{pmatrix} J & I_2 & I_3 \\ -I & J_2 & J_3 \end{pmatrix} \begin{pmatrix} \omega_1 \\ \omega_2 \\ \omega_3 \end{pmatrix}.$$

2.4. *Some global consequences of $K \equiv 1$*

Suppose now that M is connected and geodesically complete, i.e., that the vector field X_1 is complete on Σ (in both forward and backward time). Of course, if M were assumed to be compact, then Σ would be also, and the completeness of X_1 would follow from this.

Let $\Psi : \mathbb{R} \times \Sigma \to \Sigma$ be the flow of X_1 and, for brevity, let $\Psi_t : \Sigma \to \Sigma$ denote the time t flow of X_1. Since the structure equations imply

$$(2.6) \qquad \mathsf{L}_{X_1}\,\omega_1 = 0, \qquad \mathsf{L}_{X_1}\,\omega_2 = \omega_3, \qquad \mathsf{L}_{X_1}\,\omega_3 = -\omega_2,$$

it follows (letting $t : \mathbb{R} \times \Sigma \to \mathbb{R}$ denote the coordinate that is the projection on the first factor) that

$$(2.7) \qquad \begin{aligned} \Psi^*\omega_1 &= \omega_1 + \mathrm{d}t, \\ \Psi^*\omega_2 &= \cos t\,\omega_2 + \sin t\,\omega_3, \\ \Psi^*\omega_3 &= -\sin t\,\omega_2 + \cos t\,\omega_3. \end{aligned}$$

Proposition 2 (The quasi-antipodal map). *There   exists   a   unique orientation-reversing Finsler isometry $\alpha : M \to M$ such that $\alpha' = \Psi_\pi$. For any point $p \in M$, every geodesic leaving p passes through $\alpha(p)$ at distance π.*

Proof.   By (2.7), it follows that $\Psi_\pi : \Sigma \to \Sigma$ satisfies

$$(2.8) \qquad\qquad \Psi_\pi^*\omega = (\omega_1, -\omega_2, -\omega_3).$$

Hence, by Corollary 2, there is a unique orientation-reversing Finsler isometry $\alpha : M \to M$ such that $\Psi_\pi = \alpha' : \Sigma \to \Sigma$.

Since X_1 is the geodesic flow vector field, any geodesic leaving p at time 0 is of the form $\gamma(t) = \pi\big(\Psi_t(u)\big)$ for some $u \in \Sigma_p \subset T_p M$. Thus, $\gamma(\pi) = \pi\big(\Psi_\pi(u)\big) = \pi\big(\alpha'(u)\big) = \alpha(p)$, as claimed. $\square$

Now, for any fixed $p \in M$, the fiber $\Sigma_p \subset T_p M$, is diffeomorphic to a circle and is naturally oriented by taking the pullback of ω_3 to Σ_p to be a positive 1-form. Define $\ell(p) > 0$ by

$$(2.9) \qquad\qquad \ell(p) = \frac{1}{2\pi} \int_{\Sigma_p} \omega_3\,.$$

Then Σ_p can be parametrized by a mapping $\iota_p : S^1 \to \Sigma_p$ that satisfies $\iota_p^*(\omega_3) = \ell(p)\,\mathrm{d}\theta$ and that is uniquely determined once one fixes $\iota_p(0) \in \Sigma_p$. Such an ι_p will be referred to as an *angle measure* on Σ_p.

Proposition 3 (Geodesic polar coordinates). *Given $p \in M$, choose an angle measure $\iota_p : S^1 \to \Sigma_p$. There is a map $E_p : S^2 \to M$ defined by the condition*

$$(2.10)\ \ E_p(\sin t \cos\theta, \sin t \sin\theta, \cos t) = \pi\big(\Psi(t, \iota_p(\theta))\big) \qquad \text{for } 0 \le t \le \pi,$$

and it is an orientation-preserving homeomorphism that is smooth away from the two 'poles' $(0, 0, \pm 1) \in S^2$. *In particular, M is homeomorphic to the 2-sphere and its oriented diameter as a Finsler space is equal to π.*

Proof. Consider the mapping $R_p : \mathbb{R} \times S^1 \to \Sigma$ defined by

$$(2.11) \qquad R_p(t, \theta) = \Psi(t, \iota_p(\theta)).$$

The formulae (2.7), the fact that Ψ is the flow of X_1, and the defining property of ι_p then combine to show that

$$(2.12) \qquad R_p^*(\omega_1 \wedge \omega_2) = \mathrm{d}t \wedge \left(\sin t \, \ell(p) \, \mathrm{d}\theta \right) = \ell(p) \sin t \, \mathrm{d}t \wedge \mathrm{d}\theta.$$

Thus, the composition $\pi \circ R_p : \mathbb{R} \times S^1 \to M$ is a smooth map that is a local diffeomorphism away from the circles $(t, \theta) = (k\pi, \theta)$ for each integer k. Of course, $\pi(R_p(0, \theta)) = p$ and $\pi(R_p(\pi, \theta)) = \alpha(p)$ for all $\theta \in S^1$.

It now follows that the formula (2.10) does define a mapping $E_p : S^2 \to M$ that, away from $(0, 0, \pm 1)$, is smooth and an orientation-preserving local diffeomorphism. Near the two points $(0, 0, \pm 1)$, the mapping E_p is still a (not necessarily differentiable) orientation-preserving local homeomorphism.

It follows that $E_p : S^2 \to M$ is a topological covering map. Since M is orientable by assumption, it follows that E_p must be a homeomorphism and, in particular, must be one-to-one and onto.

In particular, the mapping $\pi \circ R_p : (0, \pi) \times S^1 \to M$ is a diffeomorphism onto its image, which is M minus the two points p and $\alpha(p)$. It follows that there is a unique 1-form ρ defined on M minus these two points with the property that $(\pi \circ R_p)^*(\rho) = \mathrm{d}t$. The 1-form ρ is exact and pulls back to each geodesic segment from p to $\alpha(p)$ to be the Finsler arc length along that geodesic segment. Thus, one can define $r : M \to [0, \pi]$ so that $r(p) = 0$ and $\mathrm{d}r = \rho$ and r will be continuous on M and smooth away from p and $\alpha(p)$.

Moreover, because of the convexity properties of Σ, it is not difficult to show that $\mathrm{d}r(v) \leq 1$ for all $v \in \Sigma$ (other than the two circles Σ_p and $\Sigma_{\alpha(p)}$, where $\mathrm{d}r$ is not defined), with equality if and only if v is an oriented tangent vector to a geodesic segment from p to $\alpha(p)$. Consequently, for any Σ-curve $\gamma : [a, b] \to M$ that satisfies $\gamma(a) = p$ and $\gamma(b) = \alpha(p)$, one has

$$(2.13) \qquad b - a = \int_a^b \mathrm{d}s_\gamma \geq \int_a^b \gamma^*(\mathrm{d}r) = r(\gamma(b)) - r(\gamma(a)) = \pi,$$

with equality if and only if γ is a geodesic segment of length π from p to $\alpha(p)$.

The claim about oriented diameters now follows.   □

Remark 6 (Topology of Σ). One consequence of the fact that M is diffeomorphic to S^2 is that Σ is diffeomorphic to the unit tangent bundle of S^2 endowed with any Riemannian metric. In particular, $\Sigma \simeq SO(3) \simeq \mathbb{RP}^3$.

Remark 7 (Shen's results). Propositions 2 and 3 were proved by Shen [21] in the case that Σ is reversible (see Definition 2).

Proposition 4. *Either $\alpha^2 = \mathrm{id}$ on M (in which case, all of the geodesics are closed of length 2π) or else α^2 has exactly two fixed points, say n and $\alpha(n)$.*

In the latter case, there exists a positive definite inner product on T_nM that is invariant under $(\alpha^2)'(n) : T_nM \to T_nM$ and there is an angle $\theta_n \in (0, 2\pi)$ such that $(\alpha^2)'(n)$ is a counterclockwise rotation by θ_n in this inner product.

Proof. Assume that $\alpha^2 : M \to M$ is not the identity, or else there is nothing to prove. By the Lefschetz fixed point formula, since α^2 is an orientation-preserving diffeomorphism of the 2-sphere, it must have at least one fixed point. Let n be such a fixed point. By the very definition of α, it then follows that $\alpha(n)$ is also a fixed point of α^2, one that is distinct from n by Proposition 3. It must be shown that α^2 has no other fixed points.

Consider the linear map $L = (\alpha^2)'(n) : T_nM \to T_nM$. Since α^2 is a Finsler isometry, the linear map L must preserve $\Sigma_n \subset T_nM$. Let $K_n \subset T_nM$ be the convex set bounded by Σ_n.

Define a positive definite quadratic form on T_n^*M by letting $\langle \lambda_1, \lambda_2 \rangle$ be defined for $\lambda_1, \lambda_2 \in T_n^*M$ to be the average of the quadratic function $2\lambda_1\lambda_2$ over K_n (using any translation invariant measure on K_n induced by its inclusion into the vector space T_nM). Since L is a linear map carrying K_n into itself, it must preserve this quadratic form and hence must also preserve the dual (positive-definite) quadratic form on T_nM. Since L also preserves an orientation on T_nM, it follows that, with respect to this L-invariant inner product, L must be a counterclockwise rotation by some angle $\theta_n \in [0, 2\pi)$.

If θ_n were 0, i.e., L were the identity on T_nM, then all of the geodesics through n would close at length 2π. In particular, the mapping $\Psi_{2\pi} : \Sigma \to \Sigma$ would have a fixed point and would preserve the coframing ω, forcing $\Psi_{2\pi}$ to the identity on Σ and, thus, α^2 to be the identity on M. Thus, $0 < \theta_n < 2\pi$.

Since n was an arbitrarily chosen fixed point of α^2, it follows that every fixed point of α^2 is an isolated elliptic fixed point, i.e., a fixed point of index 1. Since M is diffeomorphic to S^2, the Lefschetz fixed point formula implies that the map α^2 has exactly two fixed points. Thus α^2 has no fixed points other than n and $\alpha(n)$. □

Remark 8 (Properties of α^2). When α^2 is not the identity and n is one of its fixed points, $\theta_n + \theta_{\alpha(n)} = 2\pi$.

If θ_n is not a rational multiple of π, then the iterates of α^2 are dense in a circle of Finsler isometries of (M, Σ) that fix n and $\alpha(n)$. In such a case, (M, Σ) is rotationally symmetric about n. Moreover, it is symmetric (in an orientation-reversing sense) with respect to α.

If $\theta_n = 2\pi(p/q)$ where $0 < p \leq q$ and p and q have no common factors, then α^{2q} is the identity, so that every geodesic closes at length $2\pi q$ (though some may close sooner).

Remark 9 (The Katok examples). The second possibility in Proposition 4 does occur. The Katok examples [16] (further analyzed by Ziller [24]) have been shown by Shen [23] to have $K \equiv 1$ and are examples in which α^2 is not the identity. In fact, Katok's examples show that the number θ_n can assume any value in $(0, 2\pi)$.

For a discussion of the geometry of the geodesic flow of the Katok examples, see [19]. For a discussion of the Katok examples via the Cartan coframing formalism, see [7].

3. A Double Fibration

Throughout this section, Σ will be a Finsler structure on M (assumed diffeomorphic to the 2-sphere) satisfying $K \equiv 1$.

I begin by noting that, if all the geodesics on M close at distance 2π, then the set of oriented geodesics has the structure of a manifold in a natural way.

Proposition 5 (The space of oriented geodesics). *If α^2 is the identity, then the action*

$$(3.1) \qquad\qquad u \cdot e^{it} = \Psi(t, u)$$

defines a smooth, free, right S^1-action on Σ whose orbits are the integral curves of $\omega_2 = \omega_3 = 0$ and there exists a smooth surface Λ diffeomorphic to S^2 and a smooth submersion $\lambda : \Sigma \to \Lambda$ so that the action (3.1) makes $\lambda : \Sigma \to \Lambda$ into a principal right S^1-bundle over Λ.

Proof. If α^2 is the identity, then the flow of X_1 is periodic of period 2π, so (3.1) defines a smooth S^1-action on Σ. Since X_1 never vanishes, this action has no fixed points. Thus, if this action were not free, then there would be a $u \in \Sigma$ and an integer $k \geq 2$ such that $\Psi(2\pi/k, u) = u$. However, since $0 < 2\pi/k \leq \pi$, the equality $\Psi(2\pi/k, u) = u$ would violate Proposition 3, since then $E_{\pi(u)} : S^2 \to M$ could not be one-to-one.

Thus, the S^1-action (3.1) is free and the rest of the proposition follows by standard arguments. $\qquad\qquad\square$

Remark 10 (Double fibration and path geometries). The two mappings $\pi : \Sigma \to M$ and $\lambda : \Sigma \to \Lambda$ define a double fibration and it is easy to see that this double fibration satisfies the usual nondegeneracy axioms for double fibrations. For example, $\lambda \times \pi : \Sigma \to \Lambda \times M$ is clearly a smooth embedding. The other properties are similarly easy to verify using the structure equations. Thus, Σ defines a (generalized) path geometry on each of Λ and M.

For more background on path geometries and their invariants, see, for example, Section 2 of [5].

3.1. *Induced structures on* Λ

I will now recall some results from [5]. Throughout this subsection, I will be assuming that α^2 is the identity, so that Λ exists as a smooth manifold.

The relations (2.7) show that the quadratic form $\omega_2{}^2 + \omega_3{}^2$ is invariant under the flow of X_1. Consequently, there is a unique Riemannian metric on Λ, say g, such that

$$(3.2) \qquad\qquad \lambda^*(g) = \omega_2{}^2 + \omega_3{}^2.$$

Moreover, the 2-form $\omega_3 \wedge \omega_2$ is invariant under the flow of X_1, so it is the pullback under λ of an area 2-form for g, which will be denoted dA_g.

Now, there is an embedding $\xi : \Sigma \to T\Lambda$ defined by

$$(3.3) \qquad\qquad \xi(u) = \lambda'\big(X_3(u)\big)$$

and one sees that ξ embeds Σ as the unit circle bundle of Λ endowed with the metric g.

The structure equations (2.4) show that, under this identification of Σ with the unit circle bundle of Λ, the Levi-Civita connection form on Σ is

$$(3.4) \qquad\qquad \rho = -\omega_1 + I\,\omega_2 + J\,\omega_3\,.$$

Note that $-\omega_1$ and $I\,\omega_2 + J\,\omega_3$ are invariant under the flow of X_1.

For the next two results, which follow from the structure equations derived so far by simply unraveling the definitions, the reader may want to consult LeBrun and Mason [18] for the definition and properties of the projective structure associated to an affine connection on a surface. [They restrict themselves to the consideration of torsion-free connections, but, as they point out, this does not affect the results.]

Proposition 6. *There exists a g-compatible affine connection ∇ on Λ such that the $[\nabla]$-geodesics are the λ-projections of the integral curves of $\omega_1 = \omega_2 = 0$.* $\square$

Corollary 3. *The geodesics of the projective structure $[\nabla]$ on Λ are closed.*

Proof. By Proposition 6, the geodesics of $[\nabla]$ are the λ-projections of the integral curves of the system $\omega_1 = \omega_2 = 0$, but these integral curves are the fibers of the map $\pi : \Sigma \to M$ and hence are closed. $\square$

3.2. *Geodesic reversibility implies geodesic periodicity*

It is now time to come to the main point of this note.

Definition 2 (Reversibility). A Finsler structure $\Sigma \subset TM$ is said to be *reversible* if $\Sigma = -\Sigma$.

Definition 3 (Geodesic reversibility). A Finsler structure Σ on M is said to be *geodesically reversible* if any geodesic $\gamma : (a, b) \to M$ can be reparametrized in an orientation-reversing way so as to remain a geodesic.

Remark 11. Any reversible Finsler structure is geodesically reversible. On the other hand, the non-Riemannian Finsler examples constructed in Section 4 of [5] are geodesically reversible but not reversible.

Proposition 7. *If (M, Σ) is geodesically reversible, then α^2 is the identity on M.*

Proof. For any point $p \in M$, consider the geodesic segments of length π originating at p. By Proposition 3, they all terminate at $\alpha(p)$ but are pairwise disjoint between the two endpoints. By assumption, reversing these geodesic segments, i.e., reparametrizing them in a orientation reversing sense as Σ-curves, yields geodesic segments, now originating at $\alpha(p)$ but that, *a priori* need not be of length π.

Now, these reversed geodesic segments are pairwise disjoint until they all terminate at p. By Proposition 3, the geodesics leaving $\alpha(p)$ remain pairwise disjoint for distances between 0 and π and they all converge on $\alpha\big(\alpha(p)\big)$ at distance π.

It follows that $\alpha\big(\alpha(p)\big)$ must be p. In other words, α^2 is the identity.
□

Remark 12. The converse of Proposition 7 does not hold. The $K \equiv 1$ examples provided by Theorem 3 of [6] that are based on Guillemin's Zoll metrics have all their geodesics closed of length 2π (and hence α^2 is the identity), but none of the non-Riemannian ones are geodesically reversible.

3.3. *Geodesic reversibility implies projective flatness*

The next step is to consider the space of *unoriented* geodesics on M. This only makes sense if one assumes Σ to be geodesically reversible, so assume this for the rest of this subsection.

For each oriented geodesic $\gamma : S^1 \to M$, let $\beta(\gamma)$ denote the reversed curve, parametrized so as to be a geodesic. Obviously $\beta : \Lambda \to \Lambda$ is a fixed-point free involution of Λ, so that the quotient manifold Λ/β is diffeomorphic to $\mathbb{RP}^2$.

Proposition 8. *The path geometry on Λ defined by the geodesics of $[\nabla]$ is invariant under β and hence descends to a well-defined path geometry on Λ/β. Moreover, this path geometry is the path geometry of a projective connection on Λ/β with all of its geodesics closed.*

Proof. Since, by definition, a point p in M lies on a geodesic γ if and only if it lies on $\beta(\gamma)$, it follows that β carries each $[\nabla]$-geodesic into itself. In particular, even though β may not (indeed, most likely does not) preserve ∇, it must preserve $[\nabla]$ since the projective equivalence class of ∇ is determined by its geodesics. Thus, the claims of the Proposition are verified.
□

It is at this point that the crucial contribution of LeBrun and Mason [18] enters:

Theorem 2 (LeBrun-Mason). *Any projective structure on $\mathbb{RP}^2$ that has all of its geodesics closed is projectively equivalent to the standard (i.e., flat) projective structure.*

Corollary 4. *If* Σ *is a geodesically reversible Finsler structure on* $M \simeq S^2$ *that satisfies* $K \equiv 1$, *then the induced projective structure* $[\nabla]$ *on* Λ *is projectively flat.* □

Remark 13 (LeBrun and Mason's classification). The article [18] contains, in addition to Theorem 2, much information about *Zoll projective structures* on the 2-sphere, i.e., projective structures on the 2-sphere all of whose geodesics are closed. It turns out that, in a certain sense, there are many more of them than there are Zoll metrics on the 2-sphere.

Their results could quite likely be very useful in understanding the case of non-reversible Finsler metrics satisfying $K \equiv 1$ on the 2-sphere that satisfy $\alpha^2 = \mathrm{id}$, which is still not very well understood. It is even possible that an orbifold version of their results could be useful in the case in which α^2 is not the identity but has finite order. This may be the subject of a later article.

4. Classification

In this section, the main theorem will be proved.

4.1. *Consequences of projective flatness*

Recall from Section 2 of [5] that if a projective structure on a surface is projectively flat then its dual path geometry is projective and, moreover, projectively flat.

Proposition 9. *If* Σ *is a geodesically reversible Finsler structure on* $M \simeq S^2$ *with* $K \equiv 1$, *then the geodesics in* M *are the geodesics of a flat projective structure.*

Proof. The dual path geometry of Λ with its projective structure $[\nabla]$ is M with the space of paths being the geodesics. Now apply Corollary 4. □

Corollary 5. *Let* M *be diffeomorphic to* S^2. *Up to diffeomorphism, any geodesically reversible Finsler structure* $\Sigma \subset TM$ *with* $K \equiv 1$ *is equivalent to a member of the 2-parameter family described in Theorem 10 of* [5].

Proof. In light of Proposition 9, one can apply Theorems 9 and 10 of [5], which gives the result. □

Remark 14. It is interesting to note that each member of the 2-parameter family described in Theorem 10 of [5] is projectively flat and hence geodesically reversible.

4.2. *Reversibility*

Now for the main rigidity theorem.

Theorem 3. *Any reversible Finsler structure on* $M \simeq S^2$ *that satisfies* $K \equiv 1$ *is Riemannian and hence isometric to the standard unit sphere.*

Proof. Such a Finsler structure would be geodesically reversible and hence, by Corollary 5, a member of the family described in Theorem 10 of [5]. However, by inspection, the only member of this geodesically reversible family that is actually reversible is the Riemannian one. $\qquad\square$

Remark 15 (The argument of Foulon-Reissman). In Section 4 of [12], P. Foulon sketches an argument, due to himself and A. Reissman, that a reversible Finsler metric on the 2-sphere satisfying $K \equiv 1$ that satisfies a certain integral-geometric condition (called by them 'Radon-Gelfand') is necessarily Riemannian. Their condition holds, in particular, whenever the projective structure $[\nabla]$ on Λ is projectively flat. Thus, an alternate proof of Theorem 3 could be given by combining LeBrun and Mason's Theorem 2 with Foulon and Reissman's argument.

The proof of Theorem 3 in this article instead relies on the classification in [5].

References

[1] H. Akbar-Zadeh, *Sur les espaces de Finsler à courbures sectionnelles constantes*, Acad. Roy. Belg. Bull. Cl. Sci. (5) **74** (1988), 281–322. MR 91f:53069 96

[2] D. Bao, S.-S. Chern, and Z. Shen, *An Introduction to Riemann–Finsler Geometry*, Graduate Texts in Mathematics **200**, Springer-Verlag, New York. 2000. MR 2001g:53130   96, 99

[3] D. Bao and Z. Shen, *Finsler metrics of constant curvature on the Lie group* S^3, preprint, 2001.   96

[4] R. Bryant, *Finsler structures on the 2-sphere satisfying* $K = 1$, Finsler Geometry, Contemporary Mathematics **196** (1996), 27–42. MR 97e:53128 96

[5] ———, *Projectively flat Finsler 2-spheres of constant curvature*, Selecta Math., New Series **3** (1997), 161–204. MR 98i:53101   95, 96, 106, 107, 109, 110

[6] ———, *Some remarks on Finsler manifolds with constant flag curvature*, Houston J. Math. **28** (2002), 221–262. MR 03h:53102 108

[7] ———, *The geodesic flow on Finsler 2-spheres of constant curvature*, preprint, September 2004. 105

[8] É. Cartan, *Sur un problème d'équivalence et la théorie des espaces métriques généralisés*, Mathematica **4** (1930), 114–136. (Reprinted in *Oeuvres Complètes*, partie III, vol. 2, Éditions du CNRS, 1984.) 98, 101

[9] ———, *Les Espace Finsler*, Exposés de Géometrie, t. 79, Hermann, Paris, 1934. 96

[10] S.-S. Chern, *On the Euclidean connections in a Finsler space*, Proc. Natl. Acad. Sci. USA **29** (1943), 33–37. (Reprinted in *Shiing-shen Chern: Selected Papers, vol. II*, Springer-Verlag, New York, 1989, pp. 107–111.) MR 4,259c 96

[11] ———, *Local equivalence and Euclidean connections in Finsler spaces*, Science Reports Tsing Hua Univ. **5** (1948), 95–121. (Reprinted in *Shiing-shen Chern: Selected Papers, vol. II*, Springer-Verlag, New York, 1989, pp. 194–212. MR 11,212a 96

[12] P. Foulon, *Curvature and global rigidity in Finsler manifolds*, Houston J. Math. **28** (2002), 263–292. 110

[13] P. Funk, *Über Geometrien, bei denen die Geraden die Kürzesten sind*, Math. Annalen **101** (1929), 226–237. 96

[14] ———, *Über zweidimensionale Finslersche Räume, insbesondere über solche mit geradlinigen Extremalen und positiver konstanter Krümmung*, Math. Zeitschr. **40** (1936), 86–93. 96

[15] ———, *Eine Kennzeichnung der zweidimensionalen elliptischen Geometrie*, Österreich. Akad. Wiss. Math.-Natur. Kl. S.-B. II **172** (1963), 251–269. MR 30 #1460 96

[16] A. Katok, *Ergodic properties of degenerate integrable Hamiltonian systems*, Izs. Akad. Nauk SSSR Ser. Math. **37** (1973), 539–576. (English translation: Math. USSR Izv. **7** (1973), 535–572.) MR 48 #9758 96, 105

[17] S. Kobayashi, *Theory of connections*, Ann. Math. Pura Appl. **43** (1957), 119–194. MR 20 #2760 100

[18] C. LeBrun and L. J. Mason, *Zoll manifolds and complex surfaces*, J. Differential Geom. **61** (2002), 453–535. MR 04d:53043 95, 107, 108, 109

[19] C. Robles, *Geodesics in Randers spaces of constant curvature*, preprint 2004. 105

[20] H. Rund, *The differential geometry of Finsler surfaces*, Grundlehren der Math. Wiss., Band 101, Springer-Verlag, Berlin, 1959. MR 21 #4462 96

[21] Z. Shen, *Finsler manifolds of constant positive curvature*, In: Finsler Geometry, Contemporary Math. **196** (1996), 83–92. MR 97m:53120 96, 104

[22] ———, *Projectively flat Finsler metrics of constant flag curvature*, Trans. Amer. Math. Soc. **355** (2003), 1713–1728 (electronic). MR 03i:53111 96

[23] ———, *Two-dimensional Finsler metrics with constant flag curvature*, Manuscripta Math. **109** (2002), 349–366. MR 03k:53091 96, 105

[24] W. Ziller, *Geometry of the Katok examples*, Ergodic Theory Dynam. Systems **3** (1983), 135–157. MR 86g:58036 96, 105

Chapter 5

MULTIPLE SOLUTIONS OF THE PRESCRIBED MEAN CURVATURE EQUATION

K. C. Chang

School of Math. Sci., Peking Univ., Beijing 100871, China
kcchang@math.pku.edu.cn

Tan Zhang

Dept. Math & Stats., Murray State Univ., Murray, KY 42071, USA
tan.zhang@murraystate.edu

In memory of Professor S.S. Chern

We combine heat flow method with Morse theory, super- and sub-solution method with Schauder's fixed point theorem to show the existence of multiple solutions of the prescribed mean curvature equation under some special circumstances.

Keywords and phrases: Mean curvature equation, Morse Theory, multiple solutions.

1991 *Mathematics Subject Classification*. Primary 53A10, 58E05; Secondary 35A15, 58C30.

In this paper, we study the existence of multiple solutions of the following equation:

$$(0.1) \qquad \begin{cases} \mathcal{M}(u) = nH(x,u) & \text{in } \Omega \subset \mathbb{R}^n \\ u|_{\partial\Omega} = \varphi, \end{cases}$$

where

$$\mathcal{M}(u) := \text{div} \frac{\nabla u}{(1 + |\nabla u|^2)^{1/2}} \, .$$

It is well known in [4], that equation (0.1) has a unique solution if

$$(0.2) \qquad H_0 := \sup_{(x,u)\in\bar{\Omega}\times\mathbb{R}} |H(x,u)| < \frac{n-1}{n} \inf_{x\in\partial\Omega} H'(x) \, ,$$

where $H'(x)$ stands for the mean curvature of $\partial\Omega$ at the point x, and if

(0.3) $H_u(x, u) \geq 0$.

For simplicity, we shall henceforth consider the boundary condition $u|_{\partial\Omega} = 0$ instead. In order to find multiple solutions of equation (0.1), we must avoid assumption (0.3). However, to our knowledge, this condition has always played a crucial role in the a priori estimate for the solution of equation (0.1). It is quite difficult to study equation (0.1) for general Ω, so we shall only focus on the following special cases:

(1) For $n = 1$, we set up the Morse theory for the functional

$$I(u) := \int_\Omega (\sqrt{1 + \dot{u}^2} + \mathcal{H}(x, u))\ dx\,,$$

with $\mathcal{H}(x, u) = \displaystyle\int_0^u H(x, t)\ dt$, and provide a multiple solution result under assumptions (H1), (H2), (H3).

Because of the special feature of equation (0.1), it is well known that, critical point theory cannot be set up on the $W^{1,p}$ spaces for $p > 1$ using the above functional. But for $p = 1$ or other function spaces, the lack of Palais Smale Condition is again a major difficulty. However, as the first author has noticed in [3], sometimes the heat flow, to which the Palais Smale Condition is irrelevant, can be used as a replacement of the pseudo-gradient flow in critical point theory. We shall study the related heat equation, and use the heat flow to set up the Morse theory of isolated critical points for the above functional. Critical groups for isolated critical points are counted, and Morse relation is applied. The main result of the first section is Theorem 1.2, which asserts the existence of three nontrivial solutions.

(2) For $n > 1$, we shall further assume $H = H(r, u)$ is rotationally symmetric. Consequently, equation (0.1) can be reduced to the following O.D.E.:

$$\ddot{u} + \frac{n-1}{r}\dot{u}\ (1 + \dot{u}^2) = nH(r, u)(1 + \dot{u}^2)^{3/2}.$$

We then use the sub- and super-solution method to prove that there are at least two nontrivial solutions if assumptions (H1′), (H2′) hold, and a degree argument to prove the existence of the third nontrivial solution under the additional assumption (H3′) and (H4). The main result of the second section is Theorem 2.2.

This paper is divided into two parts. In part 1, we deal with case (1), using the heat flow method instead of the traditional pseudo-gradient flow method on some Hölder space. This approach will thereby enable us to bypass the Palais Smale Condition. In part 2, we deal with case (2). We first make some simple estimates, and then comes the crucial point in which, we construct a positive small sub-solution and a negative small super-solution of equation (0.1).

Part 1

In the case $n = 1$, we have:

$$\mathcal{M}(u) = \frac{\ddot{u}}{(1 + \dot{u}^2)^{3/2}} \cdot$$

Let $R > 0$ and let $J = [-R, R]$. Equation (0.1) is now reduced to the O.D.E.:

$$(1.1) \qquad \begin{cases} \ddot{u} = H(x, u)(1 + \dot{u}^2)^{3/2} & \text{for } x \in (-R, R) \\ u(\pm R) = 0 \,. \end{cases}$$

We assume that $H \in C^1(J \times \mathbb{R}^1, \mathbb{R}^1)$ satisfies:

$$(H1) \qquad H_0 := \sup_{(x,u) \in J \times \mathbb{R}} |H(x, u)| < \frac{1}{2R} \,.$$

It is easy to verify that

$$w_\lambda^\pm(x) = \pm(-\lambda + \sqrt{R^2 + \lambda^2 - x^2}), \qquad \forall \lambda > 0, \ |x| \le R$$

is a pair of super- and sub-solutions of equation (1.1), provided $\lambda < \sqrt{3}R$. If we also assume:

$$(H2) \qquad H(x, 0) = 0 \text{ and } -H_u(x, 0) > \left(\frac{\pi}{2R}\right)^2, \qquad \forall x \in J$$

then

$$u^\pm(x) = \pm\epsilon \cos\left(\frac{\pi x}{2R}\right), \qquad |x| \le R$$

are positive sub- and negative super-solutions of (1.1) respectively, provided $\epsilon > 0$ is sufficiently small.

In fact, we have:

$$-\mathcal{M}(u^+) = \frac{(\frac{\pi}{2R})^2 u^+}{(1 + (\dot{u}^+)^2)^{3/2}} < -H_u(x, 0)u^+ \,.$$

It follows that $-\mathcal{M}(u^+) \le -H(x, u^+)$ for $\epsilon > 0$ small, i.e., u^+ is a sub-solution. Hence, (u^+, w_λ^+) and (u^-, w_λ^-) are two pairs of sub- and super-solutions of (1.1).

Now we consider the heat equation related to (1.1):

(1.2)
$$\begin{cases} v_t - \mathcal{M}(v) = -H(x,v) \text{ in } (0,T) \times (-R,R), \\ v(t,\pm R) = 0, \forall t \in [0,T], \\ v(0,x) = v_0, \forall x \in J, \end{cases}$$

where the initial data $v_0 \in C^{1+\alpha}(J)$ with $v(\pm R) = 0$ and $\alpha \in (0,1)$.

We begin by introducing the weighted *parabolic* Hölder space $\mathbb{H}_{2+\alpha}^{(-1-\alpha)}(\Omega)$ on $\Omega := [0,T] \times J$, for $0 < \alpha < 1$, as follows:

$$\left\{ v \in C(\Omega) \Big| \sum_{\beta+2j\leq 2} d(X)^{\max\{\beta+2j-1-\alpha,0\}} |\partial_x^\beta \partial_t^j v(X)| \right.$$

$$+ d(X,Y) \left(\Sigma_{\beta+2j=2} \frac{|\partial_x^\beta \partial_t^j v(X) - \partial_x^\beta \partial_t^j v(Y)|}{|X-Y|^\alpha} \right.$$

$$\left. \left. + \frac{|\partial_x v(X) - \partial_x v(Y)|}{|X-Y|^{1+\alpha}} \right) < +\infty, \forall X,Y \in \Omega, X \neq Y \right\}$$

where $X = (t,x), Y = (\tau,y), |X-Y| = (|x-y|^2 + |t-\tau|)^{\frac{1}{2}}, d(X,Y) = \min\{d(X), d(Y)\}$, and $d(X_0) = \text{dist}\{X_0, (\partial\Omega\backslash\{t = T\}) \cap \{t < t_0\}\}$, for $X_0 = (t_0, x_0)$.

The norm of $\mathbb{H}_{2+\alpha}^{(-1-\alpha)}$ is defined by

$$||v|| = \sup_{X,Y\in\Omega, X\neq Y} \left(\sum_{\beta+2j\leq 2} d(X)^{\max\{\beta+2j-1-\alpha,0\}} |\partial_x^\beta \partial_t^j v(X)| \right.$$

$$+ d(X,Y) \left(\Sigma_{\beta+2j=2} \frac{|\partial_x^\beta \partial_t^j v(X) - \partial_x^\beta \partial_t^j v(Y)|}{|X-Y|^\alpha} \right.$$

$$\left. \left. + \frac{|\partial_x v(X) - \partial_x v(Y)|}{|X-Y|^{1+\alpha}} \right) \right).$$

According to [5], the solution $v \in \mathbb{H}_{2+\alpha}^{(-1-\alpha)}(\Omega)$ exists for any $T > 0$. Applying the Maximum Principle, we have:

$$u^+(x) \leq v(t,x) \leq w^+(x), \quad \forall(t,x) \in \Omega,$$

if

$$u^+(x) \leq v_0(x) \leq w^+(x), \quad \forall x \in J.$$

Similarly,

$$w^-(x) \leq v(t,x) \leq u^-(x), \quad \forall(t,x) \in \Omega,$$

if

$$w^-(x) \le v_0(x) \le u^-(x), \qquad \forall x \in J.$$

We now view the solution $v(t, x)$ as a flow, and consider the functional

$$I(u) := \int_J [\sqrt{1 + \dot{u}^2} + \mathcal{H}(x, u)] \, dx,$$

with $\mathcal{H}(x, u) = \int_0^u H(x, t) \, dt$, then the Euler-Lagrange equation for I is exactly equation (1.1).

Along the flow, we have:

$$\frac{d}{dt} I(v(t, \cdot)) = - \int_J \frac{\partial v}{\partial t}(t, x)[\mathcal{M}(v(t, x)) - H(x, v(t, x))] \, dx$$

$$= - \int_J \left(\frac{\partial v}{\partial t}(t, x) \right)^2 \, dx \le 0.$$

Hence, the functional I is nonincreasing.

We denote $[u, v] = \{w \in C^{2+\alpha}(J) \cap C_0(J) \mid u(x) \le w(x) \le v(x), \forall x \in J\}$, and notice that, if the initial data v_0 falls into the ordered interval $[u^+, w^+]$ (or $[w^-, u^-]$ resp.), then $v(t, x)$ is bounded, and $I(v(t, \cdot))$ is therefore bounded from below. Thus, $c := \lim_{t \to \infty} I(v(t, \cdot))$ exists, and

$$\int_0^\infty \int_J \left(\frac{\partial v}{\partial t}(t, x) \right)^2 \, dx \, dt = I(v_0) - c.$$

There must be a sequence $t_j \to +\infty$ such that

$$v_j(x) := \frac{\partial v}{\partial t}(t_j, x) \to 0 \text{ in } L^2(J).$$

Let $u_j(x) := v(t_j, x)$ for all j. Substituting these into equation (1.2), we obtain a sequence of equations:

(1.3) $$\mathcal{M}(u_j) - H(x, u_j) = v_j, \qquad \forall j.$$

We want to show that $\{u_j\}$ subconverges to a solution of (1.1). To this end, we shall prove that $\|\ddot{u}_j\|_2$ is bounded as follows.

Let

$$z_j = \frac{\dot{u}_j}{(1 + \dot{u}_j^2)^{1/2}}.$$

Since $u_j(\pm R) = 0$, there is $\xi \in J$ such that $\dot{u}_j(\xi) = 0$, i.e., $z_j(\xi) = 0$. According to equation (1.3), we have:

$$|z_j(x)| \le \int_\xi^x (|H(x, u_j(x))| + |v_j(x)|)dx$$

$$\le 2H_0 R + (2R)^{1/2}\left(\int_J |v_j|^2 dx\right)^{1/2}.$$

It follows, from $||v_j||_2 \to 0$ and (H1), that there is $\epsilon > 0$, which depends only on H_0 and R, such that $||z_j||_\infty \le 1 - \epsilon$. Hence, $||\dot{u}_j||_\infty \le \dfrac{1-\epsilon}{\epsilon}$.

By setting $M_\epsilon := (1 + (\frac{1}{\epsilon} - 1)^2)^{3/2}$, we have that $||\ddot{u}_j||_2 \le (H_0 + 1)M_\epsilon$; hence, $\exists\, u_+^* \in W^{2,2}(J)$ such that $u_j \rightharpoonup u_+^*$ in $W^{2,2}(J)$, and then $u_j \to u_+^*$ in $C^1(J)$.

Finally, $\forall \varphi \in C_0^\infty(J)$, from

$$\int_J \left[\frac{\ddot{u}_j(x)}{(1 + \dot{u}_j^2(x))^{3/2}} - H(x, u_j(x))\right]\varphi(x)\, dx \to 0,$$

it follows that

$$\int_J [\mathcal{M}(u_+^*) - H(x, u_+^*)]\varphi(x)\, dx = 0,$$

i.e., $\mathcal{M}(u_+^*(x)) = H(x, u_+^*(x))$ a.e., thus $u_+^* \in C^3(J)$.

In summary, along the heat flow, there is a subsequence $\{t_j\} \nearrow +\infty$ such that $v(t_j, \cdot) \rightharpoonup u_+^*$ in $W^{2,2}(J)$, where u_+^* is a solution of (1.1). Let K be the set of all solutions of (1.1) in $[u^+, w^+]$. According to the previous estimates, it is $W^{2,2}$ bounded, so is compact in $C^1(J)$. Since $\inf\{|(u)|u \in [u^+, w^+]\} = \inf\{|(u)|u \in K\}$, the functional I has a minimizer, again denoted by u_+^*, in the ordered interval $[u^+, w^+]$. A similar argument shows that we also have a solution $u_-^* \in C^3(\bar{J})$ which minimizes the functional I in the ordered interval $[w^-, u^-]$. Since $[w^-, u^-] \cap [u^+, w^+] = \emptyset$, $u_+^* \ne u_-^*$; they are two distinct nontrivial solutions of (1.1).

In light of the above result, we shall next seek a third nontrivial solution of (1.1). This is based on a Morse-theoretic approach, we refer to [2] and [3] for further details.

When $v_0 \in C^{1+\alpha}(J) \cap C_0(J)$, $0 < \alpha < \frac{1}{2}$, it is known ([5]) that the solution $v(t, x) \in \mathbb{H}_{2+\alpha}^{(-1-\alpha)}(\Omega)$, which defines a deformation $\eta : [0, +\infty) \times (C^{1+\alpha}(J) \cap C_0(J)) \to C^{1+\alpha}(J) \cap C_0(J)$ by $\eta(t, v_0) = v(t, \cdot)$. It is easy to verify the continuity of the mapping η by standard arguments. In particular, $\eta : [0, +\infty) \times X \to X$, where $X = [w^-, w^+] \cap C^{1+\alpha}(J) \cap C_0(J)$ is a closed convex set in the Banach space $C^{1+\alpha}(J) \cap C_0(J)$. Now, $\forall a \in \mathbb{R}^1$, we denote $I_a := \{u \in X \mid I(u) \le a\}$. The following deformation lemma holds:

Lemma 1.1. (*Deformation lemma*) *If there exists no critical point of the functional* I *in the energy interval* $I^{-1}[a, b]$, *except perhaps some isolated critical points at the level* a, *then* I_a *is a deformation retract of* I_b.

Proof. It is sufficient to prove: If the orbit $\mathbf{O}(v_0) = \{\eta(t, v_0) \,|\, t \in \mathbb{R}^1_+\} \subset I^{-1}(a, b]$, and if the limiting set $\omega(v_0)$ is isolated; then $\eta(t, v_0)$ has a weak $W^{2,2}$ limit w, (and then $\lim_{t \to +\infty} \eta(t, v_0) = w$ in $C^{1+\alpha}$, $0 < \alpha < \frac{1}{2}$) on the level $I^{-1}(a)$.

Indeed, by the previous argument, $\forall w \in \omega(v_0)$, we have a sequence $t_j \to +\infty$ such that $v(t_j, \cdot) \rightharpoonup w$ in $W^{2,2}(J)$. Thus,

$$\|v(t_j, \cdot) - w\|_{C^1} \to 0.$$

Suppose the conclusion is not true, then $\exists \epsilon_0 > 0$, $\exists t'_1 < t''_1 < t'_2 < t''_2 < \cdots$, such that

$$\|v(t'_j, \cdot) - w\|_{C^1} = \frac{\epsilon_0}{2} \leq \|v(t, \cdot) - w\|_{C^1} \leq \|v(t''_j, \cdot) - w\|_{C^1} = \epsilon_0,$$

$\forall\, t \in I_j := [t'_j, t''_j]$, $\forall\, j = 1, 2, \ldots$, and

$$\omega(v_0) \cap B_{\epsilon_0}(w) \backslash \{w\} = \emptyset.$$

The above inequalities imply the C^1- boundedness of $v(t, \cdot)$ on I_j, namely,

$$\|v(t, \cdot)\|_{C^1} \leq \|w\|_{C^1} + \epsilon_0.$$

After simple estimates, we have

$$\|v(t, \cdot)\|_{C^2} \leq C, \ \forall t \in I_j,$$

where C is a constant independent of j. But

$$\int_{I_j} \|\partial_t v(t, \cdot)\|_2^2 dt = I(v(t'_j, \cdot)) - I(v(t''_j, \cdot)) \to 0.$$

Since

$$\frac{\epsilon_0}{2} \leq \|v(t''_j, \cdot) - v(t'_j, \cdot)\|_2$$

$$\leq \int_{t'_j}^{t''_j} \|\partial_t v(t, \cdot)\|_2 dt$$

$$\leq (t''_j - t'_j)^{\frac{1}{2}} \left(\int_{t'_j}^{t''_j} \|\partial_t v(t, \cdot)\|_2^2 dt \right)^{\frac{1}{2}},$$

it follows $t''_j - t'_j \to +\infty$, and then $\exists t^*_j \in I_j$ such that $\|\partial_t v(t^*_j, \cdot)\|_2 \to 0$. Obviously, $\|v(t^*_j, \cdot)\|_{C^2} \leq C$, and $v(t^*_j, \cdot) \in B_{\epsilon_0} \backslash B_{\frac{\epsilon_0}{2}}(w)$.

Again, by the previous argument, $\exists z \in \omega(v_0)$ such that $v(t_j^*, \cdot) \to z$. Then we have: $z \in \omega(v_0) \cap B_{\epsilon_0} \backslash B_{\frac{\epsilon_0}{2}}(w)$. This is impossible, since $\omega(v_0) \cap B_{\epsilon_0}(w) \backslash \{w\} = \emptyset$. $\square$

From Lemma 1.1, the Morse relation holds for I on X. In context, critical groups $C_q(I, u_0) = H_q(U \cap I_c, (U \backslash \{u_0\}) \cap I_c)$ are defined for an isolated critical point u_0 of I, where U is an isolated neighborhood of u_0, $c = I(u_0)$, and $H_q(Y, Z)$ are the graded singular relative homology groups for $q = 0, 1, \ldots$.

In the estimation of number of solutions, we can assume, without loss of generality, that there are only finitely many critical points $\{u_1, u_2, \ldots, u_N\}$ of I. Noticing that both $u_\pm^*$ are local minimizers of I, it follows that

$$C_q(I, u_\pm^*) = \delta_{q0}.$$

On the other hand, let β_q be the qth Betti number of X, $q = 0, 1, \ldots$. Since the set X is contractible, $\beta_0 = 1$, and $\beta_q = 0$, $\forall q \geq 1$. Let M_q be the qth Morse type number of I:

$$M_q = \Sigma_{j=1}^N rank\, C_q(I, u_j), \qquad \forall q.$$

The Morse relation reads as

$$\Sigma_0^\infty (M_q - \beta_q)t^q = (1 + t)P(t),$$

where P is a formal power series with nonnegative coefficients.

Therefore, we must have at least one more critical point u^* of I. If there are critical points other than u^* and $u_\pm^*$, the conclusion follows, so we may assume u^* is the unique critical point other than $u_\pm^*$. Then we have

$$rank\, C_0(I, u^*) + rank\, C_1(I, u^*) \neq 0,$$

according to the Morse relation.

In order to distinguish u^* from θ (the trivial solution), we assume further

(H3) $-H_u(x, 0) > \left(\dfrac{\pi}{R}\right)^2.$

Since

$$d^2 I(\theta, \varphi) = \int_J (\dot{\varphi}^2 + H_u(x, 0)\varphi^2)\, dx,$$

under the assumption (H3), we have $C_q(I, \theta) = 0$, for $q = 0$ and 1. Again, this will be a contradiction, if besides $u_\pm^*$, I has only the critical point θ. Thus we have indeed established the following:

Theorem 1.2. *Assume that $H \in C^1(J \times \mathbb{R}^1, \mathbb{R}^1)$ satisfies* (H1) *and* (H2), *then equation* (1.1) *has at least two distinct nontrivial solutions, one positive and one negative. If in addition,* (H3) *is satisfied, then* (1.1) *will have at least three distinct nontrivial solutions.*

Part 2

In the case $n \geq 1$, we assume $H = H(r, u)$ is rotationally symmetric. Let $\Omega = B_R(0) \subset \mathbb{R}^n$, the ball centered at the origin of radius R. Equation (0.1) is then reduced to:

$$(2.1) \quad \begin{cases} \dfrac{1}{r^{n-1}} \dfrac{d}{dr} \left(\dfrac{r^{n-1} \frac{du}{dr}}{(1 + (\frac{du}{dr})^2)^{1/2}} \right) = nH(r, u) & \text{for } r \in (0, R) \\[4mm] \dfrac{du}{dr}(0) = u(R) = 0 \,, \end{cases}$$

and assumption (0.2) then becomes

$$(2.2) \quad H_0 := \sup_{(r,u) \in [0,R] \times \mathbb{R}} |H(r, u)| < \frac{n-1}{n} \cdot \frac{1}{R}, \qquad \text{if } n \geq 2, \text{ and}$$

$$(2.3) \quad H_0 < \frac{1}{R}, \qquad \text{if } n = 1.$$

We shall use a fixed point argument in conjunction with the super- and sub-solution method to tackle (2.1). We begin by delivering the following *a priori* estimate for solutions of (2.1):

Lemma 2.1. *There is a constant C, depending only on n, H_0, and R, such that all solutions u of* (2.1) *satisfy*

$$(2.4) \qquad \qquad \|u\|_{C^2} \leq C \,.$$

Proof. For $n \geq 2$, it suffices to show

$$(1) \quad |u_r(r)| \leq \sqrt{n} H_0 r, \text{ and}$$

$$(2) \quad |u_r(r)| \leq \sqrt{n}\left(1 - \frac{1}{n}\right).$$

For then, $u(r)$ is bounded by (2) and the boundary condition $u(R) = 0$; in the mean time, u_{rr} is bounded by (1) and the alternative expression of (2.1):

$$u_{rr} + \frac{n-1}{r} u_r (1 + u_r^2) = nH(r, u)(1 + u_r^2)^{3/2}.$$

For simplicity, we let

$$v = \frac{r^{n-1} u_r}{(1 + u_r^2)^{1/2}}.$$

It follows that

(2.5)
$$u_r = \pm \frac{v}{r^{n-1}(1 - (\frac{v}{r^{n-1}}))^{1/2}},$$

and

(2.6)
$$v_r = n r^{n-1} H.$$

It now remains to show $|v| \leq (1 - \frac{1}{n}) r^{n-1}$. If so, from (2.5), we have:

$$|u_r| \leq \left(1 - \frac{1}{n}\right) \left[1 - \left(1 - \frac{1}{n}\right)^2\right]^{-1/2} \leq \sqrt{n}\left(1 - \frac{1}{n}\right),$$

which proves (2).

Since $u_r(0) = 0$, $v(0) = 0$, by (2.6), we have:

$$|v(r)| \leq \int_0^r |v_r(t)| \, dt \leq H_0 r^n,$$

hence,

(2.7)
$$\frac{|v(r)|}{r^{n-1}} \leq H_0 r.$$

(2.7) subsequently implies

$$\frac{|v(r)|}{r^{n-1}} \leq H_0 R \leq 1 - \frac{1}{n},$$

and

$$|u_r| = \frac{(\frac{v}{r^{n-1}})}{(1 - (\frac{v}{r^{n-1}})^2)^{1/2}} \leq \sqrt{n} H_0 r.$$

For $n = 1$, assumption (2.2) is replaced by assumption (2.3), and we have $|v| \leq H_0 r \leq H_0 R$. Let $\epsilon := 1 - H_0 R$, we then have: $|u_r| \leq \frac{1 - \epsilon}{\sqrt{\epsilon}}$. This completes the proof. $\square$

Next, we construct two pairs of sub- and super-solutions of (2.1) as in Part 1. However, for $n > 1$, the construction of the second pair of such solutions is a bit more complicated. Again, we let

$$w_\lambda(r) = \pm(-\lambda + \sqrt{R^2 + \lambda^2 - r^2}), \qquad \forall \lambda > 0, \ 0 \leq r \leq R.$$

The following properties hold:

$$(w1) \quad w_\lambda(R) = 0, \ \dot{w}_\lambda(0) = 0,$$

$$(w2) \quad \mathcal{M}(w_\lambda) = -\mathcal{M}(-w_\lambda) = \frac{-n}{(R^2 + \lambda^2)^{1/2}},$$

$$(w3) \quad \dot{w}_\lambda(R) = -\frac{R}{\lambda},$$

$$(w4) \quad |w_\lambda(r)| \leq \frac{R^2 - r^2}{2\lambda}, \ |\dot{w}_\lambda| \leq \frac{r}{\lambda}.$$

These properties lead to the estimate:

$$-\mathcal{M}(w_\lambda) = \frac{n}{(R^2 + \lambda^2)^{1/2}} \geq n\left(\frac{n-1}{n}\right)\frac{1}{R} \geq -nH(r, w_\lambda),$$

provided

$$\lambda \leq \lambda_0 := \begin{cases} \dfrac{\sqrt{2n-1}}{n-1}R, \ \text{for } n > 1, \\[3mm] \epsilon R, \ \text{for } n = 1, \text{ where } 0 < \epsilon < \sqrt{\dfrac{1}{(H_0 R)^2} - 1}. \end{cases}$$

This verifies $(-w_\lambda, w_\lambda)$ is a pair of sub- and super-solutions of (2.1), when $\lambda \leq \lambda_0$. Next, we consider the functions

$$z_n(r) := r^{1-\frac{n}{2}} J_{\frac{n}{2}-1}(r), \text{ for } n = 1, 2, \ldots,$$

where $J_{\frac{n}{2}-1}(r)$ denotes the $(\frac{n}{2} - 1)$-order Bessel function.

Let μ_n be the first zero of $J_{\frac{n}{2}-1}(r)$, and let

$$v_n(r) := z_n\left(\frac{\mu_n}{R}r\right), n = 1, 2, \ldots,$$

we then have:

$$(v1) \quad v_n(R) = 0, \ \dot{v}_n(0) = 0,$$

$$(v2) \quad \ddot{v}_n + \frac{n-1}{r}\dot{v}_n = -\left(\frac{\mu_n}{R}\right)^2 v_n,$$

$$(v3) \quad \exists \ M_n > 0 \text{ such that } |\dot{v}_n(r)| \leq M_n.$$

We introduce assumption (H2′) in place of (H2) for $n > 1$:

$$(H2') \qquad H(r, 0) = 0 \text{ and } -H_u(r, 0) > \left(\frac{\mu_n}{R}\right)^2, \ \forall r \in [0, R].$$

and define

$$u_+ := \begin{cases} \delta v_n - w_\lambda, & \text{for } n > 1 \\ \delta v_1, & \text{for } n = 1. \end{cases}$$

We compute to see that

$$\mathcal{M}(u_+) = \frac{1}{(1 + (u_+)_r^2)^{3/2}} \left\{ \delta \left(\ddot{v}_n + \frac{n-1}{r} \dot{v}_n \right) - \left[\ddot{w}_\lambda + \frac{n-1}{r}(\dot{w}_\lambda + \dot{w}_\lambda^3) \right] \right.$$

$$\left. + \frac{n-1}{r} (\delta^3 \dot{v}_n^3 - 3\delta^2 \dot{v}_n^2 \dot{w}_\lambda + 3\delta \dot{v}_n \dot{w}_\lambda^2) \right\}$$

$$= \frac{-(\frac{\mu_n}{R})^2}{(1 + (u_+)_r^2)^{3/2}} \delta v_n + \left(\frac{1 + \dot{w}_\lambda^2}{1 + (u_+)_r^2} \right)^{3/2} \frac{n}{(R^2 + \lambda^2)^{1/2}} + \frac{Q(n, \delta, \lambda)(r)}{(1 + (u_+)_r^2)^{3/2}},$$

where the remainder

$$Q(n, \delta, \lambda)(r) = \frac{n-1}{r} (\delta^3 \dot{v}_n^3 - 3\delta^2 \dot{v}_n^2 \dot{w}_\lambda + 3\delta \dot{v}_n \dot{w}_\lambda^2).$$

One has

$$-\mathcal{M}(u_+) = \frac{(\frac{\mu_n}{R})^2}{(1 + (u_+)_r^2)^{3/2}} u_+ + \frac{(\frac{\mu_n}{R})^2}{(1 + (u_+)_r^2)^{3/2}} w_\lambda$$

$$- \left(\frac{1 + \dot{w}_\lambda^2}{(1 + (u_+)_r^2)} \right)^{3/2} \frac{n}{(R^2 + \lambda^2)^{1/2}} - \frac{Q(n, \delta, \lambda)(r)}{(1 + (u_+)_r^2)^{3/2}}.$$

From $(w4)$ and $(v3)$, we obtain:

$$\frac{1}{(1 + (u_+)_r^2)^{3/2}} = 1 + o(1) \text{ as } \delta \to 0, \ \lambda \to +\infty,$$

$$Q(n, \delta, \lambda)(r) = O(\delta^3 + \delta^2 \lambda^{-1} + \delta \lambda^{-2}), \text{ and}$$

$$\left(\frac{\mu_n}{R} \right)^2 w_\lambda - (1 + \dot{w}_\lambda^2)^{3/2} \frac{n}{(R^2 + \lambda^2)^{1/2}} \leq \frac{\mu_n^2}{2\lambda} \left(1 - \left(\frac{r}{R} \right)^2 \right) - \frac{n}{(R^2 + \lambda^2)^{1/2}}.$$

Let $\epsilon := H_u(r, 0) - (\frac{\mu_n}{R})^2$, we have:

$$-\mathcal{M}(u_+) < H_u(r, 0)u_+ - \frac{1}{(1 + (u_+)_r^2)^{3/2}} \left[\frac{\epsilon}{2} u_+ - \frac{\mu_n^2}{2\lambda} \left(1 - \left(\frac{r}{R} \right)^2 \right) \right.$$

$$\left. + \frac{n}{(R^2 + \lambda^2)^{1/2}} \right] + O(\delta^3),$$

with $\lambda = \delta^{-2}$ as $\delta \to 0$.

If one can choose $\delta > 0$ so small that

$$(2.8) \quad \frac{1}{2}\left[\left(\frac{\mu_n}{R}\right)^2 + \frac{\epsilon}{2}\right](R^2 - r^2)\delta^2 \leq \delta\epsilon v_n(r) + \frac{n}{(R^2 + \delta^{-4})^{1/2}},$$

then u_+ is a sub-solution of (2.1) for $n > 1$.

This is possible, since $v_n(r) > 0$ in $[0, R)$, and since the last term in the right hand side of (2.8) is of order δ^2, furthermore, it is a positive constant on $[0, R]$. So, (2.8) holds for small $\delta > 0$.

As to $n = 1$, it has already been shown in section 1.

In summary, we obtained, as in the previous section, two pairs of sub- and super-solutions of (2.1), (u_+, w_+) and (w_-, u_-) where $w_\pm = \pm w_{\lambda_0}$ and $u_- = -u_+$.

In the next step, a fixed point argument is applied to obtain a third nontrivial solution of (2.1). Let $X := \{u \in C^1([0, R]) \mid u_r(0) = u(R) = 0\}$. We define an operator $T : X \to X \cap C^2([0, R])$ via $v := Tu$, $\forall u \in X$, and if v is a solution of the linear O.D.E.:

$$\begin{cases} \ddot{v} + \dfrac{n-1}{r}\dot{v} = (1 + \dot{u}^2)^{3/2}\left(nH(r, u) - \dfrac{n-1}{r}\dot{u}^3\right) \\ \dot{v}(0) = v(R) = 0. \end{cases}$$

It is clear that u is a fixed point of T if and only if u is a solution of (2.1). According to the Maximum Principle, T maps the ordered interval $[w_-, u_-] \cap X$ and $[u_+, w_+] \cap X$ into themselves. By the Schauder's Fixed Point Theorem, there is a fixed point $u_\pm^*$ in each of these ordered intervals.

We may assume, without loss of generality, that $u_\pm^*$ is the only fixed point of T in $O_\pm$, where $O_+ = int([u_+, w_+] \cap X)$ and $O_- = int([w_-, u_-] \cap X)$, the interior of each of the corresponding set. For otherwise, there must be a third nontrivial solution.

Applying the Leray–Schauder's degree theory, we have

$$ind(I - T, u_\pm^*) = 1,$$

where $ind(I - T, u_\pm^*)$ denotes the Leray–Schauder index of $I - T$ at $u_\pm^*$, respectively.

We can then use Amann's Three Solutions Theorem [1] to confirm the existence of a third solution u^*. One may again assume, without loss of generality, that u^* is the only fixed point in $[w_-, w_+] \cap X$ other than $u_\pm^*$. By a degree computation,

$$ind(I - T, u^*) = 1,$$

provided $deg\ (I - T, int\ ([w_-, w_+] \cap X), \theta) = 1$.

As before, we should distinguish u^* from θ, the trivial solution. For this reason, stronger assumptions are imposed.

Let μ_n^j be the jth zero of the Bessel function $J_{\frac{n}{2}-1}$, and let m_n^j be the multiplicity of μ_n^j.

We assume that $\exists j_0 \geq 2$ satisfying

(H3′) $\mu_n^{j_0} < \sqrt{-H_u(r,0)}R < \mu_n^{j_0+1}, \ \forall\, r \in [0, R]\,,$

and

(H4) $m_n = \Sigma_{j=1}^{j_0} m_n^j$ is odd$\,.$

Thus,

$$ind\ (I - T, \theta) = (-1)^{m_n} = -1\,.$$

This implies $u^* \neq \theta$.

Lastly, we conclude our paper by stating the following:

Theorem 2.2. *Assume that $H \in C^1([0, R] \times \mathbb{R}^1, \mathbb{R}^1)$ satisfies (2.2) and (H2′), then equation (2.1) has at least two distinct nontrivial solutions, one positive and one negative. If in addition, (H3′) and (H4) are satisfied, then (2.1) will have at least three distinct nontrivial solutions.*

Remark. In the case $n = 1$, by combining methods used in Parts 1 and 2, we may obtain a result slightly different from both Theorem 1.2 and Theorem 2.2. Namely, if $H \in C^1(J \times \mathbb{R}^1, \mathbb{R}^1)$ satisfies

$$H_0 < \frac{1}{R},$$

$$H(x, u) = H(-x, u), \ \forall\, x \in [-R, R], \ \forall\, u \in \mathbb{R}^1\,,$$

and

$$H(x, 0) = 0 \text{ and } -H_u(x, 0) > \left(\frac{3\pi}{2R}\right)^2 \ \forall\, x \in [-R, R]\,,$$

then there exist at least three nontrivial symmetric solutions of equation (1.1).

The proof is the same as that of Theorem 1.2, with an improvement of the a priori estimates of the solutions obtained in the first few paragraphs of Part 2, under the assumption that u is symmetric.

References

1. Amann, H. On the number of solutions of nonlinear equations in ordered Banach spaces, J. Funct. Anal. 14 (1973), 346–384.
2. Chang, K. C. Infinite dimensional Morse theory and multiple solution problems, Birkhauser (1993).
3. Chang, K. C. Heat method in nonlinear elliptic equations, Topological methods, Variational methods, and their applications, (ed. by Brezis, H., Chang, K. C., Li, S. J., Rabinowitz, P.) World Sci. (2003), 65–76.
4. Gilbarg, D., Trudinger, N., Elliptic partial differential equations of second order, Grundlehren der Mathematischen Wissenschaften 224 (1983).
5. Liebermann, G. M., Second order parabolic differential equations. World Sci. (1996).

S. S. Chern and K. C. Chang at ICM, Aug. 20, 2002, Beijing.

Chapter 6

ON THE DIFFERENTIABILITY OF LIPSCHITZ MAPS
FROM METRIC MEASURE SPACES
TO BANACH SPACES

Jeff Cheeger[*]

*J.C.: Courant Institute of Mathematical Sciences,
251 Mercer Street, New York, NY 10012, USA*

Bruce Kleiner

*Mathematics Department, Yale University,
New Haven, CT 06520, USA*

August 8, 2006

We consider metric measure spaces satisfing a doubling condition and a Poincaré inequality in the upper gradient sense. We show that the results of [Che99] on differentiability of real valued Lipschitz functions and the resulting bi-Lipschitz nonembedding theorems for finite dimensional vector space targets extend to Banach space targets having what we term a *good finite dimensional approximation*. This class of targets includes separable dual spaces. We also observe that there is a straightforward extension of Pansu's differentiation theory for Lipschitz maps between Carnot groups, [Pan89], to the most general possible class of Banach space targets, those with the Radon–Nikodym property.

Contents

[*]The first author was partially supported by NSF Grant DMS 0105128 and the second by NSF Grant DMS 0505610.

1. Introduction

A classical theorem of Rademacher states that real valued Lipschitz functions on $\mathbf{R}^n$ are differentiable almost everywhere with respect to Lebesgue measure; [Rad20]. The literature contains numerous extensions of Rademacher's result in which either the domain, the target, or the class of maps is generalized. Classical examples include almost everywhere approximate differentiability of Sobolev functions, [Zie89], analysis of and on rectifiable sets, [Fed69; Mat95], almost everywhere differentiability of quasi-conformal homeomorphisms between domains in Euclidean space, [Geh61; Mos73], and differentiability of Lipschitz maps from $\mathbf{R}^k$ into certain Banach spaces.

In many of the recent results in this vein, a significant part of the achievement is to make sense of differentiation in a context where some component of the classical setting is absent e.g. the infinitesimal affine structure on the domain or target, or a good measure on the domain. Included are Pansu's differentiation theorem, [Pan89], for Lipschitz maps between Carnot groups, differentiation theorems for Lipschitz mappings between Banach spaces [BL00, Chap. 6-7], metric differentiation, [Kir94; Pau01], and the differentiation theory developed in [Che99], for Lipschitz functions on metric measure spaces which are doubling and for which a Poincaré inequality holds in the upper gradient sense.

We now discuss in more detail those developments which are most relevant to our main theme.

The Radon–Nikodym Property

While the extension of Rademacher's theorem from real valued functions to $\mathbf{R}^m$ valued functions is essentially immediate, the extension to infinite dimensional Banach space targets is not. In fact, the theorem holds for some Banach space targets and fails for others.

The following example from [Aro76] illustrates the failure when the target is L^1. Let $\mathcal{L}$ denote Lebesgue measure and let $f : (0,1) \to L^1((0,1), \mathcal{L})$ be defined by $f(t) = \chi_{(0,t)}$, where χ_A denotes the characteristic function of A. If f were differentiable at $t \in (0,1)$, then the difference quotients

$$\frac{f(t') - f(t)}{t' - t}$$

would converge in L^1 as $t' \to t$. However, this is clearly not the case, since the difference quotients converge weakly to a delta function supported at t.

A Banach space, V, for which every Lipschitz function, $f : \mathbf{R}^k \to V$, is differentiable almost everywhere, is said to have the *Radon–Nikodym property*. The following facts are taken from Chapter 5 of [BL00], which contains an extensive discussion of the Radon–Nikodym property in its various (not obviously) equivalent formulations. The term, "Radon–Nikodym property", derives from a formulation which posits the validity of the Radon–Nikodym theorem for V-valued measures.

According to an early result of Gelfand, separable dual spaces and reflexive spaces have the Radon–Nikodym property; [Gel38] and Corollary 5.12 of [BL00]. In fact, a dual space, E^*, has the Radon–Nikodym property if and only if every separable subspace of E has a separable dual; see Corollary 5.24 of [BL00]. There does exist a Banach space with the Radon–Nikodym property which is not isomorphic to a subspace of a separable dual space; see Example 5.25 of [BL00]. However, every infinite dimensional Banach space with the Radon–Nikodym property contains an infinite dimensional separable subspace which is isomorphic to a dual space; see [GM84].

Differentiating Maps between Carnot Groups

Pansu gave an extension of Rademacher's theorem in which both the domain and the target are generalized. He considered Lipschitz maps between Carnot groups i.e. nilpotent groups with certain left-invariant *subriemannian* metrics, of which the Heisenberg group, $\mathbb{H}$, with its so-called Carnot–Cartheodory metric, $d^{\mathbb{H}}$, is the simplest example; see [Pan89] and compare Section 6 below. Pansu showed for such maps, almost everywhere with respect to Haar measure, the function converges under the natural inhomogeneous rescaling to a group homorphism. This assertion makes sense only for Carnot groups.

Differentiating Real Valued Functions on Metric Measure Spaces

At a minimum, the statement of Rademacher's theorem requires that the domain carries both a metric, so that Lipschitz functions are defined, and a notion of "almost everywhere", which for finite dimensional domains, is typically provided by a measure. (In the case of infinite dimensional Banach space domains, absent an appropriate measure, other notions of "almost everywhere" have been used; see [BL00, Chap. 6-7].)

The paper [Che99] introduced a new notion of differentiability for metric measure spaces and proved a Rademacher type theorem in that setting. To explain this notion of differentiation, we begin with the following definition.

Definition 1.1. Let X denote a metric measure space, $A \subset X$ a measurable subset and $f_1, \ldots, f_k : A \to \mathbf{R}$, real-valued Lipschitz functions. For $x \in A$, we say $\{f_i\}$ *is dependent to first order at* x, if there a linear combination, $\sum_i a_i f_i$, such that for $x' \in A$, as $x' \to x$,

$$\sum_i a_i \left(f_i(x') - f_i(x) \right) = o(d^X(x', x)).$$

We say that f *is dependent on* $\{f_1, \ldots, f_k\}$ *to first order at* x, if $\{f, f_1, \ldots, f_k\}$ is *dependent* to first order at x. *Linear independence* at x is defined similarly.

If the f_i's were smooth functions on a manifold, then dependence to first order at x would be equivalent to the linear dependence of their differentials at x.

A key fact established in [Che99] is that if a metric measure space, (X, d^X, μ), is a PI space (see below for the definition) then there is an integer N such that any collection of Lipschitz functions of cardinality at least $N + 1$ is dependent to first order almost everywhere. Intuitively, this says that there is at most an N-dimensional space of possibilities for the differential of any Lipschitz function at almost every point.

Definition 1.2. A countable collection of pairs, $\{(A_\alpha, u_\alpha)\}$, with A_α a measurable set and $u_\alpha : A_\alpha \to \mathbf{R}^{n_\alpha}$ a Lipschitz map, is an *atlas* if $\bigcup_\alpha A_\alpha$ has full measure and:

1) For every Lipschitz function, $f : X \to \mathbf{R}$, and for μ-a.e. $x \in A_\alpha$ and all α, the restriction, $f \mid A_\alpha$, is dependent to first order on the component functions, $u_1, \ldots, u_{n_\alpha}$, of u_α at x.

2) For all α, the functions, $u_1, \ldots, u_{n_\alpha}$, are independent to first order at μ-a.e. $x \in A_\alpha$.

We refer to the maps $u_\alpha : A_\alpha \to \mathbf{R}^{n_\alpha}$ as *charts*.

It was shown in [Che99] that any PI space admits an atlas by using the above mentioned bound on the cardinality of set of Lipschitz functions which is independent to first order, and a selection argument (analogous to choosing a basis from a spanning set).

Property 1) leads to a definition of pointwise differentiability with respect to the chart (A_α, u_α).

Definition 1.3. A Lipschitz function is *differentiable at* $x \in A_\alpha$, *with respect to* (A_α, u_α), if there is a linear function, $\Phi : \mathbf{R}^{n_\alpha} \to \mathbf{R}$, such that for $x' \in A_\alpha$, as $x' \to x$,

$$(1.4) \qquad f(x') = f(x) + \Phi \circ (u_\alpha(x') - u_\alpha(x)) + o(d^X(x', x)).$$

Note that (1.4) generalizes the notion of first order Taylor expansion in local coordinates.

It follows readily from properties 1), 2), above, that every Lipschitz function, $f : X \to \mathbf{R}$, is differentiable almost everywhere with respect to every chart in $\{(A_\alpha, u_\alpha)\}$, and that the corresponding linear function, Φ, is uniquely determined almost everywhere, and varies measurably. Furthermore, as in the case of Lipschitz manifolds, on the overlap of any two charts one gets almost everywhere defined bounded measurable transition functions for derivatives (i.e. bounded measurable jacobians) permitting one to define the cotangent bundle T^*X. This is a measurable vector bundle which carries a natural measurable norm. The norm is characterized by the property that every Lipschitz function $f : X \to \mathbf{R}$ defines a measurable section Df whose fiberwise norm is given by its pointwise Lipschitz constant almost everywhere. The *tangent bundle*, TX, is defined as the dual bundle of T^*X, and is equipped with the dual norm.

PI Spaces

We now define the term *PI space*. We note that conditions (1.6), (1.7), below were introduced by Heinonen–Koskela; [HK96].

A measure μ is *locally doubling* there are constants R, β, such that for every point x and every $r \leq R$,

$$(1.5) \qquad \mu(B_{2r}(x)) \leq \beta \cdot \mu(B_r(x)).$$

The function, g, is called an *upper gradient* for f if for every rectifiable curve, $c : [0, \ell] \to X$, parameterized by arclength, s,

$$(1.6) \qquad |f(c(\ell)) - f(c(0))| \leq \int_0^\ell g(c(s))\, ds.$$

Put

$$f_{x,r} = \frac{1}{\mu(B_r(x))} \int_{B_{2r}(x)} d\mu.$$

The *local* $(1,p)$-*Poincaré inequality* is the condition that for some $\lambda, R, \tau < \infty$,

(1.7)

$$\frac{1}{\mu(B_r(x))} \int_{B_r(x)} |f - f_{x.r}| \, d\mu \quad \leq \quad r \cdot \tau \cdot \left(\frac{1}{\mu(B_r(x))} \int_{B_{\lambda r}(x)} g^p \, d\mu \right)^{\frac{1}{p}},$$

where $x \in X$, $r \leq R$, and f is an L^1 function for which has an upper gradient in L^p.

Definition 1.8. Let (X, d^X) denote a metric space such that closed metric balls are compact, and let μ denote a Radon measure on X which for $r \leq 1$, satisfies a local doubling condition and a local $(1,p)$-Poincaré inequality. The triple, (X, d^X, μ), will be refered to as a *PI space*.

Sometimes we will suppress the distance function, d^X, and measure, μ, and just say that X is a PI space.

Examples of PI spaces include Gromov–Hausdorff limits of sequences of Riemannian manifolds with a uniform lower Ricci curvature bound, Bourdon–Pajot spaces, Laakso spaces and Carnot groups such as the Heisenberg group $\mathbb{H}$; see [CC97], [BP99], [Laa00], [Gro96].

Unlike $\mathbf{R}^n$ or Carnot groups, general PI spaces do not have any *a priori* infinitesimal linear structure. Indeed, the Hausdorff dimension of a PI space need not be an integer and there need not exist any points for which sufficiently small neighborhoods have finite topology.

In addition to the measurable tangent bundle, there is a quite distinct notion of a *tangent cone*, at $x \in X$, which, by definition, is any rescaled (pointed, measured) Gromov–Hausdorff limit space of a sequence of rescaled spaces, $(X, x, r_i^{-1} d^X, \mu_i)$, where $r_i \to 0$ and $\mu_i = \mu/\mu(B_{r_i}(x))$. It follows from the doubling condition and Gromov's compactness theorem that tangent cones exist for all $x \in X$. Although tangent cones at x need not be unique, we use the notation, X_x, for any tangent cone at x (supressing the base point, distance function and measure). The terminology notwithstanding, X_x, need not be a cone in either the metric or topological sense. However, any tangent cone is again a PI space; see Section 9 of [Che99].

As shown in [Che99], for almost every $x \in X$, there exist natural *surjective Lipschitz maps*, $X_x \to T_x X$, for any X_x; hence

(1.9) $$\dim T_x X \leq \dim X_x,$$

where dim denotes Hausdorff dimension.

Bi-Lipschitz Nonembedding

Generalized differentiation theories have been used to show that solutions of certain mapping problems are very restricted or even fail to exist. A good example is Mostow's rigidity theorem for hyperbolic space forms, whose proof uses the differentiability of quasi-conformal mappings; see [Mos73]. The same principle is a powerful tool in the bi-Lipschitz classification of Banach spaces [BL00, Chap. 7].

Closer to our topic is the observation of Semmes that the differentiablity theory of Pansu for Lipschitz maps between Carnot groups implies that certain metric spaces do not bi-Lipschitz embed in any finite dimensional Banach space; see [Sem96].

It was shown in [Che99] using the differentiation theory for real valued Lipschitz functions on PI spaces developed there, that if X admits a bi-Lipschitz embedding into a finite dimensional normed space, then there is a full measure set of points $x \in X$ where the tangent space $T_x X$ is defined, and for every tangent cone the canonical map

$$X_x \to T_x X$$

is an isometry; see Section 14 of [Che99] and compare (1.9) above. This result unifies and generalizes earlier nonembedding theorems which were known in particular cases, the proofs of which employed special features of the domains e.g. that of [Sem96]; compare Section 6. The class of spaces to which the result of [Che99] applies includes Bourdon–Pajot spaces, Laakso spaces and Carnot groups.

For example, with respect to the Carnot–Caratheodory distance, the Hausdorff dimension of the Heisenberg group, $\mathbb{H}$, is 4; compare [Gro96]. Since any tangent cone of $\mathbb{H}$ is isometric to $\mathbb{H}$ itself, dim $\mathbb{H}_x = 4$ as well. On the other hand the measurable tangent bundle of $T\mathbb{H}$ can be seen to have dimension 2, so tangent cones are not isometric to tangent spaces.

Summary of Results

The notion of differentiability for real-valued Lipschitz functions on PI spaces introduced in [Che99] extends in a straightforward fashion to Lipschitz functions taking values in a Banach space V. For X a PI space and $\{(A_\alpha, u_\alpha)\}$ an atlas, it reads as follows:

Definition 1.10. A Lipschitz map $f : X \to V$ is *differentiable at $x \in A_\alpha$, with respect to (A_α, u_α)*, if there is a linear map, $\Phi : \mathbf{R}^{n_\alpha} \to V$, such that for $x' \in A_\alpha$, as $x' \to x$,

$$(1.11) \quad f(x') = f(x) + \Phi \circ (u_\alpha(x') - u_\alpha(x)) + o(d^X(x', x)).$$

With the aim of finding a common generalization of earlier results, one may ask:

Question 1.12. *For which pairs, (X, V), with X a PI space and V a Banach space, does one have almost everywhere differentiability for every Lipschitz map $X \to V$?*

The main result of this paper gives a partial answer to this question. We show in Theorem 4.1 that if the Banach space V satisfies the GFDA condition (see below), then a Lipschitz map from any PI space into V is differentiable almost everywhere. As a corollary, the criterion, (5.1), of [Che99], also implies the nonexistence of bi-Lipschitz embeddings into GFDA targets, see Theorem **??**.

Discussion of the Proof

From the theory of [Che99], it immediately follows that every Lipschitz map from a PI space into a Banach space V is weakly differentiable, in the sense that its composition with every $\ell \in V^*$ is differentiable almost everywhere. In general, weak differentiability does not imply differentiability, and is too weak to give rise to bi-Lipschitz nonembedding theorems. This is inevitable, since *any* metric space, X, can be canonically isometrically embedded in $L^\infty(X)$ via the *Kuratowski embedding*.[1]

The following definition identifies a property that is *not* implied by weak differentiability.

Definition 1.13. A map $f : X \to V$ is *finite dimensional to first order at* $x \in X$, if there is a finite dimensional subspace $F \subset V$ such that

$$(1.14) \qquad \limsup_{r \to 0} \frac{1}{r} \sup_{x' \in B_r(x)} d(f(x') - f(x), F) = 0,$$

where $d(\cdot, F)$ denotes the distance in V to the subspace F.

One easily checks that if a Lipschitz map from a PI space to a Banach space is finite dimensional to first order almost everywhere, then weak differentiability can be promoted to differentiability, see Corollary 4.4.

---

[1] The Kuratowski embedding of a metric space X is the map $X \to L^\infty(X)$ which assigns to each $x \in X$ the function $d(x, \cdot) - d(x, x_0)$, where $x_0 \in X$ is a basepoint.

Thus, the proof is reduced to verifying that the GFDA property for the target implies that Lipschitz maps are finite dimensional to first order almost everywhere. This in turn, is proved by combining the GFDA property of the target with weak differentiability.

The GFDA Property

The essential idea of the GFDA property is that a vector, $v \in V$, should be determined up to small error once one knows, up to small error, a *suitable* projection, $\pi(v)$, onto a finite dimensional quotient space, with the property that $\pi(v)$ has nearly the same length as v. A key point is to choose a meaning for "suitable" such that the corresponding GFDA property is strong enough to imply differentiability but weak enough to apply to large class of Banach spaces; see Definition 3.6.

Separable dual spaces have the GFDA property. This follows readily from an elementary, but extremely useful, renorming result of Kadec and Klee; see Proposition 1.b.11 of [LT77] and [Kad59], [Kle61]. It asserts that a separable dual space admits an equivalent norm, for which weak* convergence together with convergence of norms implies strong convergence i.e. $v_i \overset{w^*}{\to} v$ and $\|v_i\| \to \|v\|$, implies $\|v - v_i\| \to 0$.

Examples of separable dual spaces include the classical spaces, L^p, for $1 < p < \infty$, and l^p, for $1 \le p < \infty$. Also, if V^{**} is separable, then we can naturally regard f as taking values in the separable dual space V^{**} and the theory applies. If V is reflexive (but not necessarily separable) then every separable subspace, $\widehat{V} \subset V$ is reflexive and in particular, a dual space; see [BL00]. Thus, the theory applies to all reflexive targets, separable or not.

Radon–Nikodym Targets

It is natural to ask if the differentiation assertion of the main theorem is actually valid for Lipschitz maps into an arbitrary Banach space with the Radon–Nikodym property. This would follow if for example, the GFDA property, which implies the Radon–Nikodym property were actually equivalent to it. At present, we do not know a counter example to this statement. At least for duals, E^*, of separable spaces, E, our result is optimal, since in such a case, as indicated above, E^* has the Radon Nikodym property if and only if it is separable.

In any event, for certain special PI spaces such as Carnot groups, for which there are sufficiently nice curve families, the differentiation and bi-

Lipschitz nonembedding theorems do hold for arbitrary Radon–Nikodym targets. Since we do not have a completely general result, we will just illustrate matters by pointing out in Section 6, that in the special case of the Heisenberg group, the proof of the nonembedding theorem of [Sem96] can be quite directly carried over to Radon–Nikodym targets; see Section 6. This was also observed by Lee and Naor, [LN06].

The proof is quite short and does not require any new concepts. In contrast to the main theorems, which are valid for general PI spaces, the argument does make strong use of special properties of the Heisenberg group. Some readers might prefer to read Section 6 before proceeding to the main body of the paper.

Results Proved Elsewhere; L^1 Targets

The separable space, L^1, is not the dual space of any Banach space. Moreover, the example of [Aro76] demonstrates the failure of differentiatibility. Nonetheless, as we will show elsewhere, by using a different notion of differentiation, the Heisenberg group with its Carnot–Caratheodory metric does not admit a bi-Lipschitz embedding into L^1; see [CK06a].

On the other hand, there is a class of PI spaces which includes Laakso spaces, every member of which has a bi-Lipschitz embedding in L^1; see [CK06b]. In particular, the members of this class are examples of spaces which bi-Lipschitz embed in L^1 but not in ℓ^1.

2. Finite Dimensional Approximations and Weak Derivatives

Henceforth V will denote a Banach space and $\{W_i\}$ will denote an inverse system of finite dimensional Banach spaces indexed by the positive integers, whose projection maps, $\theta_i : W_i \to W_{i-1}$, are *quotient maps*.

A sequence, $w_1, w_2, \ldots$, with $w_k \in W_k$ is called *compatible* if $\theta_i(w_i) = w_{i-1}$, for all $i \leq j$. For future reference, we recall that the *inverse limit* Banach space, $\varprojlim W_i$, is defined to be the set of all compatible sequences, $\{w_k\}$, such that

$$\sup_i \|w_i\| < \infty \,,$$

with the obvious vector space structure and norm

$$\|\{w_i\}\| := \lim_{i \to \infty} \|w_i\| .$$

Definition 2.1. A *finite dimensional approximation* of V is a pair, $\{(W_i, \pi_i)\}$, with $\{W_i\}$ an inverse limits system as above and $\pi_i : X \to W_i$ a compatible system of quotient maps, such that for all $v \in V$, the sequence of lengths $\|\pi_i(v)\|$ converges to $\|v\|$.

It is clear that the induced map, $\pi : V \to \varprojlim W_i$, is an isometric embedding.

A finite dimensional approximation is equivalent to a choice of a suitable inverse system $\{V_i\}$ of closed finite codimensional subspaces of V. We will frequently refer to finite dimensional approximations as FDA's.

Example 2.2. Let $V = \ell^1$, and let $\pi_i : V \to \mathbf{R}^i$ be the map which sends the sequence $(a_j)_{j=1}^\infty$ to the finite sequence $(a_j)_{j=1}^i$, and let $\theta_i : \mathbf{R}^i \to \mathbf{R}^{i-1}$ be the truncation map. Then the pair, $\{(\mathbf{R}^i, \pi_i)\}$, is an FDA.

Any separable Banach space admits finite dimensional approximations:

Lemma 2.3. *A separable Banach space has a finite dimensional approximation.*

Proof. Take a sequence, v_i, of unit vectors in the separable Banach space, V, which is dense in the unit sphere. For each i, apply the Hahn–Banach theorem to obtain a unit norm linear functional, $\ell_i : V \to \mathbf{R}$, such that $\ell_i(v_i) = 1$. Let $V_i \subset X$ denote the intersection of the kernels of $\ell_1, \dots, \ell_i$, and define W_k to be the quotient Banach space $W_i := V/V_i$. $\square$

Consider a PI space, (X, d^X, μ), and an atlas $\{(A_\alpha, u_\alpha)\}$. For every Lipschitz function, $f : X \to \mathbf{R}$, the differential, Df, is defined on a full measure subset, $\mathrm{domain}(Df) \subset X$.

Let $\{(W_i, \pi_i)\}$ denote a finite dimensional approximation of a Banach space V. A Lipschitz map, $f : X \to V$, induces a compatible family of Lipschitz maps $f_i := \pi_i \circ f : X \to W_i$. The differentiation theory of [Che99] for real valued functions extends immediately to maps with finite dimensional targets, and in particular, to each f_i. Thus, for μ-a.e. $x \in X$, we obtain a compatible system of linear maps $D_x f_i : T_x X \to W_i$. For each i, this defines a measurable family of maps which is uniquely determined almost everywhere.

Definition 2.4. The *weak derivative* of f at $x \in X$ with respect to the finite dimensional approximation, $\{(W_i, \pi_i)\}$, is the induced linear mapping $\{D_x f_i\} : T_x X \to \varprojlim W_i$.

The weak derivative is defined for almost every $x \in X$. We denote by domain $(\{Df_i\}) \subset X$, the full measure subset on which the weak derivative, $\{D_x f_i\}$, is defined.

Lemma 2.5. *Weak derivatives have the following properties:*

1) *For μ-a.e. $x \in X$, and every $e \in T_x X$,*

(2.6) $$\lim_{n \to \infty} \|(D_x f_n(e)\| = \|\{D_x f_i\}(e)\|.$$

2) $(x, e) \to \|\{D_x f_i\}(e)\|$ *is a measurable function on TX.*

3) *If $X = [0, 1]$, then*

$$\int_0^1 \|\{D_t f\}\| dt = \text{length}(f).$$

Proof.

1) If $x \in \text{domain}(\{Df_i\})$, then for every $e \in T_x X$, we obtain a compatible system, $\{D_x f_n(e)\}$, which defines the element $\{D_x f_n\}(e)$. Since the norm of an element in $\varprojlim W_i$ is defined by (2.6), we are done.

2) For all $n < \infty$, the formula $(x, e) \to \|D_x f_n(e)\|$ defines a measurable function on the restriction of TX to domain$(Df_n) \subset X$. This is a monotone increasing family of functions, so its pointwise limit is measurable.

3) For $X = [0, 1]$, it is classical that

$$\text{length}(f_n) = \int_X \|Df_n\| dt.$$

If $\mathcal{P} := \{0 = t_0 < t_1 < \ldots < t_k = 1\}$ is a partition of $[0, 1]$, then

$$\text{length}(f, \mathcal{P}) = \sum_i \|f(t_i) - f(t_{i-1})\| = \sum_i \lim_{n \to \infty} \|f_n(t_i) - f_n(t_{i-1})\|$$

$$= \lim_{n \to \infty} \text{length}(f_n, \mathcal{P}) \quad \leq \limsup_{n \to \infty} \text{length}(f_n)$$

$$= \limsup_{n \to \infty} \int_{[0,1]} \|Df_n\| \quad = \int_{[0,1]} \|\{Df_n\}\|,$$

where the last equation follows from the monotonicity of the sequence of functions $\|Df_n\|$. Now, property 3) follows from

$$\text{length}(f) := \sup_{\mathcal{P}} \text{length}(f, \mathcal{P}). \qquad \square$$

3. Good Finite Dimensional Approximations

In this section we discuss GFDA's, a special class of FDA's with an additional property which is sufficient for proving differentiability and which can be verified for a large class of Banach spaces, including separable dual spaces.

Example 3.1. (Example 2.2 continued.) To motivate the the GFDA property, we examine the FDA from Example 2.2 more closely.

Note that if $v, v' \in \ell^1$, and for some i, the projections, $\pi_i(v)$, $\pi_i(v')$, are close, and their norms are close to the norms of v, v', respectively, then the vectors v and v' themselves are close. Indeed, if $\widehat{\pi}_i : \ell^1 \to \ell^1$ denotes the projection map which drops the first i terms of the sequence then

$$\|v - v'\| = \|\pi_i(v) - \pi_i(v')\| + \|\widehat{\pi}_i(v) - \widehat{\pi}_i(v')\|$$
$$\leq \|\pi_i(v) - \pi_i(v')\| + \|\widehat{\pi}_i(v)\| + \|\widehat{\pi}_i(v')\|$$
$$\leq \|\pi_i(v) - \pi_i(v')\| + \big| \|v\| - \|\pi_i(v)\| \big| + \big| \|v'\| - \|\pi_i(v')\| \big|.$$

The GFDA property is just a more technical version of this kind of statement; Remark 3.7.

To formulate the GFDA property we need the following:

Definition 3.2. Given a finite dimensional approximation, $(\{W_i\}, \{\pi_i\})$, of V, we call a positive decreasing finite sequence, $1 \geq \rho_1, \ldots, \rho_N$, ϵ-*determining* if the conditions,

(3.3)

$$\|v\| - \|\pi_i(v)\| < \rho_i \cdot \|v\|, \qquad \|v'\| - \|\pi_i(v')\| < \rho_i \cdot \|v'\|, \qquad 1 \leq i \leq N,$$

and

(3.4) $$\|\pi_N(v) - \pi_N(v')\| < N^{-1} \cdot \max(\|v\|, \|v'\|),$$

imply

(3.5) $$\|v - v'\| < \epsilon \cdot \max(\|v\|, \|v'\|).$$

Observe that by dividing by $\max(\|v\|, \|v'\|)$, it suffices to consider pairs v, v' for which $\max(\|v\|, \|v'\|) = 1$.

Definition 3.7. A finite dimensional approximation $\{(W_i, \pi_i\}$, of a Banach space V is *good* if for every $\epsilon > 0$ and every infinite decreasing sequence, $1 \geq \rho_i \searrow 0$, some finite initial segment, $\rho_1, \ldots, \rho_N$, is ϵ-determining.

A Banach space, V, which admits a good finite dimensional approximation will be called a GFDA.

Remark 3.7. In the eventual application, relation (3.5), which states that v, v' are close in norm, will be used in proving that for GFDA targets, weak derivatives are actually derivatives. In order to know that relation (3.5) holds, it is necessary to verify relation (3.3). This will follow from Egoroff's theorem, whose conclusion asserts the uniform convergence of a pointwise sequence of functions off subsets of arbitrarily small measure. Note that uniform convergence means convergence no slower than some sequence, $\rho_i \to 0$, which, as in Definition 3.7, is otherwise uncontrolled.

Lemma 3.8. *If* $\{(W_i, \pi_i)\}$ *is a GFDA of* V, *then the induced linear transformation,* $\pi : V \to \varprojlim W_i$, *is an isometry of Banach spaces.*

Proof. We already know that π is an isometric embedding, so it suffices to check that π is surjective.

Pick $\{w_i\} \in \varprojlim W_i$, and choose a decreasing sequence $\rho_i \to 0$. Let

(3.9)

$$C_i := \{v \in V \mid \pi_i(v) = w_i, \text{ and for all } 1 \leq j \leq i, \|v\| - \|w_j\| \leq \rho_j\}.$$

Since π_i is a quotient map, C_i is a nonempty, bounded, closed set for each i. The GFDA property implies that $\text{diam}\,(C_i) \to 0$. Since the family, $\{C_i\}$, is clearly nested, it follows from completeness that the intersection $\bigcap_i C_i$ is nonempty and consists of a single vector v. Then $\pi(v) = \{w_i\}$. $\square$

Remark 3.10. The above argument does not use the full strength of the GFDA assumption.

Proposition 3.11. *If* $V = E^*$ *is a separable dual space, then it is isomorphic to a GFDA space.*

Proof. We apply the following renorming procedure to E, and hence E^*.

Lemma 3.12. [LT77, p. 12] *Suppose* E *is a separable Banach space and* $F \subset E^*$ *is a separable subspace of its dual. Then* E *can be renormed so that if the sequence,* $e_i^* \in E^*$, *weak* converges to* $e_\infty^* \in F$ *and* $\|e_i^*\| \to \|e_\infty^*\|$, *then* $\|e_i^* - e_\infty^*\| \to 0$.

To continue with the proof of Proposition 3.11 we construct an inverse system $\{W_i\}$ by taking an increasing family of finite dimensional subspaces

$E_1 \subset E_2 \subset \ldots \subset E$ with dense union, and taking π_i to be the restriction mapping $E^* \to W_i := E_i^*$, where W_i is endowed with the norm dual to the norm on E_i. It follows from the Hahn–Banach theorem that $\{(W_i, \pi_i)\}$ is an FDA of $V = E^*$.

Suppose $\{(W_i, \pi_i)\}$ is not a GFDA. Then for some decreasing sequence, $\{\rho_i\} \subset (0, \infty)$, with $\rho_i \to 0$, and some $\epsilon > 0$, there are sequences, $v_k, v'_k \in V$, such that for all $k < \infty$,

$$(3.13) \qquad \|v_k\|, \|v'_k\| \leq 1,$$

$$(3.14) \quad \max\left(\|v_k\| - \|\pi_i(v_k)\|, \ \|v'_k\| - \|\pi_i(v'_k)\|\right) < \rho_i \text{ for } 1 \leq i \leq k,$$

$$(3.15) \qquad \|\pi_j(v_k) - \pi_j(v'_k)\| < \frac{1}{k},$$

$$(3.16) \qquad \|v_k - v'_k\| \geq \epsilon.$$

By the Banach–Alaoglu theorem, we can pass to weak* convergent subsequences, with respective limits v_∞ and v'_∞. Semicontinuity of norm with respect to weak* convergence implies

$$\|v_\infty\| \leq \liminf_{k \to \infty} \|v_k\| \leq 1, \quad \|v'_\infty\| \leq \liminf_{k \to \infty} \|v'_k\| \leq 1.$$

Also, by the definition of π_i and weak* convergence, for all i we have $\pi_i(x_v) \to \pi_i(v_\infty)$, $\pi_i(v'_k) \to \pi_i(v'_\infty)$. Hence, for fixed i,

$$(3.17) \quad \|v_\infty\| \geq \|\pi_i(v_\infty)\| = \lim_{k \to \infty} \|\pi_i(v_k)\| \geq \limsup_{k \to \infty} \|v_k\| - \rho_i,$$

which forces $\|v_\infty\| \geq \limsup_{k \to \infty} \|v_k\|$. Similarly $\|v'_\infty\| \geq 1$. Therefore, $\|v_k\| \to \|v_\infty\|$ and $\|v'_k\| \to \|v'_\infty\|$. By Lemma 3.12, this implies that v_k and v'_k converge strongly. However, it is clear from (3.14) that v_k and v'_k have the same weak* limit. Thus, they converge strongly to the same limit, contradicting (3.16). $\qquad \square$

Remark 3.18. By the lemma, any separable reflexive space, and ℓ^1 are isomorphic to GFDA spaces. However, $L^1([0, 1])$ is not a GFDA space. This follows from the failure of differentiability (see the example from [Aro76] mentioned in the introduction) and Theorem 4.1.

4. Differentiability for GFDA Targets

In this section (X, d^X, μ) will denote a fixed PI space, and $f : X \to V$, a Lipschitz map to a Banach space V.

Fix $\{(A_\alpha, u_\alpha)\}$, an atlas for X as in Definition 1.2. Without loss of generality, we may assume that for each α, there exists an $L_\alpha < \infty$, such that the derivative of u_α induces an L_α-bi-Lipschitz fiberwise isomorphism, $D_x u_\alpha : T_x X \to \mathbf{R}^{n_\alpha}$, for each $x \in A_\alpha$, where $T_x X$ is equipped with its canonical norm, and $\mathbf{R}^{n_\alpha}$ is given the usual norm.

Our main theorem is:

Theorem 4.1. *If V satisfies the GFDA property, then the Lipschitz map, $f : X \to V$, is differentiable at μ-a.e. $x \in X$. Moreover, the derivative is uniquely determined at μ-a.e. x and defines a bounded measurable section, Df, of the bundle $\mathrm{End}(TX, V)$.*

As indicated in the introduction, we use the next lemma and its corollary to reduce the proof of Theorem 4.1 to showing that f is finite dimensional to first order μ-a.e.

Lemma 4.2. *Let $\{(W_i, \pi_i)\}$ denote an FDA of a Banach space V. If $f : X \to V$ is finite dimensional to first order at $x \in X$, and x is a point of weak differentiability of f with respect to $\{(W_i, \pi_i)\}$, then f is differentiable at x.*

Proof. Let $F \subset V$ denote a finite dimensional subspace so that

$$(4.3) \qquad \limsup_{r \to 0} \frac{1}{r} \sup_{x' \in B_r(x)} d(f(x') - f(x), F) = 0.$$

Since F is finite dimensional, there is a bounded linear projection map $\Psi : V \to F$. Set $\widehat{f} := f(x) + \Psi \circ (f - f(x))$. By (4.3), the maps, f and $\widehat{f}$, agree to first order at x. This has two implications: that $\widehat{f}$ is also weakly differentiable at x, and that it suffices to prove that $\widehat{f}$ is differentiable at x. But $\widehat{f}$ takes values in a finite dimensional subspace, and is weakly differentiable at x, so clearly $\widehat{f}$ is differentiable at x. □

Corollary 4.4. *If a Lipschitz map, $f : X \to V$, from a PI space to an arbitrary Banach space is finite dimensional to first order μ-a.e., then f is differentiable μ-a.e.*

Proof. Since X is separable, the map f takes values in a separable subspace of V. It follows that without loss of generality, we may assume that V itself is separable. By Lemma 2.3, V admits an FDA, $\{(W_i, \pi_i)\}$, with respect to which and the weak derivative of f is defined almost everywhere. Thus, the corollary follows from Lemma 4.2. □

In preparation for the proof of Theorem 4.1, we need a two more definitions.

Definition 4.5. If F is a finite dimensional normed space and $h : X \to F$ is a Lipschitz map, then x *is an approximate continuity point of* Dh if the point, x, is a density point of A_α, and $Dh \mid A_\alpha$ is approximately continuous at x, with respect to the trivialization, $TX \mid A_\alpha \to A_\alpha \times \mathbf{R}^{n_\alpha}$, induced by u_α.

Given $\{(W_i, \pi_i)\}$, we set $f_i := \pi_i \circ f$. We recall that domain$(\{Df_i\}) \subset X$ denotes the domain of definition of the weak derivative of f; see Definition 2.4.

Definition 4.6. A point $x \in$ domain $(\{Df_i\})$ is a *weak approximate continuity point of* $\{Df_i\}$ if x is an approximate continuity point of Df_i for all i.

Clearly, the set of weak approximate continuity points of $\{Df_i\}$ has full measure.

For each $x \in$ domain$(\{Df_i\})$, and every i, define the semi-norm, $\| \cdot \|_i$, on $T_x X$, to be the pull-back,

$$\| \cdot \|_i = Df_i^*(\| \cdot \|_{W_i}) .$$

This defines a measurable Finsler pseudo-metric on TX (also denoted $\| \cdot \|_i$) which is a pointwise nondecreasing function of i, and for which we have $\| \cdot \|_i \le C \cdot \| \cdot \|_{TX}$, for some constant C. We take the pointwise limit of this sequence and put

$$\| \cdot \|_\infty := \lim_{i \to \infty} \| \cdot \|_i$$
$$\le C \cdot \| \cdot \|_{TX} .$$

For each $x \in$ domain $(\{Df_i\})$ and $i < \infty$, set

(4.7) $\qquad \nu(x, i) := \sup \{ \|\xi\|_\infty - \|\xi\|_i \mid \xi \in T_x X, \|\xi\| \le 1 \} .$

Then $\nu(\cdot, i)$ is uniformly bounded, measurable, and converges pointwise to zero almost everywhere.

Proof of Theorem 4.1. Since $\{(W_i, \pi_i)\}$ is assumed to be a GFDA, by Lemma 3.8, the map $V \xrightarrow{\pi} \varprojlim W_i$ is an isometry and hence invertible.

By Corollary 4.4, it suffices to show that f is finite dimensional to first order almost everywhere. In fact, we will show that for μ-a.e. $x \in X$, relation (1.14) holds, with $F = \text{image}(\{D_x f_i\})$, the image of the weak derivative. Here we identify $\varprojlim W_i$ with V, using the isomorphism $\varprojlim \pi_i$.

Suppose f is L-Lipschitz. It will be convenient to assume that for every α, every $x \in A_\alpha \cap \text{domain}(Du_\alpha)$, and every $\xi \in \mathbf{R}^{n_\alpha}$, with $\|\xi\| \leq 1$, we have

$$\|(D_x u_\alpha)^{-1}(\xi)\| \leq \min\left(1, \frac{1}{L}\right).$$

This can be arranged by rescaling the maps u_α. In particular, if $x \in A_\alpha \cap \text{domain}(\{Df_i\}) \cap \text{domain}(Du_\alpha)$, then

(4.8) $\qquad \|(\{D_x f_i\})(D_x u_\alpha)^{-1}(\xi)\| \leq \|\{D_x f_i\}\|\left(\frac{1}{L}\right) \leq 1.$

Pick $\epsilon, \epsilon_1, \epsilon_2, \epsilon_3 > 0$, to be further constrained later.

Choose a subset $X' \subset X$ with finite μ-measure. By Egoroff's theorem, there is a subset $S_1 \subset X'$ such that $\mu(X' \setminus S_1) < \epsilon_1$, and the quantity, $\nu(\cdot, i)$, of (4.7) converges to zero uniformly on S_1. Otherwise put, there is a sequence, $\rho_i \to 0$, such that for all $x \in S_1$,

(4.9) $\qquad\qquad\qquad\qquad \nu(x, i) < \rho_i.$

Let $S_2 \subset \text{domain}(\{Df_i\})$ denote the full measure set of weak approximate continuity points of $\{Df_i\}$, and let $S_3 := S_1 \cap S_2$.

Pick $x \in S_3$. Using the GFDA property, there exists a number N which is ϵ_2-determining for the sequence $\{\rho_i\}$. Since x is an approximate continuity point of Df_j, certainly $x \in A_\alpha$ for some α. Moreover, there exists $r_0 > 0$, such that for all $r < r_0$, there is a subset, $\Theta_r \subset B_r(x) \cap A_\alpha$, with density,

$$\frac{\mu(\Theta_r)}{\mu(B_r(x))} > 1 - \epsilon_3,$$

such that for all $x' \in \Theta_r$, $\xi \in \mathbf{R}^{n_\alpha}$, with $\|\xi\| \leq 1$, we have

(4.10) $\quad \|(D_{x'} f_j)((D_{x'} u_\alpha)^{-1}(\xi) - (D_x f_j)((D_x u_\alpha)^{-1}(\xi)\| < \frac{1}{N}.$

By the GFDA property of $\{(W_i, \pi_i)\}$ and the choice of N, one gets for $\xi \in \mathbf{R}^{n_\alpha}$, with $\|\xi\| \leq 1$, $x' \in \Theta_r$,

(4.11) $\quad \|\{D_{x'} f_i\} \circ (D_{x'} u_\alpha)^{-1}(\xi) - \{D_x f_i\} \circ (D_x u_\alpha)^{-1}(\xi)\| < \epsilon_2.$

In particular,

$$(4.12) \qquad d(\{D_{x'}f_i\} \circ (D_{x'}u_\alpha)^{-1}(\xi), \text{image}(\{D_x f_i\})) < \epsilon_2.$$

Pick $k < \infty$, and $\ell \in W_k^*$, with $\|\ell\| \leq 1$, such that ℓ annihilates image $(Df_k) \subset W_k$. Then by (4.11), for every $x' \in \Theta_r$ and every $\xi \in \mathbf{R}^{n_\alpha}$, with $\|\xi\| \leq 1$, there holds

$$(4.13) \quad |D_{x'}(\ell \circ f_k)(D_{x'}u_\alpha)^{-1}(\xi)| = |\ell\left((D_{x'}f_k)(D_{x'}u_\alpha)^{-1}(\xi)\right)| < \epsilon_2.$$

Thus, by choosing ϵ_2, ϵ_3 small enough, we can arrange that $\ell \circ f_k$ has small derivative on a subset of $B_r(x)$ with density as close to 1 as we like. By the Poincaré inequality and the fact that f is L-Lipschitz, we can choose ϵ_2, ϵ_3 small enough (independent of k, ℓ) such that when $r \leq c \cdot r_0$, for a suitable constant, c, we have

$$(4.14) \qquad\qquad \max_{B_r(x)} \ell \circ f_k - \min_{B_r(x)} \ell \circ f_k < \frac{\epsilon r}{2}.$$

It follows that for all $x' \in B_r(x)$,

$$(4.15) \qquad \begin{aligned} d(f(x') - f(x), \text{image}(\{D_z f_i\})) &= \sup_{k,\ell}\{\ell \circ f_k(x') - \ell \circ f_k(x)\} \\ &\leq \frac{\epsilon r}{2}. \end{aligned}$$

Since ϵ is arbitrary, (4.15) implies that f is finite dimensional to first order at points $x \in S_1$. Since $\mu(X' \setminus S_1) < \epsilon_1$ and ϵ_1 is arbitrary, this implies that f is differentiable on a full measure subset of X'. Since X' was arbitrary, it follows that f is differentiable on a full measure subset of X.

Uniqueness of the Differential

We claim that for each α, the equation (1.11) has a unique solution for almost every $x \in A_\alpha$.

By composing both sides with π_i, we get for $x' \in A_\alpha$, as $x' \to 0$,

$$f_i(x') = f_i(x) + (\pi_i \circ \Phi) \circ (u_\alpha(x') - u_\alpha(x)) + o(d^X(x', x)).$$

Now, from the almost everywhere uniqueness of the differential for maps into finite dimensional targets, it follows that $\pi_i \circ \Phi$ is uniquely determined almost everywhere. Since this holds for all i, it follows that Φ itself is uniquely determined almost everywhere.

Measurability of the Differential Df

From the proof above and the proof of Lemma 4.2, it follows that the differential Df agrees with the composition, $\{\pi_i\}^{-1} \circ \{D_x f_i\} : T_x X \to V$, almost everywhere, so it suffices to verify the measurability of this composition. This follows in a straightforward fashion from the measurability of Df_i, for all i, and the GFDA property of V.

$\square$

5. Bi-Lipschitz (Non)embedding for GFDA Targets

We indicate briefly how the discussion of Section 14 of [Che99], which contains nonembedding theorems for finite dimensional targets, extends directly to GFDA targets.

Let $f : X \to V$ be a Lipschitz map from a PI space into a Banach space V, which is differentiable almost everywhere; in particular, by Theorem 4.1, this is true for any Lipschitz map when V is a GFDA space. Then for μ-a.e. $x \in X$, blow-ups converge to a map $X_x \to \text{image}\,(D_x f) \subset V$ which factors through a surjective map $X_x \to T_x X$.

Now suppose f is a bi-Lipschitz embedding. Then each A_α contains a full measure subset $\widehat{A}_\alpha$ whose image under f has the property that blow-ups are equal to linear subspaces of dimension equal to n_α. This implies:

1) For almost every $x \in A_\alpha$, every tangent cone at x is bi-Lipschitz equivalent to $\mathbf{R}^{n_\alpha}$.

2) By a straightforward modification of a standard result in the finite dimensional case [Fed69], the set $f(\widehat{A}_\alpha) \subset V$ is rectifiable; in particular, the Hausdorff dimension of $\widehat{A}_\alpha$ is at most n_α.

Assertion 1) yields:

Theorem 5.1. *If X contains a positive μ-measure set of points, x, such that X_x is not bi-Lipschitz to $T_x X$, for some tangent cone, X_x then X does not admit a bi-Lipschitz embedding in any space bi-Lipschitz homeomorphic to a GDFA space.*

It follows in particular that nontrivial Carnot groups, Bourdon–Pajot spaces, and Laakso spaces do not bi-Lipschitz embed in a GFDA space.

The hypothesis of Theorem 5.1 should be compared to (1.9).

We remark that as far as we know, the n_α-dimensional Hausdorff measure of this rectifiable part $\widehat{A}_\alpha$ could be zero, and the μ-null set being discarded could have Hausdorff dimension $> n_\alpha$.

We note that for any doubling space, X, there exist injective Lipschitz maps, $f : X \to \ell_1$, for which f^{-1} has better regularity than C^α, for all $\alpha < 1$.

6. Appendix: Carnot Groups and Radon–Nikodym Targets

Here we consider the special case of the 3-dimensional Heisenberg group, $\mathbb{H}$, and general Radon Nikodym targets.

We recall that $\mathbb{H}$, is the simply connected nilpotent Lie group whose Lie algebra has a basis P, Q, Z, with the bracket relations

$$[P, Q] = Z, \quad [P, Z] = [Q, Z] = 0.$$

Let Δ denote the two dimensional distribution on $\mathbb{H}$ spanned by the left invariant vector fields P, Q. We define a riemannian metric on Δ by stipulating that P, Q are pointwise orthonormal. Then the *Carnot–Caratheodory distance* between two points $x_1, , x_2 \in \mathbb{H}$ is defined to be the infimum of the lengths of the C^1 paths joining x_1 to x_2, which are everywhere tangent to Δ. We denote the Carnot–Caratheodory distance function by $d^{\mathbb{H}}$.

Theorem 6.1. *The Heisenberg group with the Carnot–Caratheodory metric does not admit a bi-Lipschitz embedding into any Banach space satisfying the Radon–Nikodym property.*

Proof. The proof is almost identical to the proof in [Sem96]. The hypothesis on the target space is invoked only once, to deduce the almost everywhere directional differentiability of the map.

Let V denote a Banach space with the Radon–Nikodym property and let $f : \mathbb{H} \to V$ denote an L-Lipschitz map.

Step 1. The map, f, has directional derivatives $P(f)$ and $Q(f)$, almost everywhere.

It is well known that the (images of) integral curves of P and Q are isometric copies of the real line. Thus if $\gamma : \mathbf{R} \to \mathbb{H}$ is such an integral curve, the composition $f \circ \gamma : \mathbf{R} \to V$ is a Lipschitz mapping, and since V satisfies the Radon–Nikodym property, $f \circ \gamma$ is differentiable almost everywhere. Thus the set of points where the directional derivative $P(f)$ is not defined, which is easily seen to be a measurable set, intersects (the image of) each integral curve of P in a set of measure zero; as the integral curves define a smooth 1-dimensional foliation of $\mathbb{H}$, it follows from Fubini's theorem that the directional derivative $P(f)$ is defined almost everywhere. It follows by standard reasoning that $P(f)$ is measurable (it agrees almost everywhere with a pointwise limit of a sequence of measurable difference quotients). The Lipschitz condition on f implies that $\|P(f)\| \leq L$ almost everywhere. In particular, $P(f) \in L^\infty(\mathbb{H}, V)$, and similarly, $Q(f) \in L^\infty(\mathbb{H}, V)$.

Step 2. The directional derivatives $P(f)$ and $Q(f)$ are approximately continuous almost everywhere.

We recall that a measurable mapping $u : X \to Y$ from a metric measure space, (X, d^X, μ), to a metric space, Y, is *approximately continuous* at $x \in X$, if for all $\epsilon > 0$, the density of

$$S_\epsilon := \{x' \in B_r(x) \mid d(u(x'), u(x)) < \epsilon\}$$

satisfies

$$\lim_{r \to 0} \frac{\mu(S_\epsilon)}{\mu(B_r(x))} = 1 \,.$$

We observe that the Lipschitz map f takes values in a closed separable subspace $V' \subset V$. Hence, $P(f)$ and $Q(f)$ take values in V' as well. Therefore, the almost everywhere approximately continuity of $P(f)$ and $Q(f)$ follows from:

Lemma 6.2. *Let (X, d^X, μ) be a doubling metric measure space, and let $u : X \to Y$ be a measurable mapping from X into a separable metric space Y. Then u is approximately continuous for μ-a.e. $x \in X$.*

Proof. Pick $\epsilon > 0$, and let $\{U_i\}$ be a countable cover of Y by open sets of diameter $< \epsilon$. The the countable collection $\{\Omega_i\}$, where $\Omega_i := u^{-1}(U_i) \subset X$, defines a countable cover of X by measurable sets. By a standard covering argument, the set of density points $\widetilde{\Omega}_i$ in Ω_i has full measure in Ω_i, so $\bigcup_i \widetilde{\Omega}_i$ has full measure in X. By considering a countable sequence, $\epsilon_j \to 0$, this clearly suffices to complete the proof. $\square$

Step 3. If $x \in \mathbb{H}$ is an approximate continuity point of $P(f)$ and $Q(f)$, then the quantity, $d(f(x'), f(x))$ is $o(d(x', x))$, when x' lies on the integral curve $\{x \exp(tZ)\}_{t \in \mathbf{R}}$.

Pick $\epsilon > 0$. We will assume that x' is of the form $x' = x \exp t^2 Z$; the case when $x' = x \exp t^2(-Z)$ is similar.

We write, $\xi_1, \xi_2, \xi_3, \xi_4$, for $P, Q, -P, -Q$. It is a standard fact that there is an open quadrilateral $\gamma_1, \gamma_2, \gamma_3, \gamma_4$ starting at x and ending at x', where $\gamma_i : [0, t] \to G$ is an integral curve of ξ_i of elapsed time t. Furthermore, $d^{\mathbb{H}}(x', x) \geq Ct$, for some universal constant $C > 0$.

By using Fubini's theorem, and the fact that x is an approximate continuity point of the L^∞ mappings $P(f)$ and $Q(f)$, it follows that when t is sufficiently small, there exists an integral curve, $\widehat{\gamma}_i$, of ξ_i, with elapsed time t, such that $d(\widehat{\gamma}_i(s), \gamma_i(s)) < \epsilon t$, for all $s \in [0, t]$, and

$$(6.3) \qquad \frac{1}{t} \int_{\widehat{\gamma}_i} \|\xi_i(f) - \xi_i(f)(x)\| \, dt < \epsilon.$$

Therefore, since f is L-Lipschitz,

$$\begin{aligned}
\|f(x') - f(x)\| &= \|f \circ \gamma_4(t) - f \circ \gamma_1(0)\| \\
&\leq \|f \circ \widehat{\gamma}_4(t) - f \circ \widehat{\gamma}_1(0)\| + 2L\epsilon t \\
&\leq \| \sum_i (f \circ \widehat{\gamma}_i(t) - f \circ \widehat{\gamma}_i(0)) \| + 6L\epsilon t \\
&\leq \| \sum_i (\xi_i(f))(x) \, t\| + 4\epsilon t + 5L\epsilon t \\
&\leq 4\epsilon t + 6L\epsilon t \\
&= o(d^X(x', x)),
\end{aligned}$$

where in the next to last inequality we have used (6.3) and in the last, $\xi_3 = -\xi_1$ and $\xi_4 = -\xi_2$. $\qquad\square$

References

[Aro76]   N. Aronszajn. Differentiability of Lipschitzian mappings between Banach spaces. *Studia Math.*, 57(2):147–190, 1976.

[BL00]   Y. Benyamini and J. Lindenstrauss. *Geometric nonlinear functional analysis. Vol. 1*, volume 48 of *American Mathematical Society Colloquium Publications*. American Mathematical Society, Providence, RI, 2000.

[BP99]   M. Bourdon and H. Pajot. Poincaré inequalities and quasiconformal structure on the boundary of some hyperbolic buildings. *Proc. Amer. Math. Soc.*, 127(8):2315–2324, 1999.

[CC97]   J. Cheeger and T. Colding. On the structure of spaces with Ricci curvature bounded below. *J. Diff. Geom.*, 46:406–480, 1997.

[Che99]   J. Cheeger. Differentiability of Lipschitz functions on metric measure spaces. *Geom. Funct. Anal.*, 9(3):428–517, 1999.

[CK06a]   J. Cheeger and B. Kleiner. Differentiating maps to L^1 and the geometry of BV functions. Preprint, 2006.

[CK06b]   J. Cheeger and B. Kleiner. Embedding Laakso spaces in L^1. Preprint, 2006.

[CK06c]   J. Cheeger and B. Kleiner, Generalized differentiation and bi-Lipschitz nonembedding in L^1, C. R. Acad. Sci. Paris, Ser. I 343 (2006) 297–301.

[Fed69]   H. Federer. *Geometric measure theory.* Die Grundlehren der mathe-
            matischen Wissenschaften, Band 153. Springer-Verlag New York Inc.,
            New York, 1969.

[Geh61]   F. W. Gehring. Rings and quasiconformal mappings in space. *Proc.
            Nat. Acad. Sci. U.S.A.*, 47:98–105, 1961.

[Gel38]   I. M. Gelfand. Abstracte Functionen und lineare Operatoren. *Mat.
            Sbornik*, 46(4):235–284, 1938.

[GM84]   N. Ghoussoub and B. Maurey. Counterexamples to several problems
            concerning G_δ-embeddings. *Proc. Amer. Math. Soc.*, 92(3):409–412,
            1984.

[Gro96]   M. Gromov. Carnot-Caratheodory spaces seen from within. In *Sub-
            riemannian geometry*, Progr. in Math., pages 79–323. Birkhäuser,
            Basel, 1996.

[HK96]   J. Heinonen and P. Koskela. From local to global in quasiconformal
            structures. *Proc. Nat. Acad. Sci. USA*, 93:554–556, 1996.

[Kad59]   M. Ĭ. Kadeć. On the connection between weak and strong convergence.
            Dopovidi Akad. Nauk Ukrain. RSR, 1959, 1959.

[Kir94]   B. Kirchheim. Rectifiable metric spaces: local structure and regularity
            of the Hausdorff measure. *Proc. Amer. Math. Soc.*, 121(1):113–123,
            1994.

[Kle61]   V. Klee. Mappings into normed linear spaces. *Fund. Math.*, 49:25–34,
            1960/1961.

[Laa00]   T. J. Laakso. Ahlfors Q-regular spaces with arbitrary $Q > 1$ admitting
            weak Poincaré inequality. *Geom. Funct. Anal.*, 10(1):111–123, 2000.

[LN06]   J. Lee and A. Naor. L^p metrics on the Heisenberg group and the
            Goemans-Linial conjecture. Preprint, 2006.

[LT77]   J. Lindenstrauss and L. Tzafriri. *Classical Banach spaces. I.* Springer-
            Verlag, Berlin, 1977. Sequence spaces, Ergebnisse der Mathematik und
            ihrer Grenzgebiete, Vol. 92.

[Mat95]   P. Mattila. *Geometry of sets and measures in Euclidean spaces*, vol-
            ume 44 of *Cambridge Studies in Advanced Mathematics*. Cambridge
            University Press, Cambridge, 1995. Fractals and rectifiability.

[Mos73]   G. D. Mostow. *Strong rigidity of locally symmetric spaces.* Princeton
            University Press, Princeton, N.J., 1973. Annals of Mathematics Stud-
            ies, No. 78.

[Pan89]   P. Pansu. Métriques de Carnot-Carathéodory et quasiisométries des
            espaces symétriques de rang un. *Ann. of Math. (2)*, 129(1):1–60, 1989.

[Pau01]   S. D. Pauls. The large scale geometry of nilpotent Lie groups. *Comm.
            Anal. Geom.*, 9(5):951–982, 2001.

[Rad20]   H. Rademacher. Über partielle und totale Differenzierbarkeit von
            Funktionen mehrerer Variabeln. II. *Math. Ann.*, 81(1):52–63, 1920.

[Sem96]   S. Semmes. On the nonexistence of bi-Lipschitz parameterizations
            and geometric problems about A_∞-weights. *Rev. Mat. Iberoamericana*,
            12(2):337–410, 1996.

[Zie89]   W. P. Ziemer. *Weakly differentiable functions*, volume 120 of *Gradu-
            ate Texts in Mathematics*. Springer-Verlag, New York, 1989. Sobolev
            spaces and functions of bounded variation.

Chapter 7

TWO-FORMS ON FOUR-MANIFOLDS
AND ELLIPTIC EQUATIONS

S. K. Donaldson

Department of Mathematics, Imperial College, London

June 29, 2006

1. Background

Let V be a four-dimensional real vector space with a fixed orientation. Then the wedge product can be viewed, up to a positive factor, as a canonical quadratic form of signature $(3,3)$ on the six-dimensional space $\Lambda^2 V$. This gives a homomorphism from the identity component of the general linear group $GL^+(V)$ to the conformal group of the indefinite form, which is a local isomorphism. The significance of this is that geometrical structures on V can be expressed in terms of the six-dimensional space $\Lambda^2 V$ with its quadratic form. Now if X is an oriented 4-manifold we can apply this idea to the cotangent spaces of X. Many important differential geometric structures on X can fruitfully be expressed in terms of the bundle of 2-forms, with its quadratic wedge product and exterior derivative. We recall some examples

- A *conformal structure* on X is given by a 3-dimensional subbundle $\Lambda^+ \subset \Lambda^2$ on which the form is strictly positive. A *Riemannian metric* is specified by the addition of a choice of volume form.
- A (compatibly oriented) *symplectic structure* on X is a closed 2-form ω which is positive at every point, $\omega \wedge \omega > 0$.
- An *almost-complex structure* on X is given by a 2-dimensional oriented subbundle $\Lambda^{2,0} \subset \Lambda^2$ on which the form is strictly positive.
- A symplectic form ω *tames* an almost-complex structure $\Lambda^{2,0}$ if $\langle \omega \rangle + \Lambda^{2,0}$ is a 3-dimensional subbundle on which the form is positive, and

the induced orientation agrees with a standard orientation inherited from that of X. The symplectic form is *compatible* with the almost-complex structure if ω is orthogonal to $\Lambda^{2,0}$ and in this case we say that we have an *almost-Kahler* structure.

- A *complex-symplectic structure* is given by a pair of closed two forms θ_1, θ_2 such that

$$\theta_1^2 = \theta_2^2, \qquad \theta_1 \wedge \theta_2 = 0.$$

- A *complex structure* is given by an almost-complex structure $\Lambda^{2,0}$ such that in a neighbourhood of each point there is a complex-symplectic structure whose span is $\Lambda^{2,0}$.
- A *Kahler structure* is an almost-Kahler structure such that $\Lambda^{2,0}$ defines a complex structure.
- A *hyperkahler structure* is given by three closed two-forms $\theta_1, \theta_2, \theta_3$ such that

$$\theta_1^2 = \theta_2^2 = \theta_3^2, \qquad \theta_i \wedge \theta_j = 0 \text{ for } i \neq j.$$

A feature running through these examples is the interaction between pointwise, algebraic, constraints and the differential constraint furnished by the exterior derivative. In this article we introduce and begin the study of a very general class of questions of this nature, and discuss the possibility of various applications to four-dimensional differential geometry.

It is a pleasure to acknowledge the influence of ideas of Gromov and Sullivan on this article, through conversations in the 1980's which in some instances may only have been subconsciously absorbed by the author at the time.

2. A Class of Elliptic PDE

Let P be a three-dimensional submanifold in the vector space $\Lambda^2 V$, where V is as in the previous section. We say that

- P has *negative tangents* if the wedge product is a strictly negative form each tangent space TP_ω for all ω in P.
- P has *negative chords* if for all pairs $\omega, \omega' \in P$ we have $(\omega - \omega')^2 \leq 0$ with equality if and only if $\omega = \omega'$.

Clearly if P has negative chords it also has negative tangents, but the converse is not true. Now we turn to our 4-manifold X and consider a 7-dimensional submanifold $\mathcal{P}$ of the total space of the bundle Λ^2 such that the projection map induces a fibration of $\mathcal{P}$, so each fibre is a 3-dimensional submanifold, as considered before. We say that $\mathcal{P}$ has negative tangents, or negative chords, if the fibres do. In any case we can consider the pair of conditions, for a 2-form ω on X

$$\omega \subset \mathcal{P}, \qquad d\omega = 0. \tag{1}$$

Here, of course, the first condition just means that at each point $x \in X$ the value $\omega(x)$ is constrained to lie in the given 3-dimensional submanifold $\mathcal{P}_x \subset \Lambda_x^2$. The pair of conditions represent a partial differential equation over X. Now suppose that X is compact and fix a maximal positive subspace $H_+^2 \subset H^2(X; \mathbf{R})$, for example the span of the self-dual harmonic forms for some Riemannian metric. Given a class $C \in H^2(X; \mathbf{R})$, we augment Equation 1 by the cohomological condition

$$[\omega] \in C + H_+^2 \subset H^2(X; \mathbf{R}). \tag{2}$$

Proposition 1.

- If $\mathcal{P}$ has negative chords then there is at most one solution ω of the constraints (1), (2).
- If $\mathcal{P}$ has negative tangents and ω is a solution of the constraints (1), (2) then there is a neighbourhood N of ω (in, say, the C^∞ topology on 2-forms) such that there are no other solutions of the constraints in N. Further, if $\mathcal{P}^{(t)}$ is a smooth 1-parameter family of deformations of $\mathcal{P} = \mathcal{P}^{(0)}$ then we can choose N so that for small t there is a unique solution of the deformed constraint in N.

The proof of the first item is just to observe that if ω, ω' are two solutions then $(\omega - \omega')^2$ is non-positive, pointwise on X, by the negative chord assumption. On the other hand the de Rham cohomology class of $\omega - \omega'$ lies in H_+^2 so

$$\int_X (\omega - \omega')^2 \geq 0.$$

Thus $(\omega - \omega')^2$ vanishes identically and the negative chord assumption implies that $\omega = \omega'$.

For the second item we consider, for each point x of X the tangent space to the submanifold $\mathcal{P}_x$ at the point $\omega(x)$. This is a maximal negative subspace for the wedge product form and so ω determines a conformal structure, on X (the orthogonal complement of the tangent space is a maximal positive subspace). For convenience, we fix a Riemannian metric g in this conformal class. The condition that a nearby form $\omega + \eta$ lies in $\mathcal{P}$ takes the shape

$$\eta^+ = Q(\eta),$$

where η^+ is the self-dual part of η with respect to g, and Q is a smooth map with $Q(\eta) = O(\eta^2)$. We choose 2-forms representing the cohomology classes in H^2_+, so H^2_+ can be regarded as a finite-dimensional vector space of closed 2-forms on X. Then closed forms $\omega + \eta$ satisfying our cohomological constraint can be expressed as $\omega + da + h$ where $h \in H^2_+$ and where a is a 1-form satisfying the "gauge fixing" constraint $d^*a = 0$. Thus our constraints correspond to the solutions of the PDE

$$d^*a = 0, d^+a = Q(da + h) - h^+,$$

where d^+ denotes the self-dual component of d. This is not quite a $1-1$ correspondence, we need to identify the solutions $a, a + \alpha$ where α is a harmonic 1-form on X. The essential point now is that the linear operator

$$d^* \oplus d^+ : \Omega^1 \to \Omega^0 \oplus \Omega^2_+$$

is *elliptic*. This means that our problem can be viewed as solving a non-linear elliptic PDE and we can apply the implicit function theorem, in a standard fashion. The linearisation of the problem is represented by the linear map

$$L = d^* \oplus d^+ : \Omega^1/\mathcal{H}^1 \to \Omega^0/\mathcal{H}^0 \oplus \Omega^2_+/H^2_+,$$

where $\mathcal{H}^i$ denotes the space of harmonic i-forms, for $i = 0, 1$. We claim that the kernel and cokernel of L both vanish. If $d^+a \in H^2_+$ we have $dd^+a = 0$ (since the forms in H^2_+ are closed) and then

$$0 = \int_X dd^+a \wedge a = \int_X d^+a \wedge da = \int_X |d^+a|^2,$$

so $d^+a = 0$. Now we use a fundamental identity

$$0 = \int_X da \wedge da = \int_X |d^+a|^2 - |d^-a|^2, \tag{3}$$

so d^-a vanishes as well. It follows that the kernel of L is trivial. The cokernel of $d^+ : \Omega^1 \to \Omega^+$ is represented by the space of self-dual harmonic forms $\mathcal{H}^2_+$. The assertion that the cokernel of L is trivial is equivalent to the statement that L^2 projection $\pi : H^2_+ \to \mathcal{H}^2_+$ is surjective. Since these are both maximal positive subspaces for the cup product form they have the same dimension, so it is equivalent to prove that π is injective. But if a form $h \in H^2_+$ is in the kernel of π the cohomology class $[h]$ lies in the subspace $\mathcal{H}^2_- \subset H^2(X; \mathbf{R})$ defined by the anti-self dual harmonic forms, so $\int_X h^2 \leq 0$ and it follows that $h = 0$.

To sum up, for a given choice of $\mathcal{P}$ our constraints are represented by a system of nonlinear elliptic PDE whose linearisation is invertible. Now the assertions in the second item follow in a standard way from the implicit function theorem.

For the rest of this paper we will consider constraint manifolds $\mathcal{P}$ with negative tangents. Notice that we could generalise the set-up slightly by choosing a submanifold $Q \subset H^2(X; \mathbf{R})$ with the property that at each point the tangent space of Q is maximal positive subspace for the cup product form. Then we can take in place of the constraint (2) the condition $[\omega] \in Q$. The proof above goes through without essential change. We could also express things differently by considering the moduli space $\mathcal{M}$ of solutions to (1) with no cohomological constraint. Then the same proof shows that $\mathcal{M}$ is a manifold of dimension $b^2_-(X)$ and that the map $\omega \mapsto [\omega]$ defines an immersion of $\mathcal{M}$ in $H^2(X; \mathbf{R})$ whose derivative at each point takes the tangent space of $\mathcal{M}$ to a maximal *negative* subspace for the intersection form.

3. Examples

1. We fix a Riemannian metric on X and let $\mathcal{P}$ be the vector subbundle of anti-self dual forms. Then the solutions of our constraints are just the anti-self dual harmonic forms. This case is not very novel but, as we have seen in the previous section, is the model for the general situation for the purposes of local deformation theory.

2. Take $X = \mathbf{R}^4 = \mathbf{R}^3 \times \mathbf{R}$ with co-ordinates (x_1, x_2, x_3, t) and consider a case where $\mathcal{P}$ is preserved by translations in the four variables, so is determined by a single 3-manifold $P \subset \Lambda^2(\mathbf{R}^4)$. (Of course X is not compact here, but we only want to illustrate the local PDE aspects.) In the usual way, we write a 2-form as a pair of vector fields (E, B)

$$\omega = \sum E_i dx_i \wedge dt + \frac{1}{2} \sum \epsilon_{ijk} B_i dx_j \wedge dx_k \,.$$

that is, we are identifying $\Lambda^2 \mathbf{R}^4$ with $\mathbf{R}^3 \oplus \mathbf{R}^3$. The condition that a submanifold $P \subset \Lambda^2 \mathbf{R}^4$ has negative tangents implies that its tangent space at each point is transverse to the two $\mathbf{R}^3$ factors, so locally P can be written as the graph of a smooth map $F : \mathbf{R}^3 \to \mathbf{R}^3$. That is, locally around a given solution, we can write the constraint as $B = F(E)$. Consider solutions which are independent of translation in the t variable. The condition that ω is closed becomes

$$\nabla \cdot B = 0, \qquad \nabla \times E = 0.$$

We can write $E = \nabla u$ for a function u on $\mathbf{R}^3$, so our constraint is a nonlinear elliptic PDE for a function u on $\mathbf{R}^3$ of the form

$$\nabla.(F(\nabla u)) = 0.$$

Let

$$(H_{ij}) = \left(\frac{\partial F_i}{\partial \xi_j} \right),$$

be the matrix of derivatives of F. The condition that P has negative tangents becomes $H + H^T > 0$ and the linearisation of the problem is the linear elliptic PDE

$$\sum_{ij} \frac{\partial}{\partial x_i} \left(H_{ij} \frac{\partial f}{\partial x_j} \right) = 0.$$

3. The next example is the central one in this article. We fix a volume form on our 4-dimensional real vector space V and also a complex structure, i.e. a 2-dimensional positive subspace $\Lambda^{2,0} \subset \Lambda^2 V$. We define P to be the set of positive $(1,1)$-forms whose square is the given volume form. Then P is one connected component of the set

$$\{\omega \in \Lambda^2 V : \omega^2 = 1, \omega \wedge \Lambda^{2,0} = 0\},$$

(the component being fixed by orientation requirements). It is easy to see that P has negative chords. In fact if ω, ω' are two points in P we can choose complex coordinates z_1, z_2 so that

$$\omega = i dz_1 \wedge d\bar{z}_1 + i dz_2 \wedge d\bar{z}_2, \qquad \omega' = \lambda i dz_1 \wedge d\bar{z}_1 + \lambda^{-1} i dz_2 \wedge d\bar{z}_2,$$

where $\lambda > 0$. Then

$$(\omega - \omega')^2 = (\lambda - 1)(\lambda^{-1} - 1) = 2 - (\lambda + \lambda^{-1}) \leq 0,$$

with equality if and only if $\lambda = 1$.

Thus if X is a 4-manifold with a volume form and a choice of almost-complex structure we get a constraint manifold $\mathcal{P}$ with negative chords, and a problem of the kind we have been considering. In particular, we can consider the case when the almost-complex structure is integrable. Then our problem becomes the renowned Calabi conjecture (in the case of two complex variables) solved by Yau: prescribing the volume form of a Kahler metric. The solution can be most easily expressed in terms of the moduli space $\mathcal{M}$ of solutions to (1): it maps bijectively to the intersection of the Kahler cone in $H^{1,1}(X)$ with the quadric $[\omega]^2 = \mathrm{Vol}$, where Vol is the integral of the prescribed volume form. Notice however that the problem formulated in terms of the cohomological constraints (2) does *not* always have a solution. In the most extreme case, we could take X to be a complex surface which does not admit any Kahler metric. More generally, by deforming our choice of C and H_+^2 we can deform from a case when a solution exists to a case when it does not, and understanding this phenomenon is essentially the question of understanding the boundary of the Kahler cone.

The extension of the Calabi–Yau theory to the case when the almost structure is not integrable has been considered recently by Weinkove [13]. Suppose, for simplicity, that $b_+^2(X) = 1$ and suppose that ω_0 is a symplectic form compatible with the given almost-complex structure. In this case we take $C = 0$ and H_+^2 to be the 1-dimensional space spanned by ω_0. Then the cohomology class of any solution of (1), (2) is fixed by the prescribed volume form and without loss of generality we can suppose it is the same as $[\omega_0]$. Weinkove extends Yau's *a priori* estimates to prove existence under the assumption that the Nijenhius tensor of the almost complex structure is small in a suitable sense.

In general, we will say that a constraint $\mathcal{P} \subset \Lambda^2 X$ is *unimodular* if there is a volume form ρ on X such that $\omega^2 = \rho$ for any section $\omega \subset \mathcal{P}$. So for a unimodular constraint any solution of (1) is a symplectic form.

4. Our final example is in some ways a simple modification of the previous one. We take an almost-complex structure and a fixed positive $(1,1)$-form Θ and we consider the submanifold $\mathcal{P}$ of positive $(1,1)$ forms ω with $(\omega - \Theta)^2$ a given volume form on X. In each fibre this is a translate of the submanifold considered before, so again has negative chords. This example arises in the following way. Let X be a hyperkahler 4-manifold, with an orthogonal triple of closed forms $\theta_1, \theta_2, \theta_3$ and let ω be another symplectic form on X. We define three functions $\mu_i = \mu_i(\omega)$ on X by

$$\mu_i = \frac{\theta_i \wedge \omega}{\omega^2}.$$

These arose in [2] as the "moment maps" for the action of the symplectomorphism group of (X, ω) with respect to the θ_i and the triple (μ_1, μ_2, μ_3) can be regarded as a hyperkahler moment map. Now we ask the question: given three functions f_i on X, can we find a symplectic form ω with

$\mu_i(\omega) = f_i$? An obvious necessary condition is that $F = \sqrt{\sum f_i^2}$ does not vanish anywhere on X (for then the self-dual part of ω would vanish and ω^2 would be negative). Assuming this condition, we write $\sum f_i \theta_i = F\sigma$ where σ is a unit self-dual 2-form. Then σ determines an almost-complex structure on X (with $\Lambda^{2,0}$ equal to the orthogonal complement of σ in the span of $\theta_1, \theta_2, \theta_3$) and σ is a positive form of type (1,1) with respect to this structure. It is easy to check that the condition that $\mu_i(\omega) = f_i$ can be expressed in the form above, with $\Theta = \sigma/2F$ and the volume form $\sigma^2/(4F^2)$.

4. Partial Regularity Theory

We have introduced a very general class of elliptic PDE problems on 4-manifolds and shown that their behaviour with respect to deformations is straightforward. The crux of the matter, as far as proving existence results goes, is thus to obtain *a priori* estimates for solutions. Here we make some small steps in this direction, assuming an L^∞ bound. So throughout this section we assume that $\mathcal{P}$ is a constraint manifold with negative tangents and ω is a closed form in $\mathcal{P}$ with $|\omega| \leq K$ at each point. Here the norm is defined by some fixed auxiliary Riemannian metric g_0 on X. We will also use the conformal structure determined by the pair $\omega, \mathcal{P}$ which for convenience we promote to another metric g, with, say, the same volume form as g_0. Since the set $\{\Omega \in \mathcal{P}, |\Omega| \leq K\}$ is compact the metrics g, g_0 are uniformly equivalent, for fixed K.

Our estimates will depend on K and $\mathcal{P}$. More precisely, the dependence on $\mathcal{P}$ will involve local quantities that could be written down explicitly and, crucially, will be uniform with respect to continuous families of constraints $\mathcal{P}_t$ (with fixed K).

Lemma 1. *There exists* $C = C(K, \mathcal{P})$ *such that*

$$\|\nabla\omega\|_{L^2} \leq C.$$

(Here ∇ is the covariant derivative associated with the fixed metric g_0.)

It suffices to show that for any vector field v on X the Lie derivative $L_v\omega$ is bounded in L^2 by some $C(K, \mathcal{P}, v)$. For then we consider a cover of X by D coordinate patches and $4D$ vector fields such that on each patch four of the vector fields are the standard constant unit fields, for which the Lie derivative becomes the ordinary derivative. Write ω_v for $L_v\omega$. So $d\omega_v = 0$, and in fact $\omega_v = di_v(\omega)$ is exact. Imagine first that the flow generated by

v preserves the constraint manifold $\mathcal{P}$. Then if we apply the Lie derivative to the condition $\omega \subset \mathcal{P}$ we obtain an identity

$$(\omega_v)^+ = 0\,,$$

where $()^+$ denotes the self-dual part with respect to metric g defined by ω and $\mathcal{P}$. In general we will have an identity

$$(\omega_v)^+ = \rho,$$

where the L^∞ norm of ρ can be bounded in terms of of $v, \mathcal{P}$ and K. In particular, the L^2 norm of $(\omega_v)^+$ is *a priori* bounded (since the metrics g and g_0 are uniformly equivalent we can take the L^2 norm with respect to either here). Now using the metric g we have, since ω_v is exact

$$\|\omega_v\|_{L^2}^2 = 2\|(\omega_v)^+\|_{L^2}^2$$

just as in (3), and we obtain the desired L^2 bound on ω_v.

Lemma 2. *There is a constant c, depending on K and $\mathcal{P}$ such that if*

$$\int_{B(r)} |\nabla\omega|^2 \le cr^2\,,$$

for all r-balls $B(r)$ in X with $r \le r_0$ then for any p

$$\|\nabla\omega\|_{L^p} \le C\,,$$

where C depends on $p, K, \mathcal{P}, r_0$.

We use of some of the ideas developed in [3]. Fix attention on balls with a given centre, and choose local coordinates about this point such that the metric g is the standard Euclidean metric at the origin. Let $\Lambda_0^\pm$ be the space of $\pm$ self-dual forms for the Euclidean structure. The metric g is represented by a tensor $\mu \in Hom(\Lambda_0^-, \Lambda_0^+,$ vanishing at the origin, such that the g-anti-self-dual 2-forms have the shape $\sigma + \mu(\sigma)$ for σ in Λ_0^-. Now for any such tensor field μ, defined over the unit ball B in $\mathbf{R}^4$ say, consider the operator

$$d_\mu^+ = d_0^+ + \mu d_0^- : \Omega^1 \to \Omega^+\,.$$

Here d_0^+, d_0^- are the constant co-efficient operators defined by Λ_0^+, Λ_0^-. The basic point is that, for any given p, if μ is sufficiently small in L^∞ the usual Calderon-Zygmund theory can be applied to this operator, regarded as a perturbation of d_0^+. More precisely, for given p there is a $\delta = \delta(p)$ such that if $|\mu| \le \delta$ then for any 1-form α over the ball with $d^*\alpha = 0$ we have

$$\|d\alpha\|_{L^p(B/2)} \le C_p(\|d_\mu^+\alpha\|_{L^p(B)} + \|d\alpha\|_{L^2(B)})\,.$$

It follows that for closed 2-forms ρ over B we have an inequality

$$\|\rho\|_{L^p(B/2)} \leq C_p(\|\rho^{+,\mu}\|_{L^p(B)} + \|\rho\|_{L^2(B)}) \tag{4}$$

where $\rho^{+,\mu}$ denotes the self-dual part with respect to the conformal structure defined by μ. (To see this we write $\rho = d\alpha$ with $d^*\alpha = 0$.) To apply this idea in our situation we first fix some $p > 4$. For each r we rescale the ball $B(r)$ to the unit ball in $\mathbf{R}^4$ and apply (4) to the ordinary derivatives of the form corresponding to ω, defined with respect to the Euclidean coordinates, which are closed 2-forms. Since the tensor μ which arises is determined by the tangent space of $\mathcal{P}$, which varies continuously with ω, there is some $\epsilon = \epsilon(p)$ such that (4) holds provided the oscillation of ω over $B(r)$ is less than ϵ. (Here the "oscillation" of ω refers to the oscillation of the coefficients in the fixed coordinate system.) Transferring back to $B(r)$, and taking account of the rescaling behaviour of the quantities involved, we obtain an inequality of the form

$$r^{-4/p}\|\nabla\omega\|_{L^p(B(r/2))} \leq C_p r^{-2}\|\nabla\omega\|_{L^2(B(r))} + C'_p,$$

for constants C_p, C'_p depending on $p, K, \mathcal{P}$. This holds provided the oscillation of ω over $B(r)$ is less than ϵ. On the other hand the Sobolev inequalities tell us that the oscillation of ω over $B(r/2)$ is bounded by a multiple of $r^{(p-4)/p}\|\nabla\omega\|_{L^p(B(r/2))}$. So we conclude that if the oscillation of ω over $B(r)$ is less than ϵ then the oscillation of ω over $B(r/2)$ is at most

$$C\left(r^{-2}\int_{B(r)}|\nabla\omega|^2\right)^{1/2} + C'r \leq Cc^{1/2} + C'r.$$

Thus if c is sufficiently small we can arrange that the oscillation of ω over the half-sized ball is less than $\epsilon/10$, say, once r is small enough. Applying this to a pair of balls, we see that there is some r_0 such that if the oscillation of ω over all r-balls is less than ϵ, for all $r \leq r_0$, then this oscillation is actually less than $\epsilon/2$. It follows then, by a continuity argument taking $r \to 0$, that the oscillation can never exceed ϵ over balls of a fixed small size and this gives an *a priori* bound on the L^p norm of $\nabla\omega$ for this fixed p. This gives a fixed Holder bound on ω and we can repeat the discussion, starting with this, to get an L^p bound on $\nabla\omega$ for any p.

Let $B(r)$ be an embedded geodesic ball in (X, g_0). We let $\hat{\omega}$ denote the 2-form over $B(r)$ obtained by evaluating ω at the centre of the ball and extending over the ball by radial parallel transport along geodesics.

Lemma 3. *For any $c > 0$ there is a constant γ, depending on $K, \mathcal{P}, c$ such that if for all ρ-balls $B(\rho)$ (with $\rho < \rho_0$) we have*

$$\int_{B(\rho)} |\omega - \hat{\omega}|^2 \leq \gamma \rho^4$$

then the hypothesis of Lemma 2 is satisfied, i.e.

$$\int_{B(r)} |\nabla \omega|^2 \leq cr^2$$

for all $r < r_0(\rho_0, K, \mathcal{P})$.

As before, we work in small balls centred on a given point in X and standard coordinates about this point. There is no loss in supposing that the metric g_0 is the Euclidean metric in these coordinates: then the form ω_0 is just obtained by freezing the coefficients at the centre of the ball. Let v be a constant vector field in these Euclidean co-ordinates. Fix a cut-off function β equal to 1 on $[0, 1/2]$ and supported in $[0, 1)$ and let β_r be the function on X given by $\beta_r(x) = \beta(|x|/r)$ in these Euclidean coordinates. Set

$$I(r) = \int_{B(r)} \beta_r \omega_v \wedge \omega_v \,.$$

As in the proof of Lemma 1, it suffices to show that, when r is small, $I(r) \leq cr^2$ (for a different constant c). Suppose first that $r = 1$ and write $\omega = \hat{\omega} + \eta$, so $\|\eta\|^2_{L^2(B(1))} \leq \gamma$. By an L^2 version of the Poincaré inequality, as in [12], we can find a 1-form α over the ball with $\eta = d\alpha$ and $\|\nabla \alpha\|^2_{L^2(B(1))} \leq C\gamma$ for some fixed constant C. Now we have $\omega_v = \eta_v$, since the form $\hat{\omega}$ is constant in these coordinates, and $\omega_v = d\alpha_v$, where $\|\alpha_v\|^2_{L^2(B(1))} \leq C\gamma$. Then

$$I(1) = \int_{B(1)} \beta_1 \omega_v \wedge d\alpha_v = \int_{B(1)} d\beta_1 \wedge \omega_v \wedge \alpha_v \,,$$

so

$$I(1) \leq \|d\beta_1\|_{L^\infty} \|\omega_v\|_{L^2(B(1))} \|\alpha_v\|_{L^2(B(1))} \,.$$

Using the L^2 bound on α_v and the fact that the L^2 norm of ω_v on $B(1)$ is controlled by $\sqrt{I(2)}$ we obtain

$$I(1) \leq C\sqrt{\delta}\sqrt{I(2)}.$$

For general r, we scale the $2r$ ball to the unit ball and apply the same argument. Taking account of the scaling behaviour of the various quantities involved one obtains an inequality

$$I(r) \leq (C\sqrt{\gamma})r\sqrt{I(2r)}\,.$$

It is elementary to show that this implies the desired decay condition on $I(r)$ as r tends to 0. If we put $J(r) = r^{-2}I(r)$ then $J(r) \leq 2C\sqrt{\gamma}\sqrt{J(2r)}$. So if $L_n = \log J(2^{-n})$ we have

$$L_{n+1} \leq \frac{L_n}{2} + \sigma$$

where $\sigma = \log(2C\sqrt{\gamma})$. This gives

$$L_n \leq 2^{-n}(L_0 - \sigma) + 2\sigma\,.$$

Combined with the global *a priori* bound of Lemma 1, which controls L_0, this yields the desired result.

To sum up we have

Proposition 2. *Suppose $\mathcal{P}$ is a constraint manifold with negative tangents and $K > 0$. There is a $\gamma = \gamma(\mathcal{P}, K)$ with the following property. For each k, p, r_0 there is a constant $C = C(k, K, \mathcal{P}, p, r_0)$ such that if ω is any solution of (1) with $\|\omega\|_{L^\infty} \leq K$ and with*

$$\int_{B(r)} |\omega - \hat{\omega}|^2 \leq \gamma r^4$$

for all $r \leq r_0$ then $\|\omega\|_{L_k^p} \leq C$.

For $k = 1$ this is a combination of the two preceding Lemmas. The extension to higher derivatives follows from a straightforward bootstrapping argument.

Of course the hypotheses of Proposition 2 are satisfied if ω is bounded in L^∞ and has any fixed modulus of continuity, such as a Holder bound.

5. Discussion

5.1. *Motivation from symplectic topology*

We will outline the, rather speculative, possibility of applications of these ideas to questions in symplectic topology. Recall that the fundamental topological invariants of a symplectic form ω on a compact 4-manifold X are the first Chern class $c_1 \in H^2(X; \mathbf{Z})$ and the de Rham class $[\omega] \in H^2(X; \mathbf{R})$.

Question 1. *Suppose X is a compact Kähler surface with Kähler form ω_0. If ω is any other symplectic form on X, with the same Chern class and with $[\omega] = [\omega_0]$, is there a diffeomorphism f of X with $f^*(\omega) = \omega_0$?*

(McMullen and Taubes [7] have given examples of inequivalent symplectic structures on the same differentiable 4-manifold, but their examples have different Chern classes.)

A line of attack on this could run as follows. Suppose first that $b_+^2(X) = 1$. Given a general symplectic form ω we choose a unimodular constraint manifold $\mathcal{P}_1$ containing it and deform $\mathcal{P}_1$ through a 1-parameter family $\mathcal{P}_t$ for $t \in [0,1]$ to a standard Calabi–Yau constraint $\mathcal{P}_0$ containing ω_0. We suppose that $\mathcal{P}_t$ are all unimodular, with the same fixed volume form. Then we choose the cohomological constraint by taking $C = 0$ and $H_+^2 = \langle \omega \rangle$. Thus for any t the cohomology class of a solution is forced to be the fixed class $[\omega_0]$. At $t = 0$ we know that the solution ω_0 is unique. If we imagine that we have obtained suitable *a priori* estimates for solutions in the whole family of problems it would follow from the deformation result, Proposition 1, that there is a unique solution ω_t for each t, varying smoothly with t, and $\omega = \omega_1$. Then Moser's theorem would imply that ω and ω_0 are equivalent forms.

This strategy can be extended to the situation where $b_+^2 > 1$. We take $H_+^2 = \langle \omega_0 \rangle + H^{2,0}$ where $H^{2,0}$ consists of the real parts of the holomorphic 2-forms. There are two cases. If the Chern class c_1 vanishes in $H^2(X; \mathbf{R})$ then for any $\theta \in H^{2,0}$ and $s \in \mathbf{R}$ the form $\theta + s\omega_0$ is symplectic, provided θ and s are not both zero . (For the zero set of θ is either empty or a nontrival complex curve in X, and the second possibility is excluded by the assumption on c_1.) If c_1 does not vanish then the form is symplectic provided that $s \neq 0$. We follow the same procedure as before and (assuming the *a priori* estimates) construct a path ω_t from ω_0 to $\omega = \omega_1$ with $[\omega_t] = s_t[\omega_0] + [\theta_t]$ for $\theta_t \in H^{2,0}$. Now $[\omega_t]$ is not identically zero, so in the case $c_1 = 0$ we have a family of "standard" symplectic forms

$$\tilde{\omega}_t = s_t\omega_0 + \theta_t,$$

with $[\tilde{\omega}_t] = [\omega_t$. Then a version of Moser's Theorem yields a family of diffeomorphisms f_t with $f_t^*(\tilde{\omega}_t) = \omega_t$. Since, by hypothesis, $\tilde{\omega}_1 = \omega_0$ and $\omega_1 = \omega$, the diffeomorphism f_1 solves our problem. When $c_1 \neq 0$ the same argument works provided we know that s_t is never 0. But here we can use one of the deep results of Taubes [11]. If s_t were zero then there would be a class $\theta = \theta_t$ in $H^{2,0}$ which is the class of a symplectic form with the same first Chern class $c_1(X)$. But Taubes' inequality would give $c_1.[\theta] < 0$ whereas in our case $c_1.[\theta] = 0$ since c_1 is represented by a form of type (1,1) and θ has type $(2,0) + (0,2)$.

5.2. *The almost-complex case*

We have seen that, even within the standard framework of the Calabi–Yau problem on complex surfaces, solutions can blow up. In that case, everything can be understood in terms of the class $[\omega]$ and the Kahler cone. These difficulties become more acute in the nonintegrable situation. If J_0 is any almost complex structure on a 4-manifold X we can construct a 1-parameter family J_t such that there is no symplectic form, in any cohomology class, compatible with ω_1. (This is in contrast with the integrable case, where deformations of a Kahler surface are Kahler.) To do this we can simply deform J_0 in a small neighbourhood of a point so that J_1 admits a null homologous pseudo-holomorphic curve. More generally we could consider null-homologous currents T whose (1,1) part is positive. Thus if we form a 1-parameter family of constraints $\mathcal{P}_t$ using these J_t, and any fixed volume form, solutions ω_t must blow up sometime before $t = 1$, however we constrain the cohomology class $[\omega_t]$. It seems plausible that solutions blow-up at the first time t when a null-homologous (1,1)-current T appears, and become singular along the support of the T. It is a result of Sullivan [10] that the nonexistence of such currents implies the existence of a symplectic form *taming* the almost-complex structure. These considerations lead us to formulate a tentative conjecture.

Conjecture 1. *Let X be a compact 4-manifold and let Ω be a symplectic form on X. If $\mathcal{P}$ is a constraint manifold defined by an almost-complex structure which is tamed by Ω, and any smooth volume form, then there are C^∞ a priori bounds on a closed form $\omega \subset \mathcal{P}$ with $[\omega] = [\Omega]$.*

(By the results of Weinkove [13] it suffices to obtain L^∞ bounds on ω.)

This conjecture is relevant to the following problem.

Question 2. *If J is an almost-complex structure on a compact 4-manifold which is tamed by a symplectic form, is there a symplectic form compatible with J?*

If Conjecture 1 were true it would imply, by a simple deformation argument, an affirmative answer to Question 2 in the case when $b_+^2 = 1$, see the discussion in [13]. Such a result would be of interest even in the integrable case. It is a well-known fact that any compact complex surface with b_+^2 odd is Kahler. This was originally obtained from the classification theory, but more recently direct proofs have been given [1], [6]. Harvey and Lawson showed that any surface with b_+^2 odd admits a symplectic form taming the complex structure ([5], Theorem 26 and page 185), so a positive answer to the question above would yield another proof of the Kahler property, in the case when $b_+^2 = 1$. (Turning things around, one might hope that the techniques used in [1], [6] could have some bearing on Conjecture 1.)

In connection with Question 2, note first that this is special to 4-dimensions. In higher dimensions a generic almost-complex structure does not admit any compatible symplectic stucture, even locally. In another direction, the answer is known to be affirmative in the case when the taming form is the standard symplectic form on $\mathbf{CP}^2$, by an argument of Gromov. This constructs a compatible form by averaging over the currents furnished by pseudoholomorphic spheres.

5.3. *Hyperkahler structures*

Recall that complex-symplectic and hyperkahler structures on 4-manifolds can be described by, respectively, pairs and triples of orthonormal closed 2-forms. We can ask

Question 3. *Let X be a compact oriented 4-manifold*

- *Suppose there are closed two-forms θ_1, θ_2 on X such that each point θ_i span a positive 2-plane in Λ^2. Does X admit a complex-symplectic structure?*
- *Suppose there are closed two-forms $\theta_1, \theta_2, \theta_3$ on X such that at each point θ_i span a positive 3-plane in Λ^2. Does X admit a hyperkahler structure?*

The hypotheses are equivalent to saying that the symmetric matrices (2×2 and 3×3 in the two cases) $\theta_i \wedge \theta_j$ are positive definite, and the question asks whether we can find another choice of closed forms $\tilde{\theta}_i$ to make the corresponding matrix $\tilde{\theta}_i \wedge \tilde{\theta}_j$ the identity.

For simplicity we will just discuss the second version of the question, for triples of forms and in the case when X is simply connected. The hypotheses imply that the symplectic structure θ_1, say, has first Chern class zero, and a result of Morgan and Szabo [8] tells us that $b_+^2(X) = 3$. Thus the θ_i generate a maximal positive subspace H_+^2. For reasons that will appear

soon we make an additional auxiliary assumption. We suppose that there
is an involution $\sigma : X \to X$ with

$$\sigma^*(\theta_1) = \theta_1 \ , \ \sigma^*(\theta_2) = -\theta_2 \ , \ \sigma^*(\theta_3) = -\theta_3.$$

(A model case of this set-up is given by taking X to be a K3 surface double-
covering the plane, with σ the covering involution.) We show that under
this extra assumption the truth of Conjecture 1 would imply a positive
answer to our question. To see this we let U_0 be the orthogonal complement
of θ_1 in the span of the θ_i, so U_0 defines an almost-complex structure
on X, compatible with θ_1. We can chooose a smooth homotopy U_t to
the subbundle $U_1 = \mathrm{Span}(\theta_2, \theta_3)$ giving a 1-parameter family of almost-
complex structures, all tamed by θ_1. We choose a 1-parameter family of
volume forms equal to ω_1^2 when $t = 0$ and to ω_2^2 when $t = 1$. We make all
these choices σ-invariant. Thus we get a 1-parameter family of constraints
$\mathcal{P}_t$ and seek ω_t with $[\omega_t]$ in H_+^2. The form θ_1 gives a solution at time 0, so
$\omega_0 = \theta_1$. The σ-invariance of the problem means that in this case we can fix
the cohomology class of ω_t to be a multiple of the taming form θ_1 and still
solve the local deformation problem. Thus the hypotheses of Conjecture 1
are fulfilled so, assuming the conjecture, we can continue the solution over
the whole interval to obtain an ω_1 which satisfies

$$\omega_1^2 = \theta_2^2 \ , \ \omega_1 \wedge \theta_2 = \omega_1 \wedge \theta_3 = 0.$$

Now ω_1, θ_2 define a complex-symplectic stucture on X, and it well known
that under our hypotheses X must be a K3 surface and hence hyperkahler.

The only purpose of the involution σ in this argument is to fix the
cohomology class of the form in the 1-parameter family. Of course one could
hope that a suitable extension of the ideas could remove this assumption,
but our discussion is only intended to illustrate how these ideas might
possibly be useful.

5.4. *Relation with Gromov's theory*

We have seen that there is a "Calabi–Yau" constraint manifold $\mathcal{P}$ associated to an almost complex-structure J (and volume form) on a 4-manifold.
On the other hand, we have the notion of a "J-holomorphic" curve, with
renowned applications in global symplectic geometry due to Gromov and
others. These ideas can be related, and extended, as we will now explain.
Consider the Grassmannian $Gr_2(\mathbf{R}^4)$ of oriented 2-planes. It can be identified with the space of null rays in $\Lambda^2 \mathbf{R}^4$ by the map which takes a rank two

2-form to its kernel. Thus, choosing a Euclidean metric on $\mathbf{R}^4$ and hence a decomposition into self-dual and anti-self dual forms, the Grassmannian is identified with pairs (ω_+, ω_-) where $|\omega_+|^2 = |\omega_-|^2 = 1$, or in other words with the product $S_+^2 \times S_-^2$ of the unit spheres in the 3-dimensional vector spaces $\Lambda_\pm^2$. Now there is a canonical conformal structure of signature $(2, 2)$ on $Gr_2(\mathbf{R}^4)$, induced by the wedge product form and in [4] Gromov considers embedded surfaces $T \subset Gr_2(\mathbf{R}^4)$ on which the conformal structure is negative definite. A prototype for such a surface is given by $T(\omega_+) = \{\omega_+\} \times S_-^2$ for some fixed ω_+ in S_+^2. This is just the set of 2-planes which are complex subspaces with respect to the almost-complex structure defined by ω_+ (with $\Lambda^{2,0}$ the orthogonal complement of ω_+ in Λ_+^2). On the other hand, Gromov shows that *any* surface T leads to an elliptic equation generalising that defining holomorphic curves in $\mathbf{C}^2$. Slightly more generally still, let X be an oriented 4-manifold and form the bundle $Gr_2(X)$ of Grassmannians of 2-planes in the tangent spaces of X. Let $\mathcal{T}$ be a 6-dimensional submanifold of the Grassman bundle, fibering over X with each fibre $\mathcal{T}_x$ a "negative" surface in the above sense. Then we have the notion of a "T-pseudoholomorphic curve" in X: an immersed surface whose tangent spaces lie in $\mathcal{T}$. We say that a symplectic form Ω on X *tames* $\mathcal{T}$ if $\Omega(H) > 0$ for every subspace H in every $\mathcal{T}_x$. Then Gromov explains that all the fundamental results about J-holomorphic curves in symplectic manifolds extend to this more general context.

We will now relate this discussion to the rest of this article. Consider a negative submanifold $T \subset Gr_2(\mathbf{R}^4)$ as before and a map $f : [0, \infty) \times T \to \Lambda^2 \mathbf{R}^4$ with $f(r, \theta) = O(r^{-1})$, along with all its derivatives. Call the image of the map $(r, \theta) \mapsto r\theta + f(r, \theta)$ the f-deformed cone over T. Then we say that a 3-dimensional submanifold $P \subset \Lambda^2 \mathbf{R}^4$ is asymptotic to T if there is a map f as above such that, outside compact subsets, P coincides with the f-deformed cone over T. This notion immediately generalises to a pair of constraint manifolds $\mathcal{P}, \mathcal{T}$ over a 4-manifold. The prototype of this picture is to take the Calabi–Yau constraint defined by a volume form and almost-complex structure, which one readily sees is asymptotic to the submanifold of complex subspaces. Now suppose that $\mathcal{P}$ lies in the positive cone with respect to the wedge-product form, has negative chords and is asymptotic to $\mathcal{T}$. Then if $\omega \in \mathcal{P}_x$ and $\theta \in \mathcal{T}_x$ we have $(\omega - r\theta)^2 \leq O(r^{-1})$ for large r. Since $\theta^2 = 0$ and $\omega^2 > 0$ we see that $\omega \wedge \theta > 0$, which is the same as saying that ω tames $\mathcal{T}$. In other words, just as in the Calabi–Yau case, a necessary condition for there to be a solution $\omega \subset \mathcal{P}$ in a given cohomology class $h = [\omega]$ is that there is a taming form in h for $\mathcal{T}$. It is tempting then to extend Conjecture 1 to this more general situation.

Conjecture 2. *Let X be a compact 4-manifold and let Ω be a symplectic form on X. If $\mathcal{P}$ is a unimodular constraint manifold with negative chords which is asymptotic to $\mathcal{T}$, where $\mathcal{T}$ is tamed by Ω, then there are C^∞ a priori bounds on a closed form $\omega \subset \mathcal{P}$ with $[\omega] = [\Omega]$.*

(There is little hard evidence for the truth of this, so perhaps it is better considered as a question. Of course, by our results in Section 4 above, it suffices to obtain an L^∞ bound and a modulus of continuity in the "BMO sense" of Proposition 2. Going a very small way in this direction, it is easy to show that a taming form for $\mathcal{T}$ leads to an *a priori* L^1 bound on ω.)

5.5. *A counterexample*

We can use sophisticated results in symplectic topology to obtain a negative result — showing that Conjecture 2 cannot be extended to the case where $\mathcal{P}$ only has negative tangents.

Proposition 3. *There is a simply connected 4-manifold X with $b_+^2(X) = 1$ and a 1-parameter family of unimodular constraints $\mathcal{P}_t$ ($t \in [0, 1]$) on X with the following properties*

1. *For each t, $\mathcal{P}_t$ has negative tangents and is asymptotic to the manifold $\mathcal{T}$ associated with an almost-complex stucture on X which is tamed by a symplectic form Ω.*
2. *For $t < 1$ there is a closed 2-form $\omega_t \subset \mathcal{P}_t$ with $[\omega_t] \in \langle[\Omega]\rangle$.*
3. *The ω_t do not satisfy uniform C^∞ bounds as $t \to 1$.*

In fact we can take the almost-complex structure to be integrable, Ω to be a Kahler form, and arrange that, for each parameter value, $\mathcal{P}_t$ coincides with the standard Calabi–Yau constraint outside a compact set.

The proof of Proposition 3 combines some simple general constructions with results of Seidel. Let J be an almost-complex structure on a 4-manifold X and let ω be a symplectic form on X. Choose an almost-complex structure J' compatible with ω, and suppose that J and J' are homotopic, through a 1-parameter family $J(s)$. In other words, for each s we have a 2-dimensional positive subbundle $\Lambda^{2,0}(s)$. For convenience, suppose that $J(s) = J'$ for s close to 0 and $J(s) = J$ for s close to 1. For fixed small ϵ we define a subset of Λ^2 by

$$\mathcal{P}^* = \{\theta : \theta^2 = \omega^2, \theta \in \Lambda^{2,0}(\epsilon|\theta|)\},$$

where $|\theta|$ is the norm measured with respect to some arbitrary metric. This set $\mathcal{P}^*$ has two connected components, interchanged by $\theta \mapsto -\theta$, and it is easy to check that, if ϵ is sufficiently small, one of these components is a constraint manifold $\mathcal{P}$ with negative tangents, containing ω and equal to the Calabi–Yau constraint defined by J and ω^2 outside a compact set.

Now we can extend this construction to families. If ω_z is a family of symplectic forms parametrised by a compact space Z, and if the corresponding homotopy class of maps from Z to the space of almost-complex structures on X is trivial, then we can construct a family $\mathcal{P}_z$ of unimodular constraints, equal to the fixed Calabi–Yau constraint outside a compact set and with $\mathcal{P}_z$ containing ω_z. For our application we take z to be a circle, so we have a loop of symplectic forms ω_z, $z \in S^1$ with fixed cohomology class $h = [\omega_z]$. Clearly the family $\mathcal{P}_z$ for $z \in S^1$ can be extended over the disc D^2. So for each $z \in D^2$ we have a $\mathcal{P}_z$ satisfying the conditions of the Proposition. If the Proposition were false, then it would follow that we could extend the family of symplectic forms ω_z over the disc, using a simple continuity argument.

The space $\mathcal{S}_\omega$ of symplectic forms equivalent to ω can be identified with the quotient of the identity component of the full diffeomorphism group by the symplectomorphisms of (X, ω) which are isotopic to the identity. So the fundamental group of $\mathcal{S}_\omega$ can be identified with the classes of symplectomorphisms isotopic to the identity modulo symplectic isotopy. Now in [9] Seidel gives examples of Kähler manifolds (X, ω) satisfying our hypotheses and symplectomorphisms which are isotopic, but not symplectically isotopic, to the identity. Thus Seidel's results assert that there are maps from the circle to $\mathcal{S}_\omega$ which cannot be extended to the disc (although the corresponding almost-complex structures can be). This conflict with our previous argument completes the proof of Proposition 3.

The author has not yet succeeded in understanding more explicitly the blow-up behaviour which Proposition 3 asserts must occur. Seidel's symplectomorphisms are squares of generalised Dehn twists, associated to Lagrangian 2-spheres in (X, ω), and it is tempting to hope that the blow-up sets should be related to these spheres in some way. It should also be noted that Seidel's results depend crucially on fixing the cohomology class of the symplectic form.

References

1. N. Buchdahl, *On compact Kahler surfaces*, Ann. Inst. Fourier 49 (1999) 263–85.
2. S. K. Donaldson, *Moment maps and diffeomorphisms*, Asian Jour. Math. 3 (1999) 1–15.
3. S. K. Donaldson and D. P. Sullivan, *Quasiconformal four-manifolds*, Acta Math. 163 (1990) 181–252.
4. M. Gromov, *Pseudoholomorphic curves in symplectic manifolds*, Inventiones Math. 82 (1985) 307-47.
5. R. Harvey and H. B. Lawson, *An intrinsic characterisation of Kahler manifolds*, Inventiones Math. 74 (1983) 169–98.
6. A. Lamari, *Courants kahleriens et surfaces compactes*, Ann. Inst. Fourier 49 (1999) 263–85.
7. C. McMullen and C. H. Taubes, *4-manifolds with inequivalent symplectic forms and 3-manifolds with inequivalent fibrations*, Math. Res. Letters 6 (1999) 681–96.
8. J. W. Morgan and Z. Szabo, *Homotopy K3 surfaces and mod 2 Seiberg–Witten invariants*, Math. Res. Letters 4 (1997) 17–21.
9. P. Seidel, *Lectures on four-dimensional Dehn twists*, Math. Arxiv SG/030912.
10. D. P. Sullivan, *Cycles for the dynamical structure of foliated manifolds and complex manifolds*, Inventiones Math. (1976) 225–55.
11. C. H. Taubes, *The Seiberg–Witten and Gromov invariants*, Math. Res. Letters 2 (1995) 221–38.
12. K. K. Uhlenbeck, *Connections with L^p bounds on curvature*, Commun. Math. Phys. 83 (1982) 11–29.
13. B. Weinkove, *The Calabi–Yau equation on almost-complex four-manifolds*, Math. Arxiv DG/0604408, to appear in Journal of Differential Geometry.

Chapter 8

PARTIAL CONNECTION FOR p-TORSION LINE BUNDLES IN CHARACTERISTIC $p > 0$

Hélène Esnault[*]

Universität Duisburg-Essen, Mathematik, 45117 Essen, Germany
esnault@uni-due.de

August 7, 2006

To S. S. Chern, in memoriam

The aim of this brief note is to give a construction for p-torsion line bundles in characteristic $p > 0$ which plays a similar rôle as the standard connection on an n-torsion line bundle in characteristic 0.

1. Introduction

In [3] (see also [4]) we gave an algebraic construction of characteristic classes of vector bundles with a flat connection (E, ∇) on a smooth algebraic variety X defined over a field k of characteristic 0. Their value at the generic point $\mathrm{Spec}(k(X))$ was studied and redefined in [1], and then applied in [2] to establish a Riemann-Roch formula.

One way to understand Chern classes of vector bundles (without connection) is via the Grothendieck splitting principle: if the receiving groups $\oplus_n H^{2n}(X, n)$ of the classes form a cohomology theory which is a ring and is functorial in X, then via the Whitney product formula it is enough to define the first Chern class. Indeed, on the flag bundle $\pi : \mathrm{Flag}(E) \to X$, $\pi^*(E)$ acquires a complete flag $E_i \subset E_{i+1} \subset \pi^*(E)$ with E_{i+1}/E_i a line bundle, and $\pi^* : H^{2n}(X, n) \to H^{2n}(\mathrm{Flag}(E), n)$ is injective, so it is enough to construct the classes on $\mathrm{Flag}(E)$. However, if ∇ is a connection on E, $\pi^*(\nabla)$ does not stabilize the flag E_i. So the point of [3] is to show that there is a differential graded algebra $A^\bullet$ on $\mathrm{Flag}(E)$,

---

[*]Partially supported by the DFG Leibniz Preis.

together with a morphism of differential graded algebras $\Omega^\bullet_{\mathrm{Flag}(E)} \xrightarrow{\tau} A^\bullet$, so that $R\pi_* A^\bullet \cong \Omega^\bullet_X$ and so that the operator defined by the composition

$$\pi^*(E) \xrightarrow{\pi^*(\nabla)} \Omega^1_{\mathrm{Flag}(E)} \otimes_{\mathcal{O}_{\mathrm{Flag}(E)}} \pi^*(E) \xrightarrow{\tau \otimes 1} A^1 \otimes_{\mathcal{O}_{\mathrm{Flag}(E)}} \pi^*(E) \text{ stabilizes}$$

E_i. We call the induced operator $\nabla_i : E_i \to A^1 \otimes_{\mathcal{O}_{\mathrm{Flag}(E)}} E_i$ a (*flat*) τ-*connection*. So it is a k-linear map which fulfills the τ-*Leibniz formula*

$$(1.1) \qquad\qquad \nabla_i(\lambda \otimes e) = \tau d(\lambda) \otimes e + \lambda \nabla_i(e)$$

for λ a local section of $\mathcal{O}_{\mathrm{Flag}(E)}$ and e a local section of E_i. It is flat when $0 = \nabla_i \circ \nabla_i \in H^0(X, A^2 \otimes_{\mathcal{O}_X} \mathcal{E}nd(E))$, with the appropriate standard sign for the derivation of forms with values in E_i. The last point is then to find the correct cohomology which does not get lost under π^*. It is a generalization of the classically defined group

$$(1.2) \qquad\qquad \mathbb{H}^1(X, \mathcal{O}_X^\times \xrightarrow{\mathrm{dlog}} \Omega^1_X \xrightarrow{d} \Omega^2_X \xrightarrow{d} \cdots)$$

of isomorphism classes of rank one line bundles on X with a flat connection.

A typical example of such a connection is provided by a torsion line bundle: if L is a line bundle on X which is n-torsion, that is which is endowed with an isomorphism $L^n \cong \mathcal{O}_X$, then the isomorphism yields an $\mathcal{O}_X$-étale algebra structure on $\mathcal{A} = \oplus_0^{n-1} L^i$, hence a finite étale covering $\sigma : Y = \mathrm{Spec}_{\mathcal{O}_X} \mathcal{A} \to X$, which is a principal bundle under the group scheme μ_n of n-th roots of unity, thus is Galois cyclic as soon as $\mu_n(\mathbb{C}) \subset k^\times$. Since the μ_n-action commutes with the differential $d_Y : \mathcal{O}_Y \to \Omega^1_Y = \sigma^* \Omega^1_X$, it defines a flat connection $\nabla_L : L \to \Omega^1_X \otimes_{\mathcal{O}_X} L$. Concretely, if $g_{\alpha,\beta} \in \mathcal{O}_X^\times$ are local algebraic transition functions for L, with trivialization

$$(1.3) \qquad\qquad g_{\alpha,\beta}^n = u_\beta u_\alpha^{-1}, u_\alpha \in \mathcal{O}_X^\times,$$

then

$$(1.4) \qquad\qquad \left(g_{\alpha,\beta}, \frac{1}{n}\frac{du_\alpha}{u_\alpha}\right) \in (\mathcal{C}^1(\mathcal{O}_X^\times) \times \mathcal{C}^0((\Omega^1_X)_{\mathrm{clsd}}))_{\mathrm{dlog}-\delta}$$

$$\frac{dg}{g} = \delta\left(\frac{du}{u}\right)$$

is a Cech cocyle for the class

$$(1.5) \qquad\qquad (L, \nabla_L) \in \mathbb{H}^1(X, \mathcal{O}_X^\times \xrightarrow{\mathrm{dlog}} \Omega^1_X \xrightarrow{d} \Omega^2_X \xrightarrow{d} \cdots).$$

Clearly (1.4) is meaningless if the characteristic p of k is positive and divides n. The purpose of this short note is to give an Ersatz of this canonical construction in the spirit of the τ-connections explained above when p divides n.

2. A Partial Connection for p-Torsion Line Bundles

Let X be a scheme of finite type over a perfect field k of characteristic $p > 0$. Let L be a n-torsion line bundle on X, thus endowed with an isomorphism

$$(2.1) \qquad \qquad \theta : L^n \cong \mathcal{O}_X.$$

Then θ defines an $\mathcal{O}_X$-algebra structure on $\mathcal{A} = \oplus_0^{n-1} L^i$ which is étale if and only if $(p, n) = 1$. It defines the principal μ_n-covering

$$(2.2) \qquad \qquad \sigma : Y = \mathrm{Spec}_{\mathcal{O}_X} \mathcal{A} \to X$$

which is étale if and only if $(p, n) = 1$, else decomposes into

$$(2.3) \qquad \qquad \sigma : Y \xrightarrow{\iota} Z \xrightarrow{\sigma'} X$$

with σ' étale and ι purely inseparable. More precisely, if $n = m \cdot p^r$, $(m, p) = 1$, and $M = L^{p^r}$, θ defines an $\mathcal{O}_X$-étale algebra structure on $\mathcal{B} = \oplus_0^{m-1} M^i$, which defines $\sigma' : Z = \mathrm{Spec}_{\mathcal{O}_X} \mathcal{B} \to X$ as an (étale) μ_m-principal bundle. The isomorphism θ also defines an isomorphism $(L')^{p^r} \cong \mathcal{O}_Z$ as it defines the isomorphism $(\sigma')^*(M) \cong \mathcal{O}_Z$, where $L' = (\sigma')^*(L)$. So $\mathcal{C} = \oplus_0^{p^r-1} (L')^i$ becomes a finite purely inseparable $\mathcal{O}_Z$-algebra defining the principal μ_{p^r}-bundle $\iota : Y = \mathrm{Spec}_{\mathcal{O}_Z} \mathcal{C} \to Z$.

If $(n, p) = 1$, that is if $r = 0$, the formulae (1.3), (1.4) define (L, ∇) as in (1.5). We assume from now on that $(n, p) = p$. Then, as is well known, as a consequence of (1.3) one sees that the form

$$(2.4) \qquad \qquad \omega_L := \frac{du_\alpha}{u_\alpha} \in \Gamma(X, \Omega_X^1)_{\mathrm{clsd}}^{\mathrm{Cartier}=1}$$

is globally defined and Cartier invariant. Let e_α be local generators of L, with transition functions $g_{\alpha,\beta}$ with $e_\alpha = g_{\alpha,\beta} e_\beta$. The isomorphism θ yields a trivialization

$$(2.5) \qquad \qquad \sigma^* L \cong \mathcal{O}_Y$$

thus local units v_α on Y with

$$(2.6) \qquad \qquad v_\alpha \in \mathcal{O}_Y^\times, \ g_{\alpha,\beta} = v_\beta v_\alpha^{-1}$$
$$\text{so that } 1 = v_\alpha \sigma^*(e_\alpha) = v_\beta \sigma^*(e_\beta).$$

Definition 2.1. One defines the $\mathcal{O}_X$-coherent sheaf Ω_L^1 as the subsheaf of $\sigma_* \Omega_Y^1$ spanned by $\mathrm{Im}(\Omega_X^1)$ and $\frac{dv_\alpha}{v_\alpha}$.

Lemma 2.2. Ω_L^1 *is well defined and one has the exact sequence*

$$(2.7) \qquad 0 \to \mathcal{O}_X \xrightarrow{\cdot \omega_L} \Omega_X^1 \xrightarrow{\sigma^*} \Omega_L^1 \xrightarrow{s} \mathcal{O}_X \to 0$$

$$s\left(\frac{dv_\alpha}{v_\alpha}\right) = 1 \,.$$

Proof. The relation (2.6) implies

$$(2.8) \qquad \frac{dg_{\alpha,\beta}}{g_{\alpha,\beta}} = \frac{dv_\beta}{v_\beta} - \frac{dv_\alpha}{v_\alpha}$$

$$\text{so} \quad \frac{dv_\beta}{v_\beta} \equiv \frac{dv_\alpha}{v_\alpha} \in \sigma_* \Omega_Y^1 / \mathrm{Im}(\Omega_X^1) \,.$$

Hence the sheaf Ω_L^1 is well defined. If e_α' is another basis, then one has $e_\alpha = w_\alpha e_\alpha'$ for local units $w_\alpha \in \mathcal{O}_X^\times$. The new v_α are then multiplied by local units in $\mathcal{O}_X^\times$, so the surjection s is well defined. It remains to see that $\mathrm{Ker}(\sigma^*) = \mathrm{Im}(\cdot \omega_L)$. By definition, on the open of X on which L has basis e_α, one has

$$(2.9) \qquad Y = \mathrm{Spec}\, \mathcal{O}_X[v_\alpha]/(v_\alpha^n - u_\alpha).$$

This implies $\Omega_Y^1 = \langle \mathrm{Im}(\Omega_X^1), dv_\alpha \rangle_{\mathcal{O}_Y} / \langle du_\alpha \rangle_{\mathcal{O}_Y}$ on this open and finishes the proof. $\square$

Remarks 2.3.

1) Assume for example that X is a smooth projective curve of genus g, and $n = p$. Recall that $0 \neq \omega_L \in \Gamma(X, \Omega_X^1)$. In particular, if $g \geq 2$, necessarily $0 \neq \Omega_X^1 / \mathcal{O}_X \cdot \omega_L$ is supported in codimension 1. So Ω_L^1 contains a non-trivial torsion subsheaf.

2) The sheaf Ω_L^1 lies in $\sigma_* \Omega_Y^1$ but is not equal to it. Indeed, on the smooth locus of X (assuming X is reduced) the torsion free quotient of Ω_L^1 has rank equal to the dimension of X, while $\sigma_* \Omega_Y^1$ has rank $n \cdot \text{dimension } (X)$ on the étale locus of σ (which is non-empty if L itself is not a p-power line bundle).

3) The class in $\mathrm{Ext}_{\mathcal{O}_X}^2(\mathcal{O}_X, \mathcal{O}_X) = H^2(X, \mathcal{O}_X)$ defined by (2.7) vanishes. Indeed, let us decompose (2.7) as an extension of $\mathcal{O}_X$ by $\Omega_X^1 / \mathcal{O}_X \cdot \omega_L$, followed by an extension of $\Omega_X^1 / \mathcal{O}_X \cdot \omega_L$ by $\mathcal{O}_X \cdot \omega_L$. The first extension class in $H^1(X, \Omega_X^1 / \mathcal{O}_X \cdot \omega_L)$ has cocycle $\frac{dv_\beta}{v_\beta} - \frac{dv_\alpha}{v_\alpha} = \frac{dg_{\alpha,\beta}}{g_{\alpha,\beta}}$ (see (2.8)), thus is the image of the Atiyah class of L in $H^1(X, \Omega_X^1)$. Thus the second boundary to $H^2(X, \mathcal{O}_X)$ dies.

Definition 2.4. We set $\Omega_L^0 := \mathcal{O}_X$ and for $i \geq 1$ we define the $\mathcal{O}_X$-coherent sheaf Ω_L^i as the subsheaf of $\sigma_* \Omega_Y^i$ spanned by $\mathrm{Im}(\Omega_X^i)$ and $\frac{dv_\alpha}{v_\alpha} \wedge \mathrm{Im}(\Omega_X^{i-1})$.

Proposition 2.5. *The sheaf Ω_L^i is well defined. One has an exact sequence*

$$(2.10) \qquad 0 \to \omega_L \wedge \Omega_X^{i-1} \to \Omega_X^i \xrightarrow{\sigma^*} \Omega_L^i \xrightarrow{s} \Omega_X^{i-1} \to 0$$

$$s\left(\frac{dv_\alpha}{v_\alpha} \wedge \beta\right) = \beta.$$

Furthermore, the differential $\sigma_*(d_Y)$ on $\sigma_* \Omega_Y^\bullet$ induces on $\oplus_{i \geq 0} \Omega_L^i$ the structure of a differential graded algebra $(\Omega_L^\bullet, d_L)$ so that $\sigma^* : (\Omega_X^\bullet, d_X) \to (\Omega_L^\bullet, d_L)$ is a morphism of differential graded algebras.

Proof. One proves (2.10) as one does (2.7). One has to see that $\sigma_*(d_Y)$ stabilizes $\Omega_L^\bullet$. As $0 = d_X(\omega_L) \in \Omega_X^2$, $0 = d_Y(\frac{dv_\alpha}{v_\alpha}) \in \sigma_* \Omega_Y^2$, (2.10) extends to an exact sequence of complexes

$$(2.11) \quad 0 \to (\omega_L \wedge \Omega_X^{\bullet-1}, -1 \wedge d_X) \to (\Omega_X^\bullet, d_X) \xrightarrow{\sigma^*}$$

$$(\Omega_L^\bullet, d_L) \xrightarrow{s} (\Omega_X^{\bullet-1}, -d_X) \to 0.$$

This finishes the proof. $\qquad\qquad\qquad\qquad\qquad\qquad\qquad\qquad\qquad\qquad$ □

Remark 2.6. As $\frac{dg_{\alpha,\beta}}{g_{\alpha,\beta}} \in (\Omega_X^1)_{\mathrm{clsd}}$ the same proof as in Remark 2.3, 3) shows that the extension class $\mathrm{Ext}^2(\Omega_X^{\bullet-1}, \omega_L \wedge \Omega_X^{\bullet-1})$ defined by (2.11) dies.

In order to tie up with the notations of the Introduction, we set

$$(2.12) \qquad\qquad \tau = \sigma^* : \Omega_X^\bullet \to \Omega_L^\bullet.$$

Proposition 2.7. *The formula $\nabla(e_\alpha) = -\frac{dv_\alpha}{v_\alpha} \otimes e_\alpha \in \Omega_L^1 \otimes_{\mathcal{O}_X} L$ defines a flat τ-connection ∇_L on L. So (L, ∇_L) is a class in $\mathbb{H}^1(X, \mathcal{O}_X^\times \xrightarrow{\tau d \log} \Omega_L^1 \xrightarrow{d_L} \Omega_L^2 \xrightarrow{d_L} \cdots)$, the group of isomorphism classes of line bundles with a flat τ-connection.*

Proof. Formula (2.6) implies that this defines a τ-connection. Flatness is obvious. A Cech cocycle for (L, ∇_L) is $(g_{\alpha,\beta}, \frac{dv_\alpha}{v_\alpha})$. $\qquad\qquad\qquad$ □

Remarks 2.8.

1) The same formal definitions 2.1 and 2.4 of $\Omega_L^\bullet$ when $(n, p) = 1$ yield $(\Omega_L^\bullet, d_L) = (\Omega_X^\bullet, d_X)$, and the flat τ-connection becomes the flat connection defined in (1.4) and (1.5). So Proposition 2.7 is a direct generalization of it.

2) Let X be proper reduced over a perfect field k, irreducible in the sense that $H^0(X, \mathcal{O}_X) = k$, and admitting a rational point $x \in X(k)$. A generalization of torsion line bundles to higher rank bundles is the notion of Nori finite bundles, that is bundles E which are trivialized over principal bundle $\sigma : Y \to X$ under a finite flat group scheme G (see [6] for the original definition and also [5] for a study of those bundles). So for the n-torsion line bundles considered in this section, $G \cong \mu_n$. If the characteristic of k is 0, then again σ is étale, the differential $d_Y : \mathcal{O}_Y \to \sigma^* \Omega^1_X = \Omega^1_Y$ commutes with the action of G, inducing a connection $\nabla_E : E \to \Omega^1_X \otimes_{\mathcal{O}_X} E$ and characteristic classes in our groups $\mathbb{H}^i(X, \mathcal{K}^m_i \xrightarrow{d\log} \Omega^i_X \xrightarrow{d} \Omega^{i+1}_X \cdots)$ (see [3]). If the characteristic of k is $p > 0$, then σ is étale if and only if G is smooth (which here means étale), in which case one can also construct those classes. If G is not étale, thus contains a non-trivial local subgroup-scheme, then one should construct as in Proposition 2.5 a differential graded algebra $(\Omega^\bullet_E, d_E)$ with a map $(\Omega^\bullet_X, d_X) \xrightarrow{\tau} (\Omega^\bullet_E, d_E)$, so that E is endowed naturally with a flat τ-connection $\nabla_E : E \to \Omega^1_E \otimes_{\mathcal{O}_X} E$. The techniques developed in [3] should then yield classes in the groups $\mathbb{H}^i(X, \mathcal{K}^m_i \xrightarrow{\tau d\log} \Omega^i_E \xrightarrow{d_E} \Omega^{i+1}_E \cdots)$.

References

[1] Bloch, S., Esnault, H.: Algebraic Chern-Simons theory, Am. J. of Mathematics **119** (1997), 903–952.

[2] Bloch, S., Esnault, H.: A Riemann-Roch theorem for flat bundles, with values in the algebraic Chern–Simons theory, Annals of Mathematics **151** (2000), 1–46.

[3] Esnault, H.: Algebraic Differential Characters, in Regulators in Analysis, Geometry and Number Theory, Progress in Mathematics, Birkhäuser Verlag, **171** (2000), 89–117.

[4] Esnault, H.: Characteristic classes of flat bundles and determinant of the Gauß–Manin connection, Proceedings of the International Congress of Mathematicians, Beijing 2002, Higher Education Press, 471–483.

[5] Esnault, H., Hai P. H., Sun, X.: On Nori's Fundamental Group Scheme, preprint 2006, 29 pages.

[6] Nori, M.: The fundamental group scheme, Proc. Indian Acad. Sci. **91** (1982), 73–122.

Chapter 9

ALGEBRAIC CYCLES AND SINGULARITIES OF
NORMAL FUNCTIONS, II

Mark Green

Institute for Pure and Applied Mathematics, UCLA, USA

Phillip Griffiths

Institute for Advanced Study, Princeton, USA

June 13, 2006

In our previous paper [14], denoted by "I", and whose notations we shall follow here, we proposed a definition of extended normal functions (ENF), and for an ENF ν we defined its singular locus sing ν. There is a reciprocal relationship between primitive Hodge classes $\zeta \in Hg^n(X)_{\mathrm{prim}}$ and the corresponding ENF ν_ζ. Moreover, algebraic cycles Z with $[Z] = \zeta$ give rise to singularities of ν_ζ and vice versa.

In this paper we shall discuss universally defined rational maps ρ to partially compactified classifying spaces for polarized Hodge structures of odd weight and paritally compactified universal families of intermediate Jacobians over these spaces. These maps have the property that sing ν_ζ appears as a component of the inverse image of certain boundary components $\mathcal{B}$; i.e.,

$$\mathrm{sing}\,\nu_\zeta \subseteq \rho^{-1}(\mathcal{B}) \ .$$

Since the Hodge conjecture (HC) is equivalent to

$$\mathrm{sing}\,\nu_\zeta \neq \emptyset \ \text{for} \ L \gg 0 \ ,$$

and since for the fundamental class $[\mathcal{B}]$ of $\mathcal{B}$

$$\rho^*([\mathcal{B}]) \neq 0 \Rightarrow \rho^{-1}(\mathcal{B}) \neq \emptyset \ ,$$

a natural question is to investigate $\rho^*([\mathcal{B}])$. We will see that this leads to attempting to understand $\dim(\mathrm{sing}\,\nu_\zeta)$, and this in turn leads to some to us very beautiful interplay between algebraic geometry and Hodge theory

179

"at the boundary." Among other things, we find that there is transversality for $L \gg 0$ in the classical case (Lefschetz $(1,1)$ theorem), but assuming the HC there *cannot* be transversality in the higher codimensional situation. In part this lack of transversality is Hodge-theoretic, due to the infinitesimal period relation and with a very nice geometric interpretation. Additionally there is a non-zero algebro-geometric contribution to non-transversality, a contribution that we do not yet understand geometrically.

OUTLINE

I. Introduction

I(i). *Notations*

This is a continuation of [14], referred to below as "I". We shall retain the notations introduced there, and which will be recalled as needed. In addition we shall use the following notation:

- $\mathcal{M}_g$ is the moduli space of stable curves of genus g and $\overline{\mathcal{M}}_g$ is the Deligne-Mumford compactification.
- $\mathcal{C}_g \to \mathcal{M}_g$ is the universal family of genus g curves with compactification $\overline{\mathcal{C}}_g \to \overline{\mathcal{M}}_g$, with the usual understanding that these spaces must be interpreted as stacks.
- $\mathcal{A}_{\mathbf{h}}$ is the classifying space for polarized Hodge structures of weight $2n-1$ and with given Hodge numbers $\mathbf{h} = (h^{2n-1,0}, \dots, h^{n,n-1})$. When $n = 1$ and $h^{1,0} = g$, we assume that the polarization is principal and use the customary notation $\mathcal{A}_g$.
- $\mathcal{J}_{\mathbf{h}} \to \mathcal{A}_{\mathbf{h}}$ is the associated universal family of polarized complex tori, again with the stack interpretation. When $n = 1$ we shall write $\mathcal{J}_g \to \mathcal{A}_g$.
- For a choice of fan Σ, $\overline{\mathcal{A}}_{\mathbf{h},\Sigma}$ will denote the corresponding partial compactification (cf. Kato-Usui [17] and for $n = 1$ cf. Carlslon-Cattani-Kaplan [4], Cattani [5], Alexeev-Nakamura [2], Alexeev [1] and the references cited there; cf. also the note at the end of (I(i)).
- For $\mathbf{g} = (g_1, g_2)$ where

$$g = g_1 + g_2 + l - 1 \qquad (l \geqq 1)$$

we shall denote by $\Sigma_{\mathbf{g}} \subset \overline{\mathcal{M}}_g$ the boundary component of reducible curves

$$C = C_1 \cup C_2$$

where $g_i = g(C_i)$ and $l = \#(C_1 \cap C_2)$

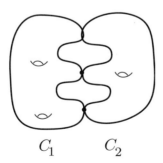

$$C_1 \qquad C_2$$

- For a rational nilpotent cone $\sigma \in \Sigma$ we shall denote by

$$\mathcal{B}_\sigma \subset \overline{\mathcal{A}}_{\mathbf{h},\Sigma}$$

the corresponding boundary component;

- We shall anticipate further work in progress by Kato-Usui and denote by

$$\overline{\mathcal{J}}_{\mathbf{h},\Sigma} \to \overline{\mathcal{A}}_{\mathbf{h},\Sigma}$$

a partial compactification of the universal family $\mathcal{J}_{\mathbf{h}} \to \mathcal{A}_{\mathbf{h}}$ of polarized complex tori (cf. the note below).

- Although the boundary component structure of $\overline{\mathcal{J}}_{\mathbf{h},\Sigma}$ is, at this time, work in progress there will be particular boundary components $\Gamma_{d,\sigma}$ of geometric interest and which will be described below.

- We shall denote by $\mathrm{Pic}^\circ(\mathcal{C}_g) \to \mathcal{M}_g$ the universal family of Jacobians and by

$$\overline{\mathrm{Pic}^\circ(\mathcal{C}_g)} \to \overline{\mathcal{M}}_g$$

a compactification, also to be described below (cf. Caparosa [3]).

- Given (X, L) where X is a smooth projective variety of dimension $2n$ and $L \to X$ is a very ample line bundle, we set $|L| = \mathbb{P}H^0(\mathcal{O}_X(L))$ and shall denote by

$$\mathcal{X}_L \subset X \times |L|$$
$$\downarrow$$
$$|L|$$

the universal family $\{X_s\}_{s \in |L|}$ of hyperplane sections relative to the embedding $X \hookrightarrow \mathbb{P}\check{H}^0(\mathcal{O}_X(L))$.

- For (X, L) as above, we shall denote by $\mathcal{M}_X$ the moduli space, assumed to exist, and by $\mathcal{X}_{\mathcal{M}} \to \mathcal{M}_X$ the corresponding family $\{X_t, L_t\}_{t \in \mathcal{M}_X}$ of polarized varieties.

- Finally, we set

$$\mathcal{L} = \bigcup_{t \in \mathcal{M}_X} |L_t|$$

and denote by

$$\mathcal{X}_{\mathcal{L}} \subset \mathcal{X}_{\mathcal{M}} \times |\mathcal{L}|$$
$$\downarrow$$
$$|\mathcal{L}|$$

the family of hyperplane sections of all the X_t, $t \in \mathcal{M}_X$.

We shall always assume that $L \to X$ has been chosen sufficiently ample so as to have properties such as

$$h^q(\mathcal{O}_{X_t}(L_t)) = 0, \qquad q > 0$$

and projective normality of the embedding given by $|L_t|$.

Note: The paper by Alexeev gives a canonical compactification $\overline{A}_g$ of A_g, one not requiring an *a priori* choice of a fan. Moreover, Alexeev's $\overline{A}_g$ coincides with the compactification associated to the second Voronoi fan. Now a fan Σ is composed of rational nilpotent cones σ in $\mathcal{G}_{\mathbb{Q}}$ satisfying certain conditions, and in this work the essential point is that Σ contain those σ's that arise from several variable degenerations of Hodge structures in the families that we shall be considering. The reason is that certain of these cones gives rise to the boundary components $\mathcal{B}(\sigma)$ that correspond to singularities of extended normal functions. For this reason we shall in this paper work with $\overline{A}_{g,\Sigma}$. At a later time we hope to more directly relate this work to Alexeev's $\overline{A}_g$.

Of course, a natural question is whether for the odd weight principally polarized case the Kato-Usui work may be extended to give a *canonical* partial compactification $\overline{A}_{\mathbf{h}}$, together with a partially compactified universal family $\overline{\mathcal{J}}_h \to \overline{A}_{\mathbf{h}}$ for which the Alexeev-Nakamura result [2] holds.

As noted above, in this paper we will anticipate some consequences of the work [17] by Kato-Usui on partial compactifications $\overline{A}_{\mathbf{h},\Sigma}$ of classifying spaces for polarized Hodge structure of odd weight. Boundary components $\mathcal{B}(\sigma(\mathbf{N}))$ of $\overline{A}_{\mathbf{h},\Sigma}$ are constructed from certain rational cones $\sigma(\mathbf{N})$ associated to commuting nilpotent elements $N_1, \ldots, N_l$ of $\mathcal{G}_{\mathbb{Q}}$. The boundary component $\mathcal{B}(\sigma(l,1))$ corresponds to the particular situation (4) in section I(ii) below. In general one may imagine something like a local system

$$(1) \qquad \qquad \mathcal{H}(\mathcal{J}) \to \overline{A}_{\mathbf{h},\Sigma}$$

given over the interior of $\mathcal{B}(\sigma(\mathbf{N}))$ by the 1st cohomology group $H^1(B^\bullet(\mathbf{N}))$ where $B^\bullet(\mathbf{N})$ is the Koszul-like complex constructed from the N_i (cf. section I.C in "I"), modulo the action of the discrete "monodromy group" associated to $\mathcal{B}(\sigma(\mathbf{N}))$. We feel that the study of the local system (1) should be interesting and could be of importance for the study of cycles.

I(ii). *Informal statements of results*

In this paper we shall study the geometry and topology of a number of rational maps arising from the linear systems $|L|$ (X fixed), $|\mathcal{L}|$ (X variable), and $\mathcal{X}_{\mathcal{L}}$. We are especially interested in these maps near a point s_0 where

(2) X_{s_0} *has ordinary double points (nodes)* p_i *with one relation among them.*[1]

---

[1]In a future work we shall study the "opposite" case of degenerations with maximally unipotent monodromy. Here the geometry is quite different and is, to us, particularly interesting.

The motivation for studying this situation is explained in "I"; cf. especially section 4.2.4. Briefly, given a primitive Hodge class $\zeta \in Hg^n(X)_{\mathrm{prim}}$ and assuming the Hodge conjecture (HC), we will have for L sufficiently ample (written $L \gg 0$) and some non-zero integer k_0

$$(3) \qquad\qquad k_0\zeta = [W - H] ,$$

where W is a smooth n-dimensional subvariety and H is a complete intersection. Moreover, a general

$$X_{s_0} \in |\mathfrak{I}_W(L)| \qquad (\mathfrak{I}_W(L) = \mathcal{O}_X(L) \underset{\mathcal{O}_X}{\otimes} \mathfrak{I}_W)$$

will have the property (1) (cf. loc. cit. and section IV(i) below for various constructions of W). There are several equivalent formulations of the condition that there be one relation among the nodes. For us the one we shall use is

(4) *For X_s smooth with s close to s_0, let $\delta_i \in H_{2n-1}(X_s, \mathbb{Z})$ be the vanishing cycle, defined up to ± 1, associated to p_i. Then there is one generating relation*

$$\sum_{i=1}^{l} \delta_i \sim 0 .$$

If ν_ζ is the extended normal function (ENF) associated to ζ with singular locus $\mathrm{sing}\, \nu_\zeta$ (loc. cit.), then (4) implies that

$$(5) \qquad\qquad s_0 \in \mathrm{sing}\, \nu_\zeta .$$

The basic observation will be the following result:

There are rational maps, a general one of which we denote here by ρ, to compactified moduli/classifying spaces, a general one of which we again here denote by $\overline{\mathcal{P}}$, such that near s_0 and with Δ denoting a suitable boundary compact of $\overline{\mathcal{P}}$ we have

$$(6) \qquad\qquad \rho^{-1}(\Delta) = \mathrm{sing}\, \nu_\zeta .$$

Some of these maps depend on ζ and some do not. Those which do not will give

$$|L| \xrightarrow{\ \rho\ } \overline{\mathcal{P}} \qquad (X \text{ fixed})$$

and

$$|\mathcal{L}| \xrightarrow{\;\rho\;} \overline{\mathcal{P}} \qquad (X \text{ variable}) .$$

In general, we will use boldface notation to signal that X is allowed to vary. If $\rho = \rho_\zeta$ does depend on ζ, then denoting by

$$\mathcal{M}_{X,\zeta} \subset \mathcal{M}_X$$

the *Noether-Lefschetz locus* where ζ remains a Hodge class (this is an algebraic subvariety by Cattani-Deligne-Kaplan [7]) and by $|\mathcal{L}|_\zeta$ the subvariety of $|\mathcal{L}|$ lying over $\mathcal{M}_{X,\zeta}$, we will have again near s_0

(7) $$\rho_\zeta^{-1}(\Delta) = \operatorname{sing} \boldsymbol{\nu}_\zeta .$$

Thus, restricting to the nodal locus

(8) $$\begin{cases} \text{(i) } \rho^{-1}(\Delta) \neq \emptyset \Rightarrow \operatorname{sing} \nu_\zeta \neq \emptyset \\ \text{(ii) } \boldsymbol{\rho}^{-1}(\Delta) \neq \emptyset \Rightarrow \operatorname{sing} \boldsymbol{\nu}_\zeta \neq \emptyset \end{cases}$$

and, assuming the inductive statement of the HC, either of these gives, by "I", the existence of an algebraic cycles Z with

$$\langle \zeta, [Z] \rangle \neq 0 ,$$

where in (ii) the cycles exist on the non-empty set of $X_t \in \mathcal{M}_X$ where

$$|L_t| \cap \boldsymbol{\rho}^{-1}(\Delta) \neq \emptyset .$$

Next we assume the "generally true" statements

(9) $$\begin{cases} \rho^*([\Delta]) \neq 0 \Rightarrow \rho^{-1}(\Delta) \neq \emptyset \\ \boldsymbol{\rho}^*([\Delta]) \neq 0 \Rightarrow \boldsymbol{\rho}^{-1}(\Delta) \neq \emptyset \end{cases}$$

where here we are leaving aside the technical issues dealing with the non-compactness of $\mathcal{M}_X$. The point here is that

A topological result gives an existence result,

here in the context of the induced map on cohomology of maps to certain universal spaces. This leads to the central question of this paper:

Assuming the Hodge conjecture, for at least some of the maps ρ do we have

$$\begin{cases} \text{(i)} & \rho^*([\Delta]) \neq 0 \\ \text{(ii)} & \boldsymbol{\rho}_\zeta^*([\Delta]) \neq 0 . \end{cases}$$

We shall see that this question leads to some to us very beautiful and unexpected geometry. Our results are very preliminary and may be informally summarized as follows:

(10) *In the case $n = 1$ when X is an algebraic surface, if ρ does not depend on ζ we have for essentially trivial reasons that $\rho^*([\Delta]) = 0$, but for $L \gg 0$ we have with certain technical assumptions that*

$$\rho_\zeta^*([\Delta]) \neq 0 .$$

We expect that these technical assumptions, the main one of which is

(11) $\operatorname{codim}_{\mathcal{M}_X}(\mathcal{M}_{X,\zeta}) = h^{2,0}(X) ,$

are not necessary.

Underlying the analysis of $\rho, \boldsymbol{\rho}$ and $\boldsymbol{\rho}_\zeta$ are transversality results. In the $n = 1$ case we shall prove that these always hold along components of $\operatorname{sing} \nu_\zeta$ and $\operatorname{sing} \boldsymbol{\nu}_\zeta$ whose general point corresponds to a nodal curve with one generating relation among the vanishing cycles (we may think of these as "principal components"). We shall also show that this transversality is not always true for the maps that do not depend on ζ, but is true for those which do provided that $L \gg 0$.

When we turn to the higher codimensional case, which in this paper will mean the case $n = 2$ where we are studying codimension 2 cycles in a fourfold, our principal finding is

(12) *the situation for $n = 2$ is completely different from the classical case.*

Some of what turns up is not unexpected. For example, concerning the Noether-Lefschetz locus corresponding to a Hodge class $\zeta \in Hg^2(X)_{\text{prim}}$ we find that we have (cf. section IV(iii))

(13) $\operatorname{codim}_{\mathcal{M}_X}(\mathcal{M}_{X,\zeta}) \leqq h^{3,1}(X) - \sigma_\zeta$

where

(14) $\begin{cases} T_\zeta = \{\theta \in H^1(\Theta_X) : \theta \cdot \zeta = 0 \text{ in } H^3(\Omega_X^1)\} \\ \sigma_\zeta = \dim(\operatorname{Image}\{H^0(\Omega_X^4) \otimes T_\zeta \to H^1(\Omega_X^3)\}) . \end{cases}$

At first glance one has

$$\operatorname{codim}_{\mathcal{M}_X}(\mathcal{M}_{X,\zeta}) \leqq h^{4,0}(X) + h^{3,1}(X)$$

since the RHS is the apparent number of conditions for a class $\zeta \in H^4(X, \mathbb{Q})$ to be of Hodge type $(2, 2)$. The refinement (13) is a consequence of the infinitesimal period relations and their integrability conditions.

As will be seen in section IV(iv) the inequality (13) is an equality in significant examples.

Somewhat more subtle, although not unexpected when one thinks about it, is that again in contrast to the $n = 1$ case

(15) *even assuming the Hodge conjecture, there are torsion and "layers of Chow group" obstructions to deforming a subvariety W along with (X, ζ) where $[W]_{\mathrm{prim}} = \zeta$, no matter how "ample" one makes W*

(cf. section IV(ii)).

Perhaps most unexpected is the following: In the $n = 1$ case for $L \gg 0$ we have

$$
(16) \quad
\begin{cases}
\text{(i) } \operatorname{codim}_{|L|}(\operatorname{sing} \nu_\zeta)_{s_0} = l - h^{2,0} & (X \text{ fixed}) \\
\text{(ii) } \operatorname{codim}_{|\mathcal{L}|}(\operatorname{sing} \boldsymbol{\nu}_\zeta)_{s_0} = l & (X \text{ variable})
\end{cases}
$$

where in the second statement we have assumed (11). This statement has the implication (cf. section III(ii))

(17) *the rational mapping*

$$
|\mathcal{L}| \overset{\Psi}{-\!\!\!-\!\!\!-\!\!\rightarrow} \overline{\mathcal{A}}_{g,\Sigma}
$$

obtained sending a smooth curve to its Jacobian is transverse along the boundary component $\mathcal{B}(\sigma(l, 1))$ corresponding to rational nilpotent cone generated by the logarithms of Picard-Lefschetz transformations corresponding to cycles δ_i satisfying (4).

Here we have (6) above where $\rho = \Psi$ and $\Delta = \mathcal{B}(\sigma(l, 1))$, which using

$$
(18) \qquad \operatorname{codim}_{\overline{\mathcal{A}}_{g,\Sigma}}(\mathcal{B}(\sigma, 1)) = l
$$

leads to the conclusion (9) in this case (cf. (10)).

In contrast, for $n = 2$ we have, still assuming the Hodge conjecture,

$$
(19) \quad
\begin{cases}
\text{(i) } \operatorname{codim}_{|L|}(\operatorname{sing} \nu_\zeta)_{s_0} \leqq l - (h^{3,1} - \sigma_\zeta) - \delta_W & (X \text{ fixed}) \\
\text{(ii) } \operatorname{codim}_{|\mathcal{L}|}(\operatorname{sing} \boldsymbol{\nu}_\zeta)_{s_0} \leqq l - \delta_W & (X \text{ variable})
\end{cases}
$$

where

$$
(20) \qquad\qquad\qquad \delta_W > 0
$$

is an algebro-geometric correction term depending on the particular subvariety W with $W \subset X_{s_0}$ and where δ_W becomes large as $L \gg 0$. This will imply that, assuming the Hodge conjecture,

(21) *the rational mapping*

$$|\mathcal{L}| \overset{\Psi}{\dashrightarrow} \overline{A}_{\mathbf{h}, \Sigma}$$

obtained by sending a smooth threefold to its intermediate Jacobian will, for $L \gg 0$, never be I-transverse to the boundary component $\mathcal{B}(\sigma(l, 1))$.

Here the concept of I-transversality will be explained; for reasons arising from the infinitesimal period relation it is the only form of transversality to have meaning for higher weight Hodge structures. That the Hodge conjecture implies a *generic non-transversality* result (cf. section V(ii)) stands in contrast to what one might naively expect and was to us a striking and thus far non-understood phenomenon.

The result and (21) depends on what to us is a very beautiful new phenomenon. Namely, in the $n = 2$ case we will show that (cf. section V(i))

(22) $\operatorname{codim}_{\overline{A}_{\mathbf{h}, \Sigma}}(\mathcal{B}(\sigma(l, 1))) = l + h^{3,0}(l - 1)$.

For a map such as

$$|\mathcal{L}| \overset{\Psi}{\dashrightarrow} \overline{A}_{\mathbf{h}, \Sigma}$$

that is regular near a point s_0, restricting to a neighborhood of s_0 one may conclude from (22) that

(23) $\operatorname{codim}_{|\mathcal{L}|} \left(\Psi^{-1}(\mathcal{B}(\sigma(l, 1))) \right) \leqq l + h^{3,0}(l - 1)$.

Eliminating the $h^{3,0}(l - 1)$ results from the infinitesimal period relation in the boundary is a special case of the following observation: Given a diagram of maps of smooth varieties

$$A \overset{F}{\longrightarrow} M$$
$$\cup \qquad \cup$$
$$B \longrightarrow N$$

where

$$F^{-1}(N) = B$$

then

(24) $$\operatorname{codim}_A(B) \leqq \operatorname{codim}_M(N) \,.$$

Under suitable assumptions on the differential F_* this may be rewritten as

(25) $$\operatorname{codim}_A(B) \leqq \operatorname{rank}\,(TM/TN) \,.$$

Now suppose there is a sub-bundle

$$I \subset TM$$

such that $F_* : TA \to I \subset TM$ and I meets TN transversely. Then (25) may be refined to

(26) $$\operatorname{codim}_A(B) \leqq \operatorname{rank}\,(I/I \cap TN) \,.$$

When applied to

$$
\begin{array}{ccc}
|\mathcal{L}| & \xrightarrow{\ \Psi\ } & \overline{A}_{\mathbf{h},\Sigma} \\[4pt]
\cup & & \cup \\[4pt]
|\mathcal{L}|_\zeta & \dashrightarrow & \mathcal{B}(\sigma(l,1))
\end{array}
$$

where $|\mathcal{L}|_\zeta$ is the subvariety of $|\mathcal{L}|$ lying over $\mathcal{M}_{X,\zeta}$ and I is the infinitesimal period relation, this leads to the elimination of the $h^{3,0}(l-1)$ term in (23). We shall refer to this phenomenon as *I-transversality*.

The result (10) follows from an excess intersection formula applied to the mapping in (17), which gives

(27) $$\Psi_\zeta^*\big([\mathcal{B}(\sigma(l,1))]\big) = \Big[\Psi_\zeta^{-1}(\mathcal{B}(\sigma(l,1)))\Big] \wedge c_{\mathrm{top}}(\mathcal{H}^{0,2})$$

where Ψ_ζ is the restriction of Ψ to $|\mathcal{L}|_\zeta$. By the Lefschetz $(1,1)$ theorem

$$\Psi_\zeta^{-1}(\mathcal{B}(\sigma(l,1))) \neq \emptyset$$

and this leads to (10) in this case.

When $n = 2$ the formula (27) holds where the "correction term" on the far right is replaced by $c_{\mathrm{top}}(V)$ where V is a vector bundle that has three components:

(i) a bundle of rank $h^{3,1} - \sigma_\zeta$ that arises for the same reason as $\mathcal{H}^{0,2}$ in the $n = 1$ case;

(ii) a bundle of rank $h^{3,0}(l-1)$ that arises from the infinitesimal period relation in the boundary of $\overline{A}_{\mathbf{h},\Sigma}$; and

(iii) a bundle of rank δ_W.

The situations (i) and (ii) are Hodge-theoretic and do not depend on the particular W, provided that a certain technical condition that holds in many examples is satisfied. The situation (iii) is algebro-geometric and represents a phenomenon that is a consequence of the HC and does not appear in the classical $n = 1$ case.

All of the dimension count phenomena described just now are quite visible and explicit in the case of the example

$$\Lambda \subset X \subset \mathbb{P}^5$$

where Λ is a $\mathbb{P}^2$, X is a smooth fourfold of degree $d \geq 6$, and

$$\zeta = [\Lambda]_{\mathrm{prim}} \, .$$

The calculations have by now the familiar polynomial flavor, with there being a few new subtleties (cf. section IV(iv)).

To conclude this introduction we remark that to us one of the more surprising and gratifying aspects of this study turned out to be the very elegant interplay between Hodge theory and geometry near and at the boundary of the partially compactified classifying spaces for Hodge structures. As is well known, in algebraic geometry a major tool is to study the degenerations of an algebraic variety, these degenerations frequently being "simpler" than the smooth variety. What we had not expected was the nice geometry that arises when these degenerations are viewed as pullbacks of the aforementioned boundary components.

II. The Classical Case of Curves on a Surface, Part A

II(i). *Maps to moduli when X is fixed*

Let (X, L, ζ) be as in the introduction where

- X is a smooth algebraic surface and $L \to X$ is a very ample line bundle;
- $\zeta \in Hg^1(X)_{\mathrm{prim}}$ is a primitive Hodge class.

We shall write

$$L_m = mH$$

where H is a fixed ample line bundle and the positive integer m varies. Then for $m_1 \gg 0$ we may find a smooth curve

$$C_1 \in |\zeta + m_1 H| \, ,$$

and for $m = m_1 + m_2$ with $m_2 \gg 0$ we may find a smooth curve

$$C_2 \in |-\zeta + m_2 H|$$

and where

$$X_{s_0} = C_1 \cup C_2 \in |L_m|$$

has ordinary nodes so that C_1, C_2 meet transversely in

(1) $$l = \#(C_1 \cap C_2) = -\zeta^2 + m_1 m_2 h, \qquad h = H^2$$

points. Denoting by ν_ζ the extended normal function associated to ζ (cf. "I"), we have

(2) **Theorem:** *For $m_2 \gg 0$, there is an irreducible component* $(\mathrm{sing}\, \nu_\zeta)_{s_0}$ *of the singular locus of* $\mathrm{sing}\, \nu_\zeta$ *passing through $s_0 \in |L|$ and*

(3) $$\mathrm{codim}_{|L|}(\mathrm{sing}\, \nu_\zeta)_{s_0} = l - h^{2,0}$$

where $h^{2,0} = h^{2,0}(X)$.

The LHS of (3) may be thought of as an algebro-geometric quantity, since it is

(4) $$\dim \left\{ \begin{array}{c} \text{deformation space} \\ \text{of } C_1 \text{ in } X \end{array} \right\} + \dim \left\{ \begin{array}{c} \text{hypersurfaces in } |L_m| \\ \text{passing through } C_1 \end{array} \right\}.$$

The RHS of (3) is Hodge-theoretic, since the number l of nodes may be read off from the local monodromy around s_0.

The above interpretation of (3) will be seen to hold for $n = 2$ — i.e., when X is a fourfold — as well as the present case, where of course writing as above

$$X_{s_0} = C = C_1 + C_2$$

the second term is

$$\{\text{deformations of } C_2 \text{ in } X\}.$$

As noted in the introduction, in contrast to the $n = 1$ case, in the $n = 2$ case there will always be a non-zero algebro-geometric "correction term."

Proof of #1: For simplicity we assume that $h^{1,0}(X) = 0$. In the general case the same argument with slightly more complicated computations will apply.

From the cohomology sequence of

$$0 \to \mathcal{O}_X \to \mathcal{O}_X(C_i) \to N_{C_i/X} \to 0 \,,$$

and using the assumption $h^1(\mathcal{O}_X) = 0$ and

$$h^q(\mathcal{O}_X(C_i)) = 0 \,, \qquad\qquad q > 0$$

which holds for $m_1 \gg 0$, $m_2 \gg 0$, we have (cf. the remark below following Proof #2)

$$\begin{aligned}\dim(\text{sing } \nu_\zeta)_{s_0} &= h^0\left(N_{C_1/X}\right) + h^0\left(N_{C_2/X}\right) \\ &= h^0\left(\mathcal{O}_X(C_1)\right) + h^0\left(\mathcal{O}_X(C_2)\right) - 2 \,.\end{aligned}$$

Now using

$$\mathcal{X}(\mathcal{O}_X(C_i)) = \frac{1}{2}\left(C_i^2 + C_i \cdot K_X\right) + \mathcal{X}(\mathcal{O}_X)$$

we have

$$h^0(\mathcal{O}_X(C_i)) = \frac{1}{2}\left(C_i^2 + C_i \cdot K_X\right) + 1 + h^{2,0} \,,$$

which gives

$$\begin{aligned}\dim(\text{sing } \nu_\zeta)_{s_0} &= \frac{1}{2}\left(C_1^2 + C_2^2 + C \cdot K_X\right) + 2h^{2,0} \\ &= \frac{1}{2}\left(C^2 + C \cdot K_X\right) - l + 2h^{2,0}\end{aligned}$$

since $C_1 \cdot C_2 = l$. Then since $\mathcal{O}_X(C) \cong L$

$$\begin{aligned}\dim |L| &= h^0(\mathcal{O}_X(C)) - 1 \\ &= \frac{1}{2}\left(C^2 + C \cdot K_X\right) + h^{2,0}\end{aligned}$$

where we have used $h^q(\mathcal{O}_X(C)) = 0$ for $q > 0$. It follows that

$$\text{codim}(\text{sing } \nu_\zeta)_{s_0} = l - h^{2,0}. \qquad\qquad \square$$

Proof #2: We set

$$\Delta = C_1 \cap C_2$$

and note that near s_0

$$(\text{sing } \nu_\zeta)_{s_0} = \left\{\begin{array}{c}\text{curves } C' \in |L| \text{ near } C \text{ such that} \\ \Delta \text{ deforms to nodes } \Delta' \subset C'\end{array}\right\}.$$

The reason is that for topological reasons, there must be one relation among the nodes Δ', and hence

$$C' = C_1' \cup C_2' \in \Sigma_{\mathbf{g}} .$$

Thus (cf. the remark below)

$$T(\operatorname{sing}\nu_\zeta)_{s_0} \cong H^0(\mathfrak{I}_\Delta \otimes L) ,$$

so that from the cohomology sequence of

$$0 \to \mathfrak{I}_\Delta \otimes L \to L \to L_\Delta \to 0$$

and $h^q(L) = 0$ for $q > 0$, we have

$$\operatorname{codim}(\operatorname{sing}\nu_\zeta)_{s_0} = \dim \big(\operatorname{Image}\{H^0(L) \to H^0(L_\Delta)\}\big)$$
$$= \deg \Delta - h^1(\mathfrak{I}_\Delta \otimes L) .$$

Now there is the usual Koszul sequence

$$0 \to \mathcal{O}_X \to \overset{2}{\underset{i=1}{\oplus}} \mathcal{O}_X(L - C_i) \to \mathfrak{I}_\Delta \otimes L \to 0$$

and since $C_1 + C_2 \in |L|$ the middle term is $\mathcal{O}_X(C_1) \oplus \mathcal{O}_X(C_2)$. Then from $h^q(\mathcal{O}_X(C_i)) = 0$ for $q > 0$ we have

$$\operatorname{codim}(\operatorname{sing}\nu_\zeta)_{s_0} = l - h^{2,0}. \qquad \square$$

(5) *Remark:* Above we have used

$$\dim(\operatorname{sing}\nu_\zeta)_{s_0} = \dim T(\operatorname{sing}\nu_\zeta)_{s_0}$$

where the RHS is the Zariski tangent space. In fact, from the above discussion we see that $(\operatorname{sing}\nu_\zeta)_{s_0}$ *is smooth and is equal to the Veronese variety*

$$|\zeta + m_1 H| \times |-\zeta + m_2 H| \subset |mH|, \qquad m = m_1 + m_2 .$$

We shall now discuss one interpretation of Theorem (2). For this we set $L = mH$ where m is sufficiently large as above and consider the rational map

(6) $$|L| \overset{\varphi}{-\!\!-\!\!\to} \overline{\mathcal{M}}_g$$

defined for a general point s by

$$\varphi(s) = X_s \in \overline{\mathcal{M}}_g$$

where as usual

$$g = \frac{1}{2}\left(L^2 + L \cdot K_X\right) + 1 .$$

The map φ is regular in a neighborhood of s_0 where

$$(7) \qquad\qquad X_{s_0} = C_1 \cup C_2$$

as above.

We denote by

$$\Sigma_{\mathbf{g}} \subset \overline{\mathcal{M}}_g, \qquad\qquad \mathbf{g} = (g_1, g_2)$$

the boundary component of all curves of the type (7) where $g_i = g(C_i)$. Then we have

$$(8) \qquad\qquad (\text{sing}\, \nu_\zeta)_{s_0} = \varphi^{-1}(\Sigma_{\mathbf{g}})_{s_0} \ .$$

where both sides denote the irreducible components passing through s_0. From

$$g = g_1 + g_2 + l - 1$$

we have

$$3g - 3 = (3g_1 - 3) + (3g_2 - 3) + l$$

which gives the well-known result that

$$(9) \qquad\qquad \text{codim}_{\overline{\mathcal{M}}_g}(\Sigma_{\mathbf{g}}) = l \ .$$

Geometrically, the l nodes of $C_1 \cup C_2$ may be smoothed independently in $\overline{\mathcal{M}}_g$. From (3) we conclude the

(10) **Corollary:** *The rational mapping (6) fails by $h^{2,0}$ to be transverse at s_0.*

From (8) and (9) we have trivially

$$\text{codim}(\text{sing}\, \nu_\zeta)_{s_0} \leqq \text{codim}_{\overline{\mathcal{M}}_g}(\Sigma_{\mathbf{g}}) = l \ .$$

As noted above, the LHS is an algebro-geometric term while the RHS is a Hodge-theoretic term. We shall write it as

$$(11) \qquad\qquad AG \leqq HT \ ,$$

and by the corollary the correction term needed to make this inequality into an equality is also a Hodge-theoretic term.

Assuming the Hodge conjecture, we will find an analogue of the inequality (11) for fourfolds, and the investigation of the correction term will lead to new and to us very interesting and only partially understood phenomena. In particular, the correction term will turn out to *not* be purely Hodge theoretic, one of a number of significant differences between the classical and higher codimensional cases.

Any rational map (or correspondence for that matter) induces in the usual way a map on cohomology by considering the fundamental class of the graph of the map and using the Künneth decomposition. In the next section we shall prove the

(12) **Proposition:** *For the induced mapping on cohomology associated to the rational mapping* (6) *we have*

$$\varphi^*([\Sigma_{\mathbf{g}}]) = 0 \ .$$

Intuitively this has to be true. If we let X vary in moduli, then a general nearby X' will have no Hodge class ζ' corresponding to ζ and therefore no curves of the type (7). The mapping φ does not depend on ζ, and if $\varphi^*([\Sigma_{\mathbf{g}}]) \neq 0$ then this would be true for the φ' corresponding to (X', L') and

$$\varphi'^*([\Sigma_{\mathbf{g}}]) \neq 0 \Rightarrow \varphi'^{-1}(\Sigma_{\mathbf{g}}) \neq \emptyset \ .$$

This reasoning is of course heuristic, but it does explain (12).

II(ii). *Maps to moduli when X varies*

We recall the notation $\{X_t\}_{t \in \mathcal{M}_X}$ for the moduli space of X, which is assumed to exist and which for simplicity we assume is irreducible, and

$$|\mathcal{L}| = \bigcup_{t \in \mathcal{M}_X} |L_t|$$

for the family of hyperplane sections $|L_t|$ of X_t. We will consider the rational map

(13) $$|\mathcal{L}| \overset{\Phi}{-\!\!\!-\!\!\!\to} \overline{\mathcal{M}}_g \ ,$$

which we think of as the rational map (6) when X varies.

We also recall the notation

$$\mathcal{M}_{X,\zeta} \subset \mathcal{M}_X$$

for the Noether-Lefshetz locus where ζ remains a Hodge class $\zeta_t \in Hg^1(X_t)_{\mathrm{prim}}$. Now $\mathcal{M}_{X,\zeta}$ is defined by the condition

(14) $$(\zeta_t)^{0,2} = 0$$

where $\zeta_t \in H^2(X_t, \mathbb{Z})$. The class ζ_t is uniquely defined for X_t close to X, and if (14) is satisfied locally then it is invariant under monodromy and therefore satisfied globally (cf. Cattani-Deligne-Kaplan [7]).

Because of (14) we may say that

(15) $\qquad\qquad$ "expected" $\mathrm{codim}_{\mathcal{M}_X} (\mathcal{M}_{X,\zeta}) = h^{2,0}$.

It is known that (cf. [14]): *In general*

(16) $\qquad\qquad\qquad \mathrm{codim}_{\mathcal{M}_X} (\mathcal{M}_{X,\zeta}) = h^{2,0}$,

but there are exceptional cases, such as when

$$\zeta = [\Lambda]_{\mathrm{prim}}$$

where $\Lambda \subset X \subset \mathbb{P}^3$ is a line on a smooth surface in $\mathbb{P}^3$ of degree at least five. In this case

$$\mathrm{codim}_{\mathcal{M}_X} (\mathcal{M}_{X,\zeta}) = h^{2,0} - h^{2,0}(-\Lambda)$$

where

$$h^{2,0}(-\Lambda) = \dim H^0 \left(\Omega_X^2(-\Lambda) \right) \ .$$

For simplicity of exposition we will assume (16) and remark that the modifications necessary in the general case may be made by considerations similar to this example.

We denote by

$$|\mathcal{L}|_\zeta \subset |\mathcal{L}|$$

the part of $|\mathcal{L}|$ lying over $\mathcal{M}_{X,\zeta}$, which we think of as complete linear systems of hyperplane sections of X_t having $\zeta_t \in Hg^1(X_t)$. By our assumption

$$\mathrm{codim}_{|\mathcal{L}|}(|\mathcal{L}|_\zeta) = h^{2,0} \ .$$

We shall denote by

$$|\mathcal{L}|_\zeta \xrightarrow{\ \Phi_\zeta\ } \overline{\mathcal{M}}_g$$

the restriction to $|\mathcal{L}|_\zeta$ of the mapping Φ in (13). We also recall our notation $\boldsymbol{\nu}_\zeta$ for the family of ν_{ζ_t}'s over $\mathcal{M}_{X,\zeta}$. Let

$$s_0 \in \mathrm{sing}\, \nu_\zeta$$

correspond to a nodal $X_{s_0} = C_1 \cup C_2$ as above. Then $s_0 \in \operatorname{sing} \boldsymbol{\nu}_\zeta$ and from Theorem (2) and its proof

$$(17) \qquad \begin{cases} \text{(i)} & (\operatorname{sing} \boldsymbol{\nu}_\zeta)_{s_0} \subset |\mathcal{L}|_\zeta \\[2mm] \text{(ii)} & \operatorname{codim}_{|\mathcal{L}|_\zeta}(\operatorname{sing} \boldsymbol{\nu}_\zeta)_{s_0} = l - h^{2,0} \\[2mm] \text{(iii)} & \operatorname{codim}_{|\mathcal{L}|}(\operatorname{sing} \boldsymbol{\nu}_\zeta)_{s_0} = l \, . \end{cases}$$

From this we may draw the following

(18) **Conclusion:** *For the mapping* Φ *in* (13) *we have*

$$\Phi^{-1}(\Sigma_{\mathbf{g}})_{s_0} = \Phi_\zeta^{-1}(\Sigma_{\mathbf{g}})_{s_0} = (\operatorname{sing} \boldsymbol{\nu}_\zeta)_{s_0}$$

and Φ *is transverse relative to* $\Sigma_{\mathbf{g}}$ *in a neighborhood of* s_0.

The infinitesimal calculation needed to establish transversality may be established by a standard sheaf cohomological computation which we shall not give here.

In the following it will be understood that we are considering irreducible components passing through s_0. We also denote by

$$\mathcal{H}^{0,2} \to |\mathcal{L}|$$

the pullback of the Hodge bundle with fibres $H^{0,2}(X_t)$ over $\mathcal{M}_X$ and we set

$$h = h^{2,0} \, .$$

(19) **Theorem:** *With the above assumptions and notations we have in* $H^{2l}(|\mathcal{L}_\zeta|)$

$$(20) \qquad \left[\Phi_\zeta^{-1}(\Sigma_{\mathbf{g}})\right] \wedge c_h(\mathcal{H}^{0,2}) = \Phi_\zeta^*([\Sigma_{\mathbf{g}}]) \, .$$

Proof: We first observe that the normal bundle of $|\mathcal{L}|_\zeta$ in $|\mathcal{L}|$ is $\mathcal{H}^{0,2}$. Here we recall our assumption that in a neighborhood of (X, ζ) the Noether-Lefschetz locus is smooth of codimension h and defining equations (14). The result then follows from standard excess intersection formula considerations (cf. Fulton [11] — in the special case one may rely on relatively elementary considerations using the Gysin map on smooth compactifications; cf. the remark below). $\qquad \square$

We remark that

(21) $$\Phi^*([\Sigma_{\mathbf{g}}]) \neq 0$$

in the following sense: Let $\overline{\mathcal{M}}_X$ be *any* smooth compactification of $\overline{\mathcal{M}}_X$ and $\overline{\Phi}$ the extension of the rational map (13) to $\overline{\mathcal{M}}_X$. Then

$$\overline{\Phi}^*([\Sigma_{\mathbf{g}}]) = [\overline{\Phi}^{-1}(\Sigma_{\mathbf{g}})]$$

where the RHS is the fundamental class of the closure in $\overline{\mathcal{M}}_X$ of the irreducible, codimension h subvariety

$$\Phi^{-1}(\Sigma_{\mathbf{g}})_{s_0} .$$

The LHS is therefore non-zero, and this is what is meant by (21).

What is obviously of more interest is to know that $\Phi_\zeta^*([\Sigma_{\mathbf{g}}]) \neq 0$, and for this there are various assumptions that will imply that the LHS of (20) is non-zero. For example, if there is another Hodge class $\zeta' \in Hg^2(X)_{\mathrm{prim}}$ such that

(22) $$\mathcal{M}_{X,\zeta'} \text{ meets } \mathcal{M}_{X,\zeta} \text{ transversely,}$$

then the LHS of (20) is the fundamental class

$$\left[\Phi_\zeta^{-1}(\Sigma_{\mathbf{g}}) \cap \mathcal{M}_{X,\zeta'} \right] ,$$

which is non-zero in the sense explained above. Thus we have the

(23) **Corollary:** *With the asssumption* (22),

$$\Phi_\zeta^*([\Sigma_{\mathbf{g}}]) \neq 0 .$$

Remark that this result is illustrative and certainly not definitive. Finally, since for the rational mapping φ in (6) we have

$$\varphi = \Phi_\zeta \big|_{|L|}$$

where $|L| \subset |\mathcal{L}|_\zeta$, we have

$$\varphi^*([\Sigma_{\mathbf{g}}]) = 0$$

since the LHS of (20) restricted to $|L|$ is zero. This proves Corollary (23).

Remark: As noted above, (20) is an elementary formula. The general situation is where we first have a diagram of maps

$$A \xrightarrow{\ F\ } M$$
$$\cup \qquad \cup \ , \quad B = F^{-1}(N)$$
$$B \longrightarrow N$$

where

$$\operatorname{codim}_A(B) = \operatorname{codim}_M(N) \, ,$$

and where for simplicity of explanation we assume that everything is smooth. Then

(24)
$$F^*([N]) = [F^{-1}(N)] = [B] \, .$$

Next suppose we have

$$B \subset C \subset A \, .$$

Denoting by $j : C \to A$ the inclusion and by $[B]_C \in H^*(C)$ the fundamental class of B in C we have

$$[B]_C = j^*([B]) \wedge c_{\text{top}}(N_{C/A}) \, .$$

Setting $F_C = F|_C$ and combining this with the formula above gives the elementary excess intersection formula

(25)
$$F_C^*([N]) = [F_C^{-1}(N)] \wedge c_{\text{top}}(N_{C/A}) \, .$$

The result (20) is the special case of this formula where

$$A = |\mathcal{L}|, \quad B = \Phi^{-1}(\Sigma_{\mathbf{g}}), \quad C = |\mathcal{L}|_\varsigma$$
$$M = \overline{A}_{\mathbf{h},\Sigma}, \quad N = \Sigma_{\mathbf{g}} \, .$$

The formula in the $n = 2$ case that will by way of contrast be given in section V(iii) will be a true excess intersection formula. To anticipate it we have in general

$$F_* : N_{B/A} \to N_{N/M}$$

and assuming F_* is injective we have the excess normal bundle

$$Q = F^{-1}(N_{N/M})/N_{B/A}$$

and then the excess intersection formula gives

(26)
$$F^*([N]) = Gy_{B/A}(c_{\text{top}}(Q))$$

where

$$Gy_{B/A} : H^*(B) \to H^{*+\operatorname{codim}_A(B)}(A)$$

denotes the Gysin map. In the situation of formula (20) we have that Q is trivial and (26) gives

(27)
$$[B] = Gy_{B/A}(1_B)$$

where $1_B \in H^0(B)$. The formula (24) then is a consequence of (27) when we insert C between B and A.

II(iii). *Maps to $\overline{\mathcal{C}}_g$*

Keeping the above notations, let us define an *interesting curve* to be an irreducible curve $C \subset X$ with

$$(28) \qquad\qquad\qquad \langle \zeta, [C] \rangle = \zeta^2 .$$

By the $(1,1)$ theorem interesting curves exist in $|\zeta + m_1 H|$ for $m_1 \gg 0$. The presence of interesting curves also implies

$$(29) \qquad\qquad \begin{array}{ll} \text{(i)} & \varphi^{-1}(\Sigma_{\mathbf{g}}) \neq \emptyset \\ \text{(ii)} & \Phi_\zeta^{-1}(\Sigma_{\mathbf{g}}) \neq \emptyset . \end{array}$$

We have seen above that $\varphi^*([\Sigma_{\mathbf{g}}]) = 0$ but under mild technical assumptions we have

$$(30) \qquad\qquad\qquad \Phi_\zeta^*([\Sigma_{\mathbf{g}}]) \neq 0 ,$$

and this implies that

$$(31) \qquad\qquad\qquad \Phi_\zeta^{-1}(\Sigma_{\mathbf{g}}) \neq \emptyset .$$

This in turn implies that there exists $(X_t, \zeta_t) \in \mathcal{M}_{X,\zeta}$ such that

$$\Phi_\zeta^{-1}(\Sigma_{\mathbf{g}}) \cap |L_t| \neq \emptyset .$$

Relabelling (X_t, ζ_t) to be just (X, ζ) we have

$$(32) \qquad\qquad\qquad \varphi^{-1}(\Sigma_{\mathbf{g}}) \neq \emptyset .$$

Theorem: *The condition (30) implies that there exist interesting curves on X.*

In other words,

> *The topological condition (30) leads to an existence theorem.*

Proof: The argument will be indirect, and we shall first explain the difficulty in giving a direct proof. We are grateful to Mark de Cataldo and Luca Migliorini for pointing this difficulty out to us.

Since Φ_ζ is only a rational map, (30) only implies the existence of a family

$$(33) \qquad\qquad\qquad \{X_s\}_{s \in \Delta} \subset |L|$$

with Δ the disc, where X_s is smooth for $s \neq 0$, and where semi-stable reduction (SSR) applied to the family (33) only produces an interesting curve on a blownup branched covering $\tilde{X}$ of X. This curve upstairs may then either contract or map to an uninteresting curve under the projection $\tilde{X} \to X$.

For the proof we shall first assume that

$$(34) \qquad\qquad h^{2,0} = 0$$

and then indicate how the argument may be modified in the general case. We consider the universal family of hyperplane sections relative to $|L|$

$$\begin{array}{c} \mathfrak{X}_{|L|} \subset X \times |L| \\ \downarrow \\ |L| \ . \end{array}$$

There is then the following diagram of mappings where $\overline{\mathcal{C}}_g$ is the compactified universal curve of genus g and the dotted arrows are rational maps

$$(35) \qquad \begin{array}{ccc} X \longleftarrow \mathfrak{X}_L & \overset{\lambda}{-\,-} \!\!\!\!\! \rightarrow & \overline{\mathcal{C}}_g \\ \Big\downarrow & & \Big\downarrow \\ |L| & \overset{\varphi}{-\,-} \!\!\!\!\! \rightarrow & \overline{\mathcal{M}}_g . \end{array}$$

We denote by $\mathcal{C}_{g_1}$ in the diagram

$$\begin{array}{ccc} \mathcal{C}_{g_1} & \subset & \overline{\mathcal{C}}_g \\ \downarrow & & \downarrow \\ \Sigma_{\mathbf{g}} & \subset & \overline{\mathcal{M}}_g \end{array}$$

the family of curves over $\Sigma_{\mathbf{g}}$ where $\mathbf{g} = (g_1, g_2)$ and whose general member is the curve component of genus g_1 lying over a general point of $\Sigma_{\mathbf{g}}$. Then

$$(36) \qquad\qquad \operatorname{codim}_{\overline{\mathcal{C}}_g}(\mathcal{C}_{g_1}) = l \ .$$

Taking components of $\lambda^{-1}(\mathcal{C}_{g_1})$ lying over a point $s_0 \in |L|$ as in the proof of Theorem (2), we have by the assumption (34)

$$(37) \qquad\qquad \operatorname{codim}_{\mathfrak{X}_L}\left(\lambda^{-1}(\mathcal{C}_{g_1})\right) = l \ .$$

Assuming for the moment that λ and φ are regular maps, we will show that

$$(38) \qquad\qquad \pi^*(\zeta) \cdot \lambda^*\left([\mathcal{C}_{g_1}]\right) \neq 0$$

in $H^{2l+2}(\mathfrak{X}_L)$. This is the main geometric step in the argument and the reason for it is as follows: Because of (36) and (37)

$$(39) \qquad \lambda^* \left([\mathcal{C}_{g_1}]\right) = \left[\lambda^{-1}(\mathcal{C}_{g_1})\right] \ .$$

Now $\lambda^{-1}(\mathcal{C}_{g_1})$ fibres over $\varphi^{-1}(\Sigma_{\mathbf{g}})$ whose general fibre is the curve C_1 in the above picture. Then

$$\langle \pi^*(\zeta), [C_1] \rangle = \zeta^2 \neq 0 \ ,$$

from which it follows that class (39) is Poincaré dual to a subvariety $Z \subset \lambda^{-1}(\mathcal{C}_{g_1})$ that maps to $\varphi^{-1}(\Sigma_{\mathbf{g}})$ as a generically finite map of degree ζ^2. Then

$$\pi^*(\zeta) \cdot \lambda^*([\mathcal{C}_{g_1}]) = [Z] \neq 0$$

which is (38).

Now from the argument used to prove (38) we see that a general fibre of $\lambda^{-1}(\mathcal{C}_{g_1}) \to \varphi^{-1}(\Sigma_{\mathbf{g}})$ maps to X onto a curve $\pi(C_1)$ with

$$\langle \zeta, [\pi(C_1)] \rangle = \zeta^2 \neq 0 \ ,$$

which then gives an interesting curve on X.

We now remove the assumption that λ and φ are regular maps. We let

$$
\begin{array}{ccc}
 & \tilde{\mathfrak{X}}_L & \\
\tilde{\pi} \swarrow & \downarrow & \searrow \tilde{\lambda} \\
X \xleftarrow{\ \pi\ } \mathfrak{X}_L & \xrightarrow{\ \lambda\ } & \overline{\mathcal{C}}_g
\end{array}
$$

be a resolution of the rational maps λ and φ, so that $\tilde{\lambda}$ and $\tilde{\varphi}$ are now regular maps. Then the above argument applies to give

$$\tilde{\pi}(\zeta) \cdot \tilde{\lambda}^* \left([C_{g_1}]\right) \neq 0 \ ,$$

which then gives as before an interesting curve $\tilde{\pi}(C_1)$ on X.

We finally remove the assumption (34). This was made in order to have the dimension count (36) leading to (39). If (34) is not satisfied then as we saw in section II(ii) we need to let (X, ζ) vary over $\mathcal{M}_{X,\zeta}$ in order to have the correct dimension counts. The argument given also then extends with the same underlying geometric idea to produce an $(X_t, \zeta_t) \in \mathcal{M}_{X,\zeta}$ such that there is an interesting curve on X_t. $\qquad \square$

II(iv). *Maps to* $\overline{\text{Pic}^{\circ}(\mathcal{C}_g)}$

The rational maps

$$|L| \overset{\varphi}{-\!\!-\!\!\to} \overline{\mathcal{M}}_g$$

$$|L| \overset{\lambda}{-\!\!-\!\!\to} \overline{\mathcal{C}}_g$$

(cf. (6) and (35)), and their analogues when X varies do not depend on a Hodge class $\zeta \in Hg^1(X)_{\text{prim}}$. What is the case is that given a ζ and at a point s_0 where

(40)
$$X_{s_0} = C_1 \cup C_2$$
$$\begin{cases} C_1 \in |\zeta + m_1 H|, & m_1 \gg 0 \\ C_2 \in |-\zeta + m_2 H|, & m_2 \gg 0 \end{cases}$$

we have

(41)
$$\varphi^{-1}(\Sigma_{\mathbf{g}})_{s_0} = (\text{sing } \nu_\zeta)_{s_0}$$

where each side denotes the component of an irreducible variety through s_0.

This is for X fixed. Letting X vary in $\mathcal{M}_X$ we have seen that

$$\Phi^{-1}(\Sigma_{\mathbf{g}})_{s_0} \text{ projects onto a component of } \mathcal{M}_{X,\zeta}$$

and

$$\Phi^{-1}(\Sigma_{\mathbf{g}})_{s_0} = (\text{sing } \boldsymbol{\nu}_\zeta)_{s_0},$$

which in particular implies that the analogue of (41) holds for all (X_t, ζ_t) near (X, ζ).

Without assuming the existence of ζ suppose we have a point s_0 such that φ is defined and regular in a neighborhood and

$$\varphi^{-1}(\Sigma_{\mathbf{g}})_{s_0} \neq \emptyset .$$

Then there exists a curve (7) where $g_i(C_i) = g_i$ are given with

$$g = g_1 + g_2 + l - 1 .$$

For suitable m_1 (which we may have to take to lie in $\mathbb{Q}$) we will have

$$(C_1 - m_1 H) \cdot H = 0 ,$$

so that ζ defined by

(42)
$$[C_1] = \zeta + [m_1 H]$$

is primitive. We note that, setting $h = H^2$,

$$\begin{cases} g_1 = \frac{1}{2}\left(\zeta^2 + m_1^2 h + \zeta \cdot K_X\right) + 1 \\ g_2 = \frac{1}{2}(\zeta^2 + m_2^2 h - \zeta \cdot K_X) + 1 \;. \end{cases}$$

(43) **Conclusion:** *If* $\varphi^{-1}(\Sigma_{\mathbf{g}}) \neq \emptyset$, *then there are Hodge classes* $\zeta \in Hg^1(X)_{\mathrm{prim}}$ *with given* ζ^2 *and* $\zeta \cdot K_X$.

In particular, there could be several ζ's corresponding to different components of $\varphi^{-1}(\Sigma_{\mathbf{g}})$.

We shall now discuss a rational map that does depend on ζ. This map is just the extended normal function viewed as a rational map

(44)
$$|L| \xrightarrow{\;\nu_\zeta\;} \overline{\mathrm{Pic}^\circ(\mathcal{C}_g)} \;.$$

For the RHS of (44) we shall take the compactification defined by Caparoso [3]. Actually the compactification in her work is $\overline{\mathrm{Pic}^k(\mathcal{C}_g)}$ where k is large relative to g. For our purposes we can set $H_s = H\,|_{X_s}$ and use the rational map

$$s \to \nu_\zeta(s) + nH_s \in \mathrm{Pic}^k(X_s)$$

for large n where $X_s \in |mH|$ and $k = mnH^2$. With this understood, for simplicity of notation and exposition we shall just consider (44). The point will be to show how the image of (44) meets certain boundary components to be described below and whose relation to (44) will be clear.

We are interested in the behaviour of $\mathrm{Pic}^\circ(X_s)$ under a specialization

$$X_s \to X_{s_0} \in \Sigma_{\mathbf{g}}$$

of a smooth curve X_s to a curve of the type (7). A line bundle $M_s \to X_s$ of degree zero will specialize to a line bundle $M_{s_0} \to X_{s_0}$ of total degree zero with

(45)
$$\begin{cases} \deg_{C_1}(M_{s_0}) = d \\ \deg_{C_2}(M_{s_0}) = -d \;. \end{cases}$$

From loc. cit. there are bounds on d in terms of g and the integer k above, but these need not concern us here. We may think of $M_{s_0} \to X_{s_0}$ as given by line bundles

(46)
$$\begin{cases} M_1 \to C_1 \\ M_2 \to C_2 \end{cases}$$

with $\deg_{C_1}(M_1) = d$, $\deg_{C_2}(M_2) = -d$ and with isomorphisms (gluing data)

(47) $$M_{1,p_i} \cong M_{2,p_i}$$

at the nodes p_i. The number of parameters in (46) and (47) is

$$g_1 + g_2 + l - 1 = g$$

where the "-1" on the LHS comes from the independent scalings of M_1 and M_2. We denote by

$$\Gamma_{d,\mathbf{g}} \subset \overline{\mathrm{Pic}^\circ(\mathcal{C}_g)}$$

the boundary component lying over $\Sigma_{\mathbf{g}}$ whose inverse image over a curve (7) is given by the data (46) and (47).

The basic observation is

(48) *For the extended normal function ν_ζ we have*

$$\nu_\zeta(s_0) \in \Gamma_{d,\mathbf{g}} , \quad d = \zeta^2 .$$

In fact

(49) $$\nu_\zeta^{-1}(\Gamma_{d,\mathbf{g}})_{s_0} = (\mathrm{sing}\, \nu_\zeta)_{s_0} .$$

As in section II(ii) we may prove that for $d = \zeta^2$

(50)
$$\nu_\zeta^*([\Gamma_{d,\mathbf{g}}]) = 0 \qquad (X \text{ fixed})$$
$$\boldsymbol{\nu}_\zeta^*([\Gamma_{d,\mathbf{g}}]) \neq 0 \qquad (X \text{ variable}) ,$$

where (50) holds under the same assumption as in Corollary (23).

Although ν_ζ does depend on ζ, the topological information in $[\nu_\zeta^{-1}(\Gamma_{d,\mathbf{g}})]$ only involves ζ^2. As will be discussed on a later occasion, we feel that ultimately one needs to understand

$$\nu_{\zeta \times \zeta'}^{-1}(\mathcal{P})$$

and its Chern classes, where $\zeta, \zeta' \in Hg^1(X)_{\mathrm{prim}}$ and $\mathcal{P}$ is an extension of the Poincaré line bundle to $\overline{\mathrm{Pic}^\circ(\mathcal{C}_g)}$ (cf. section V(D) in "I").

III. The Classical Case, Part B

III(i). *Hodge-theoretic description of $\overline{\mathcal{A}}_{g,\Sigma}$*

Thus far we have studied rational maps to compactifications of algebro-geometric moduli spaces and "universal" families over such. Now a normal function is a Hodge-theoretic object, as is an extended normal function. Moreover, although the algebro-geometric compactifications arising from algebraic curves are not generally available in the higher-dimensional case, there are recently available (partial) compactifications of many of the Hodge-theoretic players in our story.[2] This suggests studying rational maps to these Hodge-theoretic objects. In this section we shall begin this study again in the classical case of curves on a surface.

We shall use the standard notation

$$\mathcal{A}_g = \left\{ \begin{array}{l} \text{moduli space of principally polarized abelian} \\ \text{varieties (PPAV's) of dimension } g \end{array} \right\}.$$

For our purposes, $\mathcal{A}_g$ and its compactifications $\overline{\mathcal{A}}_{g,\Sigma}$ will be constructed Hodge-theoretically and we briefly review this, referring to Cattani [5], Carlson-Cattani-Kaplan [4], Alexeev-Nakamura [2], Alexeev [1] and the references cited there for details.

For the construction we assume given a pair $(H_{\mathbb{Z}}, Q)$ where $H_{\mathbb{Z}}$ is a lattice of rank $2g$ and

$$Q : H_{\mathbb{Z}} \otimes H_{\mathbb{Z}} \to \mathbb{Z}$$

is a unimodular symplectic form. A *symplectic basis* will as usual be a basis $\alpha_1, \ldots, \alpha_g; \beta_1, \ldots, \beta_g$ for $H_{\mathbb{Z}}$ relative to which

$$Q = \begin{pmatrix} 0 & I \\ -I & 0 \end{pmatrix}.$$

Setting $H = H_{\mathbb{Z}} \otimes \mathbb{C}$, a *(principally) polarized Hodge structure of weight one*[3] is given by a filtration

$$H = F^0 \supset F^1 \supset \{0\}$$

where $\dim F^1 = g$ and where

(1)
$$\begin{cases} Q(u,v) = 0 & u, v \in F^1 \\ \sqrt{-1}Q(u,\bar{u}) > 0 & 0 \neq u \in F^1. \end{cases}$$

Relative to a symplectic basis as above there is a unique basis for F^1 given by the row vectors in a $g \times 2g$ matrix

$$\Omega = (I, Z)$$

---

[2]These partial compactifications are not compact as spaces, but they are compact relative to Hodge-theoretically defined maps to them, in the sense that maps of punctured discs to them extend across the origin. For this reason, for simplicity of terminology we shall simply refer to these partial compactifications as simply compactifications.

[3]Since the polarized Hodge structures we shall consider will all be principally polarized, we shall drop the adjective "principally" in section III.

where the bilinear relations (1) are

$$(2) \qquad \begin{cases} Z = {}^t Z \\ \operatorname{Im} Z > 0 \,. \end{cases}$$

This period matrix representation will be useful in computation of examples.

We denote by D_g the set of polarized Hodge structures of weight one and set

$$\begin{cases} \Gamma_g = \operatorname{Aut}(H_{\mathbb{Z}}, Q) \\ G = \operatorname{Aut}(H_{\mathbb{R}}, Q) \end{cases}$$

where $H_{\mathbb{R}} = H_{\mathbb{Z}} \otimes R$. Then G acts transitively on D_g and the quotient

$$(3) \qquad D_g/\Gamma_g \simeq \mathcal{A}_g$$

of equivalence classes of polarized weight one Hodge structures may be identified with the moduli space of PPAV's.

Algebraic families of PPAV's over a curve may be localized to give a variation of the polarized Hodge structures of weight one (VHS)

$$(4) \qquad f : \Delta^* \to D_g/\Gamma_g$$

over the punctured disc $\Delta^* = \{t : 0 < |t| < 1\}$. Denoting $\mathcal{U} = \{z : \operatorname{Im} z > 0\}$ the upper-half-plane we may lift (4) to

$$
\begin{array}{ccc}
\mathcal{U} & \xrightarrow{\ \tilde{f}\ } & D_g \\
\downarrow & & \downarrow \\
\Delta^* & \xrightarrow{\ f\ } & D_g/\Gamma_g
\end{array}
$$

where

$$\tilde{f}(z+1) = \tilde{f}(z)T$$

with $T \in \Gamma_g$ being the monodromy transformation. It is well-known that, replacing Δ^* by a finite covering by setting $t = s^k$, we will have

$$T = I + N$$

where

$$N = \log T \in \mathcal{G}_{\mathbb{Q}}$$

with $\mathcal{G}_\mathbb{Q}$ denoting the rational subspace of the Lie algebra $\mathcal{G} \subset \mathrm{Hom}(H_\mathbb{R}, H_\mathbb{R})$ of G, and where

$$(5) \qquad\qquad\qquad\qquad N^2 = 0 \ .$$

Using this we may define the *monodromy weight filtration*

$$(6) \qquad\qquad\qquad \{0\} \subset W_0 \subset W_1 \subset W_2 = H$$

where

$$(7) \qquad\qquad\qquad \begin{cases} W_0 = \mathrm{Im}(N) \\ W_1 = \mathrm{Ker}(N) \ . \end{cases}$$

One central reason for the importance of monodromy is that the VHS (4) is asymptotic to a nilpotent orbit, as follows: Let $\check{D}_g$ be the g planes in H satisfying the first bilinear relation in (1), and define

$$\tilde{h} : \mathcal{U} \to \check{D}_g$$

by

$$\tilde{h}(z) = f(z) \cdot \exp(-zN) \ .$$

Then $\tilde{h}(z + 1) = \tilde{h}(z)$ so that $\tilde{h}$ descends to a map

$$h : \Delta^* \to \check{D}_g \ .$$

The main results are

$$(8) \quad \begin{cases} \text{(i)} \quad h \text{ extends across the origin, and we set } F_0 = h(0); \\ \text{(ii)} \quad \text{for } \mathrm{Im}\, z \gg 0,\ F_0 \cdot \exp(zN) \in D_g;\ \text{and} \\ \text{(iii)} \quad \text{in a precise sense, the nilpotent orbit} \\ \qquad\qquad F_0 \cdot \exp\left(\frac{\log t}{2\pi\sqrt{-1}} N\right) \in D_g/\{T^n\} \\ \quad is\ asymptotic\ as\ t \to 0\ to\ (4). \end{cases}$$

Remark: We may choose a symplectic basis that *over* $\mathbb{Q}$ spans W_0; say

$$W_0 = \mathrm{span}_\mathbb{Q}\{\alpha_1, \ldots, \alpha_l\} \ .$$

Then

$$N = \begin{pmatrix} 0 & \eta \\ 0 & 0 \end{pmatrix} \qquad (g \times g \text{ blocks})$$

where

$$\eta = \begin{pmatrix} \eta_{11} & 0 \\ 0 & 0 \end{pmatrix} \begin{matrix} \}l \\ \}g-l \end{matrix}$$
$$\underbrace{}_{l} \underbrace{}_{g-l}$$

where $\eta_{11} = {}^t\eta_{11}$. Then $\tilde{h}(z)$ has normalized period matrix

$$\tilde{Z}(z) - z\eta =: \tilde{W}(z) .$$

The nilpotent orbit then has normalized period matrix

$$\left(\frac{\log t}{2\pi\sqrt{-1}} \right) N + W(t)$$

which is positive definite for $0 < |t| < \epsilon$. It follows that

(9) $$\eta_{11} > 0 .$$

The above discussion extends to a localized several variable **VHS**

(10) $$f : (\Delta^*)^k \to D_g/\Gamma_g$$

(one may also include a Δ^m factor as parameters) with commuting monodromies $T_1, \ldots, T_k$, which again going to finite coverings may be assumed to satisfy

(11) $$\begin{cases} T_i = I + N_i \\ N_i^2 = 0 \\ [N_i, N_j] = 0 . \end{cases}$$

This then leads to a several variable nilpotent orbit

(12) $$F_0 \cdot \exp \left(\sum_{i=1}^k \frac{\log t_i}{2\pi\sqrt{-1}} N_i \right) .$$

Under a rescaling

$$t_i \to \exp \left(\frac{u_i}{2\pi\sqrt{-1}} \right) t_i \qquad (u_i \in \mathbb{C})$$

we have

(13) $$F_0 \to F_0 \cdot \exp \left(\sum_{i=1}^k u_i N_i \right) .$$

We set

$$\begin{cases} \mathbf{N} = (N_1, \ldots, N_k) \\ N_\lambda = \sum_{i=1}^{i} \lambda_i N_i \, , \end{cases}$$

and then by (11) we have for all λ

$$N_\lambda^2 = 0 \, .$$

A crucial fact is (cf. Cattani, loc. cit.):

(14) *For $\lambda_i \neq 0$ the monodromy weight filtration $W_k(N_\lambda)$ defined by N_λ is independent of λ.*

From that reference one also has

(i) *The filtration F_0 defines a Hodge structure of weight p on*

$$Gr_p W(N_\lambda) = W_p(N_\lambda)/W_{p-1}(N_\lambda), \qquad p = 0, 1, 2 \, .$$

(ii) *The Hodge structure on $Gr_p W(N_\lambda)$ is polarized by quadratic forms constructed from Q and N_λ.*

(15) *Remark:* From (13) and the definitions it follows that the Hodge structure on $Gr_\bullet W(N_\lambda)$ is well defined, as in the limit mixed Hodge structure on $W_1(N_\lambda)$.

We are now ready to define the boundary components that arise in the compactifications $\overline{\mathcal{A}}_{g,\Sigma}$. Let

$$W_0 \subset H$$

be an isotropic subspace, $W_1 = W_0^\perp$ and set

$$\eta(W_0) = \{N \in \mathcal{G} : \operatorname{Im} N \subset W_0\} \, .$$

Then $\ker(N) \supset W_1$ and we set

$$\eta^+(W_0) = \{N \in \eta(W_0) : Q_N > 0\}$$

where Q_N is the symmetric form on H/W_1 defined for $u, v \in H$ by

$$Q_N(u, v) = Q(u, Nv) \, .$$

The filtration $\{0\} \subset W_0 \subset W_1 \subset H$ is then the monodromy weight filtration of any $N \in \eta^+(W_0)$ as well as that of any rational cone

$$\sigma = \{\sum_{i=1}^{k} \lambda_i N_i : \lambda_i \in \mathbb{R}^+, N_i \in \overline{\eta^+(W_0)}\} \, ,$$

where we assume that σ contains an element in $\eta^+(W_0)$ and that the N_i are in $\mathcal{G}_{\mathbb{Q}}$. We set

$$B(\sigma) = \left\{ \begin{array}{c} F^1 \in \check{D}_g : (F^1, W_0) \text{ defines a mixed Hodge} \\ \text{structure polarized for every } N \in \text{Int } \sigma. \end{array} \right\}$$

(16) **Definition:** *The boundary component associated to* (W_0, σ) *is defined by*

$$\mathcal{B}(\sigma) = B(\sigma)/\exp \sigma_{\mathbb{C}} = \exp \sigma_{\mathbb{C}} \cdot D_g/\exp \sigma_{\mathbb{C}} \, .$$

The equality in the definition is a result; cf. Cattani, loc. cit.

The compactifications $\overline{A}_{g,\Sigma}$ are constructed using a fan Σ, which is a collection of rational cones as above having certain incidence properties and where the isotropic subspace is replaced by a flag

$$\{0\} \subset S_1 \subset S_2 \subset \cdots \subset S_g$$

of isotropic subspaces where $\dim S_i = i$. The details of this construction, for which we refer to Cattani and the references cited there, are not necessary for the discussion in this paper.

We conclude this section with two examples.

Example 1: We use the period matrix notation above and set

$$\eta_i = \begin{pmatrix} \delta_{ij} & 0 \\ 0 & 0 \end{pmatrix} \begin{array}{c} \}l \\ \}g-l \, . \end{array}$$
$$\underbrace{}_{l} \underbrace{}_{g-l}$$

The corresponding nilpotent orbit is

(17)
$$\begin{pmatrix} \frac{\log t_i}{2\pi\sqrt{-1}} & & 0 & \\ & \ddots & & 0 \\ 0 & & \frac{\log t_l}{2\pi\sqrt{-1}} & \\ \hline & 0 & & 0 \end{pmatrix} + \begin{pmatrix} Z_{11} & Z_{12} \\ Z_{\lambda 1} & Z_{22} \end{pmatrix}$$

where

$$(18) \quad \begin{cases} Z_{11} = {}^tZ_{11} \\ Z_{21} = {}^tZ_{12} \\ Z_{22} = {}^tZ_{22}, \quad \operatorname{Im} Z_{22} > 0 . \end{cases}$$

Under a rescaling as above we have

$$Z_{11} \to Z_{11} + \begin{pmatrix} u_1 & & 0 \\ & \ddots & \\ 0 & & u_l \end{pmatrix} .$$

Denoting this rational cone by σ_l, it follows that we may choose a unique *normalized representative* of each point in $\mathcal{B}(\sigma_l)$ to be the period matrix given by the conditions (18) together with

$$(Z_{11})_{ii} = 0 \qquad 1 \leqq i \leqq l .$$

In particular,

$$(19) \qquad \operatorname{codim}_{\overline{A}_{g,\Sigma}}(\mathcal{B}(\sigma_l)) = l .$$

Below we shall see that this boundary component corresponds under the Torelli map to the boundary component in $\overline{\mathcal{M}}_g$ given by the image of $\mathcal{M}_{g-l,2l}$.

Example 2: We define $N_1, \ldots, N_l$ by

$$\eta_i = \left(\begin{array}{cc} \delta_{ij} & 0 \\ 0 & 0 \end{array} \right) \begin{array}{l} \}l-1 \\ \}g-l+1 \end{array}$$
$$\underbrace{\qquad}_{l-1} \underbrace{\qquad}_{g-l+1}$$

$$\eta_l = \left(\begin{array}{cc} \begin{array}{c} 1\ldots 1 \\ 1\ldots 1 \end{array} & 0 \\ 0 & 0 \end{array} \right) \begin{array}{l} \}l-1 \\ \}g-l+1 . \end{array}$$
$$\underbrace{\qquad}_{l-1} \underbrace{\qquad}_{g-l+1}$$

For the corresponding nilpotent orbit of the form (12)

$$\sum_{i=1}^{l} \frac{\log t_i}{2\pi\sqrt{-1}} \eta_i + \begin{pmatrix} Z_{11} & Z_{12} \\ Z_{21} & Z_{22} \end{pmatrix} \begin{matrix} \}l-1 \\ \}g-l+1 \end{matrix}$$
$$\underbrace{\phantom{Z_{11}}}_{l-1} \quad \underbrace{\phantom{Z_{12}}}_{g-l+1}$$

we have under rescaling that

$$Z_{11} \to Z_{11} + \begin{pmatrix} u_1 & & 0 \\ & \ddots & \\ 0 & & u_{l-1} \end{pmatrix} + \begin{pmatrix} u_l & \cdots & u_l \\ \vdots & & \vdots \\ u_l & \cdots & u_l \end{pmatrix} .$$

Denoting this rational cone by $\sigma_{l,1}$, we may as above choose a unique normalized representative of each point in $\mathcal{B}(\sigma_{l,1})$ to be the period matrix given by the conditions (18) (where the block sizes are as in this example) together with

$$(20) \qquad \begin{cases} (Z_{11})_{ii} = 0 & 1 \leqq i \leqq l-1 \\ (Z_{11})_{1,l-1} = 0 \ . \end{cases}$$

In particular

$$(21) \qquad \mathrm{codim}_{\overline{\mathcal{A}}_{g,\Sigma}}(\mathcal{B}(\sigma_{l,1})) = l \ .$$

Below we shall see that this boundary component corresponds under the Torelli map to the images of all inclusions

$$\Sigma_{\mathbf{g}} \hookrightarrow \overline{\mathcal{M}}_g, \qquad g = g_1 + g_2 + l - 1 \ .$$

III(ii). *Maps to* $\overline{\mathcal{A}}_{g,\Sigma}$

Keeping the previous notation, assigning to a general curve $X_s \in |L|$ the polarized Hodge structure on $H^1(X_s)$ defines rational maps

$$(22) \qquad |L| \overset{\Psi}{-\!-\!\to} \overline{\mathcal{A}}_{g,\Sigma}$$

$$|L|_\zeta \subset |\mathcal{L}| \overset{\Psi}{-\!-\!\to} \overline{\mathcal{A}}_{g,\Sigma} \ .$$

We assume given $s_0 \in \mathrm{sing}\, \nu_\zeta$ where as above

$$X_{s_0} = C_1 \cup C_2 \in \Sigma_{\mathbf{g}} \ .$$

(23) **Theorem:** *Working in a neighborhood of* s_0, *in the diagram*

(24)
$$
\begin{array}{ccc}
|\mathcal{L}| & \xrightarrow{\ \Psi\ } & \overline{\mathcal{A}}_{g,\Sigma} \\
\cup & & \cup \\
\operatorname{sing}\boldsymbol{\nu}_\zeta & \dashrightarrow & \mathcal{B}(\sigma(l,1))
\end{array}
$$

the mapping Ψ *is generically transverse along* $\operatorname{sing}\boldsymbol{\nu}_\zeta$.

This means that

(25)
$$
\begin{cases}
\operatorname{codim}_{\mathcal{A}_{g,\Sigma}}(\mathcal{B}(\sigma(l,1))) = l & \text{(cf. section III(i))} \\
\operatorname{codim}_{|\mathcal{L}|}(\operatorname{sing}\nu_\zeta) = l & \text{(cf. section II(ii))} \\
\Psi^{-1}(\mathcal{B}(\sigma(l,1))) = \operatorname{sing}\boldsymbol{\nu}_\zeta &
\end{cases}
$$

and that an injectivity condition on the differential of Ψ is satisfied. As in section II(ii) the above result and its proof will give an excess intersection formula

(26) **Corollary:** $\Psi^{-1}(\mathcal{B}(\sigma(l,1))) \subset |\mathcal{L}|_\zeta$ *and*

$$
\Psi_\zeta^*([\mathcal{B}(\sigma(l,1))]) = [\Psi^{-1}(\mathcal{B}(\sigma(l,1)))]_\wedge c_h(\mathcal{H}^{0,2})
$$

where $\Psi_\zeta = \Psi\,|_{|\mathcal{L}|_\zeta}$.

The proof of theorem (23) and its corollary will occupy the rest of this section. Although it could be given directly, since the mapping Ψ factors

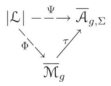

where τ is the *Torelli map* and Φ is the map previously studied in section II, we shall study τ and this will lead to the proof. The result that τ is a regular map is due to Mumford (cf. Alexeev [1] and the references cited therein). For our purposes we will need the explicit description of τ in the case of interest here.

We being by analyzing an l-parameter curve degeneration around a general point

$$
C_0 \in \mathcal{M}_{g-l,2l} \subset \overline{\mathcal{M}}_g
$$

cf. Cattani [5] and Friedman [10].

The picture is

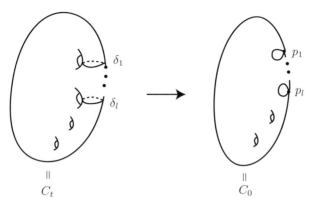

where $t = (t_1, \ldots, t_l) \in \Delta^l$ and as $t_i \to 0$ the vanishing cycle δ_i shrinks to a node p_i. We assume the family is smooth over $(\Delta^*)^l$ and the monodromy around $t_i = 0$ is the Picard-Lefschetz transformation T_{δ_i} corresponding to the vanishing cycle δ_i. We set

$$N_i = \log T_{\delta_i} = T_{\delta_i} - I \ .$$

We denote by

$$\hat{C}_0 \xrightarrow{\ \pi\ } C_0$$

the normalization with $\pi^{-1}(p_i) = q_i + r_i$ and denote by ρ_i a path connecting q_i and r_i in $\hat{C}_0$

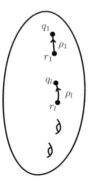

Let $\omega_1(t), \ldots, \omega_g(t)$ be a basis for $H^0(\Omega^1_{C_t})$ normalized so that the period matrix $Z(t)$ of C_t is normalized where the symplectic basis $\alpha_1, \ldots, \alpha_g$; $\beta_1, \ldots, \beta_g$ for $H_1(C_t, \mathbb{Z}) \cong H^1(C_t, \mathbb{Z})$ is chosen so that

$$\alpha_1 = \delta_1, \ldots, \alpha_l = \delta_l \ .$$

Then, again following the notations in section III(i) above, it is well known that

$$Z(t) = \begin{pmatrix} \sum_i \frac{\log t_i}{2\pi\sqrt{-1}}\eta_i + W(t) & Z_{12}(t) \\ Z_{21}(t) & Z_{22}(t) \end{pmatrix}$$

where $W(t)$, $Z_{12}(t) = {}^t Z_{21}(t)$, and $Z_{22}(t)$ are holomorphic in $t_1, \ldots, t_l$.

It is also well-known that as $t \to 0$ and for $1 \leq i \leq 1$

$$\omega_i(t) \to \eta_{q_i, r_i}$$

where η_{q_i, r_i} is a differential of the 3rd kind (dtk) on $\hat{C}_0$ with polar divisor at $q_i + r_i$ and residues ± 1. We then have the following interpretation:

(i) $Z_{22}(0)$ is the period matrix of C_0;

(ii) using the bilinear relations between differentials of 1st and 3rd kinds, the entries in $Z_{12}(0)$ are interpreted as the Abel-Jacobi images

$$AJ_{\hat{C}_0}(q_i - r_i) \in J(\hat{C}_0)$$

of the divisors $q_i - r_i$ on $\hat{C}_0$;

(iii) the off-diagonal entries in $W(0)$ are the integrals

$$\int_{\rho_i} \eta_{q_j, r_j} = \int_{\rho_j} \eta_{q_i, r_i} \, ,$$

where the equality comes from the bilinear relations for dtk's.

It may now be checked that this implies that in the diagram

$$\begin{array}{ccc} \overline{\mathcal{M}}_g & \xrightarrow{\tau} & \overline{\mathcal{A}}_{g,\Sigma} \\ \cup & & \cup \\ \mathcal{M}_{g-l,2l} & \xrightarrow{\tau} & \mathcal{B}(\sigma(l)) \end{array}$$

the Torelli map is transverse along a general l-nodal irreducible curve.

A more conceptual argument may be given as follows: The nilpotent orbit

$$(27) \qquad (u_1, \ldots, u_l) \to \begin{pmatrix} \sum_{i=1}^{l} u_i \eta_i & Z_{12} \\ Z_{21} & Z_{22} \end{pmatrix}$$

is linear in $u_1, \ldots, u_l$. Let

$$\sum_{i,j=1}^{l} a_{ij}^{\lambda} Z_{ij}, \qquad Z_{ij} = Z_{ji} \text{ and } \lambda = 1, \ldots, l(l-1)/2$$

be a basis for the linear functions on $l \times l$ symmetric matrices that annihilate the image of the top left hand block in (27). Recalling our notation $\alpha_1, \ldots, \alpha_g; \beta_1, \ldots \beta_g$ for the symplectic basis for $H_1(C_t, \mathbb{Z})$, we have

As $t \to 0$ the integrals

$$\sum_{i,j} a_{ij}^{\lambda} \int_{\beta_j} \omega_i(t)$$

remain finite and tend to the $l(l-1)/2$ quantities

$$\sum_{i,j} a_{ij}^{\lambda} \int_{\rho_j} \eta_{q_i, r_i} .$$

The quantities give the upper left hand block of

$$\lim_{t \to u} \tau(C_t) ,$$

the other parts of the period matrix being holomorphic in $t_1, \ldots, t_l$.

The point of the more conceptual argument is that it applies to a degeneration $C_t \to C_0$ where $C_0 \in \Sigma_{\mathbf{g}}$ and the degeneration arises by smoothing the l nodes on C_0. The linear combination of the $\int_{\beta_j} \omega_i(t)$ that remain finite are not quite as obvious as in the first case, but their explicit expression is not necessary to conclude that again the Torelli map is transverse in a neighborhood of C_0

Theorem (23) and its corollary (26) now follow from this discussion and that in section II(ii).

III(iii). *Maps to $\overline{\mathcal{J}}_{g,\Sigma}$*

In this section we shall give a preliminary study of the map $\mathcal{V}_\zeta$ in the situations

(28)
$$\begin{array}{ccc}
& \overline{\mathcal{J}}_{g,\Sigma} \supset \Theta(\zeta^2; l, 1) & \\
\mathcal{V}_\zeta \nearrow & \downarrow \quad \downarrow & (X \text{ fixed}) . \\
|L| \xrightarrow{\Psi} \overline{\mathcal{A}}_{g,\Sigma} \supset \mathcal{B}(\sigma(l, 1)) &
\end{array}$$

and

(29)
$$\begin{array}{ccc}
& \overline{\mathcal{J}}_{g,\Sigma} \supset \Theta(\zeta^2; l, 1) & \\
\mathcal{V}_\zeta \nearrow & \downarrow \quad \downarrow & (X \text{ variable}) . \\
|\mathcal{L}|_\zeta \xrightarrow{\Psi_\zeta} \overline{\mathcal{A}}_{g,\Sigma} \supset \mathcal{B}(\sigma(l, 1)) &
\end{array}$$

Here $\overline{\mathcal{J}}_{g,\Sigma}$ is a compactification of $\mathcal{J}_g \to \mathcal{A}_g$ corresponding to a choice of fan Σ (cf. Alexeev [1] and the references cited therein), and $\Theta(\zeta^2; l, 1)$ is a boundary component to be specified below. The objective will be to give a generic transversality result and subsequent excess intersection formula analogous to theorem (23) and corollary (26) above.

The reasons for studying (28) and (29) are

(i) the maps $\mathcal{V}_\zeta$ and $\boldsymbol{\mathcal{V}}_\zeta$ are defined purely Hodge theoretically, in contrast to the maps ν_ζ and $\boldsymbol{\nu}_\zeta$ in section II(iv) above; and

(ii) these maps depend on ζ, in contrast to the maps ψ and Ψ that were studied in section III(ii) just above.

The study is preliminary in that the boundary component $\Theta(\zeta^2; l, 1)$ will here be defined only indirectly.

Following the discussion of (28) and (29) we shall give some general remarks concerning the codimension of *any* component of $\rho^{-1}(\mathcal{B})$, where ρ is one of the rational maps studied above and $\mathcal{B}$ is the corresponding boundary component.

Rather than study the maps ν_ζ and $\boldsymbol{\nu}_\zeta$ directly, we shall study the composition of the map $\mathcal{V}_\zeta$ and $\boldsymbol{\mathcal{V}}_\zeta$ in section II(iv) with the maps in the following diagram:

(30)
$$
\begin{array}{ccc}
\overline{\mathrm{Pic}^0(\mathcal{C})} & \xrightarrow{\;\mu\;} & \overline{\mathcal{J}}_{g,\Sigma} \\
\downarrow{\scriptstyle \pi} & & \downarrow{\scriptstyle \widetilde{\omega}} \\
\overline{\mathcal{M}}_g & \xrightarrow{\;\tau\;} & \overline{\mathcal{A}}_{g,\Sigma}.
\end{array}
$$

We note that for the Torelli map τ we have

(31)
$$\Sigma_{\mathbf{g}} \xrightarrow{\;\tau\;} \mathcal{B}(\sigma(l,1)).$$

The main step in our discussion is the following

(32) **Proposition:** (i) *The number of components of $\pi^{-1}(\Sigma_{\mathbf{g}})$ is l; and* (ii) *similarly, the number of components of $\widetilde{\omega}^{-1}(\mathcal{B}(\sigma(l,1))$ is l.*

In both cases the components have dimension g, and we shall see that distinct components of $\pi^{-1}(\Sigma_{\mathbf{g}})$ map to distinct components of $\widetilde{\omega}^{-1}(\mathcal{B}(\sigma(l,1)))$. We shall define $\Theta(\zeta^2; l, 1)$ to be the component of the boundary of $\overline{\mathcal{J}}_{g,\Sigma}$ to which μ maps $\Gamma(\zeta^2, \mathbf{g})$; i.e. to have

(33)
$$\Gamma(\zeta^2; \mathbf{g}) \xrightarrow{\;\mu\;} \Theta(\zeta^2; l, 1).$$

The transversality and excess intersection formula will then follow from the results in sections II(iv) and III(ii).

Proof of the proposition: Part (i) is in Caparosa [3]. Turning to part (ii), we consider the situation of a family

$$
(34) \qquad\qquad
\begin{array}{c}
A \\
\downarrow{\scriptstyle p} \\
\Delta
\end{array}
$$

of principally polarized abelian varieties $A_t = p^{-1}(t)$, $t \neq 0$ where the total space A is smooth, p is proper, the monodromy T on $H^1(A_t, \mathbb{Z})$ is unipotent; i.e.

$$(T - I)^2 = 0 \,,$$

and the singular fibre A_{t_0} has local normal crossings with all components having multiplicity one and which is the compactification of a connected semi-abelian variety $A^0_{t_0}$ (cf. Alexeev-Nakamura [2]). We may think of (34) as $\widetilde{\omega}^{-1}(\Delta)$ where $\Delta \subset \overline{\mathcal{A}}_{g,\Sigma}$ is a general disc meeting $\mathcal{B}(\sigma(l,1))$ transversely at a general point t_0. Let $\mathbf{t} \in \Delta^*$ be a base point and

$$H^1(A_\mathbf{t}, \mathbb{Z})_{\text{ev}} \subset H^1(A_\mathbf{t}, \mathbb{Z})$$

the space of vanishing cycles. Then

$$(T - I)H^1(A_\mathbf{t}, \mathbb{Z}) \subset H^1(A_\mathbf{t}, \mathbb{Z})_{\text{ev}} \,,$$

and by Clemens-Schmid [9] the quotient

$$G = H^1(A_\mathbf{t}, \mathbb{Z})_{\text{ev}}/(T - I)H^1(A_\mathbf{t}, \mathbb{Z})$$

is a finite group. There is then a *Néron model* $\tilde{A}_{t_0}$ (cf. Clemens [8] and M. Saito [20]) which is an extension of $A^0_{t_0}$ by G

$$0 \to A^0_{t_0} \to \tilde{A}_{t_0} \to G \to 0 \,.$$

(35) **Lemma:** *(i)* $\#G = l$.
    (ii) $\#$ *(components of* $\widetilde{\omega}^{-1}(t_0)) = l$.

Proof: The second statement is a consequence of the first statement together with the general results in the literature.

For the second result we may as above choose a symplectic basis $\delta_1, \ldots, \delta_g, \gamma_1, \ldots, \gamma_g$ for $H^1(A_\mathbf{t}, \mathbb{Z})$ such that

$$T = T_{\delta_1} + \cdots + T_{\delta_{l-1}} + T_\delta$$

where the T_{δ_i} and T_δ are Picard-Lefschetz transformations and where

$$\delta = \delta_1 + \cdots + \delta_{l-1} \ .$$

Setting

$$\lambda = \delta_1 + \cdots + \delta_{l-1}$$

we have

$$\begin{cases} H^1(A_{\mathbf{t}}, \mathbb{Z})_{\text{ev}} = \text{span}_{\mathbb{Z}}\{\delta_1, \ldots, \delta_{l-1}\} \\ (T - I)H^1(A_{\mathbf{t}}, \mathbb{Z}) = \text{span}_{\mathbb{Z}}\{\delta_1 + \lambda, \ldots, \delta_{l-1} + \lambda\} \ . \end{cases}$$

In general, if we have a lattice Λ with basis γ_i and sub-lattice $\Lambda' \subset \Lambda$ with basis γ_i' where

$$\gamma_i' = \Sigma a_i^j \gamma_j, \qquad a_i^j \in \mathbb{Z} \ ,$$

then it is well-known that

$$\#(\Lambda/\Lambda') = \det \|a_i^j\| \ .$$

Taking $\Lambda = H^1(A_t, \mathbb{Z})_w$ and $\Lambda' = (T - I)H^1(A_t, \mathbb{Z})$, our result follows from

$$(\delta_1 + \lambda) \wedge \cdots \wedge (\delta_{l-1} + \lambda) = \delta_1 \wedge \cdots \wedge \delta_{l-1}$$

$$+ \sum_{i=1}^{l} \delta_1 \wedge \cdots \wedge \lambda^i \wedge \cdots \wedge \delta_{l-1}$$

$$= (1 + (l - 1))\delta_1 \wedge \cdots \wedge \delta_{l-1}$$

$$= l\delta_1 \wedge \cdots \wedge \delta_{l-1} \ . \qquad \square$$

We now turn to the question: Let

$$|L| \overset{\rho}{\relbar\joinrel\longrightarrow} \overline{\mathcal{P}} \qquad (X \text{ fixed})$$

be one of the rational maps $\varphi, \psi, \nu_\zeta, \mathcal{V}_\zeta$ considered above, $\Delta \subset \overline{\mathcal{P}}$ the corresponding boundary component. Then, recalling that $L = mH$,

Does there exist m_0 and l such that

(36) $$\text{codim}_{|L|}(\rho^{-1}(\Delta)) = l - h^{2,0}$$

*for **all** components and for $m \geq m_0$?*

Here we are taking X to be fixed — there is a corresponding question without the $h^{2,0}$ term for X variable. We shall not treat this question in full generality (non-reduced components, etc.), but shall assume that

$$C_0 = C_1 \cup C_2 \in \rho^{-1}(\Delta)$$

is a nodal curve with C_1, C_2 smooth, and ask the related question

(37) *Is there an m_0 such that for $m \geq m_0$ the deformation space of $C_0 \in |L|$ keeping the $l = C_1 \cdot C_2$ nodes is smooth and of codimension $l - h^{2,0}$?*

(38) **Theorem:** (i) *The answer to question (37) is in general **no** for the maps φ, ψ that do not depend on ζ. (ii) If*

$$\dim Hg^1(X)_{\mathrm{prim}} = 1 \, ,$$

*i.e. $Hg^1(X, \mathbb{Q})_{\mathrm{prim}} = \mathbb{Q} \cdot \zeta$, the answer is **yes** for ν_ζ and $\mathcal{V}_\zeta$.*

Proof: As in section II(ii), to have an affirmative answer to the question (37) we must in general have

(39) $\qquad h^q(\mathcal{O}_X(C_i)) = 0 \, , \qquad\qquad q > 0$ and $m \geqq m_0$.

Suppose that

$$C_1 \in |\lambda + m_1 H|$$

where $\lambda \in Hg^1(X)_{\mathrm{prim}}$. Then $C_2 \in |-\lambda + (m - m_1)H|$. We shall deal with (39) in the case of C_1, the case of C_2 being similar. For simplicity we shall assume that

(40) $\qquad\qquad\qquad K_X = H$;

the general case may be done by a similar but slightly more complicated argument.

The numerical information at hand for the mapping φ is

(41)
$$\begin{cases} g_i = \frac{1}{2}(\lambda^2 + m_i(m_1 + 1)H^2) + 1 \\ m = m_1 + m_2 \\ l = C_1 C_2 = -\lambda^2 + m_1 m_2 H^2 \\ g = g_1 + g_2 + l - 1 \, . \end{cases}$$

We set

$$C_1(k) = \lambda + kH \, .$$

According to Castelnuovo-Mumford regularity, we will have

(42) $\qquad h^q(\mathcal{O}_X(C_1(k))) = 0, \qquad\qquad q > 0$ and $k \geq k_0$

where k_0 is expressed in terms of the Hilbert polynomial

$$\chi(\mathcal{O}_X(C_1(k))) .$$

With our assumption (40), k_0 is then determined by the quantities

(43)
$$\begin{cases} C_1(0)^2 = \lambda^2 \\ \chi(\mathcal{O}_X) . \end{cases}$$

The numerical quantities given are

$$g, m, l .$$

We want to have the existence of $m_0 = m_0(g_1, g_2, l)$ such that

$$m_1 \geqq k_0 \text{ for } m \geqq m_0 .$$

From (41), (43) we see that this is not the case without an *a priori* bound on λ^2.

Next we consider the map ν_ζ and write

$$C_1 \in |a\zeta + \zeta' + m_1 H|$$

where $a \in \mathbb{Z}$, $\zeta' \in Hg^1(X)_{\text{prim}}$ and $\zeta \cdot \zeta' = 0$. With the assumption in (ii) we have $\zeta' = 0$. Then

$$\nu_\zeta(s_0) \in \overline{\text{Pic}^\circ(C_0)}$$

and the additional information is

$$\text{bidegree } (\nu_\zeta(s_0)) = (a\zeta^2, -a\zeta^2) ,$$

so that we now have an upper bound on $-\lambda^2 = -a^2\zeta^2 > 0$ and can therefore determine the k_0 for which (42) holds. We then observe from (41) that we will have an m_0 such that

$$m_1 \geqq k_0 \text{ for } k_0 \geqq k_0(m, l) \text{ and } m \geqq m_0$$

which was to be proved. More precisely, from

$$\begin{cases} m_1 m_2 \sim l \\ m_1 + m_2 = m \end{cases}$$

where "$\sim$" means modulo constants, we have

$$m_i \sim m_i(m, l) \qquad\qquad i = 1, 2 ,$$

which implies the result. $\square$

IV. Analysis of $\operatorname{sing} \nu_\zeta$ on Fourfolds

IV(i). *Structure of the space of cycles*

In the classical case $n = 1$ of curves on a surface, given

$$\zeta \in Hg^1(X)_{\text{prim}}$$

then the *principal components* of $\operatorname{sing} \nu_\zeta$, i.e. by definition those corresponding to nodal curves X_{s_0} with two components C_1, C_1 where

$$[C_1] = \zeta + m_1 H$$
$$[C_2] = -\zeta + m_2 H$$
$$L = (m_1 + m_2)H$$

are well-behaved provided

$$m_1 \gg 0, \ \ m_2 \gg 0 \ \text{relative to} \ \ \zeta \ .$$

In this case

$$H^q(\mathcal{O}_X(C_i)) = 0 \ \text{for} \ q \gg 1, \quad i = 1, 2 \ .$$

From

$$0 \to \mathcal{O}_X \to \overset{2}{\underset{i=1}{\oplus}} \mathcal{O}_X(C_i) \to \mathcal{O}_X(L) \to \mathcal{O}_X(\Delta) \to 0$$

where

$$\Delta = C_1 \cdot C_2 \ \text{is the the singular locus of} \ X_{s_0} = C_1 \cup C_2$$

and unobstructedness of $|C_1|, |C_2|$, we conclude

$$
\left\{
\begin{array}{c}
\text{codim \ in } |L| \text{ of deformations} \\
\text{of } X_{s_0} \text{ remaining reducible}
\end{array}
\right\} = \dim |L| - \dim |C_1| - \dim |C_2|
$$

$$= \dim \operatorname{Im}(H^0(\mathcal{O}_X(L)) \to H^0(\mathcal{O}(\Delta)))$$
$$= l - h^2(\mathcal{O}_X)$$
$$= l - h^{2,0}(X)$$

where

$$l = \ \text{number of nodes.}$$

Thus

$$\operatorname{codim}_{|L|}(\operatorname{sing} \nu_\zeta)_{s_0} = l - h^{2,0}(X) \ ,$$

a Hodge-theoretic quantity. Note, however, that we may expect there to be other components of $\mathrm{sing}\,(\nu_\zeta)$ where the codimension comes out differently (cf. section III(iii) above).

Note that

$$h^{2,0}(X) = \text{``expected''}\ \mathrm{codim}_{\mathcal{M}_X}(\mathcal{M}_{X,\zeta})$$

although it is only inequality the actual codimension. If equality does hold, then

$$\mathrm{codim}_{|\mathcal{L}|}(\mathrm{sing}\,\boldsymbol{\nu}_\zeta)_{s_0} = l\,.$$

In this section, *assuming the Hodge conjecture* we shall investigate the analogous questions in the $n = 2$ case of surfaces on a fourfold, both for X fixed and for (X, ζ) varying. Informally stated what we shall perphaps not non-expectedly find is

> The situation for $n = 2$ is completely different than the classical case.

For example, as will be seen in the next section, one's first guess for the Noether-Lefschetz loci is

$$\text{``expected''}\ \mathrm{codim}_{\mathcal{M}_X}(\mathcal{M}_{X,\zeta}) = h^{4,0}(X) + h^{3,1}(X)\,.$$

Because of the infinitesimal period relations this is first corrected to

$$\text{``expected''}\ \mathrm{codim}_{\mathcal{M}_X}(\mathcal{M}_{X,\zeta}) = h^{3,1}(X)\,.$$

But the integrability conditions associated to the infinitesimal period relations give an addition correction term

$$(1) \qquad\qquad \text{``expected''}\ \mathrm{codim}_{\mathcal{M}_X}(\mathcal{M}_{X,\zeta}) = h^{3,1}(X) - \sigma_\zeta$$

where σ_ζ is a Hodge-theoretic term depending on ζ. We shall also see in Section IV(iv) that (1) is sharp in an interesting set of examples.

The main general results on the structure of the space of cycles hold for all n and date to the work of Kleiman [18]:[4]

---

[4]We would like to especially thank Rob Lazarsfeld for discussions related to the work of Kleiman.

(2) *Assuming the Hodge conjecture, given $\zeta \in Hg^n(X)_{\text{prim}}$ we have for some $k \in \mathbb{Z}^*$*

$$k\zeta = [W - H]$$

where W is a smooth, codimension n subvariety and H is a smooth complete intersection of hypersurface sections;

In (2) we may assume that W is the degeneracy locus of general sections $\sigma_1, \ldots, \sigma_{r-1}$ of a rank r vector bundle $F \to X$;

Under some mild assumptions we shall also give the operations on pairs (F, S) where $S \in \text{Grass}(r - 1, H^0(F))$ that correspond to addition of cycles, modulo complete intersections.

(3) *Denoting by $\text{def}(*)_1$ the 1st order deformations of $*$, assuming that F is chosen sufficiently ample and that*

$$h^{1,0}(X) = h^{2,0}(X) = 0$$

we have

$$\dim(\text{def}_1(W)) = \dim(\text{def}_1(F)) + \dim \ \text{Grass}(r - 1, H^0(F)).$$

The assumptions are not essential, and we have given them for purposes of illustrating the essential geometric point.

(4) *For $L \gg 0$ relative to W, we may find nodal $X_{s_0} \in |L|$ with*

$$\text{(i)} \quad W \subset X_{s_0}$$

$$\text{(ii)} \quad \zeta \neq 0 \ \text{in} \ H^{2n}(\widehat{X}_{s_0})$$

where $\widehat{X}_{s_0}$ is the standard desingularization of X_{s_0} and cohomology is with $\mathbb{Q}$ coefficients;

(5) *Again for $L \gg 0$ and $W \subset X_{s_0}$ as above, to 1st order any deformation of X_{s_0} preserving the nodes carries W along with it; we write this as*

$$\text{def}_1 \left(\begin{array}{c} X_{s_0} \ \text{preserving nodes} \\ \text{and modulo } H^0(\mathcal{I}_W(L)) \end{array} \right) = \text{def}_1(W).$$

We shall also give cohomological expressions for dim and codim of $\text{def}_1(*)$ above (in fact, this is how the various results are proved). One such is

(6) $$\text{codim} \left\{ \text{def}_1 \left(\begin{array}{c} X_{s_0} \ \text{preserving} \\ \text{the nodes} \end{array} \right) \right\} = l - h^1(\mathcal{I}_{\Delta_{s_0}/X}(L))$$

where

(7) $\begin{cases} l = \# \text{ nodes} \\ \Delta_{s_0} \text{ is the set of nodes} \end{cases}$

and $\mathfrak{I}_{\Delta_{s_0}/X}(L) \subset \mathcal{O}_X(L)$ is the ideal sheaf of Δ_{s_0} in X.

All of the above is for fixed X and any n. For $n = 1$, from the second proof of Theorem (2) in section II(i) we have in the circumstances at the beginning of this section

(8) $$h^1(\mathfrak{I}_{\Delta_{s_0}/X}(L)) = h^{2,0}(X) .$$

The RHS of this equality is a Hodge-theoretic term with the geometric interpretation

$$h^{2,0}(X) = \text{"expected" } \operatorname{codim}_{\mathcal{M}_X}(\mathcal{M}_{X,\varsigma}) .$$

When $n = 2$ we will see that

$$h^1(\mathfrak{I}_{\Delta_{s_0}/X}(L)) = \text{"expected" } \operatorname{codim}_{\mathcal{M}_X}(\mathcal{M}_{X,\varsigma}) + \begin{pmatrix} \text{algebro-geometric} \\ \text{correction term} \\ \text{depending on } W \end{pmatrix}$$

(9)

where for $n = 2$ the correction term is the image of the mapping σ in the exact sequence

$$0 \to H^1(N_{W/X}) \to H^1(\mathfrak{I}_{\Delta_{s_0}/X}(L)) \xrightarrow{\sigma} H^2(\Lambda^2 N_{W/X} \otimes L^*)$$

to be derived below, and which will be seen to be non-zero in typical examples.

This discussion refers to the situation when X is fixed. When X varies, and using the notation

$$\Sigma_{X,L} = \left\{ \begin{array}{c} \text{sheaf of 1}^{\text{st}} \text{ order linear} \\ \text{differential operators on } \mathcal{O}_X(L) \end{array} \right\}$$

we will see below that there is a map

(10) $$H^1(\Sigma_{X,L}) \to H^1(\mathfrak{I}_{\Delta_{s_0}/X}(L))$$

such that

$$\operatorname{codim} \operatorname{def}_1 \begin{pmatrix} \text{pairs } (X, X_{s_0}) \\ \text{preserving} \\ \text{the nodes} \end{pmatrix} = \dim(\text{image of the map (10)}) .$$

Proof of (4). This is standard. The point is that for a section $s_0 \in H^0(L)$ such that

$$W \subset X_{s_0},$$

i.e., $s_0 = 0$ on W, the differential ds_0 is well-defined along W and gives a section

$$ds_0 \in H^0(N^*_{W/X}(L)) .$$

By choosing $L \gg 0$ relative to W, we may insure that there exists such an s_0 such that ds_0 has isolated, non-degenerate zeroes. This is equivalent to X_{s_0} having ordinary double points (nodes).

There is a standard desingularization

$$\widehat{X}_{s_0} \to X_{s_0}$$

where each node p_i is blown up to a smooth quadratic surface $Q_i \subset \widehat{X}_i$. The proper transform $\widehat{W}$ of W is smooth and the induced map $H_4(\widehat{W}) \to H_4(W)$ is an isomorphism. Since

$$\langle \zeta, [W] \rangle = k\zeta^2 \neq 0$$

it follows that the pullback

$$\hat{\zeta} \in H^4(\widehat{X}_{s_0})$$

is non-zero.

Proofs of (2) *and* (3). Although we shall give the following arguments in the $n = 2$ case they work in general. A general algebraic cycle may be written

$$Z = Z' - Z''$$

where Z' and Z'' are effective. By passing a complete intersection H of high degree through Z'' we will have

$$Z'' + Z''' = H$$

where Z''' is effective. Setting $V = Z' + Z'''$ we then have

$$Z = V - H$$

where V is effective and H is a complete intersection, which by moving in a rational equivalence class may be assumed to be smooth. We may also assume that V is reduced, since by a similar argument any multiple component of V may be moved by a rational equivalence into a sum of components of multiplicity one.

Following Kleiman [18] there is a resolution (here H is an ample line bundle)

(11) $$0 \to F \to E_3 \to E_2 \to E_1 \to E_0 \to \mathcal{O}_V \to 0$$

where

$$E_i \cong \bigoplus_j \mathcal{O}_X(-m_{ij}H)$$

where F is locally free of rank r (we identify vector bundles and locally free sheaves). It follows (cf. Fulton [11, Ex. 3.2.3]) that

$$\begin{cases} c_1(\mathcal{O}_V) = 0 \\ c_2(\mathcal{O}_V) = \pm[V] \equiv_H c_2(F) \end{cases}$$

where $\equiv_H$ denotes "congruent modulo complete intersections." Still working modulo complete intersections, we may replace F by $F(k)$ for large k and then it is well-known that

$c_2(F)$ *is represented by the degeneracy locus of* $r-1$ *general sections* $\sigma_1, \dots, \sigma_{r-1} \in H^0(F)$.

Moreover, the subvariety where the degeneracy locus drops rank by two has codimension six, so that for generic choice of $\sigma_1, \dots \sigma_{r-1}$ we may assume that

$$W = \{x \in X : \sigma_1(x) \wedge \cdots \wedge \sigma_{r-1}(x) = 0\}$$

will be a smooth surface in X. This establishes (2).

Next there is an exact Eagon-Northcut complex

(12) $$0 \to \overset{r-1}{\bigoplus} \wedge^r F^* \xrightarrow{\oplus \sigma_i} \wedge^{r-1} F^* \xrightarrow{\sigma_1 \wedge \cdots \wedge \sigma_{r-1}} \mathcal{O}_X \to \mathcal{O}_W \to 0 \ .$$

It is a general fact that for any resolution

(13) $$0 \to E_2 \to E_1 \to E_0 \to \mathcal{O}_W \to 0$$

of $\mathcal{O}_W$ by vector bundles, when restricted to W the homology sheaf of (13) is

$$\wedge^2 N_{W/X}^*, N_{W/X}^*, \mathcal{O}_W \ .$$

Applied to (12) this gives, setting $F_W^* = F^*|_W$,

$$0 \to \wedge^2 N_{W/X}^* \to \overset{r-1}{\oplus} \wedge^r F_W^* \to \wedge^{r-1} F_W^* \to N_{W/X}^* \to 0$$

or dually

$$(14) \qquad 0 \to N_{W/X} \to \wedge^{r-1} F_W \to \overset{r-1}{\oplus} \wedge^r F_W \to \wedge^2 N_{W/X} \to 0 \ .$$

Our goal is to compute $H^0(N_{W/X})$ and $H^1(N_{W/X})$; the results are:

$$(15) \quad 0 \to T \ \mathrm{Grass}(r-1, H^0(F)) \to H^0(N_{W/X})/T \ \mathrm{Aut}(F)$$

$$\to H^1(F \otimes F^*) \to 0$$

if

$$(16) \qquad\qquad h^{1,0}(X) = h^{2,0}(X) = 0 \ ,$$

and

$$(17) \qquad\qquad H^1(N_{W/X}) \cong H^2(F \otimes F^*)$$

if

$$(18) \qquad\qquad h^{1,0}(X) = h^{2,0}(X) = h^{3,0}(X) = 0 \ .$$

There are somewhat more elaborate statements that hold without assuming (16) or (18); the point is to have an illustrative result that is valid in interesting examples. We note that (15) gives (3).

We may rephrase (15) and (17) as

(19) *Under the stated assumptions and to* 1st *order, the deformations of W in X are obtained by deforming F and by varying the sections $\sigma_1, \ldots, \sigma_{r-1}$;*

(20) *The obstruction space to deforming W is isomoprphic to the obstruction space to deforming F.*

Proofs of (15) *and* (17). From (14) we have

$$(21) \qquad H^0(N_{W/X}) \cong \ker \left\{ H^0 \left(\wedge^{r-1}(F_W) \xrightarrow{\oplus \sigma_i} H^0(\oplus \wedge^r F_W) \right) \right\} \ .$$

Tensoring (12) with $\wedge^{r-1}F$ gives

$$0 \to \overset{r-1}{\oplus} F^* \to \wedge^{r-1}F^* \otimes \wedge^{r-1}F \to \wedge^{r-1}F \to \wedge^{r-1}F_W \to 0$$

$$\wr\|$$

$$F \otimes F^* \ .$$

Using the spectral sequence in cohomology and

$$\begin{cases} H^q(F^*) = 0, & q < 4 \\ H^p(\wedge^{r-1}F) = 0, & p > 0 \ . \end{cases}$$

We have

$$(22) \qquad 0 \to \frac{H^0(\wedge^{r-1}F)}{H^0(F \otimes F^*)} \to H^0(\wedge^{r-1}F_W) \to H^1(F \otimes F^*) \to 0 \ .$$

Tensoring (12) with $\wedge^r F$ gives

$$(23) \qquad 0 \to \overset{r-1}{\oplus} \mathcal{O}_X \to F \to \wedge^r F \to \wedge^r F_W \to 0 \ .$$

We shall use the notations

$$\begin{cases} V = H^0(F) \\ S = \mathrm{Im}\left\{ \overset{r-1}{\oplus} H^0(\mathcal{O}_X) \overset{\oplus\sigma_i}{\longrightarrow} H^0(F) \right\} \\ Q = V/S \end{cases}$$

so that

$$\mathrm{T\,Grass}(r-1, H^0(F)) \cong \mathrm{Hom}(S, Q) \ .$$

From the cohomology sequence of (23) we have, using the assumption (16),

$$0 \to S \to V \to H^0(\wedge^r F) \to H^0(\wedge^r F_W) \to 0$$

which gives

$$(24) \qquad 0 \to Q \to H^0(\wedge^r F) \to H^0(\wedge^r F_W) \to 0 \ .$$

The dual of (12) is

$$(25) \qquad 0 \to \mathcal{O}_X \to \wedge^{r-1}F \to \overset{r-1}{\oplus} \wedge^r F \to \wedge^2 N_{W/X} \to 0$$

which gives

$$(26) \quad 0 \to H^0(\mathcal{O}_X) \to H^0(\wedge^{r-1}F) \to H^0\left(\overset{r-1}{\oplus}\wedge^r F\right) \to H^0(\wedge^2 N_{W/X}).$$

Using this and (21) gives the commutative diagram with rows and columns

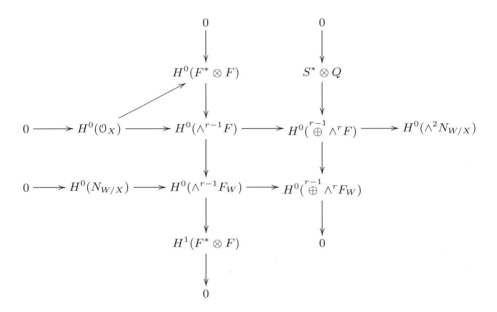

From

$$
\begin{array}{ccc}
S^* \otimes Q & & \\
\downarrow & & \\
H^0\left(\overset{r-1}{\oplus}\wedge^r F\right) & \longrightarrow & H^0(\wedge^2 N_{W/X}) \\
\downarrow & & \| \\
H^0\left(\overset{r-1}{\oplus}\wedge^r F_W\right) & \longrightarrow & H^0(\wedge^2 N_{W/X})
\end{array}
$$

we have that

$$S^* \otimes Q \to H^0(\wedge^2 N_{W/X})$$

is the zero map. This leads to the diagram with exact rows and columns

(27)

$$
\begin{array}{ccc}
& 0 & 0 \\
& \downarrow & \downarrow \\
H^0(F^* \otimes F)/H^0(\mathcal{O}_X) \to & S^* \otimes Q \\
\downarrow & \downarrow \\
0 \to H^0(\wedge^{r-1}F)/H^0(\mathcal{O}_X) \xrightarrow{\alpha} S^* \otimes H^0(\wedge^r F) \\
\downarrow & \downarrow \\
0 \to H^0(N_{W/X}) \to \quad H^0(\wedge^{r-1}F_W) \quad \to S^* \otimes H^0(\wedge^r F_W) \\
\downarrow & \downarrow \\
H^1(F^* \otimes F) & 0 \\
\downarrow \\
0 \, .
\end{array}
$$

The diagram (27) gives

(28)

$$
\begin{array}{ccc}
0 & 0 \\
\downarrow & \downarrow \\
H^0(F^* \otimes F)/H^0(\mathcal{O}_X) \to & S^* \otimes Q \quad \searrow^{\;0} \\
\downarrow & \downarrow \\
0 \to H^0(\wedge^{r-1}F)/H^0(\mathcal{O}_X) \xrightarrow{\alpha} S^* \otimes H^0(\wedge^r F) \to H^0(\wedge^2 N_{W/X}) \\
\downarrow^{r_w} & \downarrow^{r_w} & \| \\
0 \to H^0(N_{W/X}) \to \quad H^0(\wedge^{r-1}F_W) \xrightarrow{\beta} S^* \otimes H^0(\wedge^r F_W) \to H^0(\wedge^2 N_{W/X}) \\
\downarrow^{p} & \downarrow \\
H^1(F^* \otimes F) & 0 \\
\downarrow \\
0 \, .
\end{array}
$$

The row with $H^0(N_{W/X})$ in it is only exact for the first two terms. It follows that

$$
\alpha^{-1}(S^* \otimes Q) \cong S^* \otimes Q
$$

and that

$$
\mathrm{im}(\beta) \subseteq r_W(\mathrm{im}\,\alpha)
$$

It follows that the natural map $(\beta \circ p^{-1})$

$$H^1(F^* \otimes F) \to \frac{S^* \otimes H^0(\wedge^{r-1}/W)}{\operatorname{im}\beta \circ r_W}$$

is zero, because

$$\operatorname{im}(\beta \circ r_W) = \operatorname{im}(r_W \circ \alpha) \supseteq \operatorname{im}\beta .$$

Thus

$$p^{-1}(H^1(F^* \otimes F)) \subseteq \operatorname{im} r_W + \ker\beta = \operatorname{im} r_W + \operatorname{im} H^0(N_{W/X})$$

which implies the surjectivity of the map

$$H^0(N_{W/X}) \to H^1(F^* \otimes F) .$$

Now

$$\ker(H^0(N_{W/X}) \to H^1(F^* \otimes F))$$

injects to Image r_W in $H^0(\wedge^{r-1}F_W)$, and chasing the diagram leads to

$$(29) \qquad \ker(H^0(N_{W/X}) \to H^1(F^* \otimes F)) \to \frac{S^* \otimes Q}{\operatorname{im} H^0(F^* \otimes F)} .$$

Surjectivity comes because

$$S^* \otimes Q \mapsto 0$$

$\operatorname{im} H^0(\wedge^2 N_{W/X})$ by commutativity of the diagram (28) so

$$S^* \otimes Q \mapsto \operatorname{im}\alpha$$

and under r_W on the left,

$$\frac{S^* \otimes Q}{H^0(F^* \otimes F)} \mapsto \frac{\ker\beta}{\ker p} = \operatorname{im} H^0(N_{W/X}) \cap \ker p$$

$$= \ker(H^0(N_{W/X}) \to H^1(F^* \otimes F))$$

from which we get injectivity of (29). □

We have now proved (3) and (4). We shall next prove

$$(30) \qquad \operatorname{def}_1(X_{s_0} \text{ preserving the nodes } \Delta_{s_0}) \cong H^0(\mathcal{I}_{\Delta_s/X}(L)) .$$

Proof: This is a general fact. If locally

$$s(p(t), t) = 0, \qquad \frac{\partial s}{\partial z_i}(p(t), t) = 0, \qquad \text{all } i$$

where z_i are local coordinates on X, then we have

$$\dot{s}(p(0),0) + (ds(p(0),0))\rfloor \dot{p}(0) = 0$$

where $\dot{p}(0) \in T_{p(0)}X$. Since $ds(p(0),0) = 0$ we obtain $\dot{s}(p(0),0) = 0$ which gives

$$\dot{s} \in H^0(\mathfrak{I}_{\Delta_{s_0}}(L)) \,.$$

This is a 1st order condition. The 2nd order condition may be written as

$$d\dot{s}(p(0),0) + \left(d^2 s(p(0),0)\right)\rfloor \dot{p}(0) = 0$$

where $d^2 s \in S^2 T^*_{p(0)}X$ is the Hessian, which is non-degenerate. For any $d\dot{s}(p(0),0)$ this gives a unique $\dot{p}(0)$. $\qquad\square$

We next note the standard interpretation

$$(31) \qquad \begin{cases} H^1(\mathfrak{I}_{\Delta_{s_0}/X}(L)) \ \text{ is the obstruction space to} \\ \operatorname{def}_1(X_{s_0} \ \text{preserving } \Delta_{s_0}) \,. \end{cases}$$

From the cohomology sequence of

$$0 \to \mathfrak{I}_{\Delta_{s_0}/X}(L) \to \mathcal{O}_X(L) \to \mathcal{O}_{\Delta_{s_0}}(L) \to 0$$

we obtain (9) above.

$$(32) \qquad \operatorname{codim} \operatorname{def}_1(X_{s_0} \ \text{preserving } \Delta_{s_0}) = l - h^1(\mathfrak{I}_{\Delta_{s_0}/X}(L))$$

where $l = h^0(\mathcal{O}_{\Delta_{s_0}}(L))$ is the number of nodes.

We will now show that

$$(33) \qquad H^0(N_{W/X}) \cong H^0(\mathfrak{I}_{\Delta_{s_0}/X}(L)/H^0(\mathfrak{I}_{W/X}(L)) \,.$$

Proof. From the cohomology sequence of

$$0 \to \mathfrak{I}_{W/X}(L) \to \mathfrak{I}_{\Delta_{s_0}/X}(L) \to \mathfrak{I}_{\Delta_{s_0}/W}(L) \to 0 \,,$$

where $\mathfrak{I}_{W/X}(L)$ is the ideal sheaf of W tensored with $\mathcal{O}_X(L)$ and the assumption $L \gg 0$ relative to W, we obtain

$$(34) \qquad H^0(\mathfrak{I}_{\Delta_{s_0}}(L)) \cong H^0(\mathfrak{I}_{\Delta_{s_0}/X}(L))/H^0(\mathfrak{I}_{W/X}(L)) \,.$$

Next, since the p_i are ordinary nodes, the Koszul complex associated to

$$ds_0|_W \in H^0(N^*_{W/X}(L))$$

gives the exact sheaf sequence

$$(35) \qquad 0 \to \wedge^2 N_{W/X}(L^*) \to N_{W/X} \to \mathfrak{I}_{\Delta_{s_0}/W}(L) \to 0 \,.$$

Again using the assumption $L \gg 0$ relative to W we obtain

$$(36) \qquad H^0(N_{W/X}) \cong H^0(\mathcal{I}_{\Delta_{s_0}/W}(L)) \ .$$

Comparing with (34) gives (33). □

At this stage we have proved (5).

We next observe the exact cohomology sequence of (35) gives

$$0 \to H^1(N_{W/X}) \to H^1(\mathcal{I}_{\Delta_{s_0}/X}(L)) \to H^2(\wedge^2 N_{W/X} \otimes L^*)$$

where we have used

$$H^1(\mathcal{I}_{\Delta_{s_0}/W}(L)) \cong H^1(\mathcal{I}_{\Delta_{s_0}/X}(L))$$

for $L \gg 0$ relative to W. This is the sequence referred to above and may be interpreted as saying

$$\left\{ \begin{array}{c} \text{the obstruction space} \\ \text{to } \mathrm{def}_1(W \subset X) \end{array} \right\} \subset \left\{ \begin{array}{c} \text{the obstructive space to} \\ \mathrm{def}_1 \left(\begin{array}{c} X_{s_0} \text{ preserving} \\ \text{the nodes} \end{array} \right) \end{array} \right\} .$$

All of the above discussion is for X fixed. We now turn to the situation when X varies leading to (10) and the statement just below. Let then

$$\widehat{X} \xrightarrow{\epsilon} X$$

be the blowup of X at the nodes $p_i \in X_{s_0}$. Then each $\epsilon^{-1}(p_i)$ is a $\mathbb{P}^3$ that we label E_i and that meets the proper transform $\widehat{X}_{s_0}$ of X_{s_0} in a smooth quadric Q_i. We will show that

$$(37) \qquad \mathrm{def}_1(\widehat{X}, \widehat{X}_{s_0}) \cong \mathbb{H}^1(\Theta_X \to \mathcal{I}_{\Delta_{s_0}/X}(L)) \ ,$$

where the LHS denotes the 1st order deformations of the pair $\widehat{X}_{s_0} \subset \widehat{X}$.

Proof. We have

$$0 \to \epsilon_* \Theta_{\widehat{X}} \to \Theta_X \to \underset{i}{\oplus} N_{p_i/X} \to 0 \ .$$

From the diagram whose 1st column is this sequence

$$
\begin{array}{ccc}
0 & & 0 \\
\downarrow & & \downarrow \\
\epsilon_* \Theta_{\widehat{X}} & \to \epsilon_*(\epsilon^* L[-2\sum_i E_i]\big|_{\widehat{X}_{s_0}}) & \cong \mathfrak{I}^2_{\Delta_{s_0}/X}(L) \\
\downarrow & & \downarrow \\
\Theta_X & \mathfrak{I}_{\Delta_{s_0}/X}(L) & \\
\downarrow & & \downarrow \\
\underset{i}{\oplus} N_{p_i/X} & \to \epsilon_*(\epsilon^* L[-\sum_i E_i]\big|_{\widehat{X}_{s_0}}) & \cong L \otimes (\underset{i}{\oplus} N^*_{p_i/X}) \\
\downarrow & & \\
0 & &
\end{array}
$$

we infer that

(38) $$0 \to \mathbb{H}^0\left(\underset{i}{\oplus} N_{p_i/X} \xrightarrow{d^2 s_0} \oplus N^*_{p_i/X}(L)\right) \to \mathbb{H}^1(\Theta_{\widehat{X}} \to N_{\widehat{X}_{s_0}/\widehat{X}})$$

$$\to \mathbb{H}^1(\Theta_X \to \mathfrak{I}_{\Delta_{s_0}/X_{s_0}}(L)) \to 0 \ .$$

Since each p_i is a node

$$N_{p_i/X} \xrightarrow{d^2 s_0} N^*_{p_i/X}(L)$$

is an isomorphism, so the first term in (18) is zero. Since

$$\mathrm{def}_1(\widehat{X}, \widehat{X}_{s_0}) \cong \mathbb{H}^1(\Theta_{\widehat{X}} \to N_{\widehat{X}_{s_0}/\widehat{X}})$$

we may infer (17).   □

IV(ii). *The "layers of Chow" obstruction in higher codimension*

In the classical case of curves on a surface the following is well-known (and comes out of the proof of Theorem (2) in section II(iii)): Keeping the notations from that section, let (X_t, ζ_t) be a variation of (X, ζ) over an analytic neighborhood of $(X, \zeta) \in \mathcal{M}_{X,\zeta}$. Then there exists an m_0 with the following property:

(39) *For $m \geqq m_0$ and* **any** *smooth curve $W \subset X$ with*

$$[W] = \zeta + m[H] \ ,$$

there exists a smooth deformation W_t with

$$[W_t] = \zeta_t + m[H_t] \ .$$

Here, H and H_t are hyperplane sections of X and X_t, respectively. In words we may say

(40) *Any smooth representative of a sufficiently ample Hodge class deforms with the Hodge class.*

The example of Soulé-Voisin (cf. section 4.4.1 in "I") shows that, because of torsion phenomena, in higher codimension (42) will in general be false even in situations where the HC holds. So minimally the statement might be:

(41) *There exists m_0, k_0 such that for $m \geqq m_0$ and* any *smooth n-fold W with*

$$[W] = k_0 \zeta + m[H]$$

there exists a smooth deformation W_t with

$$[W_t] = k_0 \zeta_t + m[H_t] \,.$$

In fact, the HC implies that there exists k_0 and a W as above that deforms with (X_t, ζ_t). But as we shall now see

(42) **Theorem:** *Assuming the HC and for any k_0, there exist examples such that no matter how ample we make W it will not deform with (X_t, ζ_t).*

In other words, assuming the HC we will have (40) for **some**, but not **any**, W.

To construct examples we first observe: To prove theorem (42), it will suffice to find an algebraic cycle $Z \in Z^n(X)$ and a direction $\theta \in T_X \mathcal{M}_{X,\zeta}$ such that

(43) $\begin{cases} \text{(i)} & [Z] = 0 \\ \\ \text{(ii)} & \textit{for any } m, \ Z + mH \textit{ does not deform} \\ & \textit{to } 1^{\text{st}} \textit{ order in the direction } \theta \,. \end{cases}$

The reason is this: Suppose that W deforms with (X_t, ζ_t), and for any k_0 consider

$$W' = k_0 W + Z \,.$$

Suppose that the rational equivalence class of W'

$$\{W'\} \in CH^n(X) \otimes \mathbb{Q}$$

deforms to 1st order in the direction θ. Then the same would be true of

$$\{Z\} \in CH^n(X) \otimes \mathbb{Q} \,.$$

In the examples to be constructed

(44) $\{Z\}$ *will not deform to* 1st *order in the direction* θ.

It follows that $\{W'\}$ will also not deform to 1st order in the direction θ. But by the proof of (3) in section IV(i), for $m \gg 0$ we see that $\{W'\}$ will contain a smooth subvariety W'' with

$$[W''] = k_0 \zeta + m[H]$$

and which does not deform to 1st order in the direction θ.

The geometric idea behind the construction of cycles Z as in (43) is the following:

Denoting by $CH^p(X)$ the Chow group *considered modulo torsion*, according to the conjecture of Bloch and Beilinson there exists a filtration $F^k CH^p(X)$ such that

(i) $Gr^k CH^p(X)$ has a Hodge-theoretic description;
(ii) $F^0 CH^p(X) = CH^p(X)$ and $F^1 CH^p(X)$ is the kernel of the fundamental class mapping;
(iii) $F^2 CH^p(X)$ is the kernel of the Abel-Jacobi mapping;
(iv) $F^{p+1} CH^p(X) = 0$; and
(v) if X is defined over $\overline{\mathbb{Q}}$, then

$$F^2 CH^p(X(\overline{\mathbb{Q}})) = 0 \ .$$

In [12] we have given a Hodge-theoretic construction of such a filtration, assumming (v) and the GHC. Other proposed constructions are given in [16] and [19].

In our paper [13], we have given a construction of the formal tangent spaces $TF^k CH^p(X)$ with the infinitesimal version of the properties (i)–(v), and that is what we use here. Specifically, given a class

$$\{Z\} \in F^k CH^p(X)$$

and a tangent to moduli $\theta \in H^1(\Theta_X)$, we constructed an obstruction class

(45) $$\mathcal{O}_k(\{Z\}, \theta) \in H^{p-k-1, p+1}(X) \otimes \Omega^k_{\mathbb{C}/\mathbb{Q}}$$

to formally deforming the graded piece

$$\{Z\}_k \in Gr^k CH^p(X)$$

to 1st order in the direction θ. In particular:

(46) *If* $\mathcal{O}_k(\{Z\}, \theta) \neq 0$, *then* $\{Z\}$ *does not deform to* 1st *order in the direction* θ.

Note that for $k = p$, for degree reasons

$$\mathfrak{O}_p(\{Z\}, \theta) = 0$$

so that in the $p = 1$ case the only obstruction to deforming $\{Z\}$ to 1st order is that it remain of Hodge type $(1, 1)$. More generally, there are p independent conditions — the first being to remain of Hodge type (p, p) — to deforming $\{Z\} \in CH^p(X)$ to first order.

For $p = 2$ and $\{Z\} \in F^1 CH^2(X)$ the obstruction

$$\mathfrak{O}_1(\{Z\}, \theta) \in H^{0,3}(X) \otimes \Omega^1_{\mathbb{C}/\mathbb{Q}},$$

which suggests that we look for examples when $n = 2$, $\{Z\} \in F^1 CH^2(X)$ and $H^3(\mathfrak{O}_X) \neq 0$.

In general, one has the *arithmetic cycle class* (cf. loc. cit.)

$$[Z]_a \in H^p(\Omega^p_{X/\mathbb{Q}}).$$

There is a mapping

$$\Theta_{X,\mathbb{C}} \to \Theta_{X/\mathbb{Q}}$$

which gives a pairing (setting $\Theta_X = \Theta_{X/\mathbb{C}}$)

(47) $$H^1(\Theta_X) \otimes H^p(\Omega^p_{X/\mathbb{Q}}) \to H^{p+1}(\Omega^{p-1}_{X/\mathbb{Q}})$$

and in loc. cit. it is proved that

(48) *If $\{Z\}$ deforms to 1st order in the direction θ, then*

$$\theta \cdot [Z]_a = 0$$

in $H^{p+1}(\Omega^{p-1}_{X/\mathbb{Q}})$.

At the infinitesimal level one has that the formal tangent space to the Chow groups is given by

$$TCH^p(X) \cong H^{p+1}(\Omega^{p-1}_{X/\mathbb{Q}}).$$

The filtration $F^k TCH^p(X)$ results from the filtration induced on $TCH^p(X)$ by the filtration

$$F^k \Omega^{p-1}_{X/\mathbb{Q}} = \text{image} \left\{ \Omega^k_{\mathbb{C}/\mathbb{Q}} \otimes \Omega^{p-1-k}_{X/\mathbb{Q}} \to \Omega^{p-1}_{X/\mathbb{Q}} \right\},$$

and this is the basis for (45).

In the $p = 2, k = 1$ case the pairing (46) is (omitting the $\Omega^1_{\mathbb{C}/\mathbb{Q}}$ factor and setting $\Omega^1_X = \Omega^1_{X/\mathbb{C}}$)

$$H^1(\Theta_X) \otimes H^2(\Omega^1_X) \to H^3(\mathfrak{O}_X).$$

Suppose that X is defined over $\overline{\mathbb{Q}}$, that the algebraic part $J^2_{\mathrm{alg}}(X)$ of the intermediate Jacobian is non-zero and parameterized by algebraic cycles, and that for

$$TJ^2_{\mathrm{alg}}(X) \subset H^{1,2}(X)$$

the pairing

$$H^1(\Theta_X) \otimes TJ^2_{\mathrm{alg}}(X) \to H^3(\mathcal{O}_X)$$

is non-zero. Then there will be directions $\theta \in H^1(\Theta_X)$ and non-zero classes

$$\{Z\} \in F^1 CH^2(X)$$

that do not deform to 1st order in the direction θ.

Explicit examples may be constructed from Fermat hypersurfaces, as e.g. in the various works of Shioda [21].

Summary:

(i) *The HC implies that there exists $k_0 \in \mathbb{Z}^*$ and a smooth subvariety W with*

(49) $$[W] = k_0 \zeta + mH, \qquad m \gg 0 \, ;$$

which deforms with (X, ζ);

(ii) *Still assuming the HC, there exist smooth W satisfying (49) and which do not deform with (X, ζ), no matter what k_0 is and how large we take m.*

In (ii), there are (at least) two types of obstructions, *torsion obstructions* as in the Kollár-Soulé-Voisin example (cf. section 4.4.1 of "I"), and *layers of Chow obstructions* as described above.

IV(iii). *Codimension of Noether-Lefschetz loci for fourfolds*

As noted above, for an algebraic surface X and Hodge class $\zeta \in Hg^1(X)$, the Noether-Lefschetz locus $\mathcal{M}_{X,\zeta}$ in the moduli space $\mathcal{M}_X$ where ζ remains a Hodge class is an algebraic variety with

(50) $$\mathrm{codim}_{\mathcal{M}_X}(\mathcal{M}_{X,\zeta}) \leqq h^{2,0} \, .$$

As also noted above the inequality (50) is equality in many examples and one may informally say that

(51) $$\text{``expected''} \ \mathrm{codim}_{\mathcal{M}_X}(\mathcal{M}_{X,\zeta}) = h^{2,0} \, .$$

For X a fourfold, as will now be explained the situation is quite different. For $\zeta \in Hg^2(X)$ we have for the Zariski tangent space

(52) $$T_X \mathcal{M}_{X,\zeta} \overset{\text{def}}{=} T_\zeta = \ker\{H^1(\Theta_X) \overset{\zeta}{\longrightarrow} H^{1,3}(X)\} .$$

(53) **Definition:** *We set*

$$\sigma_\zeta = \dim\left(\text{Image}\{T_\zeta \otimes H^{4,0}(X) \to H^{3,1}(X)\}\right) .$$

(54) **Theorem:** *For the Noether-Lefschetz locus we have*

$$\operatorname{codim}_{\mathcal{M}_X}(\mathcal{M}_{X,\zeta}) \leqq h^{3,1} - \sigma_\zeta .$$

We shall prove this statement for the Zariski tangent space, which is sufficient. As will be seen in the next section, we will see that equality holds in (52) in significant examples. For this reason we propose to use the terminology

$$\text{"expected"} \ \operatorname{codim}_{\mathcal{M}_X}(\mathcal{M}_{X,\zeta}) = h^{3,1} - \sigma_\zeta .$$

Proof: Working in an analytic neighborhood of $X \in \mathcal{M}_X$, the Noether-Lefschetz locus is defined by the equations

$$(\zeta_t)^{(1,3)+(0,4)} = 0 .$$

Thus it would seem at first glance that

(55) $$\text{"expected"} \ \operatorname{codim}_{\mathcal{M}_X}(\mathcal{M}_{X,\zeta}) = h^{4,0} + h^{3,1} .$$

However, if $X = X_{t_0}$ since $\zeta_{t_0} \in H^{2,2}(X_{t_0})$ we have from the infinitesimal period relation

(56) $$\text{to } 1^{\text{st}} \text{ order in any direction, } (\zeta'_{t_0})^{0,4} = 0 .$$

Thus (55) might seem to need to be amended to

(57) $$\text{"expected"} \ \operatorname{codim}_{\mathcal{M}_X}(\mathcal{M}_{X,\zeta}) = h^{3,1} .$$

Now (56) is based on the infinitesimal period relation, which may be viewed as an exterior differential system (EDS) on the classifying space for polarized Hodge structures. As for any EDS there are integrability conditions, and these need to be taken into account along integral manifolds, such as $\mathcal{M}_{X,\zeta}$ for the EDS (56). This is what we shall now do.

Denoting as usual the Gauss-Manin correction by ∇, for any

$$\begin{cases} \theta \in H^1(\Theta_X) \\ \omega \in H^{4,0}(X) \end{cases}$$

we may write (56) as

$$\langle \omega, \theta \cdot \zeta \rangle = 0$$

where the pairing is

$$H^{p,q}(X) \otimes H^{4-p,4-q}(X) \to \mathbb{C} \, .$$

Now suppose that

$$\theta \in T_\zeta \subset H^1(\Theta_X)$$

and let $\theta' \in H^1(\Theta_X)$. Then

$$\begin{aligned}
\langle \theta \cdot \omega, \theta' \cdot \zeta \rangle &= \langle \omega, \theta\theta'\zeta \rangle \\
&= \langle \omega, \theta'\theta\zeta \rangle \\
&= 0
\end{aligned}$$

since $\theta\zeta \in H^{2,2}(X)$. $\square$

The crucial step in this calculation is commuting the cup-product actions of θ and θ', which is a reflection of the integrability.

Discussion: Let $D_\mathbf{h}$ be the classifying space for the set of equivalence classes of polarized Hodge structures on a lattice $H_\mathbb{Z}$ with a non-degenerate symmetric form Q and with Hodge numbers

$$\mathbf{h} = (h^{4,0}, h^{3,1})$$

where $h^{1,3} = h^{3,1}$, $h^{0,4} = h^{4,0}$, $h^{2,2} = \operatorname{rank} H_\mathbb{Z} - 2(h^{4,0} + h^{3,1})$ and

$$\text{signature } Q = 2h^{4,0} + h^{2,2} - 2h^{1,3} \, .$$

For $\zeta \in H_\mathbb{Z}$ denote by

$$D_{\mathbf{h},\zeta} \subset D_\mathbf{h}$$

the subvariety where ζ is of type $(2,2)$ (cf. the heuristic argument for (4.17) on page 29 in "I"). We have the period mapping

$$\begin{array}{ccc}
\mathcal{M}_X & \xrightarrow{P} & D_\mathbf{h} \\
\cup & & \cup \quad , \quad \mathcal{M}_{X,\zeta} = P^{-1}(D_{\mathbf{h},\zeta}) \\
\mathcal{M}_{X,\zeta} & \xrightarrow{P_\zeta} & D_{\mathbf{h},\zeta}
\end{array}$$

and $P_\zeta = P|_{\mathcal{M}_{X,\zeta}}$. Now

$$\operatorname{codim}_{D_\mathbf{h}}(D_{\mathbf{h},\zeta}) = h^{4,0} + h^{3,1}$$

so that

$P(\mathcal{M}_X)$ does not meet $D_{\mathbf{h},\zeta}$ transversely.

If we let $I \subset TD_{\mathbf{h},\varsigma}$ be the exterior differential system (EDS) given by the infinitesimal period relation, then the differential

$$P_* : T\mathcal{M}_X \to I \subset TD_{\mathbf{h}} .$$

Now I meets $TD_{\mathbf{h},\varsigma}$ transversely along $D_{\mathbf{h},\varsigma}$, and one might wonder if the period mapping is *I-transversal* along $\mathcal{M}_{X,\varsigma}$ in the sense that

$$\mathrm{codim}_{\mathcal{M}_X}(\mathcal{M}_{X,\varsigma}) = \mathrm{rank}(I_\varsigma / I_\varsigma \cap TD_{\mathbf{h},\varsigma})$$

where $I_\varsigma \subset TD_{\mathbf{h}}|_{D_{\mathbf{h},\varsigma}}$ is the restriction of I to $D_{\mathbf{h},\varsigma}$? In fact, this does not hold due to the integrability conditions, which contribute the σ_ς term.

It is plausible that there is more sophisticated a concept of I-transversality based on the subvariety

$$V(I) \subset \prod_k \mathrm{Grass}(k, TD_{\mathbf{h}})$$

of integral elements of various dimensions, but we shall not pursue this here.

We do note that the notion of I-transversality appeared in the introduction in the discussion of (19) there, where one does in fact have such a result "at the boundary."

IV(iv). *Analysis of an example*

In this section we shall study the example

(58) $$\Lambda \subset X \subset \mathbb{P}^5$$

where X is a smooth hypersurface of degree $d \geqq 6$ and Λ is a 2-plane. We set

(59) $$\varsigma = [\Lambda]_{\mathrm{prim}}$$

and denote by

(60) $$\mathcal{M}_{X,\Lambda} \subset \mathcal{M}_X$$

the local moduli space of deformations X' of X for which Λ moves to a (unique) 2-plane $\Lambda' \subset X'$.[5] We note that since $h^0(N_{\Lambda/X}) = 0$

$$\mathcal{M}_{X,\Lambda} \subseteq \mathcal{M}_{X,\varsigma} .$$

---

[5] More properly, one could define $\mathcal{M}_{X,\Lambda}$ to be the moduli space of pairs (X, Λ) as in (58) and then there is an étalè map $\mathcal{M}_{X,\Lambda} \to \mathcal{M}_X$. For the purpose of dimension counts we shall just work locally in moduli where we have (60).

In this section we shall prove the following statements (61)–(66):

(61) $\mathcal{M}_{X,\Lambda} = \mathcal{M}_{X,\zeta}$.

Geometrically, *any deformation of X preserving the Hodge class ζ is obtained by deforming the pair (X, Λ) where (58) holds.*

(62) $\operatorname{codim}_{\mathcal{M}_X}(\mathcal{M}_{X,\zeta}) = h^{3,1} - \sigma_\zeta$.

Thus, in this case the inequality in Theorem (54) is sharp. The result (62) will be proved by showing:

For $\varphi \in H^{3,1}(X)$,

(63) $\langle \varphi, \zeta \cdot H^1(\Theta_X) \rangle = 0 \Leftrightarrow \varphi \in T_\zeta \cdot H^{4,0}(X)$.

The inequality in theorem (54) results from the implication "$\Leftarrow$", which always holds. Below we shall by dimension count prove "$\Rightarrow$". We shall also show that

(64) $h^{3,1} - \sigma_\zeta = h^1(N_{\Lambda/X})$.

Next, setting $L = \mathcal{O}_X(1)$ and denoting by l the number of nodes in general with $\Lambda \subset X_{s_0}$ we shall show that

(65) $\operatorname{codim}_{|L|}\big(\operatorname{sing} \nu_\zeta\big) = l - (h^{3,1} - \sigma_\zeta) - h^2(\Lambda^2 N_{\Lambda/X}(-1))$.

This is for fixed X. Now letting (X, ζ) vary — which by (61) is the same as letting (X, Λ) vary — we shall show that

(66) $\operatorname{codim}_{|\mathcal{L}|}(\operatorname{sing} \boldsymbol{\nu}_\zeta) = l - h^2\big(\Lambda^2 N_{\Lambda/X}(-1)\big)$.

We note that

(67) $h^2\big(\Lambda^2_{\Lambda/X}(-1)\big) = h^0\big(\mathcal{O}_\Lambda(d-5)\big) = \binom{d-3}{2} \neq 0$.

Discussion: This example (for which the VHC holds) illustrates two fundamental differences between the classical $n = 1$ case and curves on a surface and the $n = 2$ case (surfaces on a fourfold). Namely

  A. *Both the infinitesimal period relations and its integrability conditions enter in the codimension of the Noether-Lefschetz locus, which in general* (not just in special cases) *depends on the particular Hodge class.*
  B. *There is a non-zero correction term $h^2(\Lambda^2 N_{\Lambda/X}(-1))$ to the codimension of* sing $\boldsymbol{\nu}_\zeta$. *In particular, the to be defined rational map*

$$|\mathcal{L}| \overset{\Psi}{-\!\!-\!\!\to} \overline{\mathcal{A}}_{\mathbf{h},\Sigma}$$

always fails to be transverse along the boundary component $\mathcal{B}(\sigma(l, 1))$ *with*

$$\Psi^{-1}(\mathcal{B}(\sigma(l, 1))) = \operatorname{sing} \boldsymbol{\nu}_\zeta .$$

Assuming the VHC we have noted the phenomenon B in general but with the assumption that $L \gg 0$ relative to W. Here of course we see it in a concrete special case.

We now turn to the proofs of the assertions (61)–(66) above. These are based on explicit computations based on the well-known polynomial description of the relevant cohomology groups (cf. [15]). We suppose that

$$\mathbb{P}^5 = \mathbb{P}V^*$$

where $\dim V = 6$ and set $V^k = \operatorname{Sym}^k V$. Then X is given by

$$F = 0 , \qquad F \in V^d .$$

We denote by

$$\mathcal{J}^k \subset V^k$$

for the kth homogeneous piece of the *Jacobian ideal* $\mathcal{J}_\bullet = \{F_{z_0}, \ldots, F_{z_5}\}$ where $z_0, \ldots z_5$ give a basis for V and $F_{z_i} = \partial F / \partial z_i$, and we set

(68) $$R^k = V^k / \mathcal{J}^k .$$

Then it is well known that

(69) $$H^1(\Theta_X) \cong R^d$$
(70) $$H^q(\Omega_X^p)_{\text{prim}} \cong R^{(q+1)d-6} \qquad (p + q = 4) .$$

We may suppose that

$$\Lambda = \{z_0 = z_1 = z_2 = 0\}$$

so that from $\Lambda \subset X$ we have

$$F = z_0 G_0 + z_1 G_1 + z_2 G_2 .$$

Then denoting by $V_\Lambda^\cdot, \mathcal{J}_\Lambda^\cdot, R_\Lambda^\cdot$ the restrictions of $V^\cdot, \mathcal{J}^\cdot, R^\cdot$ to Λ we have

$$\mathcal{J}_\Lambda^\bullet = \{G_0, G_1, G_2\} .$$

For later use we note the identifications given by the vertical maps in the diagram

$$0 \to N_{\Lambda/X} \to N_{\Lambda/\mathbb{P}^5} \to N_{X/\mathbb{P}^5}|_\Lambda \to 0$$

(71) $$\qquad\qquad \| \qquad\quad \wr\| \qquad\qquad \|\wr$$

$$0 \to N_{\Lambda/X} \to \overset{3}{\oplus} \mathcal{O}_\Lambda(1) \to \mathcal{O}_{\Lambda(d)} \to 0 .$$

We shall now show that:

(72) $$R_\Lambda^{3d-6} = V_\Lambda^{3d-6}/(G_0, G_1, G_2) \cong \mathbb{C}$$

and the surjective map

$$R^{3d-6} \to R_\Lambda^{3d-6}$$

gives a natural element of $H^2(\Omega_X^2)_{\mathrm{prim}}$ which is just

$$\zeta = [\Lambda]_{\mathrm{prim}} .$$

Up to a non-zero constant this is

$$\det\|\partial G_i/\partial z_j\| \qquad 0 \le i,\ j \le 2 .$$

For simplicity of notation we set

$$P = \det\|\partial G_i/\partial z_j\| \in R^{3d-6} .$$

Proof: From the cohomology arising from the Koszul complex for $\{G_0, G_2, G_2\}$

$$O \to \mathcal{O}_\Lambda(-3) \to \overset{3}{\oplus} \mathcal{O}_\Lambda(d-4) \to \overset{3}{\oplus} \mathcal{O}_\Lambda(2d-5) \to \mathcal{O}_\Lambda(3d-6) \to 0$$

and

$$H^2(\mathcal{O}_\Lambda(-3)) \cong H^2(\Omega_\Lambda^2) \cong \mathbb{C}$$

we conclude (72). The various identifications may be verified by standard compuations — what they amount to is that for $H \in R^{3d-6} \cong H^2(\Omega_\Lambda^2)_{\mathrm{prim}}$

$$\int_\Lambda H = \text{ image of } H \text{ in } H^2(\Omega_\Lambda^2) . \qquad \square$$

For later use we note that

(73) $$H^1(N_{\Lambda/X}) \cong V_\Lambda^d/(G_0, G_1, G_2)$$

(74) $$H^1(N_{\Lambda/X} \otimes K_X) \cong V_\Lambda^{2d-6}/(G_0, G_1, G_2) .$$

The first follows from the cohomology sequence in the bottom row of (71), and the second from the cohomology sequence arising from tensoring the bottom row with $K_X \cong \mathcal{O}_X(d-6)$.

We now turn to the proof of (61). The 1st order deformation theory of pair (X, W) is given by the hypercohomology group $\mathbb{H}^1(\Theta_X \to N_{W/X})$, from which we conclude

(75) $$\left(\begin{array}{c} \text{deformations of } X \\ \text{preserving } W \end{array} \right) = \ker\left\{ H^1(\Theta_X) \to H^1/N_{W/X} \right\}$$

In the case at hand, using the identifications above the RHS of (75) is the kernel of

$$V^d/\mathfrak{I}^d \to V^d_\Lambda/\mathfrak{I}^d_\Lambda \cong V^d_\Lambda/(G_0, G_1, G_2)$$

This kernel is just

(76) $$I^d_\Lambda/I^d_\Lambda \cap \mathfrak{I}^d$$

where $I^\bullet_\Lambda = \{z_0, z_1, z_2\}$ is the ideal of Λ.

From the commutative diagram

$$
\begin{array}{ccc}
H^1(\Theta_X) & \xrightarrow{\ \zeta\ } & H^3(\Omega^1_X) \\[4pt]
\wr\| & & \wr\| \\[4pt]
R^d & \xrightarrow{\ P\ } & R^{3d-6}
\end{array}
$$

we have

$$T_\zeta = [\mathfrak{I}^\bullet : P]^d = \{R \in R^d : RP \in \mathfrak{I}^{3d-6}\}\ .$$

It also follows from the discussion above that we have a factorization

$$
\begin{array}{ccc}
R^{2d-6} & \xrightarrow{\ P\ } & R^{4d-12} \cong \mathbb{C} \\[6pt]
& \searrow \quad \nearrow & \\
& R^{2d-6}_\Lambda & .
\end{array}
$$

The basic diagram in the story is

(77)
$$
\begin{array}{ccc}
H^1(\Theta_X) \otimes H^1(\Omega^3_X) & \longrightarrow & H^2(\Omega^2_X) \\[4pt]
\downarrow & & \downarrow \\[4pt]
H^1(\Theta_X|_\Lambda) \otimes H^1(\Omega^3_X|_\Lambda) & \longrightarrow & H^2(\Omega^2_X|_\Lambda) \\[4pt]
\downarrow & & \downarrow \\[4pt]
\end{array}
$$

$$H^1(N_{\Lambda/X}) \otimes H^1(N_{\Lambda/X} \otimes K_X) \to H^2(\Lambda^2 N_{\Lambda/X} \otimes K_X) \cong H^2(K_\Lambda) \cong \mathbb{C}$$

with a diagonal arrow labelled ζ from $H^2(\Omega^2_X|_\Lambda)$.

where we have used

$$
\begin{cases}
\Omega^3_X|_\Lambda \cong K_X \otimes \Theta_X|_\Lambda \to K_X \otimes N_{\Lambda/X} \\
\Omega^2_X|_\Lambda \cong K_X \otimes \Lambda^2\Theta_X|_\Lambda \to K_X \otimes \Lambda^2 N_{\Lambda/X}\ .
\end{cases}
$$

From this diagram we have that

$$\left\langle \zeta \cdot \ker\left\{H^1(\Theta_X) \to H^1(N_{\Lambda/X}), H^1(\Omega^3_X)\right\}\right\rangle = 0$$

and thus

(78) $$T_\zeta \subseteq \ker \left\{ H^1(\Theta_X) \to H^1(N_{\Lambda/X}) \right\} .$$

Geometrically this is clear: If Λ deforms to 1st order in the direction $\theta \in H^1(\Theta_X)$, then so does its fundamental class. It is the converse, which is equality in (61), that we want to establish.

The basic observation is that the map

(79) $H^1(\Omega_X^3) \to H^1(N_{\Lambda/X} \otimes K_X) \cong H^1(N_{\Lambda/X}^* \otimes K_\Lambda) \cong H^1(N_{\Lambda/X})^*$

is surjective; this follows from the above identifications and the surjectivity of

$$R^{2d-6} \to R_\Lambda^{2d-6} .$$

It follows from this surjectivity that the left kernel T_ζ in the map

$$H^1(\Theta_X) \otimes H^1(\Omega_X^3) \to H^2(\Omega_X^2) \xrightarrow{\zeta} \mathbb{C}$$

(77) is equal to the left kernel of the vertical column in (77), which is just

$$\ker \left\{ H^1(\Theta_X) \to H^1(N_{\Lambda/X}) \right\} .$$

This establishes equality in the inclusion (78), and with it proves (61).

We next turn to (63). Working on the other side of the $\otimes$ in (77) we have

(80) $$(\operatorname{Im} \zeta)^\perp = \ker \left\{ H^1(\Omega_X^3) \to H^1(N_{\Lambda/X} \otimes K_X) \right\}$$
$$= I_\Lambda^{2d-6} / R_\Lambda^{2d-6} \cap \mathfrak{J}^{2d-6} .$$

On the other hand we have seen that

$$T_\zeta = \text{image} \left\{ I_\Lambda^d / R_\Lambda^d \cap \mathfrak{J}^d \to V^d / \mathfrak{J}^d \right\} .$$

From the diagram

$$T_\zeta \otimes H^0(\Omega_X^4) \quad \to \quad (\text{Image } \zeta)^\perp$$
$$\wr\| \qquad\qquad\qquad\qquad \wr\|$$
$$(I_\Lambda^d / I_\Lambda^d \cap \mathfrak{J}^2) \otimes V^{d-6} \to I_\Lambda^{2d-6} / I_\Lambda^{2d-6} \cap \mathfrak{J}^{2d-6}$$

in which the bottom row is surjective we conclude the surjectivity of the top row, which is equivalent to (63).

We now also have (62), since

$$\sigma_\zeta = \dim \left(\operatorname{coker} \left\{ H^1(\Theta_X) \xrightarrow{\zeta} H^3(\Omega_X^1) \right\} \right)$$

which implies that

$$\operatorname{codim} T_\zeta = h^{3,1} - \sigma_\zeta \,,$$

while we have seen above that also

$$\operatorname{codim} T_\zeta = h^1(N_{\Lambda/X}) \,.$$

Turning to (65) we have from the Koszul resolution of the double point locus restricted to Λ the sequence

$$(81) \qquad 0 \to \Lambda^2 N_{\Lambda/X}(-1) \to N_{\Lambda/X} \to \mathcal{O}_\Lambda(1) \to \mathcal{O}_\Delta(1) \to 0 \,.$$

From the resulting spectral sequence and $h^0(\mathcal{O}_\Delta(1)) = l$ we obtain

$$l = \underbrace{h^0(\mathcal{O}_\Lambda(1)) + h^1(N_{\Lambda/X})} + h^2(\Lambda^2 N_{\Lambda/X}(-1))$$

and the term over the brackets is just

$$\operatorname{codim}_{|\mathcal{L}|}(\operatorname{sing} \boldsymbol{\nu}_\zeta) \,.$$

This gives (65), and (66) is clear since in this case

$$\operatorname{codim}_{|L|}(\operatorname{sing} \nu_\zeta) = h^0(\mathcal{O}_\Lambda(1)) \,.$$

V. Maps in the Case of Fourfolds

V(i). *Structure of nodal boundary components in* $\overline{\mathcal{A}}_{\mathbf{h},\Sigma}$ *when* $\mathbf{h} = (h^{3,0}, h^{2,1})$ *with* $h^{3,0} \neq 0$

Kato-Usui [17] have defined a partial compactification for classifying spaces that arise for general variations of Hodge structures. Although not an analytic variety in the usual sense, these spaces do have a "logarithmic structure" and much of the classical theory when the classifying space is a bounded symmetric domain seems to extend in a useable manner. For us the relevant fact is that variations of Hodge structures defined over punctured polycylinders and with unipotent monodromies give maps that extend to the Kato-Usui partial compactifications.

As in the classical case boundary components are associated to certain rational cones σ of commuting nilpotent elements in a Lie algebra. Here we will describe two boundary components

$$(1) \qquad \mathcal{B}(\sigma(l)) \leftrightarrow \left\{ \begin{array}{c} \text{cone generated by logarithms } N_i \\ \text{of Picard-Lefschetz transformation} \\ \text{corresponding to } l \text{ independent nodes} \end{array} \right\}$$

and

$$(2) \quad \mathcal{B}(\sigma(l,1)) \leftrightarrow \left\{ \begin{array}{c} \text{cone generated by logarithms } N_i \\ \text{of Picard-Lefschetz transformations} \\ \text{corresponding to } l \text{ nodes with one relation} \end{array} \right\} .$$

We shall elucidate the structure of these boundary components in a way that is amenable to algebro-geometric calculations and shall show that

$$(3) \qquad \operatorname{codim}_{\overline{\mathcal{A}}_{h,\Sigma}}(\mathcal{B}(\sigma(l)) = l + l \cdot h^{3,0}$$

and

$$(4) \qquad \operatorname{codim}_{\overline{\mathcal{A}}_{h,\Sigma}}(\mathcal{B}(\sigma(l,1)) = l + (l-1)h^{3,0} .$$

These are at first glance surprising, since for example in case (1) when $l = 1$, acquiring a single node generally imposes one condition on a family of varieties. What the proofs of (3) and (4) will show is that, because of the infinitesimal period relation, when $h^{3,0} \neq 0$ period mappings in a family of threefolds acquiring nodes that are either independent or have one relation can never be transverse in the usual sense to the corresponding boundary component. However, in the next section we shall show that because of refined notion of dimension counts arising where one is considering integral manifolds of an exterior differential system, and may drop the $lh^{3,0}$ and $(l-1)h^{3,0}$ terms in (3) and (4) when estimating codimensions of the pullbacks of these boundary components under period mappings.

We shall now discuss the boundary component $\mathcal{B}(\sigma(l))$ in (1). For this we assume given $(H_\mathbb{Z}, Q)$ where $H_\mathbb{Z}$ is a lattice of rank $2h$ and Q is a unimodular symplectic form on $H_\mathbb{Z}$. We may describe $\mathcal{A}_\mathbf{h}$ as the set of polarized Hodge structures on $(H_\mathbb{Z}, Q)$ of weight three and with given Hodge numbers $h^{3,0}$ and $h^{2,1}$ where $h^{3,0} + h^{2,1} = h$. As usual we have

$$\mathcal{A}_\mathbf{h} \subset \check{\mathcal{A}}_\mathbf{h}$$

where the dual classifying space $\check{\mathcal{A}}_\mathbf{h}$ is the set of filtrations $F^3 \supset F^2$ with

$$\begin{cases} \dim F^3 = h^{3,0} \\ \dim F^2/F^3 = h^{2,1} \\ Q(F^2, F^2) = 0 . \end{cases}$$

We set $F^1 = F^{3\perp}$.

Setting $G_\mathbb{C} = \operatorname{Aut}(H_\mathbb{C}, Q)$, in general boundary components are described as a quotient of several variable nilpotent orbits. Specifically we assume given

$$N_1, \ldots, N_l = \mathcal{G}_\mathbb{Q}$$

satisfying

$$\begin{cases} N_i^4 = 0 \\ [N_i, N_j] = 0 \ , \end{cases}$$

together with certain other conditions discussed in [7] and Kato-Usui (loc. cit.). Let

$$\sigma_{\mathbb{C}} = \left\{ \sum_i \lambda_i N_i : \lambda = (\lambda_1, \dots, \lambda_l) \in \mathbb{C}^l \right\}$$

be the nilpotent subalgebra of $\mathcal{G}_{\mathbb{C}}$ generated by the N_i. We set

$$\sigma = \left\{ \exp\left(\sum_i \lambda_i N_i \right) : \lambda_i > 0 \right\}$$

and then as a set the corresponding boundary component is

$$\mathcal{B}(\alpha) = \exp(\sigma_{\mathbb{C}}) \cdot \mathcal{A}_{\mathbf{h}} / \exp(\sigma_{\mathbb{C}}) \ .$$

As will be explained below, the necessity for passing to the quotient is because of rescaling.

We choose a symplectic basis $\delta_1, \dots, \delta_h; \gamma_1, \dots, \gamma_h$ for $H_{\mathbb{Z}}^*$, and shall write elements of $H_{\mathbb{Z}}^*$ relative to this basis as column vectors. Thus

$$\delta_1 = \begin{pmatrix} 1 \\ 0 \\ \cdot \\ \\ 0 \end{pmatrix}, \dots, \gamma_h = \begin{pmatrix} 0 \\ \cdot \\ \cdot \\ 0 \\ 1 \end{pmatrix} .$$

Elements of $H_{\mathbb{C}}$ will then be written as row vectors

$$\delta_1^* = (1, 0, \dots, 0)$$
$$\vdots$$
$$\gamma_h^* = (0, \dots, 0, 1) \ ,$$

and transformation $N \in \mathcal{G}_{\mathbb{C}}$ will be realized as $2h \times 2h$ matrices multiplying row vectors on the right. Points $F^\bullet \in \check{\mathcal{A}}_{\mathbf{h}}$ will be written as *period matrices* obtained by choosing adapted bases for $F^3 \subset F^2$; pictorially

$$\Omega = \underbrace{\begin{pmatrix} * & \cdots & * \\ \vdots & & \vdots \\ * & \cdots & * \end{pmatrix}}_{2h} \begin{matrix} \}h^{3,0} \\ \\ \}h^{2,1} \end{matrix} \ .$$

Changing the adapted basis occurs by left multiplication by a non-singular matrix of the form

$$
(5) \qquad \begin{array}{c} \left(\begin{array}{cc} * & * \\ 0 & * \end{array} \right) \begin{array}{l} \}h^{3,0} \\ \}h^{2,1} \end{array} . \\ \underbrace{}_{h^{3,0}} \; \underbrace{}_{h^{2,1}} \end{array}
$$

Specifically, in case (1) above we may choose our symplectic basis so that

$$
\delta_i^* N_j = \delta_i^j \gamma_j^* \qquad\qquad 1 \leqq i, j \leqq l
$$

and all other terms are zero. Thus

$$
N_i = \begin{array}{c} \left(\begin{array}{cc} 0 & \eta_i \\ 0 & 0 \end{array} \right) \begin{array}{l} \}h \\ \}h \end{array} \\ \underbrace{}_{h} \; \underbrace{}_{h} \end{array}
$$

where

$$
\eta_i = \begin{array}{c} \overbrace{}^{l} \\ \left(\begin{array}{c|c} \begin{array}{ccc} 0 & & \\ \cdot & 1 & \\ \cdot & & \cdot \; 0 \\ 0 & & \end{array} & \begin{array}{c} 0 \\ \vdots \\ \vdots \\ 0 \end{array} \end{array} \right) \begin{array}{l} \\ \}l \\ \\ \end{array} \end{array}
$$

has 1 in the (i, i) spot and is zero elsewhere. Below we shall show that

$$
(6) \qquad\qquad l \leqq h^{2,1}
$$

and we shall correspondingly write period matrices in the block form

$$
(7) \qquad \Omega = \begin{pmatrix} \Omega_1 \\ \Omega_2 \\ \Omega_3 \end{pmatrix} = \begin{array}{c} \left(\begin{array}{cccc} \Omega_{11} & \Omega_{12} & \Omega_{13} & \Omega_{14} \\ \Omega_{21} & \Omega_{22} & \Omega_{23} & \Omega_{24} \\ \Omega_{31} & \Omega_{32} & \Omega_{33} & \Omega_{34} \end{array} \right) \begin{array}{l} \}h^{3,0} \\ \}l \\ \}\hat{h}^{2,1} \end{array} \\ \underbrace{\phantom{\Omega_{11}}}_{l} \; \underbrace{\phantom{\Omega_{12}}}_{\hat{h}} \; \underbrace{\phantom{\Omega_{13}}}_{l} \; \underbrace{\phantom{\Omega_{14}}}_{\hat{h}} \end{array}
$$

where we have set

$$
\hat{h} = h - l
$$
$$
\hat{h}^{2,1} = h^{2,1} - l .
$$

We shall now use period matrices to describe a general VHS $\Omega(t)$ with parameters $(t_1, \ldots, t_l) \in (\Delta^*)^l$ and whose monodromy logarithm around $t_i = 0$ is N_i. This monodromy assumption implies that

$$\tilde{\Omega}(t) \stackrel{\text{def}}{=} \Omega(t) \exp\left(-\sum_i \frac{\log t_i}{2\pi\sqrt{-1}} N_i\right)$$

is single-valued in $(t_1, \ldots t_n)$. We may assume that $\tilde{\Omega}(t)$ is holomorphic in Δ^l and that $\tilde{\Omega}(0) \in \check{A}_{\mathbf{h}}$. We then may write

$$(8) \qquad \Omega(t) = \begin{pmatrix} \Omega_{11}(t) & \Omega_{12}(t) & L(\Omega_{11}(t)) + H_1(t) & \Omega_{14}(t) \\ \Omega_{21}(t) & \Omega_{22}(t) & L(\Omega_{21}(t)) + H_2(t) & \Omega_{24}(t) \\ \Omega_{31}(t) & \Omega_{32}(t) & L(\Omega_{31}(t)) + H_3(t) & \Omega_{34}(t) \end{pmatrix}$$

and where

$$(9) \qquad L(\Omega_{21}(t)) = \Omega_{21}(t)\left(\sum_{i=1}^{l} \frac{\log t_i}{2\pi\sqrt{-1}} \eta_i\right)$$

where we are here considering η_i as an $l \times l$ matrix. The reason for (9) is that analytic contribution of $\Omega_{23}(t)$ around $t_i = 0$ is given by

$$\Omega_{23}(t) \to \Omega_{23}(t) + \Omega_{21}(t) ,$$

and therefore

$$\Omega_{23}(t) - \Omega_{21}\left(\sum_{i=1}^{l} \frac{\log t_i}{2\pi\sqrt{-1}} \eta_i\right) = H_2(t)$$

is single-valued and holomorphic.

The main step in the argument is the following

(10) **Lemma:** $\Omega_{11}(t) = 0 \mod \{t_1, \ldots, t_l\}$.

Proof: The idea is to use the infinitesimal period relation

$$(11) \qquad \Omega_1'(t) = A(t)\Omega_1(t) + B(t)\Omega_2(t) + C(t)\Omega_3(t)$$

where " $'$ " is any derivative and A, B, C are holomorphic matrices of the appropriate size. To see the essential point suppose first that $l = 1$ and $h^{3,0} = 1$. Then

$$\Omega_{11}(t) = (2\pi\sqrt{-1})(c + tf(t))$$

where c is a constant and $f(t)$ is a holomorphic function. From the proof of (8) we have

$$\Omega_{13}(t) = c \log t + t f(t) \log t + h(t)$$

where $h(t)$ is again a holomorphic function. Then

$$\Omega'_{13}(t) = \frac{c}{t} + f(t) + f'(t) t \log t + h'(t)$$

from which we conclude that the constant $c = 0$, i.e.

$$\Omega_{11}(0) = 0 .$$

This argument extends as follows: Any non-zero constant term in an entry in $\Omega_{11}(t)$ will produce a term

$$c \log t_i, \qquad c \neq 0$$

in $L(\Omega_{11})$ and hence a non-zero polar term in Ω'_{13}. $\qquad\square$

This lemma has the inequality (6) as a corollary, since at $t_1 = \cdots = t_l = 0$ the first column

$$\begin{pmatrix} 0 \\ \Omega_{21}(0) \\ \Omega_{31}(0) \end{pmatrix}$$

of $\Omega(0)$ must have rank l. By a suitable basis change (5) we may make

$$(12) \qquad\qquad \Omega_{21}(t) = I_l ,$$

and then by a further basis change we may arrange that

$$(13) \qquad\qquad \Omega_{31}(t) = 0 .$$

It is known that any VHS over $(\Delta^*)^l$ and with unipotent mondromy may be approximated by a nilpotent orbit. In this case the nilpotent orbit is

$$\tilde{\Omega}(t) \overset{\text{def}}{=} \tilde{\Omega} \cdot \exp\left(\sum_i \frac{\log t_i}{2\pi\sqrt{-1}} N_i \right)$$

where

$$\tilde{\Omega} = \tilde{\Omega}(0) .$$

The boundary component $\mathcal{B}(\sigma(l))$ may be intuitively thought of as all "limits of VHS's $\Omega(t)$" as above, modulo the ambiguity resulting from rescaling the t_i. Since

$$\text{"}\lim \Omega(t)\text{"} = \text{"}\lim \tilde{\Omega}(t)\text{"}$$

our strategy to compute $\dim \mathcal{B}(\sigma(l))$ will be to count the number of parameters in $\tilde{\Omega}$ and subtract off the number resulting from rescaling.

Now using (13)

$$\tilde{\Omega}(t) = \begin{pmatrix} \Omega_{11}(t) & \Omega_{12}(t) & H_1(t) & \Omega_{14}(t) \\ I & \Omega_{22}(t) & H_2(t) & \Omega_{24}(t) \\ 0 & \Omega_{32}(t) & H_3(t) & \Omega_{34}(t) \end{pmatrix}$$

where we note that in (8)

$$\begin{cases} L(\Omega_{21}(t)) = \sum_i \frac{\log t_i}{2\pi\sqrt{-1}}\eta_i \\ L(\Omega_{31}(t)) = 0 \,. \end{cases}$$

From (10) we have at $t = 0$

$$(14) \qquad \tilde{\Omega} = \begin{pmatrix} 0 & \Omega_{12} & H_1 & \Omega_{11} \\ I & \Omega_{22} & H_2 & \Omega_{21} \\ 0 & \Omega_{32} & H_3 & \Omega_{31} \end{pmatrix} \,.$$

Restricting the VHS to a disc $\Delta^* \subset (\Delta^*)^l$ with "slope" $\lambda_1, \ldots, \lambda_l$ where all $\lambda_i \neq 0$ gives a 1-parameter VHS with a monodromy weight filtration, which by [5] is independent of the slope, and limiting MHS given by (14). Denoting the logarithm of monodromy by N_λ, the monodromy weight filtration is

$$\{0\} = W_0 = W_1 \subset \underset{\substack{\| \\ \operatorname{Im} N_\lambda}}{W_2} \subset \underset{\substack{\| \\ \operatorname{Ker} N_\lambda}}{W_3} \subset W_4 = W_5 = H$$

The weight three Hodge structure on $Gr_3 W_0 = W_3/W_2$ corresponds to a polarized complex torus $\hat{\mathfrak{J}}$ with Hodge numbers

$$\begin{cases} \hat{h}^{3,0} = h^{3,0} \\ \hat{h}^{2,1} = h^{2,1} - l \,. \end{cases}$$

To see this we note that the monodromy weight filtration and limiting Hodge filtration may be pictured as

$$
(15) \qquad \tilde{\Omega} = \begin{pmatrix} 0 & \Omega_{12} & H_1 & \Omega_{14} \\ I & \Omega_{22} & H_2 & \Omega_{24} \\ 0 & \Omega_{32} & H_3 & \Omega_{34} \end{pmatrix} \begin{array}{l} \left.\vphantom{\begin{matrix}a\\a\end{matrix}}\right\} F^3 \\ \end{array} \left.\vphantom{\begin{matrix}a\\a\\a\end{matrix}}\right\} F^2 .
$$

$$
\underbrace{\phantom{\Omega_{32}H_3}}_{W_2}
$$

$$
\underbrace{\phantom{\Omega_{12}\ H_1\ \Omega_{14}}}_{W_3}
$$

Note that

$$
F^3 \subset W_3 ,
$$

i.e., under a VHS of the type given above

> *the Hodge number $h^{3,0}$ does not change in the limit, and the Hodge number $h^{3,1}$ drops by $\dim Gr_4 = \dim Gr_2$.*

In addition we have

> *for the limiting MHS we have $F^2 \cap W_{2,\mathbb{C}} = 0$; i.e., the Hodge structure on W_2 is "of Tate type".*

Degeneration of this type, which are very special, have been studied by Clemens [8]. We are indebted to Gregory Pearlstein for showing us interesting examples where $N^2 = 0$ but $F^2 \cap W_{2,\mathbb{C}} \neq 0$.

The mixed Hodge structure on W_3 gives rise to an extension

$$
(16) \qquad 0 \to (\mathbb{C}^*)^l \to \mathcal{J}_e \to \hat{\mathcal{J}} \to 0 .
$$

where $\mathcal{J}_e$ is a complex Lie group of dimension h, which may be thought of as the intermediate Jacobian analogue of a quasi-abelian variety, and $\hat{\mathcal{J}}$ is a polarized complex torus of dimension $\hat{h}$. The blocks in the period matrix (15) have the following interpretations:

$$
(17) \qquad \begin{pmatrix} \Omega_{12} & \Omega_{14} \\ \Omega_{32} & \Omega_{34} \end{pmatrix} \text{ is the period matrix of } \hat{\mathcal{J}}
$$

$$
(18) \qquad \Omega_{24} \text{ gives the extension data for (16).}
$$

We note that

$$
(19) \qquad \#\ \text{parameters in } \hat{\mathcal{J}} = \frac{\hat{h}(\hat{h}+1)}{2} + \hat{h}^{3,0}\hat{h}^{2,1}
$$

$$
(20) \qquad \#\ \text{parameters in } \Omega_{24} = l\hat{h} .
$$

Thus far we have not used the Riemann-Hodge bilinear relations, which it seems must enter in any significant result. The first bilinear relations are

(i) $\quad \Omega_{12}{}^t\Omega_{14} = \Omega_{14}{}^t\Omega_{12}$

(ii) $\quad {}^tH_2 + \Omega_{22}{}^t\Omega_{24} = H_2 + \Omega_{24}{}^t\Omega_{22} \Rightarrow H_2 = {}^tH_2$

(iii) $\quad \Omega_{32}{}^t\Omega_{34} = \Omega_{34}{}^t\Omega_{32}$

(iv) $\quad \Omega_{12}{}^t\Omega_{24} = H_1 + \Omega_{34}{}^t\Omega_{22}$

(v) $\quad \Omega_{12}{}^t\Omega_{34} = \Omega_{14}{}^t\Omega_{32}$

(vi) $\quad \Omega_{24}{}^t\Omega_{32} = H_3 + {}^t\Omega_{34}\Omega_{22}.$

We note that

(i), (iii), (v) *give the* 1st *bilinear relations for the polarized Hodge structure on* Gr_3W.

Next we turn to rescaling. Under

$$t_i \rightarrow \exp(2\pi\sqrt{-1}u_i)t_i$$

we see from (8) that all the block entries in $\tilde{\Omega}$ are unchanged except that

$$H_2 \rightarrow H_2 + \begin{pmatrix} u_1 & & 0 \\ & \ddots & \\ 0 & & u_l \end{pmatrix}.$$

Since from the bilinear relation (ii) we have that H_2 is symmetric, we conclude that

(21) $\qquad$ # essential parameters in $H_2 = l(l-1)/2$.

For the last step we will use the remaining basis change (5) that preserves the normalization made thus far. We note that

$$\text{rank}\begin{pmatrix} \Omega_{12} \\ \Omega_{32} \end{pmatrix} = \hat{h}$$

and that we may add linear combinations of the rows of Ω_{12} to the rows in Ω_{22} and Ω_{32}, and similarly we may add linear combinations of the rows in Ω_{32} to those in Ω_{22}. Write

$$\Omega_{22} = \big(\underbrace{\Omega'_{12}}_{\hat{h}^{3,0}}, \underbrace{\Omega''_{22}}_{\hat{h}^{2,1}} \big)\big\}\hat{h}^{3,0}$$

$$\Omega_{32} = \big(\underbrace{\Omega'_{32}}_{\hat{h}^{3,0}}, \underbrace{\Omega''_{32}}_{\hat{h}^{2,1}} \big)\big\}\hat{h}^{2,1} .$$

An open set in $\breve{A}_{\mathbf{h}}$ is given by the conditions

$$\det \Omega'_{12} \neq 0, \quad \det \Omega''_{32} \neq 0,$$

and for the purposes of parameter counts it will suffice to work in that set. Then we may normalize to have

(22)
$$\begin{cases} \Omega'_{12} = I_{\hat{h}^{3,0}} \\ \Omega'_{32} = 0, \ \Omega''_{32} = I_{\hat{h}^{2,1}} . \end{cases}$$

Then by a transformation as described above we may arrange that

(23)
$$\Omega_{22} = 0 .$$

From the bilinear relations (iv) and (vi) we then find

H_1 *and* H_3 *are determined by the data* (17) *and* (18).

We are now ready for the parameter count:

$$\dim A_{\mathbf{h}} = \frac{h(h+1)}{2} + h^{3,0}h^{2,1}$$

$$= \frac{(\hat{h}+l)(\hat{h}+l+1)}{2} + \hat{h}^{3,0}(l+\hat{h}^{2,1})$$

$$= \underbrace{\frac{\hat{h}(\hat{h}+1)}{2} + \hat{h}^{3,0}\hat{h}^{2,1}}_{\overset{\|}{(19)}} + \underbrace{\hat{h}l}_{\overset{\|}{(20)}} + \underbrace{\frac{l(l-1)}{2}}_{\overset{\|}{(21)}}) + (l+lh^{3,0}) .$$

The last term gives the codimension of the boundary component $\mathcal{B}(\sigma(l))$, thus establishing (3).   □

We now turn to the boundary component $\mathcal{B}(\sigma(l,1))$ and the proof of (4). The argument will be analogous to that given above for the proof of (3). In this case we choose a symplectic basis

$$\delta_1, \ldots, \delta_{l-1}, \delta_l, \ldots \delta_h, \quad \gamma_1, \ldots, \gamma_{l-1}, \gamma_l, \ldots, \gamma_h$$

for $H_{\mathbb{Z}}^*$ and set

$$\delta = \delta_1 + \cdots + \delta_{l-1} .$$

We define N_i for $1 \leq i \leq l$ by

(24)
$$\begin{cases} \delta_i^* N_j = \delta_i^j \gamma_j^* & 1 \leq i, j \leq l-1 \\ \delta_i^* N_l = \gamma_1^* + \cdots + \gamma_{l-1}^* , \end{cases}$$

where N_l is modeled on a Picard-Lefschetz transformation corresponding to the vanishing cycle δ. In analogy to (6) above we will see that

(25) $$l - 1 \leqq h^{2,1} .$$

Using the above notations we have

$$N_i = \begin{pmatrix} 0 & \eta_i \\ 0 & 0 \end{pmatrix} \begin{matrix} \}h \\ \}h \end{matrix}$$
$$\underbrace{}_{h} \quad \underbrace{}_{h}$$

where

$$\eta_i = \left(\begin{array}{ccc|c} 0 & \cdots & 0 & 0 \\ \vdots & 1 & \vdots & \cdot \\ 0 & \cdots & 0 & \cdot \\ \hline 0 & \cdot & \cdot & 0 \end{array}\right) \begin{matrix} \left.\begin{matrix} \\ \\ \end{matrix}\right\} l-1 \\ \}\hat{h}=h-(l-1) \end{matrix} \qquad 1 \leqq i \leqq l-1$$

$$\eta_l = \left(\begin{array}{ccc|c} 1 & \cdots & 1 & 0 \\ \vdots & & \vdots & \cdot \\ 1 & \cdots & 1 & \cdot \\ \hline 0 & \cdot & \cdot & 0 \end{array}\right) \begin{matrix} \left.\begin{matrix} \\ \\ \end{matrix}\right\} l-1 \\ \}\hat{h}=h-(l-1). \end{matrix}$$

The nilpotent orbits will arise from

$$\sigma_{\mathbb{C}} = \left\{ \sum_i \lambda_i N_i : \lambda = (\lambda_1, \ldots, \lambda_l) \in \mathbb{C}^l \right\} .$$

The same arguments as above, especially lemma (10), lead to the partially normalized period matrix

(26)
$$\Omega(t) = \left(\begin{array}{cccc} \Omega_{11}(t) & \Omega_{12}(t) & L(\Omega_{11}(t)) + H_1(t) & \Omega_{14}(t) \\ I & \Omega_{22}(t) & L(t) + H_2(t) & \Omega_{24}(t) \\ 0 & \Omega_{32}(t) & H_3(t) & \end{array}\right) \begin{matrix} \}\hat{h}^{3,0} \\ \}l-1 \\ \}\hat{h}^{2,1}=h^{2,1}-(l-1) \end{matrix}$$
$$\underbrace{}_{l-1} \underbrace{}_{\hat{h}=h-(l-1)} \underbrace{}_{l-1} \underbrace{}_{\hat{h}=h-(l-1)}$$

where

$$(27) \quad L(t) = \begin{pmatrix} \frac{\log t_1}{2\pi\sqrt{-1}} & & 0 \\ & \ddots & \\ 0 & & \frac{\log t_{l-1}}{2\pi\sqrt{-1}} \end{pmatrix} + \begin{pmatrix} \frac{\log t_l}{2\pi\sqrt{-1}} & \cdots & \frac{\log t_l}{2\pi\sqrt{-1}} \\ \vdots & & \vdots \\ \frac{\log t_l}{2\pi\sqrt{-1}} & \cdots & \frac{\log t_l}{2\pi\sqrt{-1}} \end{pmatrix}$$

and where

$$\Omega_{11}(0) = 0 .$$

For $\tilde{\Omega}$ given by $\Omega(t) \exp\left(-\sum_{i=1}^{1} \frac{\log t_i}{2\pi\sqrt{-1}} N_i\right)$ at $t = 0$ we then have

$$\tilde{\Omega} = \begin{pmatrix} 0 & \Omega_{12} & H_1 & \Omega_{14} \\ I & \Omega_{22} & H_2 & \Omega_{24} \\ 0 & \Omega_{32} & H_3 & \Omega_{34} \end{pmatrix} .$$

As before we may further normalize so as to have

$$\tilde{\Omega} = \begin{pmatrix} 0 & I & \Omega_{12}'' & H_1 & \Omega_{14} \\ I & 0 & 0 & H_2 & \Omega_{24} \\ 0 & 0 & I & H_3 & \Omega_{34} \end{pmatrix} \begin{matrix} \}\hat{h}^{3,0} = h^{3,0} \\ \}l-1 \\ \}\hat{h}^{2,1} = h^{2,1} - (l-1) \end{matrix}$$

$$\underbrace{}_{l-1} \; \underbrace{}_{\hat{h}^{3,0}} \; \underbrace{}_{\hat{h}^{2,1}} \; \underbrace{}_{l-1} \; \underbrace{}_{\hat{h}}$$

where

$$(28) \qquad \begin{matrix} \hat{h}^{3,0}\{ \\ \hat{h}^{2,1}\{ \end{matrix} \begin{pmatrix} I & \Omega_{12}'' & \Omega_{14} \\ 0 & I & \Omega_{34} \end{pmatrix} \leftrightarrow \hat{\jmath} \in \mathcal{A}_{\hat{\mathbf{h}}}$$

$$\underbrace{}_{\hat{h}^{3,0}} \; \underbrace{}_{\hat{h}^{2,1}} \; \underbrace{}_{\hat{h}}$$

$$(29) \qquad\qquad \Omega_{24} \qquad\qquad \leftrightarrow e \in \mathrm{Ext}^1(\hat{\jmath}, (\mathbb{C}^*)^{l-1}) .$$

Also as before H_1 and H_3 are uniquely determined by the 1st bilinear relation from the data (28), (29) which also gives

$$H_2 = {}^t H_2 .$$

The main new point is that under rescaling

$$(30) \qquad H_2 \to H_2 + \begin{pmatrix} u_1 & & 0 \\ & \ddots & \\ 0 & & u_{l-1} \end{pmatrix} + \begin{pmatrix} u_1 & \cdots & u_l \\ \vdots & & \vdots \\ u_1 & \cdots & u_l \end{pmatrix} .$$

Thus

(i) # of parameters in (28) = $\frac{\hat{h}(\hat{h}+1)}{2} + \hat{h}^{3,0}\hat{h}^{2,1}$

(ii) # of parameters in (29) = $(l-1)\hat{h}$

and by (30)

(iii) # of parameters in $H_2 = \frac{(l-1)(l)}{2} - l = \frac{l(l-3)}{2}$.

Then

$$\dim \mathcal{A}_{\mathbf{h}} = \frac{h(h+1)}{2} + h^{3,0}h^{2,1}$$

$$= \frac{(\hat{h}+l-1)(\hat{h}+l)}{2} + \hat{h}^{3,0}(\hat{h}^{2,1}+l-1)$$

$$= \underbrace{\frac{\hat{h}(\hat{h}+1)}{2} + \hat{h}^{3,0}\hat{h}^{2,1}}_{(i)} + \underbrace{\hat{h}(l-1)}_{(ii)} + \underbrace{\frac{l(l-3)}{2}}_{(iii)} + l + h^{3,0}(l-1)$$

where the last term on the right results from

$$\frac{l(l-1)}{2} = \frac{l(l-3)}{2} + l .$$

In conclusion, since $\dim \mathcal{B}(\sigma(l,1)) = (i) + (ii) + (iii)$

$$\dim \mathcal{A}_{\mathbf{h}} = \dim \mathcal{B}(\sigma(l,1)) + l + h^{3,0}(l-1)$$

as desired. □

V(ii). *Non-transversality of the VHS at the boundary components* $\mathcal{B}(\sigma(l))$ *and* $\mathcal{B}(\sigma(l,1))$

For the situation

$$|L| \xrightarrow{\ \psi\ } \overline{\mathcal{A}}_{\mathbf{h},\Sigma}$$

of interest in this paper, all we can *a priori* conclude from (3) in Section V(i) is that

(31) $$\operatorname{codim}_{|L|}\left(\psi^{-1}(\mathcal{B}(\sigma(l)))\right)_{s_0} \leqq l + lh^{3,0}$$

near where X_{s_0} has l independent nodes. But we know that, e.g. when $l = 1$, we have

(32) $$\operatorname{codim}_{|L|}\left(\psi^{-1}(\mathcal{B}(\sigma(l)))\right)_{s_0} \leqq l ,$$

with equality frequently holding at least for small l. Geometrically, for $L \gg 0$ the first few nodes impose independent conditions on $|L|$. In fact, we shall now show

The inequality (32) always holds.

In other words,

(33) *The mapping ψ always fails, by at least in dimension equal to $lh^{3,0}$, to be transverse along $\mathcal{B}(\sigma(l))$.*

We shall see that (33) is a general VHS result. For the proof it will be notationally convenient to interchange the 2nd and 3rd row and column blocks in the period matrix (14) and write it as follows

(34)
$$
\tilde{\Omega} = \left.\begin{pmatrix} I & \omega_{12} & 0 & \omega_{14} & \omega_{15} & K_1 \\ 0 & I & 0 & \omega_{24} & \omega_{25} & K_2 \\ 0 & 0 & I & \omega_{34} & \omega_{35} & K_3 \end{pmatrix}\right\} \begin{matrix} \hat{h}^{3,0} \\ \hat{h}^{2,1} \\ l \end{matrix} \, .
$$
$$
\underbrace{}_{\hat{h}^{3,0}} \; \underbrace{\phantom{\omega_{12}}}_{\hat{h}^{2,1}} \; \underbrace{}_{l} \; \underbrace{\phantom{\omega_{14}}}_{\hat{h}^{3,0}} \; \underbrace{\phantom{\omega_{15}}}_{\hat{h}^{2,1}} \; \underbrace{}_{l}
$$

Then

$$
\begin{pmatrix} I & \omega_{12} & \omega_{14} & \omega_{15} \\ 0 & I & \omega_{24} & \omega_{24} \end{pmatrix} \quad \text{corresponds to } \hat{\jmath} \in \mathcal{A}_{\hat{\mathbf{h}}}
$$

and

$$
\left(\omega_{34}, \omega_{35}\right) \quad \text{corresponds to the extension class } e \in \operatorname{Ext}^1(\hat{\jmath}, (\mathbb{C}^*)^l)
$$

for the MHS given by $(W_3, F^\bullet)$.

The 1st bilinear relations give

$$
\begin{aligned}
&\text{(i)} && \omega_{14} + \omega_{15}{}^t\omega_{12} = {}^t\omega_{14} + \omega_{12}{}^t\omega_{15} \\
&\text{(ii)} && \omega_{15} = {}^t\omega_{24} + \omega_{12}{}^t\omega_{25} \\
&\text{(iii)} && \omega_{25} = {}^t\omega_{25} \\
&\text{(iv)} && K_1 = {}^t\omega_{34} + \omega_{12}{}^t\omega_{35} \\
&\text{(v)} && K_2 = {}^t\omega_{35} \\
&\text{(vi)} && K_3 = {}^tK_3 \, .
\end{aligned}
$$

We note that (i), (ii), (iii) are the 1st bilinear relations for $\hat{\jmath}$, and (iv), (v) show that K_1 and K_2 are determined by $(e, \hat{\jmath})$. (vi) corresponds to the symmetry (ii) in section V(i).

We write the new blocks of $\tilde{\Omega}$ as $\omega_1, \omega_2, \omega_3$

$$\tilde{\Omega} = \begin{pmatrix} \omega_1 \\ \omega_2 \\ \omega_3 \end{pmatrix} .$$

Then for a VHS given by a map

(35) $$S \to \overline{\mathcal{A}}_{\mathbf{h},\Sigma}$$

we have along the inverse image of $\mathcal{B}(\sigma(l))$ that

(36) $$\omega_1' = A\omega_1 + B\omega_2 + C\omega_3$$

where $'$ is any derivative along a direction mapping to $T(\overline{\mathcal{B}}(\sigma(l)))$. From (34) and $\omega_{11}' = 0$ we infer that

$$A = 0 .$$

From $\omega_{13}' = 0$ we obtain

$$C = 0 .$$

Then (35) gives

(37) $$B = \omega_{12}' .$$

The formulas for ω_{14}' and ω_{15}' simply say that W_3 gives an admissable VMHS (cf. [20]). The remaining relations in (36) together with (37) give

(38) $$\left({}^t\omega_{34} + \omega_{12}{}^t\omega_{25}\right)' = \omega_{12}'{}^t\omega_{35} .$$

This is a matrix equation of size $\hat{h}^{3,0} \times l$ and may be interpreted as measuring the amount of *non*-transversality in the intersection of the image of TS and $T\mathcal{B}(\sigma(l))$. Put another way

> $\omega_{12}, \omega_{34}, \omega_{35}$ *are free to move arbitrarily in the image of a general map* (35), *but are subject to* (38) *for a VHS*

which is what was to be shown. □

The same argument just given, when directly adapted to calculations given above in section V(i) for the proof of (4) there, give the conclusion

(39) *We have*

$$\mathrm{codim}_{|L|}\, \psi^{-1}(\mathcal{B}(\sigma(l,1))) \leqq l .$$

In words, the mapping ψ always fails, by at least in dimension equal to $h^{3,0}(l-1)$, to be transverse along $\mathcal{B}(\sigma(l,1))$.

The general geometric reason for this argument was explained in the introduction. The above are the specific calculations that apply in the two cases of interest here.

V(iii). *The excess intersection formula in the* $n = 2$ *case*

We will apply the considerations of the remark at the end of the section II(ii) where we had the general setup

$$
\begin{array}{ccc}
A & \xrightarrow{F} & M \\
\cup & & \| \\
C & \xrightarrow{F} & M \qquad F^{-1}(N) = B \\
\cup & & \cup \\
B & \xrightarrow{F} & N
\end{array}
$$

(40)

to the case where

$$
A = |\mathcal{L}|, \quad C = |\mathcal{L}|_\zeta, \quad B = \operatorname{sing} \nu_\zeta
$$
$$
M = \overline{A}_{\mathbf{h},\Sigma}, \quad N = \mathcal{B}(\sigma(l,1))
$$

and

$$
\begin{cases}
F = \Psi \\
\Psi_\zeta = \Psi|_{|\mathcal{L}|_\zeta} \, .
\end{cases}
$$

For simplicity of notation we set

$$
\begin{cases}
\mathcal{B} = \mathcal{B}(\sigma(l,1)) \\
\mathcal{S}_\zeta = (\operatorname{sing} \nu_\zeta)_{s_0} \, .
\end{cases}
$$

The diagram (40) is then

$$
\begin{array}{ccc}
|\mathcal{L}| & \overset{\Psi}{\dashrightarrow} & \overline{A}_{\mathbf{h},\Sigma} \\
\cup & & \| \\
|\mathcal{L}|_\zeta & \overset{\Psi_\zeta}{\dashrightarrow} & \overline{A}_{\mathbf{h},\Sigma} \qquad \mathcal{S}_\zeta = \Psi^{-1}(\mathcal{B}) \, . \\
\cup & & \cup \\
\mathcal{S}_\zeta & \dashrightarrow & \mathcal{B}
\end{array}
$$

(41)

As before we are considering only the component $(\operatorname{sing} \nu_\zeta)_{s_0}$ of $\Psi^{-1}(\mathcal{B})$. Keeping the notations from section IV(i), we make the additional assumptions

(42)
$$
\begin{cases}
\operatorname{codim}_{\mathcal{M}_X}(\mathcal{M}_{X,\zeta}) = h^{3,1} - \sigma_\zeta \\
H^1(\Theta_X) \twoheadrightarrow H^1(N_{W/X}) \, .
\end{cases}
$$

These assumptions are satisfied in our basic example (section IV(iv)) and conceivably are satisfied "in general". With these assumptions the codimensions in (41) are

$$(43) \quad \begin{cases} \text{(i)} & \text{codim}_{|\mathcal{L}|}\left(\Psi^{-1}(\mathcal{B})\right) = l - \delta \\ \text{(ii)} & \text{codim}_{|\mathcal{L}|_\zeta}\left(\Psi_\zeta^{-1}(\mathcal{B})\right) = l - \delta - (h^{3,1} - \sigma_\zeta) \\ \text{(iii)} & \text{codim}_{\overline{A}_{\mathbf{h},\Sigma}}(\mathcal{B}) = l + h^{3,0}(l-1) \end{cases}$$

where

$$\delta = \dim\left(\ker\left\{H^2(\Lambda^2 N_{W/X}(L^*)) \xrightarrow{ds_0} H^2(N_{W/X})\right\}\right)$$

so that $\delta \gg 0$ for $L \gg 0$ relative to W.

There are two excess intersection formulas, one for

$$(44) \qquad \Psi^*([\mathcal{B}]) \in H^{2(l+h^{3,0}(l-1))}(|\mathcal{L}|),$$

and one for

$$(45) \qquad \Psi_\zeta^*([\mathcal{B}]) \in H^{2(l+h^{3,0}(l-1))}(|\mathcal{L}|_\zeta).$$

For (44) the excess normal bundle

$$\mathcal{E} \to \mathcal{S}_\zeta$$

has two components which we denote by $\mathcal{E}'$ and $\mathcal{E}''$ (this means that topologically $\mathcal{E} = \mathcal{E}' \oplus \mathcal{E}''$). Their fibres are given by

$$(46) \qquad \mathcal{E}'_{s_0} = I_{\Psi(s_0)}/(I \cap T\mathcal{B})_{\Psi(s_0)}$$

where $I \subset T\overline{A}_{\mathbf{h},\Sigma}$ is the infinitesimal period relation, and

$$(47) \qquad \mathcal{E}''_{s_0} = \ker\left\{H^2(\Lambda^2 N_{W/X}(L^*)) \xrightarrow{ds_0} H^2(N_{W/X})\right\}.$$

Here, we are *not* claiming that $I \subset T\overline{A}_{\mathbf{h},\Sigma}$ has been defined in general, but only that it is defined along the locus in which we are interested, if one wishes by the explicit computations in section V(i). With this understood we have the

(48) Theorem: *With the above notations and assumptions we have*

$$\Psi^*([\mathcal{B}]) = Gys_{\mathcal{S}_\zeta/|\mathcal{L}|}\left(c_{\text{top}}(\mathcal{E}') \cdot c_{\text{top}}(\mathcal{E}'')\right).$$

For the excess intersection formula corresponding to (45), we have an additional component $\mathcal{E}'''$ to the excess normal bundle whose fibres are

$$(49) \qquad \mathcal{E}'''_{s_0} = N_{\mathcal{M}_{X,\zeta}/\mathcal{M}_X}.$$

(50) **Theorem:** *With again keeping our notations and assumptions we have*

$$\Psi_\zeta^*([\mathcal{B}]) = Gys_{\mathcal{S}_\zeta/|\mathcal{L}|_\zeta} \left(c_{\text{top}}(\mathcal{E}') \cdot c_{\text{top}}(\mathcal{E}'') \cdot c_{\text{top}}(\mathcal{E}''') \right)$$

where the fibres of $\mathcal{E}', \mathcal{E}'', \mathcal{E}'''$ are given by (46), (47), and (49) respectively.

The ranks of $\mathcal{E}', \mathcal{E}'', \mathcal{E}'''$ are given by

$$\begin{cases} \text{rank } \mathcal{E}' = h^{3,0}(l-1) \\ \text{rank } \mathcal{E}'' = \delta \\ \text{rank } \mathcal{E}''' = h^{3,1} - \sigma_\zeta \ , \end{cases}$$

and using (43) one may verify that both sides of the formulas (48) and (49) are in $H^{2(l+h^{3,0}(l-1))}(|\mathcal{L}|)$ and $H^{2(l+h^{3,0}(l-1))}(|\mathcal{L}|_\zeta)$ respectively.

Our story is not complete in that we have not described $\mathcal{E}', \mathcal{E}'', \mathcal{E}'''$ globally over $\mathcal{S}_\zeta$; all we have done is give their fibre at a general point. We expect to take up this consideration together with the computation of an example in a subsequent work.

VI. Conclusions

A first conclusion is that

(1) *When considering questions of Hodge classes and extended normal functions one should let (X, ζ) vary in moduli.*

The basic diagram governing the situation is

(2)
$$\begin{array}{ccc} & & \overline{\mathcal{J}}_{\mathbf{h},\Sigma} \\ & \nu_\zeta \nearrow \!\!\!\!\! \nearrow & \downarrow \\ |\mathcal{L}| - \overset{\Psi}{-} \!\!\! \longrightarrow & \overline{\mathcal{A}}_{\mathbf{h},\Sigma} \end{array}$$

Here, $\mathcal{J}_{\mathbf{h},\Sigma}$ is a partial compactification of the universal family of intermediate Jacobians. An issue is

(3) *The family $\overline{\mathcal{J}}_{\mathbf{h},\Sigma}$ has not yet been defined in general.*

In this paper we have explicitly given the set-theoretic boundary components of the partial compactification $\overline{\mathcal{A}}_{\mathbf{h},\Sigma}$ by describing the limiting MHS's in the nodal case where there is at most one relation among the vanishing cycles, and Kato-Usui have some preliminary handwritten notes that describe the part of $\overline{\partial}_{\mathbf{h},\Sigma}$ lying over the components of $\overline{\mathcal{A}}_{\mathbf{h},\Sigma}$ around which the monodromy satisfies $(T-I)^2 = 0$, which includes the nodal case mentioned above.[6] Of special interest may be the boundary components where the local systems $\mathcal{H}(\mathcal{I})$ with "fibres" $H^1(B^\bullet(\mathbf{N}))$ are supported (cf. the end of section I(ii)).

From sections V(i), V(ii) we would suggest that the geometry of the infinitesimal period relations and resulting "I-transversality" may be of particular interest.

Restricting to nodal loci, one has the basic observation

(4) *sing* ν_ζ *is equal to components of* $\Psi^{-1}(\mathcal{B}(\sigma(l,1)))$.

This suggests that one should study

(5) $$\operatorname{codim}_{|\mathcal{L}|}(\operatorname{sing}\nu_\zeta)$$

and

$$\Psi^*([\mathcal{B}(\sigma(l,1)])\,.$$

From (4) and general excess intersection formula considerations one expects a relation of the form

(6) $$[\operatorname{sing}_{|\mathcal{L}|}(\nu_\zeta)] = \Psi_\zeta^*([\mathcal{B}(\sigma(l,1)])_\wedge K$$

where K is a "correction" term. Much of this paper has been devoted to an initial study of (5) and (6), and perhaps the major issue raised in this work is to

(7) *Define* $\overline{\partial}_{\mathbf{h},\Sigma}$ *and a set of boundary components* $\mathcal{B}_\lambda \subset \overline{\partial}_{\mathbf{h},\Sigma}$ *such that for*
   $L \gg 0$

(8) $$\operatorname{sing}\nu_\zeta = \underset{\lambda}{\cap}\,\nu_\zeta^{-1}(\mathcal{B}_\lambda)\,.$$

References

1. V. Alexeev, Compactified Jacobians and the Torelli map, *Publ. RIMS Kyoto Univ.* **40** (2004), 1241–1265.
2. V. Alexeev and I. Nakamura, On Mumford's construction of degenerating abelian varieties, *Tohoku Math. J.* **51** (1999), 399–420.

---

[6] In work in progress since this paper was written, we have begun the study of boundary components of maximally unipotent monodromy. The mix of Hodge theory and geometry here is completely unlike the $N^2\mathcal{J} = 0$ case and appears to be much richer.

3. L. Caporaso, A compactification of the universal Picard variety over the moduli space of stable curves, *Jour. AMS* **7** (1994), 589–560.

4. J. Carslon, E. Cattani, and A. Kaplan, Mixed Hodge structures and compactifications of Siegel's space, *Journées de géometrie algébrique d'Angers*, Sijthoff & Hoordhoff, 1980, pp. 1–43.

5. Eduardo Cattani, Mixed Hodge structures, compactifications and monodromy weight filtration, Chapter IV in *Topics in Transcendental Algebraic Geometry, Annals of Math. Studies* **106** (1984), Princeton Univ. Press, Princeton, NJ.

6. E. Cattani, A. Kaplan, and W. Schmid, Degeneration of Hodge structures, *Ann. of Math.* **123** (1986), 457–535.

7. E. Cattani, P. Deligne, and A. Kaplan, On the locus of Hodge classes, *J. Amer. Math. Soc.* **8** (1995), 483–506.

8. C. H. Clemens, The Neron model for families of intermediate Jacobians acquiring "algebraic" singularities, *Publ. Math. I.H.E.S.* **58** (1983), 5–18.

9. _____, Degeneration of Kähler manifolds, *Duke Math. J.* **44** (1977).

10. Friedman, The period map at the boundary of moduli, Chapter X in *Topics in Transcendental Algebraic Geometry, Annals of Math. Studies* **106** (1984), Princeton Univ. Press, Princeton, NJ.

11. W. Fulton, *Intersection Theorem, Results in Mathematics and Related Areas, 3rd Series, A Series of Modern Surveys in Math.* **2**, Springer-Verlag, Berlin, 1998.

12. M. Green and P. Griffiths, Hodge theoretic invariants for algebraic cycles, *IMRN* **9** (2003), 477–510.

13. _____, Formal deformation of Chow groups, in *The Legacy of Niels Henrik Abel*, 467–509, Springer-Verlag, Berlin, 2004.

14. _____, Algebraic cycles and singularities of normal functions, I.

15. M. Green, J. Murre, and C. Voisin, Algebraic cycles and Hodge theory, *Proc. C.I.M.E., Torino, 1993*, A. Albano and F. Bardelli, eds., *Lecture Notes in Math.* **1593**, Springer-Verlag, Berlin, 1994.

16. U. Jannsen, Equivalence relations on algebraic cycles, *NATO Sci. Ser. C Math. Phys. Sci.* **548** (2000), Kluwer Acad. Publ. Dordrecht.

17. K. Kato and S. Usui, Classifying spaces of degenerating polarized Hodge structures, preprint.

18. S. Kleiman, Geometry on Grassmannians and applications to splitting bundles and smoothing cycles, *Publ. Math. IHES* **36** (1969), 281–297.

19. J. Murre, On a conjectural filtration on the Chow groups of an algebraic variety, I. The general conjectures and some examples, *Indag. Math.* **4** (1993), 177–188.

20. M. Saito, Admissible normal functions, *J. Alg. Geom.* **5** (1996), 235–276.

21. T. Shioda, Algebraic cycles on certain hypersurfaces, *Alg. Geom. (Tokyo/ Kyoto, 1982)*, 271–294, *Lect. Notes in Math.* **1016**, Springer-Verlag, Berlin, 1983.

Chapter 10

PLANAR WEB GEOMETRY THROUGH ABELIAN RELATIONS AND SINGULARITIES

Alain Hénaut

Institut de Mathématiques de Bordeaux, 351, cours de la Libération,
F-33405 Talence, Cedex, France

May 29, 2006

À la mémoire du Professeur S. S. Chern

1. Introduction and Basic Objects

Web geometry is devoted to the study of families of foliations which are in general position. The subject is of interest in any dimension and webs consisting of foliations of different codimensions can be considered. However in the following we essentially restrict ourselves to the plane situation, locally in the neighborhood of the origin in $\mathbb{C}^2$ or globally on $\mathbb{P}^2 = \mathbb{P}^2(\mathbb{C})$, with $d \geq 1$ complex analytic foliations of curves in general position. On the other hand we indicate how inherent singularities of these configurations can be solicited for some roles. Basic and new results are also presented in the nonsingular case. To illustrate methods and open problems, we emphasize all along the three sections on introductory examples.

A series of over sixty papers under the general title "*Topologische Fragen der Differentialgeometrie*" initiated the subject around the 1930's. Most of them were written by Wilhelm Blaschke and his students or coworkers in Hamburg, notably Gerhard Thomsen and Gerrit Bol (*cf.* [B-B] and [B] for a classical and enlarged summary of the results).

Shiing-Shen Chern defended his thesis in 1936, under the direction of Blaschke, with the papers T 60 and T 62 (respectively [C1] and [C2] in the references). In the first paper, extending methods "à la Poincaré-Blaschke", he proved that the sharp upper bound for the dimension of the abelian relations of a d-web of codimension 1 in $\mathbb{C}^k$ is the Castelnuovo number for the maximum genus of a nondegenerate reduced algebraic curve of degree d in $\mathbb{P}^k$. Already familiar to Cartan-Kähler theory and using an affine connection, he introduced in the second paper basic invariants of a 3-web of codimension n in $\mathbb{C}^{2n}$ and gave geometrical interpretation of some of them. Then, with the future that we know, Chern spent a postdoctoral year in Paris with Élie Cartan.

He returns to this subject before the 1980's with Phillip A. Griffiths (*cf.* [C-G1], [C-G2] and [C-G3]). Rather close to the paper T 60, the cycle "*Variations on a Theorem of Abel*" carry on walking with emblematic resonances (for a part of substantial chords, *cf.* [G2], [G3] and [G4]). The common work of Chern and Griffiths brings a resurgence of interest in web geometry. This revival is also supported with papers by Maks A. Akivis and Vladislav V. Goldberg (*cf.* for instance the survey [A-G]). A part of recent references are included below, but for results and applications of web geometry in various domains, refer also to Isao Nakai's introduction and all papers in [W].

Chern's influence on web geometry never failed and goes on. Not only by his thesis, survey papers, problems, interviews and even comments on his mathematical education (*cf.* for instance [C3] and [C4]), but above all through an *attitude*. This one notably occurs to us all along his general contribution on "*Geometry of Characteristic Classes*" which prepares the ground, favors the analogies and stimulates the extensions. In particular, the use of adapted extrinsic constructions coupled with algebraic manipulations which give rise to geometrical intrinsic properties and conceptual tools.

With no surprise according to the previous observations, abelian relations (that is essentially relations between the normals of the leaves) and connections not necessarily affines are used to study the geometry of planar webs in the following. Singularities as objects for generating forms as phrased by René Thom, come on stage on one's toes, but with the assurance to offer some new interests in the subject. This part of a work in progress completes the present text which illustrates the interplay between differential geometry and algebraic geometry.

Let $\mathcal{O} := \mathbb{C}\{x, y\}$ be the ring of convergent power series in two variables. A (germ of a) nonsingular *d-web* $\mathcal{W}(d) = \mathcal{W}(F_1, \ldots, F_d)$ in $(\mathbb{C}^2, 0)$ is defined by a family of *leaves* which are germs of level sets $\{F_i(x, y) = const.\}$ where $F_i \in \mathcal{O}$ can be chosen to satisfy $F_i(0) = 0$ and such that $dF_i(0) \wedge dF_j(0) \neq 0$ for $1 \leq i < j \leq d$ from the assumption of general position.

The classification of such $\mathcal{W}(d)$ is a widely open problem and we are looking for invariants. These can be either numerical invariants or more geometrical ones such as curvatures and of course local models of planar webs $\mathcal{W}(d)$. In other words, we collect a breeding ground of *forms* to approach a classification.

From the local inverse theorem, the study of possible configurations for the different $\mathcal{W}(d)$ is interesting only for $d \geq 3$ in the nonsingular case. Up to the rectified model $\mathcal{W}(x, y, F_3)$ we extend the warp and weft from the basic textile! However such a web is in general not equivalent to the hexagonal model $\mathcal{H}(x, y, x + y)$ and this problem and above all the great variety of solutions appear like the birth certificate of the planar web geometry subject.

Let $F(x, y, p) = a_0(x, y) \cdot p^d + a_1(x, y) \cdot p^{d-1} + \cdots + a_d(x, y)$ be an element of $\mathcal{O}[p]$ without multiple factor, not necessarily irreducible and such that $a_0 \neq 0$. We denote by $\overline{R} = (-1)^{\frac{d(d-1)}{2}} a_0 \cdot \overline{\Delta}$ the *p-resultant* of F where $\overline{\Delta} \in \mathcal{O}$ is its *p-discriminant*. In a neighborhood of $(x_0, y_0) \in \mathbb{C}^2$ such that $\overline{R}(x_0, y_0) \neq 0$, a theorem of Cauchy asserts that the d integral curves of the differential equation of the first order

$$F(x, y, y') = 0$$

are the leaves of a nonsingular web $\mathcal{W}(d)$ in $(\mathbb{C}^2, (x_0, y_0))$. Every such $F \in \mathcal{O}[p]$, up to an invertible element in $\mathcal{O}$, gives rise to an *implicit d-web* $\mathcal{W}(d)$ in $(\mathbb{C}^2, 0)$ which is generically nonsingular. Inversely a nonsingular *d-web* $\mathcal{W}(d) = \mathcal{W}(F_1, \ldots, F_d)$ in $(\mathbb{C}^2, 0)$ defines a differential equation of the previous type, through their different slopes, after a linear change of coordinates; to $\mathcal{W}(d)$ corresponds d vector fields $X_i = A_i \partial_x + B_i \partial_y$ with $X_i(F_i) = 0$ and one may assume $A_i(0) \neq 0$ for $1 \leq i \leq d$. According to the assumption of general position, it can be noted that in general the zeros of $\overline{\Delta} = \prod_{1 \leq i < j \leq d}(A_i B_j - A_j B_i)^2$ are the *singular points* of the planar web $\mathcal{W}(d)$ since $F = \prod_{i=1}^d (A_i\, p - B_i) \in \mathcal{O}[p]$.

This implicit form of a planar web, where no leaf is prefered, presents a natural setting for the study of planar webs and their singularities. Moreover, using the web viewpoint, this approach enlarges methods to investigate the geometry of the differential equation $F(x, y, y') = 0$.

Global webs can be defined on any analytic surface by using its pro-
jectivized tangent space. The study of global webs worn by a surface is
of interest for the geometry of both. For a *polynomial* planar web, that
is with $F \in \mathbb{C}[x, y, p]$, the associated web extends to $\mathbb{P}^2$ with a singular
locus corresponding to the zeros of the p-discriminant $\overline{\Delta}$ of F. In this case
$F(x, y, y') = 0$ is an algebraic differential equation of the first order with
degree d. Incomplete ones $F(y, y') = 0$ include those related to the elliptic functions through the Briot-Bouquet's theorem. For polylogarithmic
webs partially discussed below, leaves are level sets of rational functions
$F_i \in \mathbb{C}(x, y)$, but relations between the F_i involve very special transcendental functions which generalize the logarithm.

Coming back to the classical geometric study of differential equations
$F(x, y, y') = 0$, we shall confirm even locally, in the nonsingular case, how
some basic objects attached to the surface S defined by F with the projec-
tion $\pi : S \longrightarrow (\mathbb{C}^2, 0)$ induced by $(x, y, p) \longmapsto (x, y)$ govern the geometry
of the planar web $\mathcal{W}(d)$ associated with F. Moreover in the singular case,
meromorphic connection methods will come into use. The archetypes are
the local monodromy and regularity theorems related to periods of integrals for a suitable family of algebraic curves by means of the associated
Picard-Fuchs equation or the Gauss-Manin connection.

A basic model of planar webs comes from algebraic geometry. Let $C \subset$
$\mathbb{P}^2$ be a reduced algebraic curve of degree d, not necessarily irreducible
and possibly singular. For a generic line $l(x, y) \in G(1, \mathbb{P}^2) = \check{\mathbb{P}}^2$, we have
$l(x, y) \cap C = \sum_{i=1}^d \check{p}_i(x, y)$ as 0-cycles in C. By duality, one can get a special
linear d-web $\mathcal{L}_C(d)$ in $\check{\mathbb{P}}^2$ implicitly given by the F_i where $\check{p}_i = (F_i, \xi_i(F_i))$
and called the *algebraic web* associated with $C \subset \mathbb{P}^2$. This web is singular
and its leaves are family of straight lines. It corresponds, in a suitable local
coordinate system, to a differential equation of the previous form given by
$F(x, y, p) = P(y - px, p)$ where $P(s, t) = \prod_{i=1}^d (t - \xi_i(s)) = 0$ is an affine
equation for C (*cf.* for instance [H1] for details). If C contains no straight
lines, the leaves of $\mathcal{L}_C(d)$ are essentially the tangents of the dual curve
$\check{C} \subset \check{\mathbb{P}}^2$ of $C \subset \mathbb{P}^2$; otherwise, they belong to the corresponding pencils of
straight lines.

As usual in geometry, we begin with the nonsingular case. But, as
we shall see with the implicit approach, we do not forget the singularities. One of the main invariant of a nonsingular planar web $\mathcal{W}(d) =$
$\mathcal{W}(F_1, \ldots, F_d)$ in $(\mathbb{C}^2, 0)$ is related to the notion of abelian relation. A

d-uple $\big(g_1(F_1), \ldots, g_d(F_d)\big) \in \mathcal{O}^d$ satisfying

$$\sum_{i=1}^{d} g_i(F_i) dF_i = 0$$

where $g_i \in \mathbb{C}\{t\}$ is called an *abelian relation* of $\mathcal{W}(d)$. Such a relation links the normals of leaves with constant coefficients along the leaves. From the differential equation $F(x, y, y') = 0$ viewpoint, it is after integration an addition relation between the generic solutions. Using the above component presentation these abelian relations form a $\mathbb{C}$-vector space $\mathcal{A}(d)$ and the following *optimal inequality* holds:

$$\operatorname{rk} \mathcal{W}(d) := \dim_{\mathbb{C}} \mathcal{A}(d) \leq \pi_d := \frac{1}{2}(d-1)(d-2).$$

This bound is classic and the integer $\operatorname{rk} \mathcal{W}(d)$ called the *rank* of $\mathcal{W}(d)$ defined above is an invariant of $\mathcal{W}(d)$ which does not depend on the choice of the functions F_i. Abelian relations can be considered as "differential 1-forms" of the web $\mathcal{W}(d)$ and after integration correspond to "functions" of $\mathcal{W}(d)$, that is d-uple $\big(\widetilde{g}_1(F_1), \ldots, \widetilde{g}_d(F_d)\big) \in \mathcal{O}^d$ with $\widetilde{g}_i \in \mathbb{C}\{t\}$ such that $\sum_{i=1}^{d} \widetilde{g}_i(F_i) = const$. A natural generalization of these objects as a complex $(\mathcal{A}^\bullet, \delta)$ of $\mathbb{C}$-vector space $\mathcal{A}^p$ for $0 \leq p \leq n$ exists for a d-web $\mathcal{W}(d)$ of codimension n in $(\mathbb{C}^N, 0)$, where δ is induced by the usual exterior derivative; it looks like a de Rham complex for such a $\mathcal{W}(d)$. For $N = kn$ with $k \geq 2$, optimal upper bounds $\pi_p(d, k, n)$ appear for $0 \leq p \leq n$ and generalize the Castelnuovo's ones (*cf.* for instance [B-B], [C1], [C-G1], [C-G2], [C-G3] and [H2]). For planar d-webs, we recover $\mathcal{A}^1 = \mathcal{A}(d)$ and $\pi_1(d, 2, 1) = \pi_d$ with the following exact sequence

$$0 \longrightarrow \mathbb{C}^d \longrightarrow \mathcal{A}^0 \overset{\delta}{\longrightarrow} \mathcal{A}^1 \longrightarrow 0.$$

For example in the planar case, the rank of the hexagonal web $\mathcal{H}(x, y, x+y)$ is maximum equal to 1 because $dx + dy - d(x+y) = 0$. Conversely, if $\operatorname{rk} \mathcal{W}(3) = 1$, using a nontrivial abelian relation and after integration there exists a local isomorphism which carries $\mathcal{W}(3)$ to $\mathcal{H}(x, y, x+y)$.

Another basic result in nonsingular planar web geometry is related to *linear* webs $\mathcal{L}(d)$ (*i.e.* all leaves of $\mathcal{L}(d)$ are straight lines, not necessarily parallel). The following assertions are *equivalent*:

(i) The linear web $\mathcal{L}(d)$ is algebraic, that is $\mathcal{L}(d) = \mathcal{L}_C(d)$ where $C \subset \mathbb{P}^2$ is a reduced algebraic curve of degree d, not necessarily irreducible and possibly singular;

(ii) The rank of $\mathcal{L}(d)$ is maximum, that is equal to π_d.

This equivalence and some variants play a fundamental role in the foundation of web geometry and explains the terminology. Indeed, the implication i) $\Rightarrow$ ii) is a special case of Abel's theorem and asserts that in fact the following equalities hold:

$$\text{rk}\,\mathcal{L}_C(d) = \dim_{\mathbb{C}} H^0(C, \omega_C) = \pi_d\,.$$

With the notations used before, these equalities come from the description of elements of $H^0(C, \omega_C)$ as $\omega = r(s,t)\frac{ds}{\partial_t(P)}$ where $r \in \mathbb{C}[s,t]$ with $\deg r \leq d-3$ and since we have a $\mathbb{C}$-isomorphism $H^0(C, \omega_C) \longrightarrow \mathcal{A}(\mathcal{L}_C(d))$, where $\mathcal{A}(\mathcal{L}_C(d))$ is the abelian relation space of $\mathcal{L}_C(d)$. It is given by $\omega \longmapsto (g_i(F_i))_i$ where $\omega = g_i(s)ds$ in the neighborhood of $\check{p}_i(0,0)$, since we have globally on $\mathbb{P}^2$ the relation $\text{Trace}(\omega) := \sum_{i=1}^d \check{p}_i^*(\omega) = \sum_{i=1}^d g_i(F_i)dF_i = 0$. The implication ii) $\Rightarrow$ i) is a kind of converse to Abel's theorem. In the case $d = 4$, it has been initiated by Lie's theorem on surfaces of double translation (*cf.* for instance [C3]) and deeply generalized, for $d \geq 3$ and higher codimension questions, by Griffiths (*cf.* [G2]).

In connection with the previous result, let us quote Griffiths from [G3]:

> " ... *the presence of* addition theorems *(loosely interpreted) or* functional equations *may have rather striking consequences — both local and global — on a geometric configuration.*"

Classical proofs of the implication ii) $\Rightarrow$ i) in the equivalence before use extension process coupled with the GAGA principle of Jean-Pierre Serre. In fact with the connection (E, ∇) associated with any $\mathcal{W}(d)$ defined in the next section, it is possible for a linear web $\mathcal{L}(d)$ to prove this equivalence (*cf.* [H3]) by using only differential systems as Chern asks in [C3]. For a linear web $\mathcal{L}(d)$ with maximum rank, we have from the previous result $F(x, y, p) = P(y - px, p)$, up to an invertible element in $\mathcal{O}$, but with $P \in \mathbb{C}[s,t]$. Thus, some "normal forms" for the implicit differential equation $F(x, y, y') = 0$ can be expected in the general case with the help of abelian relations $\sum_{i=1}^d g_i(F_i)dF_i = 0$, notably in the maximum rank case.

The abelian relations space $\mathcal{A}(d)$ is a *local system* of $\mathbb{C}$-vector spaces. In the second section, we essentially deal with the geometric study of any planar web $\mathcal{W}(d)$ with the help of the previous generically finite morphism $\pi : S \longrightarrow (\mathbb{C}^2, 0)$ and a not necessarily integrable connection (E, ∇) of rank π_d associated with $\mathcal{W}(d)$, where possible horizontal sections, however, are identified with $\mathcal{A}(d)$ (*cf.* [H3] for details). Following some recents results obtained by Olivier Ripoll from the connection (E, ∇), new invariants of a planar web $\mathcal{W}(d)$ and effective methods to completely determine its rank are also presented (*cf.* [R1] and [R2] for details). Moreover, for singular planar webs, these possible abelian relations are in general *singular*, for instance "multivalued" in the g_i. For example the 4-web $\mathcal{W}(x, y, xy, \frac{x}{y}) = \mathcal{W}(F_1, F_2, F_3, F_4)$ is of maximum rank equal to 3 with the following "singular" basis of abelian relations:

$$\begin{cases} \dfrac{1}{F_1} dF_1 + \dfrac{1}{F_2} dF_2 - \dfrac{1}{F_3} dF_3 = 0 \,, \\[2ex] \dfrac{1}{F_1} dF_1 - \dfrac{1}{F_2} dF_2 - \dfrac{1}{F_4} dF_4 = 0 \,, \\[2ex] \dfrac{\log F_1}{F_1} dF_1 + \dfrac{\log F_2}{F_2} dF_2 - \dfrac{\log F_3}{2F_3} dF_3 - \dfrac{\log F_4}{2F_4} dF_4 = 0 \,. \end{cases}$$

From the logarithm relation, all extracted 3-webs of this 4-web are of maximum rank equal to 1. Defined in [H3] the *weave* of this 4-web, which is a refinement of its rank related to the kind of components which generate abelian relations, is $(2, 1)$.

These different observations explain the emergence of the singular approach for planar webs which ends the present paper. The next two sections are also motivated by what follows.

Web geometry for nonsingular planar webs of maximum rank is larger in extent than algebraic geometry of plane curves. Indeed, there exist *exceptional* webs $\mathcal{E}(d)$ in $(\mathbb{C}^2, 0)$. Such a web $\mathcal{E}(d)$, sometimes called exotic by Chern, is of maximum rank π_d and cannot be made algebraic up to a local analytic isomorphism of $\mathbb{C}^2$, that is by the previous result $\mathcal{E}(d)$ is not linearizable.

For $\mathcal{E}(d)$ one knows that necessary $d \geq 5$ and the first example (1936) is Bol's 5-web $\mathcal{B}(5)$ in $\mathbb{P}^2$ with rank 6. This exceptional web $\mathcal{B}(5)$ is formed by 4 pencils of lines whose vertices are in general position and the fifth leaf through a generic z is the only conic passing through the four vertices and z. A basis of abelian relations of $\mathcal{B}(5)$ is given by 5 relations on 3 terms completed with the relation on 5 terms satisfied by "the" dilogarithm ϕ_2 coming back to $\sum_{n \geq 1} \frac{z^n}{n^2}$. With the Rogers' modified version, it is precisely

$$\phi_2(x) - \phi_2(y) + \phi_2\left(\frac{y}{x}\right) - \phi_2\left(\frac{1-y}{1-x}\right) + \phi_2\left(\frac{x(1-y)}{y(1-x)}\right) = \frac{\pi^2}{6}$$

which extends the logarithm relation $\phi_1(x) - \phi_1(y) + \phi_1\left(\frac{y}{x}\right) = 0$. The weave of $\mathcal{B}(5)$ is $(5, 0, 1)$. Also from a Bol's result, $\mathcal{B}(5)$ is the only hexagonal d-web, that is all extracted 3-web of $\mathcal{B}(5)$ are with rank one, except the ones generated by pencils of d points in $\mathbb{P}^2$.

The next exceptional web expected was Spence-Kummer's 9-web $\mathcal{SK}(9)$ in $\mathbb{P}^2$ related to the functional relation with 9 terms of the trilogarithm (*cf.* [H1]). It contains three incarnations of $\mathcal{B}(5)$ and Gilles Robert proved in 2001 that $\mathcal{SK}(9)$ is indeed exceptional and he found "on the road" some other $\mathcal{E}(d)$ (*cf.* [Rob]). Independently Luc Pirio recovered and extented these results. Moreover, in 2004, he has given *three* examples of new and nonequivalent $\mathcal{E}(5)$ very simple with explicit basis of abelian relations (*cf.* for instance [P]):

$$\mathcal{W}\left(x, y, x+y, x-y, \begin{array}{c} x^2 + y^2 \\ xy \\ e^x + e^y \end{array}\right)$$

where the last two were known for projective differential properties, with the exception of their ranks! Thus the very rich polylogarithmic web vein for general properties of configuration must be enlarged for these questions of rank. Recently, Pirio and Jean-Marie Trépreau insert the three previous $\mathcal{E}(5)$ in an exceptional family of $\mathcal{E}(5)$ with parameter k:

$$\mathcal{W}(x, y, x+y, x-y, \mathrm{sn}_k\, x \cdot \mathrm{sn}_k\, y)$$

where the fifth family of leaves is given by the product of the same Jacobi elliptic function in x and y (*cf.* [P-T]).

The geometric structure of exceptional planar webs is of interest; even for $d = 5$, these questions are widely open and attractive as already noticed by Chern. Not only for web geometry with natural generalization in any dimension and codimension, but in other domains like for example number theory and algebraic K-theory or perhaps noncommutative geometry (*cf.* for instance [Gol], [G4], [Gon] and [Man]).

Let $\mathcal{W}(d, k, 1)$ be a d-web of codimension 1 in $(\mathbb{C}^k, 0)$ with maximum 1-rank $\pi_1(d, k, 1)$. Contrary to the planar case, it can be indicated to close this large introduction that $\mathcal{W}(d, 3, 1)$ is linearizable for $d \neq 5$ by a theorem of Bol, and in fact algebraizable by a converse of Abel's theorem (*cf.* [B-B] for $\mathcal{W}(d, 3, 1)$ and for $k \geq 3$ and $d \geq 2k$, [C-G2] for "normal" $\mathcal{W}(d, k, 1)$ and Trépreau's paper in the present volume for the general case).

2. From Abelian Relations to Connections

Let $\mathcal{W}(d)$ be a planar web in $(\mathbb{C}^2, 0)$ implicitly presented by $F \in \mathcal{O}[p]$. We suppose that the p-resultant $\overline{R} \in \mathcal{O}$ of F satisfies $\overline{R}(0) \neq 0$, thus $\mathcal{W}(d)$ is nonsingular in $(\mathbb{C}^2, 0)$. For a large part of the results presented in this section we refer to [H3].

The surface S defined by F is nonsingular and the projection $\pi : S \longrightarrow (\mathbb{C}^2, 0)$ is a covering map of degree d with local branches $\pi_i(x, y) = (x, y, p_i(x, y))$. Thus, we have

$$F(x, y, p) = a_0(x, y) \cdot \prod_{i=1}^{d} (p - p_i(x, y)) \, .$$

Locally on the surface S, we have the de Rham complex $(\Omega_S^\bullet, d)$ where as usual $\Omega_S^\bullet = \Omega_{\mathbb{C}^3}^\bullet / (dF \wedge \Omega_{\mathbb{C}^3}^{\bullet - 1}, F\Omega_{\mathbb{C}^3}^\bullet)$. An element $\omega = r \cdot \frac{dy - p\, dx}{\partial_p(F)} \in \Omega_S^1$ if and only if there exist elements $(r, r_p, t) \in \mathcal{O}_S^3$ such that

$$(\star) \qquad r \cdot \big(\partial_x(F) + p\, \partial_y(F)\big) + r_p \cdot \partial_p(F) = \big(\partial_x(r) + p\, \partial_y(r) + \partial_p(r_p) - t\big) \cdot F$$

and especially with this relation, it can be checked that $d\omega = t \cdot \frac{dx \wedge dy}{\partial_p(F)} \in \Omega_S^2$.

Let $\mathfrak{a}_F$ be the $\mathbb{C}$-vector space of particular 1-forms defined by

$$\mathfrak{a}_F := \left\{ \omega = r \cdot \frac{dy - p\, dx}{\partial_p(F)} \in \pi_*(\Omega_S^1); r \in \mathcal{O}[p] \text{ with } \deg r \leq d - 3 \text{ and } d\omega = 0 \right\},$$

we have the following basic result:

Theorem ($\mathfrak{a}_F$). *There exists a $\mathbb{C}$-isomorphism*

$$\mathcal{A}(d) \longrightarrow \mathfrak{a}_F$$

$$\big(g_i(F_i)\big)_i \longmapsto \omega := \left(F \cdot \sum_{i=1}^{d} \frac{g_i(F_i)\partial_y(F_i)}{p - p_i} \right) \cdot \frac{dy - p\, dx}{\partial_p(F)}$$

such that $\mathrm{Trace}_\pi(\omega) := \sum_{i=1}^{d} \pi_i^*(\omega) = \sum_{i=1}^{d} g_i(F_i)dF_i = 0.$

The proof of this theorem uses the ubiquitous Lagrange interpolation formula and some calculus influenced by those of Gaston Darboux in his solution of the problem of surfaces of double translation (*cf.* [Dar]). From this result any abelian relation of $\mathcal{W}(d)$ is interpreted as the vanishing trace of an element belonging to $\mathfrak{a}_F$. Already in the Abel's theorem used before and as a publicity slogan let us say "nothing vanishes without a trace"!

Differential 1-forms in $\mathfrak{a}_F$ are closed by definition and using the relation $(\star)$ every element $r = b_3 \cdot p^{d-3} + b_4 \cdot p^{d-4} + \cdots + b_d$ which appears in $\mathfrak{a}_F$ is explicitly given by an analytic solution of a homogeneous linear differential system $\mathcal{M}(d)$ with $d - 1$ equations. To be concrete for $d = 3$, it has the following form:

$$\mathcal{M}(3) \qquad \begin{cases} \partial_x(b_3) + A_1\, b_3 = 0 \\ \partial_y(b_3) + A_2\, b_3 = 0 \end{cases}$$

and for $d = 4$, we get

$$\mathcal{M}(4) \qquad \begin{cases} \partial_x(b_4) \qquad\quad + A_{1,1}\, b_3 + A_{1,2}\, b_4 = 0 \\ \partial_x(b_3) + \partial_y(b_4) + A_{2,1}\, b_3 + A_{2,2}\, b_4 = 0 \\ \partial_y(b_3) \qquad\quad + A_{3,1}\, b_3 + A_{3,2}\, b_4 = 0\,. \end{cases}$$

In general, we obtain the following system:

$$\mathcal{M}(d) \qquad \begin{cases} \partial_x(b_d) \qquad\qquad + A_{1,1}\, b_3 + \cdots + A_{1,d-2}\, b_d = 0 \\ \partial_x(b_{d-1}) + \partial_y(b_d) + A_{2,1}\, b_3 + \cdots + A_{2,d-2}\, b_d = 0 \\ \qquad\qquad\qquad\qquad\qquad\qquad \vdots \\ \partial_x(b_3) + \partial_y(b_4) + A_{d-2,1}\, b_3 + \cdots + A_{d-2,d-2}\, b_d = 0 \\ \partial_y(b_3) \qquad\qquad + A_{d-1,1}\, b_3 + \cdots + A_{d-1,d-2}\, b_d = 0 \end{cases}$$

where the matrix $(A_{i,j})$ of elements in $\mathcal{O}$ is obtain by identification, only from the coefficients of F by Cramer formulas. The classical formula of Sylvester for the p-resultant $\overline{R} = (-1)^{\frac{d(d-1)}{2}} a_0 \cdot \overline{\Delta}$ of F appears indeed in the matrix form of this identification and the result comes from the hypothesis $\overline{R}(0) \neq 0$. Moreover, it can be checked that in general $A_{i,j} \in \mathcal{O}[1/\Delta]$, that is poles of $A_{i,j}$ are on the *discriminant locus* of F or the singular locus of $\mathcal{W}(d)$, where Δ is the reduced divisor associated with $\overline{\Delta}$. This observation will be especially interesting in the study of singular planar webs which ends the present paper.

From the previous results, *we have the following identifications*:

$$\mathcal{A}(d) = \mathfrak{a}_F = \mathrm{Sol}\,\mathcal{M}(d)$$

where $\mathrm{Sol}\,\mathcal{M}(d)$ denotes the $\mathbb{C}$-vector space of analytic solutions of $\mathcal{M}(d)$. Moreover, using the special form of the symbols of $\mathcal{M}(d)$, basic results in algebraic analysis prove that $\mathrm{Sol}\,\mathcal{M}(d)$ is a *local system* with rank $\mathrm{rk}\,\mathcal{W}(d)$, which is bounded by $\pi_d := \frac{1}{2}(d-1)(d-2)$. In other words, $\mathfrak{a}_F$ is a locally constant sheaf of $\mathbb{C}$-vector spaces with finite dimensional fibers with dimension bounded by π_d.

At this stage, let us mention a couple of general remarks which specifies the differential system $\mathcal{M}(d)$ and the geometric interest of the previous identifications:

1. The special 1-form $\mathfrak{n} := \frac{dy - p\,dx}{\partial_p(F)} \in \Omega_S^1$ induces a natural *normalisation* for the planar web $\mathcal{W}(d)$ as follows. With the previous notations, the 1-forms $\mathfrak{n}_i = \pi_i^*(\mathfrak{n})$ for $1 \le i \le d$ define $\mathcal{W}(d)$ and we have

$$\mathrm{Trace}_\pi(p^k \cdot \mathfrak{n}) = \sum_{i=1}^{d} p_i^k \cdot \mathfrak{n}_i = 0 \quad \text{for} \quad 0 \le k \le d - 3$$

by the Lagrange interpolation formula; moreover, from the previous explicit expression of $d : \Omega_S^1 \longrightarrow \Omega_S^2$ we have the following equalities:

$$d(p^k \cdot \mathfrak{n}) = \left(A_{d-1,d-2-k} \cdot p^{d-2} + A_{d-2,d-2-k} \cdot p^{d-3} + \cdots + A_{1,d-2-k} \right)$$
$$\cdot \frac{dx \wedge dy}{\partial_p(F)(x,y,p)} \, .$$

2. By construction the matrix $(A_{i,j})$ depends on the implicit presentation of $\mathcal{W}(d)$ given by F. Ripoll in [R2] enlightens some results of [H3] by checking that $(A_{i,j})$ is generated from F only by the *linearization polynomial* $P_{\mathcal{W}(d)} \in \mathcal{O}[p]$ where $\deg P_{\mathcal{W}(d)} \le d - 1$ which is associated with $\mathcal{W}(d)$ and its *fundamental 1-form* $\alpha \in \Omega^1$.

The first object is constructed in such a way that the leaves of the planar web $\mathcal{W}(d)$ satisfy the second order differential equation $y'' = P_{\mathcal{W}(d)}(x, y, y')$ and plays a central role for $d \ge 4$ in the linearisation problem (*cf.* for instance [H1]). The "fundamental" 1-form α of $\mathcal{W}(d)$ is defined with the previous notations by $\alpha = A_{1,d-2}\,dx + A_{2,d-2}\,dy$ for $d \ge 4$ and $\alpha = A_1\,dx + A_2\,dy$ for $d = 3$. For any $g \in \mathcal{O}^*$, that is an invertible element in $\mathcal{O}$, it can be proved that F and $g \cdot F$ give the same $P_{\mathcal{W}(d)}$ and the same 2-form $d\alpha$, contrary to α. In this way $P_{\mathcal{W}(d)}$ and $d\alpha$ are geometric invariants of the planar web $\mathcal{W}(d)$ in the sense that these objects do not depend of its implicit presentation.

The integrability conditions for $\mathcal{M}(d)$ must be clarified, especially from the web geometric point of view, except for $d = 3$ with the help of these remarks. Indeed, for a planar web $\mathcal{W}(3)$ the integrability condition is only $d\alpha = 0$. Then it is sufficient to recall that the *Blaschke curvature* $d\gamma$ of $\mathcal{W}(3)$ does not depend on the normalisation $\omega_1 + \omega_2 + \omega_3 = 0$ used to define it, contrary to the 1-form γ such that $d\omega_i = \gamma \wedge \omega_i$ for $1 \le i \le 3$ where the ω_i correspond to the leaves of $\mathcal{W}(3)$. In particular, since here we have $d\gamma = d\alpha = \left(\partial_x(A_2) - \partial_y(A_1) \right) dx \wedge dy$ by using the properties of the special

1-form $\mathfrak{n}$, we recover the classical result: $d\gamma$ depends only of $\mathcal{W}(3)$ and

$$\operatorname{rk} \mathcal{W}(3) = 1 \quad \text{if and only if} \quad d\gamma = 0.$$

To understand the integrability conditions of the differential system $\mathcal{M}(d)$ for general $d \geq 3$, we go back to the trends of ideas "à la Cartan-Spencer".

The exterior differential on the surface S induces a linear differential operator $\varrho : \mathcal{O}^{d-2} \longrightarrow \mathcal{O}^{d-1}$ defined with the previous notations by $\varrho(b_3, \ldots, b_d) = (t_2, \ldots, t_d)$ where the second member is given by the $d-1$ equations of $\mathcal{M}(d)$ from the bottom up. Its corresponding $\mathcal{O}$-morphism $p_0 : J_1(\mathcal{O}^{d-2}) \longrightarrow \mathcal{O}^{d-1}$ satisfies $p_0 \circ j_1 = \varrho$ where $j_1(b) = \big(b, \partial_x(b), \partial_y(b)\big)$ as jets and has prolongations p_k with $R_k = \operatorname{Ker} p_k \subseteq J_{k+1}(\mathcal{O}^{d-2})$. Moreover, we have the first Spencer complex associated with p_k

$$0 \longrightarrow \operatorname{Sol} \mathcal{M}(d) \xrightarrow{j_{k+1}} R_k \xrightarrow{D} \Omega^1 \otimes_{\mathcal{O}} R_{k-1} \xrightarrow{D} \Omega^2 \otimes_{\mathcal{O}} R_{k-2} \longrightarrow 0$$

which is exact at R_k with injective j_{k+1}.

Using again the special form of the symbols of $\mathcal{M}(d)$, the properties of the operator D and some diagram chasing, we obtain from the theorem $\mathfrak{a}_F$ and the previous identifications the following result:

Theorem 1. *There exists a $\mathbb{C}$-vector bundle $E := R_{d-3} \subseteq J_{d-2}(\mathcal{O}^{d-2})$ of rank π_d on $(\mathbb{C}^2, 0)$ equipped with a connection $\nabla : E \longrightarrow \Omega^1 \otimes_{\mathcal{O}} E$ such that its $\mathbb{C}$-vector space of horizontal sections $\operatorname{Ker} \nabla$ is isomorphic to the abelian relation space $\mathcal{A}(d)$ of the planar web $\mathcal{W}(d)$. Moreover, there exists an adapted basis (e_ℓ) of (E, ∇) such that its curvature $K : E \longrightarrow \Omega^2 \otimes_{\mathcal{O}} E$ has the following matrix:*

$$\begin{pmatrix} k_1 & k_2 & \ldots & k_{\pi_d} \\ 0 & 0 & \ldots & 0 \\ \vdots & \vdots & & \vdots \\ 0 & 0 & \ldots & 0 \end{pmatrix} dx \wedge dy.$$

In particular, $\operatorname{rk} \mathcal{W}(d)$ is maximum equal to π_d if and only if $K = 0$. Moreover for $d = 3$, the 2-form $k_1\, dx \wedge dy$ is the Blaschke curvature of $\mathcal{W}(3)$.

Let (e_ℓ) be a not necessarily adapted basis of E where $1 \leq \ell \leq \pi_d$. In this basis, the matrix of 1-forms $\gamma = (\gamma_{k\ell})$ such that $\nabla(e_\ell) = \sum_k \gamma_{k\ell} \otimes e_k$ represents the connection (E, ∇) and the matrix of 2-forms $d\gamma + \gamma \wedge \gamma$ represents its curvature K; horizontal sections $f = (f_\ell) \in \operatorname{Ker} \nabla$ are defined, in matrix form, by

$$df + \gamma \cdot f = 0.$$

We denote by $(\det E, \det \nabla)$ the *determinant connection* associated with (E, ∇); the line bundle $\det E$ is defined by $\bigwedge^{\pi_d} E$, its connection $\det \nabla$ is represented by $\operatorname{tr}(\gamma)$ in the basis $e_1 \wedge \ldots \wedge e_{\pi_d}$ and its curvature is precisely $\operatorname{tr}(K)$. Moreover, from the usual properties of change of basis in connection, we have $k_1 \, dx \wedge dy = \operatorname{tr}(K)$.

For a linear and nonsingular web $\mathcal{L}(d)$ in $(\mathbb{C}^2, 0)$, we have $P_{\mathcal{L}(d)} = 0$ and some simplifications appear in the previous relation ($\star$) and in the connection (E, ∇); in particular it satisfies $\operatorname{tr}(K) = \pi_d \cdot d\alpha$. A part of the results obtained in the linear case is summarized in the following:

Corollary. *For a linear and nonsingular planar web $\mathcal{L}(d)$ in $(\mathbb{C}^2, 0)$, implicitly presented by $F(x, y, p) = a_0(x, y) \cdot p^d + a_1(x, y) \cdot p^{d-1} + \cdots + a_d(x, y)$, the following conditions are equivalent:*

(i) *The rank of $\mathcal{L}(d)$ is maximum equal to π_d;*
(ii) *The connection (E, ∇) associated with $\mathcal{L}(d)$ is integrable;*
(iii) *The fundamental 1-form α of $\mathcal{L}(d)$ satisfies $d\alpha = \partial_y^2\left(\frac{a_1}{a_0}\right) dx \wedge dy = 0$;*
(iv) *The linear web $\mathcal{L}(d)$ is algebraic, that is $\mathcal{L}(d) = \mathcal{L}_C(d)$ with $C \subset \mathbb{P}^2$.*

The next results specify the connection (E, ∇) associated with any nonsingular planar web $\mathcal{W}(d)$ in $(\mathbb{C}^2, 0)$. They are related to the trace $\operatorname{tr}(K)$ called the *Blaschke-Chern curvature* of $\mathcal{W}(d)$ on one hand and the determination of its rank on the other hand as follows:

Theorem 2 (Ripoll). *With the previous notations, for a nonsingular planar web $\mathcal{W}(d)$ in $(\mathbb{C}^2, 0)$, we have the following results:*

1. *Trace formula: We have $\operatorname{tr}(K) = \sum_{k=1}^{\binom{d}{3}} d\gamma_k$ where the $d\gamma_k$ are the Blaschke curvatures of the extracted 3-webs of $\mathcal{W}(d)$;*

2. *Rank determination: From essentially the connection (E, ∇) and an adapted basis, there exists by derivation a matrix $(k_{m\ell}) : \mathcal{O}^{\pi_d} \longrightarrow \mathcal{O}^{\pi_d}$ whose first line is $(k_1, \ldots, k_{\pi_d})$ which satisfy, via local systems, $\operatorname{Ker}(k_{m\ell}) = \mathcal{O} \otimes_{\mathbb{C}} \operatorname{Ker} \nabla$; thus we have $\operatorname{rk} \mathcal{W}(d) = \operatorname{corank}(k_{m\ell})$ and in particular $\operatorname{rk} \mathcal{W}(d) \geq 1$ if and only if $\det(k_{m\ell}) = 0$.*

In germ in [H3], the 2-form $\operatorname{tr}(K)$ plays the classic role of the Chern class for E, *but for the planar web $\mathcal{W}(d)$ itself*. Indeed, $\operatorname{tr}(K)$ depends only of $\mathcal{W}(d)$ from the trace formula, which also justifies the terminology of Blaschke-Chern curvature. The curvature K itself probably depends only on $\mathcal{W}(d)$ and not on the implicit presentation of $\mathcal{W}(d)$; it is at least true for $3 \leq d \leq 5$ (*cf.* [H3] and [R2]).

Let $\mathcal{W}(d)$ be a nonsingular planar web in $(\mathbb{C}^2, 0)$ implicitly presented by F. From the trace formula on curvature, it can be checked that if F gives γ as connection matrix for (E, ∇), then the implicit presentation $g \cdot F$ gives $^g\gamma$ for $g \in \mathcal{O}^*$ with the following *relation on traces*:

$$\operatorname{tr}(^g\gamma) = \operatorname{tr}(\gamma) - \pi_d \cdot \frac{dg}{g}.$$

Let k be an integer such that $1 \leq k \leq \binom{d}{3}$. We denote by $\mathcal{W}(d)_k$ the extracted 3-web of $\mathcal{W}(d)$ and (E_k, ∇_k) its line bundle equipped with its connection naturally induced from the matrix $(A_{i,j})$ by deletion of leaves; from the theorem 1, its curvature is the Blaschke curvature $d\gamma_k$ of $\mathcal{W}(d)_k$. Then, *up to a change of implicit presentation* of $\mathcal{W}(d)$ by multiplication with an element of $\mathcal{O}^*$, it can be proven from the previous results that we have a *connection isomorphism*

$$(\det E, \det \nabla) \overset{\sim}{\longrightarrow} \left(\bigotimes_{k=1}^{\binom{d}{3}} E_k, \bigotimes_{k=1}^{\binom{d}{3}} \nabla_k \right).$$

Ripoll proves in [R2] (*cf.* also [R1]) the previous theorem and these consequences in any case for the rank determination and only in the case $3 \leq d \leq 6$ for the trace formula with the help of explicit calculations. Following some observations by Robert, the connection (E, ∇) has been revisited recently in [H-2R] from the viewpoint of the complex $(\mathcal{A}^\bullet, \delta)$ and the trace formula has been completed essentially by induction on d. It must be noted that Pirio in [P] digs up and interprets in the connection language some results and announcements by Alexandru Pantazi and Nicolae N. Mihăileanu on planar webs rank (*cf.* [Pan1], [Pan2] and [Mih]). In particular he checked a kind of equivalent trace formula, by computation for $3 \leq d \leq 5$.

Before we end this section we give some effective examples, essentially from [R2]. They illustrate how to use the connection (E, ∇) associated with a nonsingular planar web $\mathcal{W}(d)$ to determine certain of their geometric invariants such as $\operatorname{rk} \mathcal{W}(d)$.

Examples

1. Let $\mathcal{W}(3)$ be the planar 3-web implicitly presented by

$$F = p^3 + a \quad \text{where } a \in \mathcal{O} \text{ is without multiple factor.}$$

With the previous notations, we have $\overline{\Delta} = -27a^2$ with reduced divisor $\Delta = a$,

$$P_{\mathcal{W}(3)} = \frac{-\partial_y(a)}{3a} \cdot p^2 + \frac{-\partial_x(a)}{3a} \cdot p \text{ and } \alpha = \frac{-2\partial_x(a)}{3a} \, dx + \frac{-\partial_y(a)}{3a} \, dy \,.$$

Its Blaschke curvature is $d\gamma = d\alpha = \frac{a\partial_x\partial_y(a) - \partial_x(a)\partial_y(a)}{3a^2} \, dx \wedge dy$ and can be noted $\frac{1}{3}\partial_x\partial_y(\log a) \, dx \wedge dy$. In particular, we obtain for this special $\mathcal{W}(3)$ that $\operatorname{rk} \mathcal{W}(3) = 1$ if and only if $a(x,y) = u(x) \cdot v(y)$.

2. Let $\mathcal{W}(4)$ be the planar 4-web implicitly presented by

$$F = (x^4 - 2x^3 + x^2) \cdot p^4 + (-2x^2y^2 + 2xy^2 - y^2) \cdot p^2 + y^4 \,.$$

With the previous notations $\overline{\Delta} = 16x^2y^{12}(x-1)^2(2x-1)^4$ and $\Delta = xy(x-1)(2x-1)$,

$$P_{\mathcal{W}(4)} = \frac{-x(x-1)}{y^2(2x-1)} \cdot p^3 + \frac{1}{y} \cdot p^2 + \frac{-1}{2x-1} \cdot p \text{ and } \alpha = \frac{-1}{2x-1} \, dx + \frac{-2}{y} \, dy.$$

There exists an adapted basis of (E, ∇) such that its connection matrix is

$$\gamma = \begin{pmatrix} \dfrac{-1}{2x-1} \, dx + \dfrac{-1}{y} \, dy & 0 & \dfrac{-x^2+x-1}{y^2(2x-1)^2} \, dy \\[3ex] -dx & \dfrac{-2}{2x-1} \, dx + \dfrac{-1}{y} \, dy & \dfrac{-x(x-1)}{y^2(2x-1)} \, dy \\[3ex] -dy & 0 & \dfrac{-1}{2x-1} \, dx + \dfrac{-2}{y} \, dy \end{pmatrix}$$

with rank determination matrix

$$(k_{m\ell}) = \begin{pmatrix} 0 & 0 & \dfrac{3}{y^2(2x-1)^3} \\[3ex] 0 & 0 & \dfrac{-15}{y^2(2x-1)^4} \\[3ex] \dfrac{3}{y^2(2x-1)^3} & 0 & 0 \end{pmatrix} \,.$$

In particular for this $\mathcal{W}(4)$ we have $\operatorname{tr}(\gamma) = \frac{-4}{2x-1} \, dx + \frac{-4}{y} \, dy$, thus the trace curvature of (E, ∇) satisfy $\operatorname{tr}(K) = 0$ and moreover we obtain $\operatorname{rk} \mathcal{W}(4) = 1$ with weave $(0,1)$.

3. Let $\mathcal{W}(4)$ be the planar 4-web implicitly presented by

$$F = p(p^3 + y^2 \cdot p - y) \,.$$

With the previous notations, we have $\overline{\Delta} = -y^4(27 + 4y^4)$ with $\Delta = y(27 + 4y^4)$,

$$P_{\mathcal{W}(4)} = \frac{-12}{27 + 4y^4} \cdot p^3 + \frac{9 + 4y^4}{y(27 + 4y^4)} \cdot p^2 + \frac{-8y^2}{27 + 4y^4} \cdot p \text{ and}$$

$$\alpha = \frac{-2(9 + 4y^4)}{y(27 + 4y^4)} \, dy \,.$$

There exists an adapted basis of (E, ∇) such that its connection matrix is

$$\gamma = \begin{pmatrix} \frac{-9 - 4y^4}{y(27 + 4y^4)} \, dy & \frac{-16y(-27 + 4y^4)}{(27 + 4y^4)^2} \, dy & \frac{96y^2}{(27 + 4y^4)^2} \, dy \\[2mm] -dx & \frac{-8y^2}{27 + 4y^4} \, dx + \frac{-9 - 4y^4}{y(27 + 4y^4)} \, dy & \frac{-12}{27 + 4y^4} \, dy \\[2mm] -dy & 0 & \frac{-2(9 + 4y^4)}{y(27 + 4y^4)} \, dy \end{pmatrix}$$

with rank determination matrix

$$(k_{m\ell}) = \begin{pmatrix} \frac{-16y(-27 + 4y^4)}{(27 + 4y^4)^2} & \frac{-128y^3(-27 + 4y^4)}{(27 + 4y^4)^3} & 0 \\[2mm] \frac{-128y^3(-27 + 4y^4)}{(27 + 4y^4)^3} & \frac{-1024y^5(-27 + 4y^4)}{(27 + 4y^4)^4} & 0 \\[2mm] \frac{64(243 - 306y^4 + 8y^8)}{(27 + 4y^4)^3} & \frac{512y^2(243 - 306y^4 + 8y^8)}{(27 + 4y^4)^4} & 0 \end{pmatrix}.$$

In particular for this $\mathcal{W}(4)$ we have $\text{tr}(\gamma) = \frac{-8y^2}{27 + 4y^4} \, dx + \frac{-4(9 + 4y^4)}{y(27 + 4y^4)} \, dy$ with trace curvature $\text{tr}(K) = \frac{-16y(-27 + 4y^4)}{(27 + 4y^4)^2} \, dx \wedge dy$ for $q(E, \nabla)$ and moreover we obtain $\text{rk}\,\mathcal{W}(4) = 2$ with weave $(0, 2)$. It can be also noted that the rank of the planar 3-web implicitly presented by the factor $p^3 + y^2 \cdot p - y$ of the present F is 0 since its Blasche curvature is again $\text{tr}(K)$!

Let $\mathcal{W}(d)$ be a nonsingular planar web in $(\mathbb{C}^2, 0)$. A generalization for $d \geq 3$ of the geometric closure or hexagonality condition, initially due to Thomsen for the $\mathcal{W}(3)$, can be founded in [R2]; it specifies the geometric meaning of the trace formula for $\mathcal{W}(d)$.

Some partial results exist, but it would be probably interesting to investigate in general the *total weave* of such a $\mathcal{W}(d)$, that is the weave of all extracted subwebs of $\mathcal{W}(d)$, from the associated connection viewpoint and as a combinatorial object.

Explicit relations exist between the connection (E, ∇) and the linearization problem of such a $\mathcal{W}(d)$ for $d \geq 4$. Indeed, using suitable adapted basis, the first line of the curvature matrix $d\gamma + \gamma \wedge \gamma$ appears with the following form:

$$(k_1, \partial_x(k_1) + \widetilde{k_2}, \partial_y(k_1) + \widetilde{k_3}, \partial_x^2(k_1) + \widetilde{k_4}, \partial_x\partial_y(k_1) + \widetilde{k_5}, \partial_y^2(k_1) + \widetilde{k_6}, \dots)$$

where $k_1 \, dx \wedge dy = \operatorname{tr}(K)$ and the $\widetilde{k_j}$ for $2 \leq j \leq \pi_d$ involve special differential expressions obtained from the coefficients of the implicit presentation F of $\mathcal{W}(d)$ and its linearization polynomial $P_{\mathcal{W}(d)}$. With these observations and using only differential system methods, Ripoll recovers in [R2] some results of the author obtained before "à la Poincaré" by geometric methods (*cf.* for instance [H1]).

3. Singularities of Webs Come on Stage

Before we begin the study of polynomial webs globally defined on $\mathbb{P}^2$, let us start with some results in the local singular case. Let $\mathcal{W}(d)$ be a planar web in $(\mathbb{C}^2, 0)$ implicitly presented by $F \in \mathcal{O}[p]$ with the possibility that $\Delta(0) = 0$, where its singular locus $\Delta = \prod_q \Delta_q$ is the reduced divisor associated with the p-discriminant $\overline{\Delta} = \prod_q \Delta_q^{m_q}$ of F; we denote also by Δ the analytic germ in $0 \in \mathbb{C}^2$, with irreducible components Δ_q, which is defined by $\Delta \in \mathcal{O}$. The components Δ_q are not necessarily "solutions" of the differential equation $F(x, y, y') = 0$, but Δ contains the singularities of the leaves of such a $\mathcal{W}(d)$.

From the previous section, since the matrix $(A_{i,j})$ has coefficients in $\mathcal{O}[1/\Delta]$, there exists a *meromorphic connection* (E, ∇) on $(\mathbb{C}^2, 0)$ of rank π_d with poles on Δ, which is associated with the planar web $\mathcal{W}(d)$; the connection (E, ∇) is not necessarily integrable, but *incarnates* by construction, outside Δ, a local system $\operatorname{Ker} \nabla$ which is identified with the abelian relations of the nonsingular planar web generated by F.

For instance, a $\mathcal{W}(3)$ implicitly presented by $F = a_0 \cdot p^3 + a_1 \cdot p^2 + a_2 \cdot p + a_3$ gives rise to a meromorphic connection (E, ∇) represented, from the relation $(\star)$ and the system $\mathcal{M}(3)$ defined in the previous section, by $\gamma = A_1 \, dx + A_2 \, dy \in \Omega^1[1/\Delta]$ where we have explicitly, before simplification by a_0

$$a_0 \cdot \overline{\Delta} \cdot A_1 = \begin{vmatrix} \partial_y(a_0) & a_0 & -a_0 & 0 & 0 \\ \partial_x(a_0) + \partial_y(a_1) & a_1 & 0 & -2a_0 & 0 \\ \partial_x(a_1) + \partial_y(a_2) & a_2 & a_2 & -a_1 & -3a_0 \\ \partial_x(a_2) + \partial_y(a_3) & a_3 & 2a_3 & 0 & -2a_1 \\ \partial_x(a_3) & 0 & 0 & a_3 & -a_2 \end{vmatrix}$$

and

$$a_0 \cdot \overline{\Delta} \cdot A_2 = \begin{vmatrix} 0 & \partial_y(a_0) & -a_0 & 0 & 0 \\ a_0 & \partial_x(a_0) + \partial_y(a_1) & 0 & -2a_0 & 0 \\ a_1 & \partial_x(a_1) + \partial_y(a_2) & a_2 & -a_1 & -3a_0 \\ a_2 & \partial_x(a_2) + \partial_y(a_3) & 2a_3 & 0 & -2a_1 \\ a_3 & \partial_x(a_3) & 0 & a_3 & -a_2 \end{vmatrix}.$$

The meromorphic connection (E, ∇) associated with any planar web $\mathcal{W}(3)$ has probably, at most, *simple poles* on Δ. It is the case for a $\mathcal{W}(3)$ implicitly presented by $F = a_0 \cdot p^3 + a_2 \cdot p + a_3$. Indeed, we may always assume that $a_0 \equiv 1$ from the relation on traces and use generically the logarithmic differential of $\overline{\Delta} = -27a_3^2 - 4a_2^3$. Thus it is sufficient to verify that we have

$$6a_2 \, \overline{\Delta} \, A_1 + 2a_2 \, \partial_x(\overline{\Delta}) - 3a_3 \, \partial_y(\overline{\Delta}) + 6\partial_y(a_3) \, \overline{\Delta} = 0$$

and

$$18a_3 \, \overline{\Delta} \, A_2 - 2a_2 \, \partial_x(\overline{\Delta}) + 3a_3 \, \partial_y(\overline{\Delta}) + 6\partial_x(a_2) \, \overline{\Delta} = 0.$$

These observations are probably related to classical properties of the partial derivatives of the determinant $\overline{\Delta}$ with respect to the coefficients a_i.

For $d \geq 3$ however, as the previous examples show, the meromorphic connection (E, ∇) associated with a planar web $\mathcal{W}(d)$ is in general not with simple poles on its singular locus Δ. However, it can be noted in the examples above that the associated determinant connection $(\det E, \det \nabla)$ has simple poles on Δ.

In order to begin the study of the singularities of (E, ∇) and following Pierre Deligne and Bernard Malgrange, we recall some definitions (*cf.* [D], and for instance [M] with the chapter of André Haefliger and its bibliography which is published in the same volume).

We say that the meromorphic connection (E, ∇) with poles on Δ is *regular* if for any morphism germ $u : (\mathbb{C}, 0) \longrightarrow (\mathbb{C}^2, a)$ transversal to the smooth part of Δ, so in particular $u(0) = a \in \Delta_q - \{0\}$ if $\Delta \neq \emptyset$, the pullback connection $(u^*E, u^*\nabla)$ which is automatically integrable has a regular singularity at $0 \in \mathbb{C}$ in the usual sense in dimension one.

If the matrix of 1-forms γ with coefficients in $\Omega^1[1/\Delta]$ represents the connection (E, ∇) in a not necessarily adapted basis (e_ℓ) of E for $1 \leq \ell \leq \pi_d$, the connection $(u^*E, u^*\nabla)$ is represented by the matrix $u^*(\gamma)$ in the basis $(u^*(e_\ell))$.

Moreover we recall, from complements of basic results by Lazarus Fuchs, that $(u^*E, u^*\nabla)$ has a regular singularity at $0 \in \mathbb{C}$ if for instance, one of the following four equivalent conditions holds:

(i) Up to a meromorphic change of basis, there exists a basis (ε_ℓ) of u^*E such that $u^*\nabla$ is represented by a meromorphic matrix $(m_{k\ell})$ with coefficient in $\mathbb{C}\{t\}[1/t]$, with at most a simple pole in $0 \in \mathbb{C}$, where by definition $u^*(\nabla)(\varepsilon_\ell) = \sum_{k=1}^{\pi_d} m_{k\ell} \, dt \otimes \varepsilon_k$;

(ii) Up to a meromorphic change of basis, there exists a basis of u^*E such that $u^*\nabla$ is represented by $\Gamma \frac{dt}{t}$ with a constant matrix $\Gamma \in M(\pi_d \times \pi_d, \mathbb{C})$ where only, up to conjugation in $\mathrm{GL}(\pi_d, \mathbb{C})$, the *monodromy matrix* $T = \exp(-2i\pi\Gamma)$ of $(u^*E, u^*\nabla)$ is uniquely determined;

(iii) The horizontal sections $f = (f_\ell)$ of $(u^*E, u^*\nabla)$ have moderate growth at $0 \in \mathbb{C}$, that is in a suitable sector satisfy for $1 \leq \ell \leq \pi_d$ the following estimate:

$$|f_\ell(t)| \leq c \cdot |t|^{-N} \text{ for some constants } c > 0 \text{ and } N \in \mathbb{N}^*;$$

(iv) The horizontal sections $f = (f_\ell)$ of $(u^*E, u^*\nabla)$ are in the Nilsson class with precisely, up to refinement using Jordan blocks, the following form:

$$f = \sum_{\ell=1}^{\pi_d} \sum_{k=0}^{\pi_d - 1} h_{\ell,k}(t) \cdot t^{\alpha_\ell} (\log t)^k$$

where $h_{\ell,k} \in \mathbb{C}\{t\}^{\pi_d}$ and $\alpha_\ell \in \mathbb{C}$ is such that $\exp(-2i\pi\alpha_\ell)$ is an eigenvalue of the monodromy T.

If the meromorphic connection (E, ∇) with poles on Δ is regular, its associated determinant connection $(\det E, \det \nabla)$ is also regular. For instance, as we have noticed before, the three explicit examples above have a regular associated determinant connection $(\det E, \det \nabla)$.

Algebraic webs give rise to special examples as follows. Let $\mathcal{L}_C(d)$ be an algebraic web associated by duality with a reduced algebraic curve $C \subset \mathbb{P}^2$. Its singular locus Δ essentially corresponds to the dual curve $\check{C} \subset \check{\mathbb{P}}^2$ of C and its associated connection (E, ∇) is regular. Indeed, using the previous equivalences it is sufficient to check that in this case, there exists an adapted basis (e_ℓ) of E such that its connection matrix γ has the very particular following form: $\gamma = 0$ for $d = 3$,

$$\gamma = - \begin{pmatrix} 0 & 0 & 0 \\ dx & 0 & 0 \\ dy & 0 & 0 \end{pmatrix} \text{ for } d = 4, \gamma = - \begin{pmatrix} 0 & 0 & 0 & 0 & 0 & 0 \\ dx & 0 & 0 & 0 & 0 & 0 \\ dy & 0 & 0 & 0 & 0 & 0 \\ 0 & dx & 0 & 0 & 0 & 0 \\ 0 & dy & dx & 0 & 0 & 0 \\ 0 & 0 & dy & 0 & 0 & 0 \end{pmatrix} \text{ for } d = 5, \text{ etc.}$$

We recall that the exact sequence of K. Saito-Aleksandrov

$$0 \longrightarrow \Omega^\bullet \longrightarrow \Omega^\bullet(\log \Delta) \overset{\mathrm{res}_\Delta}{\longrightarrow} \omega_\Delta^{\bullet-1} \longrightarrow 0$$

links three complex of $\mathcal{O}$-modules with the help of the exterior derivative d and the *residue morphism* of Poincaré-Leray. The *logarithmic complex* $\left(\Omega^\bullet(\log \Delta), d\right)$ where $\Omega^\bullet(\log \Delta) \subset \Omega^\bullet[1/\Delta]$ contains the usual de Rham complex $\left(\Omega^\bullet, d\right)$. It is defined by $\omega \in \Omega^\bullet(\log \Delta)$ if $\Delta\,\omega$ *and* $\Delta\,d\omega$ belong to $\Omega^\bullet$. The third is the Barlet complex $\left(\omega_\Delta^\bullet, d\right)$ defined in [Bar] with "regular" meromorphic forms on Δ where essentially we have $\mathrm{res}_\Delta(\omega) = \frac{\xi}{h}$ since for any $\omega \in \Omega^\bullet(\log \Delta)$ exists a representation $h\,\omega = \frac{d\Delta}{\Delta} \wedge \xi + \eta$ where $h \in \mathcal{O}$, ξ and η are in $\Omega^\bullet$ with $\dim_{\mathbb{C}} \Delta \cap \{h = 0\} = 0$.

From the previous exact sequence, it can be noted that if Δ is irreducible and $\omega = \frac{\theta_x\,dx + \theta_y\,dy}{\Delta} \in \Omega^1(\log \Delta)$ with $d\omega = 0$, then $\mathrm{res}_\Delta(\omega) = \frac{\theta_x}{\partial_x(\Delta)} = \frac{\theta_y}{\partial_y(\Delta)} \in \mathbb{C}$.

For planar webs $\mathcal{W}(d)$, these definitions generate *logarithmic* meromorphic connections (E, ∇) with poles on Δ, that is those represented in a basis of E by a matrix of 1-forms γ such that, in matrix form, $\Delta\,\gamma \in \Omega^1$ and $\Delta\,d\gamma \in \Omega^2$. From the previous equivalences, a logarithmic connection (E, ∇) is necessarily regular.

In general, from the usual properties of change of basis, the matrix connection of the associated determinant connection $(\det E, \det \nabla)$ is only defined up to a logarithmic differential. Using eventually here a change of implicit presentation of $\mathcal{W}(d)$ and the relation on traces, one can bypass this indetermination under some natural hypothesis, especially filled in the maximum rank case if the connection (E, ∇) is regular. The result is the following:

Proposition. *Let $\mathcal{W}(d)$ be a planar web in $(\mathbb{C}^2, 0)$ implicitly presented by $F \in \mathcal{O}[p]$ with singular locus Δ. Suppose that its associated meromorphic connection (E, ∇) with poles on Δ has a zero Blaschke-Chern curvature, that is $\operatorname{tr}(K) = 0$ or the associated determinant connection $(\det E, \det \nabla)$ is integrable. Then the following conditions are equivalent:*

(i) *$\operatorname{tr}(\gamma) \in \Omega^1(\log \Delta)$ in a basis of E, that is only here $\Delta \operatorname{tr}(\gamma) \in \Omega^1$;*

(ii) *Up to a change of implicit presentation of $\mathcal{W}(d)$ by multiplication with an element of $\mathcal{O}^*$, we have the following "determinant" formula for $\mathcal{W}(d)$:*

$$\det \mathcal{W}(d) := \sum_q \operatorname{res}_{\Delta_q}[\det \mathcal{W}(d)] \cdot \frac{d\Delta_q}{\Delta_q}$$

with constant residues $\operatorname{res}_{\Delta_q}[\det \mathcal{W}(d)] := \operatorname{res}_{\Delta_q}(\operatorname{tr}(\gamma)) \in \mathbb{C}$ which depend only on $\mathcal{W}(d)$ and the irreducible components Δ_q of Δ;

(iii) *The meromorphic determinant connection $(\det E, \det \nabla)$ is regular.*

Proof. (i) $\Rightarrow$ (ii): Since $d \operatorname{tr}(\gamma) = \operatorname{tr}(K) = 0$ and from the previous observations the different residues $\operatorname{res}_{\Delta_q}(\operatorname{tr}(\gamma))$ are complex numbers. Moreover, these depend only on $\mathcal{W}(d)$ and the Δ_q from the relation on traces. The closed 1-form defined by $\omega := \operatorname{tr}(\gamma) - \sum_q \operatorname{res}_{\Delta_q}(\operatorname{tr}(\gamma)) \cdot \frac{d\Delta_q}{\Delta_q}$ belongs to $\Omega^1(\log \Delta)$ and verifies $\operatorname{res}_\Delta(\omega) = 0$, thus $\omega \in \Omega^1$ and we get $\omega = d\varphi$ from the Poincaré lemma. Using again the relation on traces, the element $g := \exp(\frac{\varphi}{\pi_d})$ gives rise to a presentation $g \cdot F$ of $\mathcal{W}(d)$ such that $\operatorname{tr}(^g\gamma) = \sum_q \operatorname{res}_{\Delta_q}(\operatorname{tr}(\gamma)) \cdot \frac{d\Delta_q}{\Delta_q}$. This proves the implication and justifies the definition.

(ii) $\Rightarrow$ (iii): This implication follows from the fact that a logarithmic connection is automatically regular.

(iii) $\Rightarrow$ (i): It is sufficient to verify that $\operatorname{tr}(\gamma)$ has simple poles on Δ. This can be done on a generic transversal line and comes from the classical criterion of Fuchs for a linear differential equation, in the first order case since $(\det E, \det \nabla)$ is of rank one. $\square$

Under the equivalent conditions of the lemma, we will say that the complex numbers $\operatorname{res}_{\Delta_q}[\det \mathcal{W}(d)]$ are the *residues of the determinant* of the planar web $\mathcal{W}(d)$. Nonsingular planar webs $\mathcal{W}(d)$ have no residue and from previous remarks all the residues of an algebraic web $\mathcal{L}_C(d)$ are zero.

Examples of determinants

1. Let $\mathcal{W}(3)$ be the planar 3-web implicitly presented by

$$F = p^3 + x^m y^n \text{ where } (m,n) \in \mathbb{N}^2 .$$

Using the previous notations, we have $\overline{\Delta} = -27x^{2m}y^{2n}$ and $\Delta = \Delta_1 \cdot \Delta_2 := xy$, $P_{\mathcal{W}(3)} = \frac{n}{3y} \cdot p^2 + \frac{m}{3x} \cdot p$ with $\mathrm{rk}\,\mathcal{W}(3) = 1$ since $d\gamma = 0$ where $\gamma = \frac{-2m}{3x} dx + \frac{-n}{3y} dy$, or as determinant $\det \mathcal{W}(3) = -\frac{2m}{3} \cdot \frac{d\Delta_1}{\Delta_1} - \frac{n}{3} \cdot \frac{d\Delta_2}{\Delta_2}$ with a residue sequence equal to $\left(\mathrm{res}_{\Delta_q}[\det \mathcal{W}(3)] \right)_q = \left(-\frac{2m}{3}, -\frac{n}{3} \right)$.

2. Let $\mathcal{C}(3)$ be Cauchy's planar 3-web. It comes from an example of a differential equation $F(x, y, y') = 0$ which showed the inadequacy of the "first" definition of a *singular* solution and is implicitly presented by

$$F = p^3 + 4xy \cdot p - 8y^2.$$

Using the previous notations $\overline{\Delta} = -64y^3(27y + 4x^3)$ and $\Delta = \Delta_1 \cdot \Delta_2 := y(27y + 4x^3)$, $P_{\mathcal{C}(3)} = \frac{1}{2y} \cdot p^2$ with $\mathrm{rk}\,\mathcal{C}(3) = 1$ since $d\gamma = 0$ where $\gamma = \frac{-1}{2y} dy$, or as determinant $\det \mathcal{C}(3) = -\frac{1}{2} \cdot \frac{d\Delta_1}{\Delta_1}$ with the residue sequence $\left(\mathrm{res}_{\Delta_q}[\det \mathcal{C}(3)] \right)_q = \left(-\frac{1}{2}, 0 \right)$.

3. Let $\mathcal{W}(4) = \mathcal{W}(x - y, x + y, x^2 - y^2, \frac{x-y}{x+y})$ be the planar 4-web implicitly presented by

$$F = xy \cdot p^4 - (x^2 + y^2) \cdot p^3 + (x^2 + y^2) \cdot p - xy$$

which comes from the rectified model $\mathcal{W}(x, y, xy, \frac{x}{y})$ partially discussed in the introduction. With the previous notations, we have

$$\overline{\Delta} = 4(x - y)^6(x + y)^6 \quad \text{and} \quad \Delta = \Delta_1 \cdot \Delta_2 := (x - y)(x + y),$$

$$P_{\mathcal{W}(4)} = \frac{-1}{(x - y)(x + y)} \left(x \cdot p^3 - y \cdot p^2 - x \cdot p + y \right)$$

and $\alpha = \frac{-x}{(x-y)(x+y)} dx$. There exists an adapted basis of (E, ∇) such that its connection matrix is

$$\gamma = \begin{pmatrix} \dfrac{-x\,dx + y\,dy}{(x - y)(x + y)} & \dfrac{x(x\,dx - y\,dy)}{(x - y)^2(x + y)^2} & \dfrac{-y(x\,dx - y\,dy)}{(x - y)^2(x + y)^2} \\[2em] -dx & \dfrac{y\,dy}{(x - y)(x + y)} & \dfrac{-x\,dy}{(x - y)(x + y)} \\[2em] -dy & \dfrac{y\,dx}{(x - y)(x + y)} & \dfrac{-x\,dx}{(x - y)(x + y)} \end{pmatrix}$$

with $d\gamma + \gamma \wedge \gamma = 0$. Thus $\mathrm{rk}\,\mathcal{W}(4) = 3$ and $\mathrm{tr}(\gamma) = \det \mathcal{W}(4) = -\frac{d\Delta_1}{\Delta_1} - \frac{d\Delta_2}{\Delta_2}$ as determinant for $\mathcal{W}(4)$ with the residue sequence $\left(\mathrm{res}_{\Delta_q}[\det \mathcal{W}(4)]\right)_q = (-1, -1)$.

4. Let $\mathcal{W}(4)$ be the planar 4-web implicitly presented by

$$F = (x^4 - 2x^3 + x^2) \cdot p^4 + (-2x^2y^2 + 2xy^2 - y^2) \cdot p^2 + y^4.$$

This web has been already studied before (example 2 in the second section), we have $\mathrm{tr}(K) = 0$, $\mathrm{rk}\,\mathcal{W}(4) = 1$ and moreover $\Delta = \prod_{q=1}^4 \Delta_q :=$ $xy(x-1)(2x-1)$, $\mathrm{tr}(\gamma) = \frac{-4}{2x-1}\,dx + \frac{-4}{y}\,dy = \det \mathcal{W}(4) = -4 \cdot \frac{d\Delta_2}{\Delta_2} - 2 \cdot \frac{d\Delta_4}{\Delta_4}$ as determinant with the residue sequence $\left(\mathrm{res}_{\Delta_q}[\det \mathcal{W}(4)]\right)_q = (0, -4, 0, -2)$.

The analogy noted before between the Blaschke-Chern curvature $\mathrm{tr}(K)$ for the planar web $\mathcal{W}(d)$ and the Chern class of E goes on in the case of $K = 0$ with the Cheeger-Chern-Simons class for (E, ∇). Indeed, if U is a sufficient small connected open set which contains $0 \in \mathbb{C}^2$, then the element induced by $\det \mathcal{W}(d) = \sum_q \mathrm{res}_{\Delta_q}[\det \mathcal{W}(d)] \cdot \frac{d\Delta_q}{\Delta_q}$ in $H^1(U - \Delta, \mathbb{C}/\mathbb{Z})$ plays the analogous role for the planar web $\mathcal{W}(d)$.

One can notice that in the four examples above the associated meromorphic determinant connections $(\det E, \det \nabla)$ is regular and all the residues $\mathrm{res}_{\Delta_q}[\det \mathcal{W}(d)]$ are rational numbers. It is also the case for the exceptional Bol's web $\mathcal{B}(5)$ defined in $\mathbb{P}^2$ from four vertices in general position with precisely the same residue $\mathrm{res}_{\Delta_q}[\det \mathcal{B}(5)] = -2$ on the six lines Δ_q generated by these vertices (*cf.* [R2]).

More generally the study of the local monodromy and the regularity for the meromorphic connection (E, ∇) associated with a planar web $\mathcal{W}(d)$ in $(\mathbb{C}^2, 0)$ are of interest. Especially in the maximum rank case to which we will restrict from now on to simplify the presentation.

By construction, the horizontal sections $f = (f_\ell) \in \mathrm{Ker}\,\nabla$ of the meromorphic connection (E, ∇) correspond to closed 1-form

$$a = (b_3 \cdot p^{d-3} + b_4 \cdot p^{d-4} + \cdots + b_d) \cdot \frac{dy - p\,dx}{\partial_p(F)} \in \mathfrak{a}_F.$$

From the theorem $\mathfrak{a}_F$ they give rise to a *singular basis* of abelian relations of $\mathcal{W}(d)$, that is with possibly multivalued sections.

Examples of singular basis of abelian relations

Referring to examples 1, 2 and 3 previously presented for determinants, we obtain from the theorem $\mathfrak{a}_F$ the corresponding results.

1. For the $\mathcal{W}(3)$ presented by $F = p^3 + x^m y^n$ where $(m, n) \in \mathbb{N}^2$, the space $\mathfrak{a}_F$ is generated by the closed 1-form $a = x^{\frac{2m}{3}} y^{\frac{n}{3}} \cdot \frac{dy - p\, dx}{3p^2}$.

2. For the $\mathcal{C}(3)$ presented by $F = p^3 + 4xy \cdot p - 8y^2$, the space $\mathfrak{a}_F$ is generated by the closed 1-form $a = y^{\frac{1}{2}} \cdot \frac{dy - p\, dx}{3p^2 + 4xy}$.

3. For the $\mathcal{W}(4) = \mathcal{W}(x - y, x + y, x^2 - y^2, \frac{x-y}{x+y})$ it can be checked that the horizontal sections $f = \begin{pmatrix} f_1 \\ f_2 \\ f_3 \end{pmatrix} \in \mathrm{Ker}\,\nabla$ have the following fundamental matrix solution:

$$\begin{pmatrix} 1 & 0 & 2 + \log(x - y) + \log(x + y) \\ x & y & (x - y)\log(x - y) + (x + y)\log(x + y) \\ y & x & -(x - y)\log(x - y) + (x + y)\log(x + y) \end{pmatrix}$$

which gives the generators $a = (-f_2(x, y) \cdot p + f_3(x, y)) \cdot \frac{dy - p\, dx}{\partial_p(F)}$ for the space $\mathfrak{a}_F$. From this previous fundamental matrix solution and again the theorem $\mathfrak{a}_F$ since $g_i(F_i)dF_i = \pi_i^*(a)$ for $1 \le i \le d = 4$, we also explicitly recover the basis of the abelian relations space of $\mathcal{W}(4)$ corresponding to the one given in the introduction for its rectified model. This matrix gives also that $h = x^2 - y^2 = \Delta_1 \cdot \Delta_2$ is a generator of $\mathrm{Ker}(\det \nabla)$ since $dh + \mathrm{tr}(\gamma) \cdot h = 0$.

Let $\mathcal{W}(d)$ be a polynomial planar web in $\mathbb{P}^2$, that is defined by $F \in \mathbb{C}[x, y, p]$, with maximum rank π_d and singular locus Δ, not necessarily with normal crossings. The associated meromorphic connection (E, ∇) as well as its determinant connection $(\det E, \det \nabla)$ are *algebraic* by construction and integrable from the rank hypothesis. These connections correspond to a representation of the fundamental group

$$\rho : \pi_1(\mathbb{P}^2 - \Delta, z) \longrightarrow \mathrm{GL}(\pi_d, \mathbb{C})$$

and its associated determinant $\det \rho : \pi_1(\mathbb{P}^2 - \Delta, z) \longrightarrow \mathbb{C}^*$, where $\Delta \subset \mathbb{P}^2$ is a reduced algebraic curve and $z \notin \Delta$ is a base point. We may suppose that a generic line $L \in \check{\mathbb{P}}^2$ intersects transversally Δ such that $\Sigma = \Delta \cap L = \{t_1, \ldots, t_{\deg \Delta}\}$ and gives an exact sequence

$$\pi_1(L - \Sigma, z) \longrightarrow \pi_1(\mathbb{P}^2 - \Delta, z) \longrightarrow 1$$

from a result due to Oscar Zariski. The connection (E, ∇) induces on L a connection (E_L, ∇) and we may suppose that 0 is one of the points in Σ. Corresponding to a cyclic element of the connection (E_L, ∇), locally around $t = 0 \in L$, the linear differential equation

$$\xi^{(\pi_d)} + r_1(t) \cdot \xi^{(\pi_d - 1)} + \cdots + r_{\pi_d - 1}(t) \cdot \xi' + r_{\pi_d}(t) \cdot \xi = 0$$

with rational coefficients in t is relevant, notably from the fuchsian viewpoint.

In this context, at least two results on the connection (E, ∇) with poles on Δ can be expected: a quasi-unipotent local monodromy induced on generic lines L and the regularity. In particular, these properties imply that the local residues of the determinant associated with the polynomial planar web $\mathcal{W}(d)$ are rational numbers. It can be checked for instance that the two previous expected results are true for the three examples presented just before.

To strengthen these examples and above all some perspectives offered in the study of planar web geometry through singularities and some associated tools, let us end by an observation. It is based on the fact, already used before, that any abelian relation of $\mathcal{W}(d)$ comes from a closed 1-form $a \in \mathfrak{a}_F$. In specific cases such a form a induces locally on L a horizontal section of a Gauss-Manin connection (H_L, D). These cases, which may be not completely satisfying without an embedded resolution of Δ in $\mathbb{P}^2$, are conditioned by the existence of suitable pencils of curves associated with F. Nevertheless in these cases, with the help of the local monodromy and regularity theorems for (H_L, D) (*cf.* for instance [G1]), the two previous expected results hold.

References

[A-G] M. A. AKIVIS and V. V. GOLDBERG, *Differential geometry of webs* in Handbook of differential geometry, Ed. F. J. E. Dillen and L. C. A. Verstralen, vol. 1, North-Holland, Amsterdam, 2000, 1–152.

[Bar] D. BARLET, *Le faisceau $\omega_X^\bullet$ sur un espace analytique X de dimension pure*, in Fonctions de Plusieurs Variables Complexes III, Séminaire F. Norguet, Lect. Notes in Math. **670**, Springer, Berlin, 1978, 187–204.

[B] W. BLASCHKE, *Einführung in die Geometrie der Waben*, Birkhäuser, Basel, 1955.

[B-B] W. BLASCHKE und G. BOL, *Geometrie der Gewebe*, Springer, Berlin, 1938.

[C1] S. S. CHERN, *Abzählungen für Gewebe*, Abh. Hamburg **11** (1936), 163–170.

[C2] S. S. CHERN, *Eine Invariantentheorie der 3-Gewebe aus r-dimensionalen Mannigfaltigkeiten im R_{2r}*, Abh. Hamburg **11** (1936), 333–358.

[C3] S. S. CHERN, *Web Geometry*, Bull. Amer. Math. Soc. **6** (1982), 1–8.

[C4] S. S. CHERN, *My Mathematical Education* in S. S. Chern, A Great Geometer of the Twentieth Century. Expanded Edition, Ed. S.-T. Yau, International Press, Hong Kong, 1998, 1–15.

[C-G1] S. S. CHERN and P. A. GRIFFITHS, *Linearization of webs of codimension one and maximum rank*, in Proc. Int. Symp. on Algebraic Geometry, Kyoto, 1977, Ed. M. Nagata, Kinokuniya, Tokyo, 1978, 85–91.

[C-G2] S. S. CHERN and P. A. GRIFFITHS, *Abel's Theorem and Webs*, Jahresber. Deutsch. Math.-Verein. **80** (1978), 13-110 and *Corrections and Addenda to Our Paper: Abel's Theorem and Webs*, Jahresber. Deutsch. Math.-Verein. **83** (1981), 78–83.

[C-G3] S. S. CHERN and P. A. GRIFFITHS, *An Inequality for the Rank of a Web and Webs of Maximum Rank*, Ann. Scuola Norm. Sup. Pisa **5** (1978), 539–557.

[Dar] G. DARBOUX, *Leçons sur la théorie générale des surfaces*, Livre I, seconde édition, Gauthier-Villars, Paris, 1914.

[D] P. DELIGNE, *Équations différentielles à points singuliers réguliers*, Lect. Notes in Math. **163**, Springer, Berlin, 1970.

[Gol] V. V. GOLDBERG, *Theory of Multicodimensional (n + 1)-Webs*, Kluwer, Dordrecht, 1988.

[Gon] A. B. GONCHAROV, *Regulators*, in Handbook of K-theory, Ed. E. M. Friedlander and D. R. Grayson, Vol. 1, Springer, Berlin, 2005, 295–349.

[G1] P. A. GRIFFITHS, *Periods of integrals on algebraic manifolds: Summary of main results and discussion of open problems*, Bull. Am. Math. Soc. **75** (1970), 228–296.

[G2] P. A. GRIFFITHS, *Variations on a Theorem of Abel*, Invent. Math. **35** (1976), 321–390.

[G3] P. A. GRIFFITHS, *On Abel's Differential Equations*, in Algebraic Geometry, The Johns Hopkins Centennial Lectures, Ed. J.-I. Igusa, Baltimore, 1977, 26–51.

[G4] P. A. GRIFFITHS, *The Legacy of Abel in Algebraic Geometry*, in The Legacy of Niels Henrik Abel, The Abel Bicentennial, Oslo, 2002, Ed. O. A. Laudal and R. Piene, Springer, Berlin, 2004, 179–205.

[H1] A. HÉNAUT, *Analytic Web Geometry*, in Web Theory and Related Topics, Toulouse, 1996, Ed. J. Grifone and É. Salem, World Scientific, Singapore, 2001, 6–47.

[H2] A. HÉNAUT, *Formes différentielles abéliennes, bornes de Castelnuovo et géométrie des tissus*, Comment. Math. Helv. **79** (2004), 25–57.

[H3] A. HÉNAUT, *On planar web geometry through abelian relations and connections*, Ann. of Math. **159** (2004), 425–445.

[H-2R] A. HÉNAUT, O. RIPOLL et G. F. ROBERT, *Formule de la trace pour la connexion d'un tissu du plan*, writing in progress.

[M] B. MALGRANGE, *Regular connections, after Deligne*, in Algebraic D-modules, A. Borel et al., Academic Press, Boston, 1987, 151–172.

[Man] Y. I. MANIN, *Real Multiplication and Noncommutative Geometry*, in The Legacy of Niels Henrik Abel, The Abel Bicentennial, Oslo, 2002, Ed. O. A. Laudal and R. Piene, Springer, Berlin, 2004, 685–727.

[Mih] N. N. MIHĂILEANU, *Sur les tissus plans de première espèce*, Bull. Math. Soc. Roum. Sci. **43** (1941), 23–26.

[Pan1] A. PANTAZI, *Sur la détermination du rang d'un tissu plan*, C. R. Acad. Sci. Roumanie, **2** (1938), 108–111.

[Pan2] A. PANTAZI, *Sur une classification nouvelle des tissus plans*, C. R. Acad. Sci. Roumanie, **4** (1940), 230–232.

[P] L. PIRIO, *Équations fonctionnelles abéliennes et géométrie des tissus*, Thèse de doctorat, Université Paris 6, 2004.

[P-T] L. PIRIO et J.-M. TRÉPREAU, *Tissus plans exceptionnels et fonction thêta*, Ann. Inst. Fourier, **55** (2005), 2209–2237.

[R1] O. RIPOLL, *Détermination du rang des tissus du plan et autres invariants géométriques*, C. R. Acad. Sci. Paris, **341** (2005), 247–252.

[R2] O. RIPOLL, *Géométrie des tissus du plan et équations différentielles*, Thèse de doctorat, Université Bordeaux 1, 2005.

[Rob] G. F. ROBERT, *Relations fonctionnelles polylogarithmiques et tissus plans*, Prépub. Université Bordeaux 1, 2002.

[W] J. GRIFONE and É. SALEM (Ed.), *Web Theory and Related Topics*, Toulouse, 1996, World Scientific, Singapore, 2001.

Chapter 11

TRANSITIVE ANALYTIC LIE PSEUDO-GROUPS

Niky Kamran[*]

Department of Mathematics and Statistics,
McGill University, 805 Sherbrooke Street West,
Montreal, QC, H3A 2K6, Canada,
nkamran@math.mcgill.ca

The geometric formulation of the theory of analytic Lie pseudo-groups of infinite type in terms of G-structures, due to Chern,[4] has played a key role in the development of the subject. Cartan's structure equations and the fundamental theorems of Lie appear naturally in the framework developed by Chern, and make it possible in principle to compute the invariants and structure equations of any given Lie pseudo-group. The construction of an infinite-dimensional manifold that would act as a "parameter space" for transitive Lie pseudo-groups of infinite type is a question which has not yet been fully resolved. Our purpose in this paper is to give an overview of the classical theory of Cartan and Chern, and to present some recent progress on the parametrization problem.

1. Introduction

André Weil, in his famous 1947 article on the future of mathematics,[15] described the theory of infinite Lie groups, pioneered by Lie and Cartan, as an impenetrable jungle in need of clearing; to quote his exact words: *"Sur la théorie, si importante sans doute mais pour nous si obscure des "groupes de Lie infinis", nous ne savons rien que ce qui se trouve dans les mémoires de Cartan, première exploration à travers une jungle presque impénétrable; mais celle-ci menace de se refermer sur les sentiers déjà tracés, si l'on ne procède bientôt à un indispensable travail de défrichement."* Fortunately, much of the clearing process that was called for by Weil did take place, thanks to the important works of Chern,[4] Kuranishi,[9,10] Spencer,[14] Singer

---

[*]Research supported by NSERC Discovery Grant RGPIN 105490-2004.

and Sternberg,[13] Malgrange,[11] Olver and Pohjanpelto,[12] and many others. The paper by Chern,[4] where infinite Lie pseudo-groups are presented in terms G-structures, is probably the closest in spirit to the original papers of Cartan.[2,3] (It is also written in excellent French.) This paper provides a broad and beautiful perspective on the subject, and charts the course for many of the subsequent developments in the field. I first read Chern's paper when I was a graduate student, and was struck by its elegance and depth. I have been going back to it regularly ever since, and it has been a wonderful source of knowledge and inspiration for me, as I am sure it has been for many others.

My purpose in this paper is to recall some of the essentials of the classical theory of Lie pseudo-groups according to Cartan and Chern, and to review some recent results on the problem of assigning a suitable infinite-dimensional manifold structure to Lie pseudogroups of infinite type. The latter results, which first appeared in,[8] are joint work with Thierry Robart. This paper is dedicated to the memory of Professor Chern, with profound admiration and respect.

2. Invariant Forms and the Cartan Structure Equations

This section is meant to give a brief overview of Cartan's structure theory for Lie pseudogroups[3] based on the characterization of Lie pseudogroups by invariant 1-forms and generalized Maurer-Cartan equations.

Recall that a pseudogroup of transformations of an n-dimensional manifold M is a collection Γ of local diffeomorphisms of M satisfying the following set of conditions:[6]

(i) Every f in Γ is a diffeomorphism of an open set of M onto an open set of M.

(ii) Let $U = \cup_{i \in I} U_i$, where each U_i is an open set of M. A diffeomorphism f from U onto an open set of M belongs to Γ if an only if $f|_{U_i}$ belongs to Γ for all $i \in I$.

(iii) The identity transformation id_U belongs to Γ for every open set U of M.

(iv) If $f \in \Gamma$, then $f^{-1} \in \Gamma$.

(v) If $f \in \Gamma$ is a diffeomorphism from U to V and $g \in \Gamma$ is a diffeomorphism from $U' \to V'$ with $V \cap U' \neq \varnothing$, then the diffeomorphism $g \circ f : f^{-1}(V \cap U') \to g(V \cap U')$ is in Γ.

Of particular interest are the pseudogroups whose elements are solutions of a system of differential equations. Indeed, when the system is such that its solutions are parametrized by arbitrary constants, the corresponding pseudogroup is a local Lie group. A pseudogroup Γ is thus said to be a Lie pseudogroup if there exists an involutive differential system $\mathcal{D} \subset J^k(M, M)$ such that elements of Γ form the general solution of $\mathcal{D}$. Two classical examples of Lie pseudogroups are the following. Take $M = \mathbb{R}$ and consider the differential system $\mathcal{D} \subset J^3(\mathbb{R}, \mathbb{R})$ given by

$$\frac{d^3 X}{dx^3} - \frac{3}{2} \left(\frac{d^2 X}{dx^2} \right)^2 \left(\frac{dX}{dx} \right)^{-1} = 0 \,.$$

The Lie pseudogroup Γ defined by $\mathcal{D}$ corresponds to the set of unimodular fractional linear transformations of the real line, and is parametrized by three arbitrary real constants, which can be thought of as local coordinates on the Lie group $SL(2, \mathbb{R})$. For our second example, we take $M = \mathbb{R}^2$, and $\mathcal{D} \subset J^1(\mathbb{R}^2, \mathbb{R}^2)$ to be the system associated to the Cauchy-Riemann equations

$$X_x = Y_y, \; X_y = -Y_x, \; X_x^2 + X_y^2 \neq 0 \,.$$

The pseudogroup Γ defined by this differential system corresponds to the set of conformal local diffeomorphisms of the plane, and is parametrized by two arbitrary analytic functions of one variable. In the classical terminology, our first example belongs to the class of Lie pseudogroups of finite type, meaning that their elements are locally parametrized by arbitrary constants, while our second example belongs to the class of Lie pseudogroups of infinite type, meaning now that their elements depend on arbitrary functions as well as possibily some arbitrary constants. Lie pseudogroups of finite type are the ancestors of the modern finite-dimensional abstract Lie groups.

Cartan's structure theory for Lie pseudogroups is based on a set of structure equations which generalize the Maurer-Cartan equations satisfied by the right- (or left-) invariant 1-forms on any finite-dimensional Lie group. Cartan showed that given an involutive analytic differential system $\mathcal{D} \subset J^k(M, M)$, giving rise to a Lie pseudogroup Γ, every element of Γ has a natural lift to a local diffeomorphism of $\mathcal{D}$ (which we assume to be a submanifold embedded in $J^k(M, M)$). Furthermore, Cartan showed by a constructive procedure that there exists locally on $\mathcal{D}$ linearly independent 1-forms $\omega^1, \ldots, \omega^r, \pi^1, \ldots, \pi^s$ which satisfy structure equations of the form

$$d\omega^i = \sum_{j=1}^{r} \sum_{\rho=1}^{s} a_{j\rho}^i \pi^\rho \wedge \omega^j + \frac{1}{2} \sum_{j,k=1}^{r} C_{jk}^i \omega^j \wedge \omega^k, \; 1 \leq i \leq r \,.$$

An element f of Γ will be such that its lift $\tilde{f}$ to $\mathcal{D}$ leaves the 1-forms ω^i invariant,

$$\tilde{f}^*\omega^i = \omega^i,\ 1 \leq i \leq r\,.$$

In the structure equations, the coefficients $a^i_{j\rho}$ and C^i_{jk} will in general be functions of the essential scalar invariants of of Γ. For any fixed values of these invariants, the $a^i_{j\rho}$ will form an involutive tableau[1] and give a faithful matrix representation of a finite-dimensional Lie algebra $\mathfrak{g}$ of dimension s. In the transitive case, which will be our main concern in this paper, the coefficients $a^i_{j\rho}$ and C^i_{jk} are constants, and are related by a set of algebraic constraints generalizing the Jacobi indentities. These constraints are easily derived from the structure equations. Indeed, if for $1 \leq \rho \leq s$, we have

$$d\pi^\rho = \frac{1}{2}\sum_{\sigma,\tau=1}^{s} \gamma^\rho_{\sigma\tau}\pi^\rho \wedge \pi^\tau + \sum_{\sigma=1}^{s}\sum_{i=1}^{r} v^\rho_{\sigma i}\pi^\sigma \wedge \omega^i + \frac{1}{2}\sum_{i,j=1}^{r} v^\rho_{ij}\omega^i \wedge \omega^j\,.$$

From $d^2 = 0$, we obtain

$$\sum_{i=1}^{r}(a^k_{\sigma i}a^i_{\rho j} - a^k_{\rho i}a^i_{\sigma j}) = \sum_{\tau=1}^{s} a^k_{\tau j}\gamma^\tau_{\sigma\rho}\,,$$

$$\sum_{i=1}^{r}(-C^k_{im}a^i_{\rho l} + C^k_{il}a^i_{\rho m} + C^i_{lm}a^k_{\rho i}) + \sum_{\sigma=1}^{s}(a^k_{\sigma l}v^\sigma_{\rho m} - a^k_{\sigma m}v^\sigma_{\rho l}) = 0\,,$$

$$\sum_{i=1}^{r}(C^k_{ij}C^i_{lm} + C^k_{il}C^i_{mj} + C^k_{im}C^i_{jk}) + \sum_{\rho=1}^{s}(a^k_{\rho j}v^\rho_{lm} + a^k_{\rho l}v^\rho_{mj} + a^k_{\rho m}v^\rho_{jl}) = 0\,.$$

When $\pi^\rho = 0$, $1 \leq \rho \leq s$, we obtain the classical Maurer-Cartan equations

$$d\omega^i = \frac{1}{2}\sum_{j,k=1}^{r} C^i_{jk}\omega^j \wedge \omega^k,\ 1 \leq i \leq r\,,$$

which characterize Lie pseudogroups of finite type, and the constraints given above reduce to the standard Jacobi identities. Cartan also studied the second and third fundamental theorems of Lie for transitive Lie pseudogroups on the basis of the generalized Jacobi identities. The content of these theorems is that the invariant 1-forms and the structure equations provide a characterization of Lie pseudogroups. We wish to emphasize that in contrast with the case of Lie pseudogroups of finite type, the hypothesis of analyticity is essential when dealing with Lie pseudogroups of infinite type from the perspective of Cartan's theory, since the Cartan-Kähler Theorem[1] is needed to prove the existence of solutions to the involutive differential system $\mathcal{D}$, as well as the fundamental theorems of Lie.

We now sketch Cartan's construction[3] of the invariant 1-forms associated to a Lie pseudogroup, which we assume to be analytic in the infinite case. To keep the notation simple, we assume that $\mathcal{D}$ is given by second order equations, and that the locus cut out by $\mathcal{D}$ in $J^2(M, M)$ is an embedded submanifold. We also assume for simplicity that the pseudogroup does not admit any invariant functions. If $(x^1, \ldots, x^n, X^1, \ldots, X^n)$ denote a system of local coordinates on $J^0(M, M)$, then the module $\Omega^{(2)}(M, M)$ of contact 1-forms on $J^2(M, M)$ is generated by

$$\eta^i := dX^i - \sum_{j=1}^{n} X^i{}_j dx^j, \ \zeta^i{}_j := dX^i{}_j - \sum_{k=1}^{n} X^i{}_{jk} dx^k.$$

The pull-back of $\Omega^{(2)}(M, M)$ to $\mathcal{D}$ is a Pfaffian system $\mathcal{I}_\mathcal{D}$ generated by 1-forms which we also denote by η^i and $\zeta^i{}_j$, but where the coefficients $X^i{}_{jk}$ are no longer independent functions. Every solution f of $\mathcal{D}$ lifts by prolongation to the 2-jets to a local diffeomorphism $\tilde{f}$ of $J^2(M, M)$, and this lift maps the submanifold $\mathcal{D}$ to itself. Equivalently, $\tilde{f}$ pulls back $\mathcal{I}_\mathcal{D}$ to itself. This can be verified explicitly for the examples given above; indeed, the Schwarzian derivative is scaled by a non-zero factor under unimodular fractional linear transformations of the independent variable, and the Cauchy-Riemann equations are invariant under conformal local diffeomorphisms of the plane. Consider now on $\mathcal{D}$ the 1-forms

$$\omega^i := \sum_{j=1}^{n} X^i{}_j dx^j, \ 1 \le i \le n.$$

These 1-forms are invariant under the lift of every solution of $\mathcal{D}$ to a local diffeomorphism of $\mathcal{D}$ and semi-basic with respect to the restriction to $\mathcal{D}$ of the fibration of $J^2(M, M)$ over the source. In particular, we have

$$d\omega^i = \sum_{j=1}^{n} \pi^i{}_j \wedge \omega^j.$$

Choosing a maximal linearly independent set $\{\pi^1, \ldots, \pi^r\}$ amongst the 1-forms $\pi^i{}_j$, we have

$$\pi^\rho \equiv \sum_{i=1}^{n} A^\rho{}_i \omega^i + \sum_{i,j=1}^{n} B^\rho{}_i{}^j \zeta^i{}_j \ \mod\{\omega^1, \ldots, \omega^n\}.$$

The lift $\hat{f}$ to $\mathcal{D}$ of every solution f of $\mathcal{D}$ will satisfy

$$\hat{f}^*\left(\sum_{i=1}^{n} A^{\rho}{}_{i}\omega^{i} + \sum_{i,j=1}^{n} B^{\rho}{}_{i}{}^{j}\zeta^{i}{}_{j}\right) \equiv \sum_{i=1}^{n} A^{\rho}{}_{i}\omega^{i} + \sum_{i,j=1}^{n} B^{\rho}{}_{i}{}^{j}\zeta^{i}{}_{j} \mod\{\omega^{1},\ldots,\omega^{n}\}.$$

But we know that $\hat{f}$ pulls back $\mathcal{I}_{\mathcal{D}}$ to itself, so that the preceding congruence must actually be an equality. We therefore obtain $r' \leq r$ new invariant forms. This process will terminate, and by suitable relabeling of the indices, we obtain the Cartan structure equations for the Lie pseudogroup Γ associated to $\mathcal{D}$,

$$d\omega^{i} = \sum_{j=1}^{r}\sum_{\rho=1}^{s} a^{i}_{j\rho}\pi^{\rho} \wedge \omega^{j} + \frac{1}{2}\sum_{j,k=1}^{r} C^{i}_{jk}\omega^{j} \wedge \omega^{k}, \; 1 \leq i \leq r,$$

where the $a^{i}_{j\rho}$ form an involutive tableau as a consequence of the involutivity of the differential system $\mathcal{D}$ defining Γ.

We now illustrate Cartan's construction by deriving explicit expressions for 1-forms $(\omega^{1},\ldots,\omega^{r},\pi^{1},\ldots,\pi^{s})$ for the two simple examples of pseudogroups that we considered above. For the pseudogroup Γ of finite type defined by the local diffeomorphisms of the real line with zero Schwarzian derivative, we parametrize $\mathcal{D}$ by jet coordinates (x, X, X', X''), in which

$$\mathcal{I}_{\mathcal{D}} = \left\{ dX - X'dx, dX' - X''dx, dX'' - \frac{3}{2}\frac{X''}{(X')^{2}}dx \right\}.$$

We therefore let

$$\omega^{1} = X'dx,$$

and compute $d\omega^{1}$,

$$d\omega^{1} = \omega^{2} \wedge \omega^{1}, \; \omega^{2} = \frac{1}{X'}(dX - X'dx).$$

Now, we have

$$d\omega^{2} = \omega^{3} \wedge \omega^{1}, \; \omega^{3} = \frac{1}{2}\frac{(X'')^{2}}{(X')^{3}}dx + \frac{X''}{(X')^{3}}dX' - \frac{1}{(X')^{2}}dX''.$$

By construction, the 1-forms ω^{1}, ω^{2} and ω^{3} are invariant under the lift to $\mathcal{D}$ of the group of unimodular fractional linear transformations acting on x. This can also be verified by direct calculation. They satisfy the Maurer-Cartan equations of $\mathfrak{sl}(2,\mathbb{R})$. There are no 1-forms $\pi^{1},\ldots,\pi^{s}$ in this example, and we therefore have a Lie pseudogroup of finite type, whose elements are parametrized locally by three arbitrary real constants.

For the pseudogroup Γ of infinite type defined by the Cauchy-Riemann equations, we parametrize $\mathcal{D}$ by the jet coordinates (x, y, u, v, X_x, X_y), so that

$$\mathcal{I}_{\mathcal{D}} = \{\, dX - X_x dx - X_y dy, dY + X_y dx - X_x dy \,\}.$$

By construction, the 1-forms

$$\omega^1 = u_x dx + u_y dy, \ \omega^2 = -u_y dx + u_x dy,$$

are invariant under the lift to $\mathcal{D}$ of the pseudogroup of local conformal diffeomorphisms of the (x, y) plane. (Again, this can be verified by direct calculation as well.) We have

$$d\omega^1 = \pi^1 \wedge \omega^1 + \pi^2 \wedge \omega^1, \ d\omega^2 = -\pi^2 \wedge \omega^1 + \pi^1 \wedge \omega^2.$$

where

$$\pi^1 \equiv \frac{1}{X_x^2 + X_y^2}(X_x dX_x + X_y dX_y), \ \pi^2 \equiv \frac{1}{X_x^2 + X_y^2}(-X_y dX_x + X_x dX_y),$$

and where the congruence is $\mathrm{mod}\,\omega^1, \omega^2$. We have

$$(a^i_{j\rho}\pi^\rho) = \begin{pmatrix} \pi^1 & \pi^2 \\ -\pi^2 & \pi^1 \end{pmatrix},$$

and it is easily verified that the $a^i_{j\rho}$ form an involutive tableau with characters given by $\sigma_1 = 2$ and $\sigma_{1+a} = 0$ for all $a \geq 0$. We thus have a Lie pseudogroup parametrized locally by two arbitrary functions of one variable. Note that the Lie algebra $\mathfrak{g}$ is the Lie algebra of the group of similarities of the plane.

Chern's notion of an involutive G-structure gives a very natural geometric framework for understanding the structure equations of a wide class of Lie pseudogroups. Recall that a G-structure B_G on a manifold M is by definition a reduction of the frame bundle $\mathcal{F}_M$ of M to a subgroup G. The restriction of the canonical vector valued 1-form of $\mathcal{F}_M$ to B_G gives a vector valued 1-form $\omega = (\omega^1, \dots, \omega^r)$ which satisfies structure equations of the form

$$d\omega^i = \sum_{j=1}^{r}\sum_{\rho=1}^{s} a^i_{j\rho}\pi^\rho \wedge \omega^j + \frac{1}{2}\sum_{j,k=1}^{r} C^i_{jk}\omega^j \wedge \omega^k, \ 1 \leq i \leq r,$$

where the $a^i_{j\rho}$ are constants, and the C^i_{jk} are functions on B_G. By construction, the set of local automorphisms of B_G is a Lie pseudogroup, which is intransitive in general. The method of equivalence of Cartan makes it in principle possible to determine the structure of the Lie pseudogroup of local automorphisms of any G-structure, in particular to determine whether it is of finite or infinite type, and also whether it is transitive or intransitive. In the latter case, the method of equivalence leads to complete sets of local differential invariants for the problem that is being considered. This approach has been applied with great success in a wide array of geometric situations. The monograph by Gardner[5] provides a good exposition of the method of equivalence.

3. An Infinite-Dimensional Lie Group Approach

A conceptual question that is not addressed when working from the classical perspective that we have just sketched is to know to what extent one can associate an infinite dimensional Lie group structure to a Lie pseudogroup of infinite type, which could be thought of as a "parameter space" for the Lie pseudogroup. This question was dealt with successfully by Kuranishi[9,10] in the context of formal analytic mappings, but it is still open in convergent analytic case. There are however some recent results available for isotropy subgroups of transitive Lie pseudogroups, which we now present. Further details can be found in the paper.[8]

Let V be an n-dimensional real vector space which we identify with $\mathbb{R}^n$ by a choice of basis, and let $S^k(V^\star)$ the k-th symmetric power of the dual space $V^\star$. A vector field X_k whose coefficients are all homogeneous polynomials of degree k can be written as $X_k = \sum_{i=1}^n X_k^i \partial_i$ where X_k^i can be identified with an element of $V \otimes S^k(V^\star)$. We choose the vector fields

$$\left\{ \frac{\mid \alpha \mid!}{\alpha!} x^\alpha \partial_i \right\} \text{ where } i = 1, \ldots, n \text{ and } \mid \alpha \mid = k \,,$$

as a basis for $V \otimes S^k(V^\star)$. With this choice of basis, we write

$$X_k = \sum X_{k,\alpha}^i \frac{\mid \alpha \mid!}{\alpha!} x^\alpha \partial_i \,,$$

and endow each space $V \otimes S^k(V^\star)$ with the norm

$$\|X_k\|_k = \max_{\alpha,i} |X_{k,\alpha}^i| \,.$$

Let $\chi(V)$ denote the Lie algebra of formal vector fields based at the origin 0 of V, and $\chi_q(V)$ the Lie subalgebra of formal vector fields tangent up to order $q \in \mathbb{N}$ to the zero vector field. A formal vector field X in $\chi_q(V)$ can thus be written as

$$X = \sum_{i=1}^{n} \sum_{|\alpha| > q} (X_\alpha^i x^\alpha) \partial_i \, .$$

For the remainder of this paper, we will consider Lie pseudogroups Γ defined near the origin in $V \simeq \mathbb{R}^n$. From the work of Singer and Sternberg,[13] we know that one can associate to any such pseudogroup Γ a Lie algebra $\mathcal{L}(\Gamma)$ of formal vector fields. We let $\mathcal{L}_q(\Gamma) = \mathcal{L}(\Gamma) \cap \chi_q(V)$. We have $\mathcal{L}(\Gamma)/\mathcal{L}_0(\Gamma) \simeq V$ whenever Γ^ω is transitive. To the formal Lie algebra $\mathcal{L}(\Gamma)$ one associates the flat Lie algebra $L(\Gamma)$ defined by

$$L(\Gamma) = \oplus_{q=-1}^{\infty} \mathcal{L}_q(\Gamma)/\mathcal{L}_{q+1}(\Gamma) \, .$$

By definition, a transitive Lie pseudogroup Γ defined near the origin in V is flat if the Lie algebra $\mathcal{L}(\Gamma)$ is isomorphic to $L(\Gamma)$. A classical example of a flat Lie pseudogroup is the pseudogroup of local symplectomorphisms of $V \simeq \mathbb{R}^n$, where n is even, and V is endowed with the standard symplectic structure.

We now restrict our attention to the class of analytic vector fields. Recall that formal vector field

$$X = \sum_{k=0}^{\infty} X_k, \quad X_k \in V \otimes S^k(V^\star) \, ,$$

is analytic if its coefficients satisfy the boundedness condition

$$\limsup_k \|X_k\|_k^{\frac{1}{k}} < \infty \, .$$

We denote the set of analytic vector fields in $\mathcal{L}(\Gamma)$ by $\mathcal{L}^\omega(\Gamma)$. This space is naturally endowed with the structure of a topological Lie algebra, where the topology is that of a locally convex strict inductive limit of Banach spaces, defined as follows. Let $\rho > 0$ be a real number and let $\mathcal{L}_\rho^\omega(\Gamma)$ denote the subspace of $\mathcal{L}^\omega(\Gamma)$ consisting of those vector fields V for which

$$\limsup \|X_k\|_k / \rho^k < +\infty \, .$$

We have

$$\mathcal{L}^\omega(\Gamma) = \bigcup_{\rho > 0} \mathcal{L}_\rho^\omega(\Gamma) \, .$$

Each space $\mathcal{L}_\rho^\omega(\Gamma)$ is thus a Banach space for the norm

$$\|V\|_\rho = \sup_k \frac{\|X_k\|_k}{\rho^k}.$$

One shows that for $\rho < \rho'$ the injection $\mathcal{L}_\rho^\omega(\Gamma) \hookrightarrow \mathcal{L}_{\rho'}^\omega(\Gamma)$ is continuous and compact, so that $\mathcal{L}^\omega(\Gamma)$ is a complete Hausdorff locally convex topological vector space. Its associated topology is the locally convex strict inductive limit topology

$$\mathcal{L}^\omega(\Gamma) = \varinjlim_{\rho \in \mathbb{N}} \mathcal{L}_\rho^\omega(\Gamma).$$

This endows $\mathcal{L}^\omega(\Gamma)$ with a topological Lie algebra structure. We denote by $\mathcal{L}_{\rho,q}^\omega(\Gamma)$ the intersection

$$\mathcal{L}_{\rho,q}^\omega(\Gamma) = \mathcal{L}_\rho^\omega(\Gamma) \cap \chi_q(V),$$

and by $\mathcal{L}_{\rho,q,M}^\omega(\Gamma)$ the subset of $\mathcal{L}_{\rho,q}^\omega(\Gamma)$ consisting of those analytic vector fields X satisfying

$$\|X\|_\rho \leq M.$$

We are now ready to state the infinite-dimensional version of the classical second and third fundamental theorems of Lie for isotropy subgroups of analytic Lie pseudogroups. Before stating the results, we recall that from[7] that the group $G_0^\omega(V)$ of analytic local diffeomorphisms of V fixing the origin is Gâteaux-analytic Lie group. Our theorems reads:

Theorem 1.1 (Lie II). *Let Γ be a Lie pseudogroup of analytic transformations defined near the origin in V. Then the isotropy subgroup of the origin is integrable into a unique connected subgroup H embedded in $G_0^\omega(V)$. This subgroup H is a regular analytic Lie group of countable order.*

Theorem 1.2 (Lie III). *The isotropy Lie algebra $\mathcal{L}_0^\omega(\Gamma)$ is integrable into a unique connected and simply connected regular analytic Lie group Γ_0 of the second kind and countable order.*

The proofs of these theorems were first given in.[8] The idea is to construct a chart given as an infinite product of exponentials of vector fields chosen in a suitable bounded filtered basis of $\mathcal{L}_0^\omega(\Gamma)$. The boundedness of this basis gives the estimates needed to prove the convergence of the infinite product. As explained in,[8] the existence of this basis is a corollary of a fundamental estimate of Malgrange[11] which is central to his proof of the Cartan-Kähler Theorem. The Cartan-Kähler Theorem is thus one of the cornerstones of our approach, just as in the classical theory. In the remainder of this section, we will sketch the main steps of the construction of the chart.

We start from a Lie algebra $\mathcal{L}^\omega(\Gamma) \subset \chi(V)$ of local analytic vector fields defined in a neighborhood of the origin of V, and we denote by $\mathcal{L}_q^\omega(\Gamma)$ the subalgebra of vector fields in $\mathcal{L}^\omega(\Gamma)$ contained in $\chi_q(V)$. We will be interested in sections Σ_q of

$$\mathcal{L}_{q-1}^\omega(\Gamma) \to \mathcal{L}_{q-1}^\omega(\Gamma)/\mathcal{L}_q^\omega(\Gamma),$$

which satisfy a boundedness condition that will be specified below. Letting

$$E_q = \Sigma_q(\mathcal{L}_{q-1}^\omega(\Gamma)/\mathcal{L}_q^\omega(\Gamma)),$$

the isotropy Lie algebra $\mathcal{L}_0^\omega(\Gamma)$ decomposes as a direct sum

$$\mathcal{L}_0^\omega(\Gamma) = \oplus_{q=1}^{+\infty} E_q.$$

The following lemma gives the existence of uniformly bounded sections:

Lemma 1.1. *There exists a $\rho_0 > 0$ and sections $\Sigma_q : \mathcal{L}_{q-1}^\omega(\Gamma)/\mathcal{L}_q^\omega(\Gamma) \to \mathcal{L}_{q-1}^\omega(\Gamma)$ taking values in the Banach subspace $\mathcal{L}_{\rho_0}^\omega(\Gamma)$ of $\mathcal{L}^\omega(\Gamma)$, such that the norms of the continuous operators $\Sigma_q : \mathcal{L}_{q-1}^\omega(\Gamma)/\mathcal{L}_q^\omega(\Gamma) \to \mathcal{L}_{q-1}^\omega(\Gamma)$ are uniformly bounded.*

It is useful to think of these uniformly bounded sections in terms of an adapted filtered basis of the Lie algebra $\mathcal{L}^\omega(\Gamma)$. A basis $\mathcal{B}$ of the Lie algebra $\mathcal{L}^\omega(\Gamma)$ is filtered if the subset $\mathcal{B}_q$ of $\mathcal{B}$ consisting of the basis elements belonging to $\mathcal{L}_q^\omega(\Gamma)$ forms a basis of $\mathcal{L}_q^\omega(\Gamma)$. We have

Lemma 1.2. *If $\mathcal{L}^\omega(\Gamma)$ satisfies the boundedness condition of Lemma 1.1, then it admits a bounded filtered basis $\mathcal{B}$, meaning that there exists a constant $\rho_0 > 0$ such that for every element X of $\mathcal{L}_{q-1}^\omega(\Gamma)/\mathcal{L}_q^\omega(\Gamma)$,*

$$\|\tilde{X}\|_{\mathcal{L}_{\rho_0}} \le \|X\|_q,$$

where

$$\tilde{X} = \Sigma_q(X).$$

To prove Theorem 1.2 it is sufficient[7] to demonstrate that any local transformation Φ of Γ of the form

$$\Phi = I - \sum_{i=1}^{n} \sum_{|\alpha| \ge 2} \phi_\alpha^i x^\alpha$$

with I being the identity transformation, can be written uniquely as an infinite product

$$\cdots \circ Exp\, X_n \circ \cdots \circ Exp\, X_2 \circ Exp\, X_1$$

where $X_n \in (\mathcal{L}_{2^n}^{\omega}(\Gamma)/\mathcal{L}_{2^{n+1}}^{\omega}(\Gamma))_{|\Sigma}$. The existence and uniqueness of a formal solution $(X_n)_{n \in \mathbb{N}}$ is easily established by iteration. The main difficulty of the proof is to establish the analyticity of the series $\sum_n X_n$.

The key step is to effect a decomposition of the "free part" of any vector field with respect to the chosen section Σ. Let $Z = Z_{p+1} + Z_{p+2} + \cdots$ be an element of $\mathcal{L}_{\rho,p,M}^{\omega}(\Gamma)$, where each Z_{p+i} represents the homogeneous part of Z of degree $p + i$. For each integer i, we let

$$\tilde{Z}_{p+i} = \Sigma_{p+i}(Z_{p+i})\,.$$

We rewrite Z as

$$Z = \tilde{Z}_{p+1} + (Z_{p+2} - \tilde{Z}_{p+1}^{p+2}) + (Z_{p+3} - \tilde{Z}_{p+1}^{p+3}) + \cdots$$

where, for each integer $l > i$, $\tilde{Z}_{p+i}^{p+l}$ is the homogeneous part of degree $p + l$ of $\tilde{Z}_{p+i}$.

Since $Z \in \mathcal{L}_{\rho,p,M}^{\omega}(\Gamma)$, its homogeneous part of degree $p + 1$ satisfies the bound

$$\|Z_{p+1}\|_{p+1} \leq M\rho^p\,.$$

Therefore, we have

$$\|\tilde{Z}_{p+1}^{p+l}\|_{p+l} \leq M\rho^p \rho_0^{l-1} \leq M\rho^p$$

for all $l \geq 1$ whenever ρ_0 is smaller than 1. Denote now by $\tilde{Z}_{p+2}$ the image under Σ of the homogeneous part of degree $p + 2$ of Z, that is $Z_{p+2} - \tilde{Z}_{p+1}^{p+2}$. We have

$$Z = \tilde{Z}_{p+1} + \tilde{Z}_{p+2} + \sum_{l=3}^{+\infty} \left(Z_{p+l} - \tilde{Z}_{p+1}^{p+l} - \tilde{Z}_{p+2}^{p+l} \right),$$

and, for $l \geq 2$, the estimate

$$\|\tilde{Z}_{p+2}^{p+l}\|_{p+2} \leq M\rho^{p+1} + M\rho^p\,.$$

By repeating this procedure, we obtain a decomposition $Z = X + Y$ of Z into a free part X and a remainder Y, where

$$X = \tilde{Z}_{p+1} + \tilde{Z}_{p+2} + \cdots + \tilde{Z}_{2p}\,,$$

and

$$Y = \sum_{l=p+1}^{+\infty} \left(Z_{p+l} - \tilde{Z}_{p+1}^{p+l} - \tilde{Z}_{p+2}^{p+l} - \cdots - \tilde{Z}_{2p}^{p+l} \right),$$

with the estimate

$$\|Y_{n+1}\|_{n+1} \le M\rho^n + M\rho^p + (M\rho^{p+1} + M\rho^p) + \cdots + (M\rho^{2p-1} + \cdots + M\rho^p).$$

Hence if we start at $\rho \ge 4$ each $\|Y_{n+1}\|_{n+1}$ is bounded by $M(\rho^n + 2\rho^{2p-1})$. The norm of the remainder Y in $\mathcal{L}_\rho(\Gamma)$ is bounded by

$$\|Y\|_{\mathcal{L}_{\rho(\Gamma)}} \le M \limsup_{k \ge 1} \left\{ 1 + \frac{2}{\rho^k} \right\} \le \frac{3}{2}M.$$

Since $X = Z - Y$, this shows that the decomposition splits any $Z \in \mathcal{L}_{\rho,p,M}^\omega(\Gamma)$ into $Z = X + Y$ where $X \in \mathcal{L}_{\rho,p,M'=\frac{5}{2}M}^\omega(\Gamma)$ and $Y \in \mathcal{L}_{\rho,2p,M''=\frac{3}{2}M}^\omega(\Gamma)$.

We can use this to prove our main theorem. The iteration described above takes us from $x - Z$ to $(x - X) \circ e^X - Y \circ e^X = x - W$. By an estimate proved in[8] we obtain the following bound on the "error" W,

$$\|W\|_{\mathcal{L}_{\sigma\rho}^\omega} \le \frac{2M'}{\sqrt{\pi}\sigma^p} \frac{\gamma^2}{1 - \gamma^2} + \|Y\|_{\mathcal{L}_{\sigma\rho}^\omega},$$

where

$$\gamma = \frac{\sqrt{(p+1)M'}}{\sigma^{\frac{p-2}{2}} \ln \sigma},$$

can be chosen with no loss of generality to be less than 1. Since $Y \in \mathcal{L}_{\rho,2p,M''}^\omega(\Gamma)$, we have also

$$\|Y\|_{\mathcal{L}_{\sigma\rho}^\omega} \le \frac{M''}{\sigma^{2p}}.$$

Finally

$$\|W\|_{\mathcal{L}_{\sigma\rho}^\omega} \le \frac{5M}{\sqrt{\pi}\sigma^p} \frac{\gamma^2}{1 - \gamma^2} + \frac{3M}{2\sigma^{2p}}.$$

The iteration sends $\mathcal{L}_{\rho,p,M}^\omega(\Gamma)$ to $\mathcal{L}_{\sigma\rho,2p,\tilde{M}}^\omega(\Gamma)$ where

$$\tilde{M} = \frac{M}{\sigma^p} \left(\frac{5}{\sqrt{\pi}} \frac{\gamma^2}{1 - \gamma^2} + \frac{3}{2\sigma^p} \right).$$

We set the value γ to $1/2$ and use the "LambertW" function solve the equation

$$\sigma^{\frac{p-2}{2}} \ln \sigma = \sqrt{10(p+1)M}$$

for σ as a function of p and M. This function is defined as the principal branch of the solution of the equation $z = w\,exp(w)$ for w as a function of z. This gives

$$\sigma = e^{2\frac{LW\left(\frac{p-2}{2}\sqrt{10(p+1)M}\right)}{(p-2)}}.$$

The iteration takes us from $\mathcal{L}^{\omega}_{\rho, n_i, M_i}(\Gamma)$ to $\mathcal{L}^{\omega}_{\rho_{i+1}, n_{i+1}, M_{i+1}}(\Gamma)$ where $n_{i+1} = 2n_i$, $\rho_{i+1} = \sigma_i \rho_i$,

$$M_{i+1} = \frac{M_i}{\sigma_i^{n_i}}\left(\frac{5}{3\sqrt{\pi}} + \frac{3}{2\sigma_i^{n_i}}\right)$$

and

$$\sigma_i = e^{2\frac{LW\left((n_i/2-1)\sqrt{10(n_i+1)M_i}\right)}{(n_i-2)}}.$$

We have $n_i = 2^i n_0$, $\frac{5}{3\sqrt{\pi}} \approx 0.94$ and $\sigma_i > 1$ for all i, which implies that $M_i \leq (\frac{5}{2})^i M_0$. Therefore

$$\rho_{i+1} = \sigma_i \sigma_{i-1} \cdots \sigma_0 \rho_0$$

is less than

$$\rho_{i+1} \leq e^{\sum_{k=0}^{i} 2\frac{LW\left((2^{k-1}n_0-1)\sqrt{10(2^k n_0+1)(5/2)^k M_0}\right)}{(2^k n_0-2)}} \rho_0.$$

The asymptotics of LW are similar to those of the classical logarithmic function, and the infinite sum

$$\sum_{k=0}^{+\infty} 2\frac{LW\left((2^{k-1}n_0 - 1)\sqrt{10(2^k n_0 + 1)(5/2)^k M_0}\right)}{(2^k n_0 - 2)}$$

converges. This concludes our sketch of the proof of the convergence of our iterative scheme, and therefore of Theorem 1.2.

References

1. R. Bryant, S. S. Chern, R. Gardner, H. Goldschmidt, and P. A. Griffiths, *Exterior differential systems*, Mathematical Sciences Research Institute Publications, **18** (Springer-Verlag, New York, 1991).
2. E. Cartan, Sur la structure des groupes infinis de transformations, *Ann. Ec. Normale* **21**, 219–308 (1904).
3. E. Cartan, La structure des groupes infinis, Seminaire de Math., exposé G, 1er mars 1937, reprinted in *Elie Cartan, Oeuvres complètes, Vol. II* (Editions du CNRS, Paris, 1984).
4. S. S. Chern, Pseudogroupes continus infinis, in *Géométrie différentielle, Colloques Internationaux du Centre National de la Recherche Scientifique, Strasbourg* (Centre National de la Recherche Scientifique, Paris, 1953).

5. R. Gardner, *The method of equivalence and its applications*, CBMS-NSF Regional Conference Series in Applied Mathematics, **58** (Society for Industrial and Applied Mathematics (SIAM), Philadelphia, PA, 1989).

6. S. Kobayashi, *Transformation groups in differential geometry*, Ergebnisse der Mathematik und ihrer Grenzgebiete, **70** (Springer-Verlag, New York-Heidelberg, 1972).

7. N. Kamran, T. Robart, A manifold structure for analytic isotropy Lie pseudogroups of infinite type, *Journal of Lie Theory* **11** no. 1, 57–80 (2001).

8. N. Kamran and T. Robart, An infinite-dimensional manifold structure for analytic Lie pseudogroups of infinite type, *Int. Math. Res. Not.* **34**, 1761–1783 (2004).

9. M. Kuranishi, On the local theory of continuous infinite pseudo groups, I, *Nagoya Math. J.* **15**, 225–260 (1959).

10. M. Kuranishi, On the local theory of continuous infinite pseudo groups, II, *Nagoya Math. J.* **19**, 55–91 (1961).

11. B. Malgrange, Equations de Lie I & II, *J. Differential Geometry* **6**, 503–522 & **7**, 117–141 (1972).

12. P. Olver and J. Pohjanpelto, Maurer-Cartan forms and the structure of Lie pseudo-groups, *Selecta Math. (N.S.)* **11**, no. 1, 99–126 (2005).

13. I.M. Singer, S. Sternberg, The infinite groups of Lie and Cartan. I. The transitive groups, *J. Analyse Math.* **15**, 1–114 (1965).

14. D. Spencer, Deformation of structures on manifolds defined by transitive, continuous pseudogroups I & II, *Ann. of Math.* **76**, 306–445 (1962).

15. A. Weil, L'avenir des mathématiques, in *Les grands courants de la pensée mathématique*, ed. F. Le Lionnais (Cahiers du Sud, Paris, 1947).

Chapter 12

STABILITY OF CLOSED CHARACTERISTICS ON COMPACT CONVEX HYPERSURFACES

Yiming Long[*,†,‡] and Wei Wang[*,§]

Chern Institute of Mathematics, †Key Lab of Pure Mathematics and Combinatorics of Ministry of Education, Nankai University, Tianjin 300071, The People's Republic of China
‡*longym@nankai.edu.cn*
§*wangwei03@mail.nankai.edu.cn*

June 2006

Dedicated to the memory of Professor Shiing-Shen Chern

In this paper, for every compact convex hypersurface Σ in $\mathbf{R}^{2n}$ with $n \geq 2$ we prove that there exist at least two non-hyperbolic closed characteristics on Σ provided the number of geometrically distinct closed characteristics on Σ is finite.

Keywords: Compact convex hypersurfaces, closed characteristics, Hamiltonian systems, Morse theory, mean index identity, stability.

AMS Subject Classification: 58E05, 37J45, 37C75.

1. Introduction and Main Results

In this paper, let Σ be a fixed C^3 compact convex hypersurface in $\mathbf{R}^{2n}$, i.e., Σ is the boundary of a compact and strictly convex region U in $\mathbf{R}^{2n}$. We denote the set of all such hypersurfaces by $\mathcal{H}(2n)$. Without loss of generality, we suppose U contains the origin. We consider closed characteristics

‡Partially supported by the 973 Program of MOST, Yangzi River Professorship, NNSF, MCME, RFDP, LPMC of MOE of China, S. S. Chern Foundation, and Nankai University.
§Partially supported by NNSF, RFDP of MOE of China.

(τ, y) on Σ, which are solutions of the following problem

$$\begin{cases} \dot{y} = JN_\Sigma(y), \\ y(\tau) = y(0), \end{cases} \tag{1.1}$$

where $J = \begin{pmatrix} 0 & -I_n \\ I_n & 0 \end{pmatrix}$, I_n is the identity matrix in $\mathbf{R}^n$, $\tau > 0$, $N_\Sigma(y)$ is the outward normal vector of Σ at y normalized by the condition $N_\Sigma(y) \cdot y = 1$. Here $a \cdot b$ denotes the standard inner product of $a, b \in \mathbf{R}^{2n}$. A closed characteristic (τ, y) is *prime*, if τ is the minimal period of y. Two closed characteristics (τ, y) and (σ, z) are *geometrically distinct*, if $y(\mathbf{R}) \neq z(\mathbf{R})$. We denote by $\mathcal{J}(\Sigma)$ and $\tilde{\mathcal{J}}(\Sigma)$ the set of all closed characteristics (τ, y) on Σ with τ being the minimal period of y and the set of all geometrically distinct ones respectively.

Let $j : \mathbf{R}^{2n} \to \mathbf{R}$ be the gauge function of Σ, i.e., $j(\lambda x) = \lambda$ for $x \in \Sigma$ and $\lambda \geq 0$, then $j \in C^3(\mathbf{R}^{2n} \setminus \{0\}, \mathbf{R}) \cap C^0(\mathbf{R}^{2n}, \mathbf{R})$ and $\Sigma = j^{-1}(1)$. Fix a constant $\alpha \in (1, 2)$ and define the Hamiltonian function $H_\alpha : \mathbf{R}^{2n} \to [0, +\infty)$ by

$$H_\alpha(x) = j(x)^\alpha, \qquad \forall x \in \mathbf{R}^{2n}. \tag{1.2}$$

Then $H_\alpha \in C^3(\mathbf{R}^{2n} \setminus \{0\}, \mathbf{R}) \cap C^0(\mathbf{R}^{2n}, \mathbf{R})$ is convex and $\Sigma = H_\alpha^{-1}(1)$. It is well known that the problem (1.1) is equivalent to the following given energy problem of the Hamiltonian system

$$\begin{cases} \dot{y}(t) = JH'_\alpha(y(t)), & H_\alpha(y(t)) = 1, \quad \forall t \in \mathbf{R}, \\ y(\tau) = y(0). \end{cases} \tag{1.3}$$

Denote by $\mathcal{J}(\Sigma, \alpha)$ the set of all solutions (τ, y) of (1.3) where τ is the minimal period of y and by $\tilde{\mathcal{J}}(\Sigma, \alpha)$ the set of all geometrically distinct solutions of (1.3). Note that elements in $\mathcal{J}(\Sigma)$ and $\mathcal{J}(\Sigma, \alpha)$ are one to one correspondent to each other.

Let $(\tau, y) \in \mathcal{J}(\Sigma, \alpha)$. The fundamental solution $\gamma_y : [0, \tau] \to \mathrm{Sp}(2n)$ with $\gamma_y(0) = I_{2n}$ of the linearized Hamiltonian system

$$\dot{w}(t) = JH''_\alpha(y(t))w(t), \qquad \forall t \in \mathbf{R}, \tag{1.4}$$

is called the *associate symplectic path* of (τ, y). The eigenvalues of $\gamma_y(\tau)$ are called *Floquet multipliers* of (τ, y). By Proposition 1.6.13 of [Eke3], the Floquet multipliers with their multiplicities of $(\tau, y) \in \mathcal{J}(\Sigma)$ do not depend on the particular choice of the Hamiltonian function in (1.3). For any $M \in \mathrm{Sp}(2n)$, we define the *elliptic height* $e(M)$ of M to be the total algebraic multiplicity of all eigenvalues of M on the unit circle $\mathbf{U} = \{z \in \mathbf{C} | |z| = 1\}$ in the complex plane $\mathbf{C}$. Since M is symplectic, $e(M)$ is even and $0 \leq e(M) \leq 2n$. As usual a $(\tau, y) \in \mathcal{J}(\Sigma)$ is *elliptic*, if $e(\gamma_y(\tau)) = 2n$. It is *non-degenerate*, if 1 is a double Floquet multiplier of it. It is *hyperbolic*, if 1 is a double Floquet multiplier of it and $e(\gamma_y(\tau)) = 2$. It is well known that these concepts are independent of the choice of $\alpha > 1$.

The study on closed characteristics in the global sense started in 1978, when the existence of at least one closed characteristic on any $\Sigma \in \mathcal{H}(2n)$ was first established by P. Rabinowitz in [Rab1] (for star-shaped hyper-surfaces) and A. Weinstein in [Wei1] independently. In I. Ekeland and L. Lassoued's [EkL1], I. Ekeland and H. Hofer's [EkH1] of 1987, and A. Szulkin's [Szu1] of 1988, $^\#\tilde{\mathcal{J}}(\Sigma) \geq 2$ was proved for any $\Sigma \in \mathcal{H}(2n)$ when $n \geq 2$. In [HWZ1] of 1998, H. Hofer-K. Wysocki-E. Zehnder proved that $^\#\tilde{\mathcal{J}}(\Sigma) = 2$ or $+\infty$ holds for every $\Sigma \in \mathcal{H}(4)$. In [LoZ1] of 2002, Y. Long and C. Zhu further proved $^\#\tilde{\mathcal{J}}(\Sigma) \geq [\frac{n}{2}]+1$ for any $\Sigma \in \mathcal{H}(2n)$. In [LLZ1] of 2002, C. Liu, Y. Long and C. Zhu proved $^\#\tilde{\mathcal{J}}(\Sigma) \geq n$ for every $\Sigma \in \mathcal{H}(2n)$ provided $\Sigma = -\Sigma$. Recently in [WHL1], W. Wang, X. Hu and Y. Long proved $^\#\tilde{\mathcal{J}}(\Sigma) \geq 3$ for every $\Sigma \in \mathcal{H}(6)$.

On the stability problem, in [Eke2] of I. Ekeland in 1986 and [Lon2] of Y. Long in 1998, for any $\Sigma \in \mathcal{H}(2n)$ the existence of at least one non-hyperbolic closed characteristic on Σ was proved provided $^\#\tilde{\mathcal{J}}(\Sigma) < +\infty$. I. Ekeland proved also in [Eke2] the existence of at least one elliptic closed characteristic on Σ provided $\Sigma \in \mathcal{H}(2n)$ is $\sqrt{2}$-pinched. In [DDE1] of 1992, Dell'Antonio, G., B. D'Onofrio and I. Ekeland proved the existence of at least one elliptic closed characteristic on Σ provided $\Sigma \in \mathcal{H}(2n)$ satisfies $\Sigma = -\Sigma$. In [Lon3] of 2000, Y. Long proved that $\Sigma \in \mathcal{H}(4)$ and $^\#\tilde{\mathcal{J}}(\Sigma) = 2$ imply that both of the closed characteristics must be elliptic. In [LoZ1] of 2002, Y. Long and C. Zhu further proved when $^\#\tilde{\mathcal{J}}(\Sigma) < +\infty$, there exists at least one elliptic closed characteristic and there are at least $[\frac{n}{2}]$ geometrically distinct closed characteristics on Σ possessing irrational mean indices, which are then non-hyperbolic.

The following is our main result in this paper:

Theorem 1.1. *There exist at least two non-hyperbolic closed characteristics in $\tilde{\mathcal{J}}(\Sigma)$ for every $\Sigma \in \mathcal{H}(2n)$ satisfying $^\#\tilde{\mathcal{J}}(\Sigma) < +\infty$.*

Note that by the above mentioned results of [Lon3] and [LoZ1], Theorem 1.1 holds already for $n = 2$ and $n \geq 4$ respectively. Hence our Theorem 1.1 is new for the case of $n = 3$.

Our proof of Theorem 1.1 uses mainly three ingredients: the mean index identity for closed characteristics established in [WHL1] recently, Morse inequality and the index iteration theory developed by Y. Long and his coworkers, specially the common index jump theorem of Y. Long and C. Zhu (Theorem 4.3 of [LoZ1], cf. Theorem 11.2.1 of [Lon4]). In Section 2, we review briefly the equivariant Morse theory and the mean index identity for closed characteristics on compact convex hypersurfaces in $\mathbf{R}^{2n}$ developed in the recent [WHL1]. The proof of Theorem 1.1 is given in Section 3.

In this paper, let $\mathbf{N}$, $\mathbf{N}_0$, $\mathbf{Z}$, $\mathbf{Q}$, $\mathbf{R}$, and $\mathbf{R}^+$ denote the sets of natural integers, non-negative integers, integers, rational numbers, real numbers, and positive real numbers respectively. Denote by $a \cdot b$ and $|a|$ the standard inner product and norm in $\mathbf{R}^{2n}$. Denote by $\langle \cdot, \cdot \rangle$ and $\| \cdot \|$ the standard L^2 inner product and L^2 norm. For an S^1-space X, we denote by X_{S^1} the homotopy quotient of X module the S^1-action, i.e., $X_{S^1} = S^\infty \times_{S^1} X$. We define the functions

$$\begin{cases} [a] = \max\{k \in \mathbf{Z} \,|\, k \le a\}, \quad E(a) = \min\{k \in \mathbf{Z} \,|\, k \ge a\}, \\ \varphi(a) = E(a) - [a], \end{cases} \tag{1.5}$$

Specially, $\varphi(a) = 0$ if $a \in \mathbf{Z}$, and $\varphi(a) = 1$ if $a \notin \mathbf{Z}$. In this paper we use only $\mathbf{Q}$ coefficients for all homological modules. For a $\mathbf{Z}_m$-space pair (A, B), let

$$H_*(A, B)^{\pm \mathbf{Z}_m} = \{\sigma \in H_*(A, B) \,|\, L_*\sigma = \pm\sigma\}, \tag{1.6}$$

where L is a generator of the $\mathbf{Z}_m$-action.

2. Equivariant Morse Theory for Closed Characteristics

In the rest of this paper, we fix a $\Sigma \in \mathcal{H}(2n)$ and assume the following condition on Σ:

(F) **There exist only finitely many geometrically distinct closed characteristics on Σ.**

In this section, we review briefly the equivariant Morse theory for closed characteristics on Σ developed in [WHL1] which will be needed in Section 3 of this paper. All the details of proofs can be found in [WHL1].

Let $\hat{\tau} = \inf\{\tau_i | \, 1 \le i \le k\}$. Then for any $a > \hat{\tau}$, we can construct a function $\varphi_a \in C^\infty(\mathbf{R}, \mathbf{R}^+)$ which has 0 as its unique critical point in $[0, +\infty)$ such that φ_a is strictly convex for $t \ge 0$. Moreover, $\frac{\varphi_a'(t)}{t}$ is strictly decreasing for $t > 0$ together with $\lim_{t \to 0+} \frac{\varphi_a'(t)}{t} = 1$ and $\varphi_a(0) = 0 = \varphi_a'(0)$. Define the Hamiltonian function $H_a(x) = a\varphi_a(j(x))$ and consider the fixed period problem

$$\begin{cases} \dot{x}(t) = JH_a'(x(t)), \\ \quad x(1) = x(0). \end{cases} \tag{2.1}$$

Then $H_a \in C^3(\mathbf{R}^{2n} \setminus \{0\}, \mathbf{R}) \cap C^1(\mathbf{R}^{2n}, \mathbf{R})$ is strictly convex. Solutions of (2.1) are $x \equiv 0$ and $x = \rho y(\tau t)$ with $\frac{\varphi_a'(\rho)}{\rho} = \frac{\tau}{a}$, where (τ, y) is a solution of (1.1). In particular, nonzero solutions of (2.1) are one to one correspondent to solutions of (1.1) with period $\tau < a$.

In the following, we use the Clarke-Ekeland dual action principle. As usual, let G_a be the Legendre transform of H_a defined by

$$G_a(y) = \sup\{x \cdot y - H_a(x) \mid x \in \mathbf{R}^{2n}\}. \tag{2.2}$$

Then $G_a \in C^2(\mathbf{R}^{2n} \setminus \{0\}, \mathbf{R}) \cap C^1(\mathbf{R}^{2n}, \mathbf{R})$ is strictly convex. Let

$$L_0^2(S^1, \mathbf{R}^{2n}) = \left\{ u \in L^2([0, 1], \mathbf{R}^{2n}) \,\middle|\, \int_0^1 u(t)dt = 0 \right\}. \tag{2.3}$$

Define a linear operator $M : L_0^2(S^1, \mathbf{R}^{2n}) \to L_0^2(S^1, \mathbf{R}^{2n})$ by

$$\frac{d}{dt} Mu(t) = u(t), \quad \int_0^1 Mu(t)dt = 0. \tag{2.4}$$

The dual action functional on $L_0^2(S^1, \mathbf{R}^{2n})$ is defined by

$$\Psi_a(u) = \int_0^1 \left(\frac{1}{2} Ju \cdot Mu + G_a(-Ju) \right) dt. \tag{2.5}$$

Then the functional $\Psi_a \in C^{1,1}(L_0^2(S^1, \mathbf{R}^{2n}), \mathbf{R})$ is bounded from below and satisfies the Palais-Smale condition. Suppose x is a solution of (2.1). Then $u = \dot{x}$ is a critical point of Ψ_a. Conversely, suppose u is a critical point of Ψ_a. Then there exists a unique $\xi \in \mathbf{R}^{2n}$ such that $Mu - \xi$ is a solution of (2.1). In particular, solutions of (2.1) are in one to one correspondence with critical points of Ψ_a. Moreover, $\Psi_a(u) < 0$ for every critical point $u \neq 0$ of Ψ_a.

Suppose u is a nonzero critical point of Ψ_a. Then following [Eke3] the formal Hessian of Ψ_a at u is defined by

$$Q_a(v, v) = \int_0^1 (Jv \cdot Mv + G_a''(-Ju)Jv \cdot Jv)dt, \tag{2.6}$$

which defines an orthogonal splitting $L_0^2 = E_- \oplus E_0 \oplus E_+$ of $L_0^2(S^1, \mathbf{R}^{2n})$ into negative, zero and positive subspaces. The index of u is defined by $i(u) = \dim E_-$ and the nullity of u is defined by $\nu(u) = \dim E_0$. Let $u_a = \dot{x}_a$ be the critical point of Ψ_a such that x_a corresponds to the closed characteristic (τ, y) on Σ. Then the index $i(u_a)$ and the nullity $\nu(u_a)$ defined above coincide with the Ekeland indices defined by I. Ekeland in [Eke1] and [Eke3]. Specially $1 \leq \nu(u_a) \leq 2n - 1$ always holds.

We have a natural S^1 action on $L_0^2(S^1, \mathbf{R}^{2n})$ defined by

$$\theta \cdot u(t) = u(\theta + t), \quad \forall \theta \in S^1, \ t \in \mathbf{R}. \tag{2.7}$$

Clearly Ψ_a is S^1-invariant. For any $\kappa \in \mathbf{R}$, we denote by

$$\Lambda_a^\kappa = \{u \in L_0^2(S^1, \mathbf{R}^{2n}) \mid \Psi_a(u) \leq \kappa\}. \tag{2.8}$$

For a critical point u of Ψ_a, we denote by

$$\Lambda_a(u) = \Lambda_a^{\Psi_a(u)} = \{w \in L_0^2(S^1, \mathbf{R}^{2n}) \mid \Psi_a(w) \leq \Psi_a(u)\}. \tag{2.9}$$

Clearly, both sets are S^1-invariant. Since the S^1-action preserves Ψ_a, if u is a critical point of Ψ_a, then the whole orbit $S^1 \cdot u$ is formed by critical points of Ψ_a. Denote by $crit(\Psi_a)$ the set of critical points of Ψ_a. Note that by the condition (F), the number of critical orbits of Ψ_a is finite. Hence as usual we can make the following definition.

Definition 2.1. *Suppose u is a nonzero critical point of Ψ_a, and $\mathcal{N}$ is an S^1-invariant open neighborhood of $S^1 \cdot u$ such that $crit(\Psi_a) \cap (\Lambda_a(u) \cap \mathcal{N}) = S^1 \cdot u$. Then the S^1-critical modules of $S^1 \cdot u$ is defined by*

$$C_{S^1,\, q}(\Psi_a, \ S^1 \cdot u) = H_q((\Lambda_a(u) \cap \mathcal{N})_{S^1}, \ ((\Lambda_a(u) \setminus S^1 \cdot u) \cap \mathcal{N})_{S^1}). \tag{2.10}$$

We have the following proposition for critical modules.

Proposition 2.2. (cf. Proposition 3.2. of [WHL1]) *The critical module $C_{S^1,\, q}(\Psi_a, \ S^1 \cdot u)$ is independent of a in the sense that if x_i are solutions of (2.1) with Hamiltonian functions $H_{a_i}(x) \equiv a_i \varphi_{a_i}(j(x))$ for $i = 1$ and 2 respectively such that both x_1 and x_2 correspond to the same closed characteristic (τ, y) on Σ. Then we have*

$$C_{S^1,\, q}(\Psi_{a_1}, \ S^1 \cdot \dot{x}_1) \cong C_{S^1,\, q}(\Psi_{a_2}, \ S^1 \cdot \dot{x}_2), \quad \forall q \in \mathbf{Z}. \tag{2.11}$$

Now let $u \neq 0$ be a critical point of Ψ_a with multiplicity $mul(u) = m$, i.e., u corresponds to a closed characteristic $(m\tau, y) \subset \Sigma$ with (τ, y) being prime. Hence $u(t + \frac{1}{m}) = u(t)$ holds for all $t \in \mathbf{R}$ and the orbit of u, namely, $S^1 \cdot u \cong S^1/\mathbf{Z}_m \cong S^1$. Let $f : N(S^1 \cdot u) \to S^1 \cdot u$ be the normal bundle of $S^1 \cdot u$ in $L_0^2(S^1, \mathbf{R}^{2n})$ and let $f^{-1}(\theta \cdot u) = N(\theta \cdot u)$ be the fibre over $\theta \cdot u$, where $\theta \in S^1$. Let $DN(S^1 \cdot u)$ be the ϱ disk bundle of $N(S^1 \cdot u)$ for some $\varrho > 0$ sufficiently small, i.e., $DN(S^1 \cdot u) = \{\xi \in N(S^1 \cdot u) \mid \|\xi\| < \varrho\}$ and let $DN(\theta \cdot u) = f^{-1}(\theta \cdot u) \cap DN(S^1 \cdot u)$ be the disk over $\theta \cdot u$. Clearly, $DN(\theta \cdot u)$ is $\mathbf{Z}_m$-invariant and we have $DN(S^1 \cdot u) = DN(u) \times_{\mathbf{Z}_m} S^1$ where the Z_m action is given by

$$(\theta, v, t) \in \mathbf{Z}_m \times DN(u) \times S^1 \mapsto (\theta \cdot v, \ \theta^{-1} t) \in DN(u) \times S^1.$$

Hence for an S^1-invariant subset Γ of $DN(S^1 \cdot u)$, we have $\Gamma/S^1 = (\Gamma_u \times_{\mathbf{Z}_m} S^1)/S^1 = \Gamma_u/\mathbf{Z}_m$, where $\Gamma_u = \Gamma \cap DN(u)$. Since Ψ_a is not C^2 on $L_0^2(S^1, \mathbf{R}^{2n})$, we need to use a finite dimensional approximation introduced by I. Ekeland in order to apply Morse theory. More precisely, we can construct a finite dimensional submanifold $\Gamma(\iota)$ of $L_0^2(S^1, \mathbf{R}^{2n})$ which admits a $\mathbf{Z}_\iota$-action with $m|\iota$. Moreover Ψ_a and $\Psi_a|_{\Gamma(\iota)}$ have the same critical points. $\Psi_a|_{\Gamma(\iota)}$ is C^2 in a small tubular neighborhood of the critical orbit $S^1 \cdot u$ and the Morse index and nullity of its critical points coincide with those of the corresponding critical points of Ψ_a. Let

$$D_\iota N(S^1 \cdot u) = DN(S^1 \cdot u) \cap \Gamma(\iota), \quad D_\iota N(\theta \cdot u) = DN(\theta \cdot u) \cap \Gamma(\iota). \quad (2.12)$$

Then we have

$$C_{S^1, *}(\Psi_a, S^1 \cdot u) \cong H_*(\Lambda_a(u) \cap D_\iota N(u), (\Lambda_a(u) \setminus \{u\}) \cap D_\iota N(u))^{\mathbf{Z}_m}. \quad (2.13)$$

Now we can apply the results of D. Gromoll and W. Meyer in [GrM1] to the manifold $D_{p\iota} N(u^p)$ with u^p as its unique critical point, where $p \in \mathbf{N}$. Then $mul(u^p) = pm$ is the multiplicity of u^p and the isotropy group $\mathbf{Z}_{pm} \subseteq S^1$ of u^p acts on $D_{p\iota} N(u^p)$ by isometries. According to Lemma 1 of [GrM1], we have a $\mathbf{Z}_{pm}$-invariant decomposition of $T_{u^p}(D_{p\iota} N(u^p))$

$$T_{u^p}(D_{p\iota} N(u^p)) = V^+ \oplus V^- \oplus V^0 = \{(x_+, x_-, x_0)\}$$

with $\dim V^- = i(u^p)$, $\dim V^0 = \nu(u^p) - 1$ and a $\mathbf{Z}_{pm}$-invariant neighborhood $B = B_+ \times B_- \times B_0$ for 0 in $T_{u^p}(D_{p\iota} N(u^p))$ together with two Z_{pm}-invariant diffeomorphisms

$$\Phi : B = B_+ \times B_- \times B_0 \to \Phi(B_+ \times B_- \times B_0) \subset D_{p\iota} N(u^p)$$

and

$$\eta : B_0 \to W(u^p) \equiv \eta(B_0) \subset D_{p\iota} N(u^p)$$

such that $\Phi(0) = \eta(0) = u^p$ and

$$\Psi_a \circ \Phi(x_+, x_-, x_0) = |x_+|^2 - |x_-|^2 + \Psi_a \circ \eta(x_0), \quad (2.14)$$

with $d(\Psi_a \circ \eta)(0) = d^2(\Psi_a \circ \eta)(0) = 0$. As [GrM1], we call $W(u^p)$ a local *characteristic manifold* and $U(u^p) = B_-$ a local negative disk at u^p, by the proof of Lemma 1 of [GrM1], $W(u^p)$ and $U(u^p)$ are $\mathbf{Z}_{pm}$-invariant. Then we have

$$H_*(\Lambda_a(u^p) \cap D_{p\iota} N(u^p), (\Lambda_a(u^p) \setminus \{u^p\}) \cap D_{p\iota} N(u^p))$$
$$= H_*(U(u^p), U(u^p) \setminus \{u^p\})$$
$$\otimes H_*(W(u^p) \cap \Lambda_a(u^p), (W(u^p) \setminus \{u^p\}) \cap \Lambda_a(u^p)), \quad (2.15)$$

where

$$H_q(U(u^p), U(u^p) \setminus \{u^p\}) = \begin{cases} \mathbf{Q}, & \text{if } q = i(u^p), \\ 0, & \text{otherwise.} \end{cases} \tag{2.16}$$

Now we have the following proposition.

Proposition 2.3. (cf. Proposition 3.10. of [WHL1]) *Let $u \neq 0$ be a critical point of Ψ_a with $mul(u) = 1$. Then for all $p \in \mathbf{N}$ and $q \in \mathbf{Z}$, we have*

$$C_{S^1, q}(\Psi_a, S^1 \cdot u^p)$$
$$\cong \left(H_{q-i(u^p)}(W(u^p) \cap \Lambda_a(u^p), (W(u^p) \setminus \{u^p\}) \cap \Lambda_a(u^p)) \right)^{\beta(u^p)\mathbf{Z}_p}, \tag{2.17}$$

where $\beta(u^p) = (-1)^{i(u^p)-i(u)}$. In particular, if u^p is non-degenerate, i.e., $\nu(u^p) = 1$, then

$$C_{S^1, q}(\Psi_a, S^1 \cdot u^p) = \begin{cases} \mathbf{Q}, & \text{if } q = i(u^p) \text{ and } \beta(u^p) = 1, \\ 0, & \text{otherwise.} \end{cases} \tag{2.18}$$

We make the following definition

Definition 2.4. *Let $u \neq 0$ be a critical point of Ψ_a with $mul(u) = 1$. Then for all $p \in \mathbf{N}$ and $l \in \mathbf{Z}$, let*

$$k_{l,\pm 1}(u^p) = \dim \left(H_l(W(u^p) \cap \Lambda_a(u^p), (W(u^p) \setminus \{u^p\}) \cap \Lambda_a(u^p)) \right)^{\pm \mathbf{Z}_p}, \tag{2.19}$$

$$k_l(u^p) = \dim \left(H_l(W(u^p) \cap \Lambda_a(u^p), (W(u^p) \setminus \{u^p\}) \cap \Lambda_a(u^p)) \right)^{\beta(u^p)\mathbf{Z}_p}. \tag{2.20}$$

$k_l(u^p)s$ are called critical type numbers of u^p.

We have the following properties for critical type numbers

Proposition 2.5. (cf. Proposition 3.13. of [WHL1]) *Let $u \neq 0$ be a critical point of Ψ_a with $mul(u) = 1$. Then there exists a minimal $K(u) \in \mathbf{N}$ such that*

$$\nu(u^{p+K(u)}) = \nu(u^p), \quad i(u^{p+K(u)}) - i(u^p) \in 2\mathbf{Z}, \tag{2.21}$$

and $k_l(u^{p+K(u)}) = k_l(u^p)$ for all $p \in \mathbf{N}$ and $l \in \mathbf{Z}$. We call $K(u)$ the minimal period of critical modules of iterations of the functional Ψ_a at u.

For a closed characteristic (τ, y) on Σ, we denote by $y^m \equiv (m\tau, y)$ the mth iteration of y for $m \in \mathbf{N}$. Let $a > \tau$ and choose φ_a as above. Determine

ρ uniquely by $\frac{\varphi'_a(\rho)}{\rho} = \frac{\tau}{a}$. Let $x = \rho y(\tau t)$ and $u = \dot{x}$. Then we define the index $i(y^m)$ and nullity $\nu(y^m)$ of $(m\tau, y)$ for $m \in \mathbf{N}$ by

$$i(y^m) = i(u^m), \qquad \nu(y^m) = \nu(u^m). \qquad (2.22)$$

These indices are independent of a when a tends to infinity. Now the mean index of (τ, y) is defined by

$$\hat{i}(y) = \lim_{m \to \infty} \frac{i(y^m)}{m}. \qquad (2.23)$$

Note that $\hat{i}(y) > 2$ always holds which was proved by I. Ekeland and H. Hofer in [EkH1] of 1987 (cf. Corollary 8.3.2 and Lemma 15.3.2 of [Lon4] for a different proof).

By Proposition 2.2, we can define the critical type numbers $k_l(y^m)$ of y^m to be $k_l(u^m)$, where u^m is the critical point of Ψ_a corresponding to y^m. We also define $K(y) = K(u)$. Then we have

Proposition 2.6. *We have $k_l(y^m) = 0$ for $l \notin [0, \nu(y^m)-1]$ and it can take only values 0 or 1 when $l = 0$ or $l = \nu(y^m) - 1$. Moreover, the following properties hold (cf. Lemma 3.10 of [BaL1], [Cha1] and [MaW1]):*

(i) *$k_0(y^m) = 1$ implies $k_l(y^m) = 0$ for $1 \le l \le \nu(y^m) - 1$.*

(ii) *$k_{\nu(y^m)-1}(y^m) = 1$ implies $k_l(y^m) = 0$ for $0 \le l \le \nu(y^m) - 2$.*

(iii) *$k_l(y^m) \ge 1$ for some $1 \le l \le \nu(y^m) - 2$ implies $k_0(y^m) = k_{\nu(y^m)-1}(y^m) = 0$.*

(iv) *If $\nu(y^m) \le 3$, then at most one of the $k_l(y^m)$s for $0 \le l \le \nu(y^m) - 1$ can be non-zero.*

(v) *If $i(y^m) - i(y) \in 2\mathbf{Z} + 1$ for some $m \in \mathbf{N}$, then $k_0(y^m) = 0$.*

Proof. By Definition 2.4 we have

$$k_l(y^m) \le \dim H_l(W(u^m) \cap \Lambda_a(u^m), \, (W(u^m) \setminus \{u^m\}) \cap \Lambda_a(u^m)) \equiv \eta_l(y^m). \qquad (2.24)$$

Then from Corollary 5.1 of [Cha1] or Corollary 8.4 of [MaW1], (i)–(iv) hold.

For (v), if $\eta_0(y^m) = 0$, then (v) follows directly from Definition 2.4.

By Corollary 8.4 of [MaW1], $\eta_0(y^m) = 1$ if and only if u^m is a local minimum in the local characteristic manifold $W(u^m)$. Hence $(W(u^m) \cap \Lambda_a(u^m), \, (W(u^m) \setminus \{u^m\}) \cap \Lambda_a(u^m)) = (\{u^m\}, \, \emptyset)$. By Definition 2.4, we have:

$$k_{0,+1}(u^m) = \dim H_0(W(u^m) \cap \Lambda_a(u^m), \ (W(u^m) \setminus \{u^m\}) \cap \Lambda_a(u^m))^{+\mathbf{Z}_m}$$
$$= \dim H_0(\{u^m\})^{+\mathbf{Z}_m}$$
$$= 1.$$

This implies $k_0(u^m) = k_{0,-1}(u^m) = 0$. $\qquad\qquad\square$

For a closed characteristic (τ, y) on Σ, define

$$\hat{\chi}(y) = \frac{1}{K(y)} \sum_{\substack{1 \le m \le K(y) \\ 0 \le l \le 2n-2}} (-1)^{i(y^m)+l} k_l(y^m). \qquad (2.25)$$

In particular, if all y^ms are non-degenerate, then by (2.18) we have

$$\hat{\chi}(y) = \begin{cases} (-1)^{i(y)}, & \text{if } i(y^2) - i(y) \in 2\mathbf{Z}, \\ \dfrac{(-1)^{i(y)}}{2}, & \text{otherwise.} \end{cases} \qquad (2.26)$$

We have the following mean index identity for closed characteristics on every $\Sigma \in \mathcal{H}(2n)$ when $^\#\tilde{\mathcal{J}}(\Sigma) < +\infty$.

Theorem 2.7. (cf. Theorem 1.2 of [WHL1]) *Suppose $\Sigma \in \mathcal{H}(2n)$ satisfies $^\#\tilde{\mathcal{J}}(\Sigma) < +\infty$. Denote all the geometrically distinct closed characteristics by $\{(\tau_j, y_j)\}_{1 \le j \le k}$. Then the following identity holds*

$$\sum_{1 \le j \le k} \frac{\hat{\chi}(y_j)}{\hat{i}(y_j)} = \frac{1}{2}. \qquad (2.27)$$

Let Ψ_a be the functional defined by (2.5) for some $a \in \mathbf{R}$ large enough and let $\varepsilon > 0$ be small enough such that $[-\varepsilon, +\infty) \setminus \{0\}$ contains no critical values of Ψ_a. Denote by I_a the greatest integer in $\mathbf{N}_0$ such that $I_a < i(\tau, y)$ hold for all closed characteristics (τ, y) on Σ with $\tau \ge a$. Then by Section 5 of [WHL1], we have

$$H_{S^1, q}(\Lambda_a^{-\varepsilon}) \cong H_{S^1, q}(\Lambda_a^{\infty}) \cong H_q(CP^{\infty}), \quad \forall q < I_a. \qquad (2.28)$$

For any $q \in \mathbf{Z}$, let

$$M_q(\Lambda_a^{-\varepsilon}) = \sum_{1 \le j \le k, \, 1 \le m_j < a/\tau_j} \dim C_{S^1, q}(\Psi_a, \ S^1 \cdot u_j^{m_j}). \qquad (2.29)$$

Then the equivariant Morse inequality for the space $\Lambda_a^{-\varepsilon}$ yields

$$M_q(\Lambda_a^{-\varepsilon}) - M_{q-1}(\Lambda_a^{-\varepsilon}) + \cdots + (-1)^q M_0(\Lambda_a^{-\varepsilon})$$
$$\ge b_q(\Lambda_a^{-\varepsilon}) - b_{q-1}(\Lambda_a^{-\varepsilon}) + \cdots + (-1)^q b_0(\Lambda_a^{-\varepsilon}), \qquad (2.30)$$

where

$$b_q(\Lambda_a^{-\varepsilon}) = \dim H_{S^1, q}(\Lambda_a^{-\varepsilon}) \,.$$

Now we have the following Morse inequality for closed characteristics.

Theorem 2.8. *Let* $\Sigma \in \mathcal{H}(2n)$ *satisfy* $^\#\tilde{\mathcal{J}}(\Sigma) < +\infty$. *Denote all the geometrically distinct closed characteristics by* $\{(\tau_j, \, y_j)\}_{1 \le j \le k}$. *Let*

$$M_q = \lim_{a \to +\infty} M_q(\Lambda_a^{-\varepsilon}), \quad \forall q \in \mathbf{Z}, \tag{2.31}$$

$$b_q = \lim_{a \to +\infty} b_q(\Lambda_a^{-\varepsilon}) = \begin{cases} 1, & \text{if } q \in 2\mathbf{N}_0, \\ 0, & \text{otherwise.} \end{cases} \tag{2.32}$$

Then we have

$$M_q - M_{q-1} + \cdots + (-1)^q M_0 \ge b_q - b_{q-1} + \cdots + (-1)^q b_0, \quad \forall \, q \in \mathbf{Z}. \tag{2.33}$$

Proof. As we have mentioned before, $\hat{i}(y_j) > 2$ holds for $1 \le j \le k$. Hence the Ekeland index satisfies $i(y_j^m) = i(u_j^m) \to \infty$ as $m \to \infty$ for $1 \le j \le k$. Note that $I_a \to +\infty$ as $a \to +\infty$. Now fix a $q \in \mathbf{Z}$ and a sufficiently great $a > 0$. By Propositions 2.2, 2.3 and (2.29), $M_i(\Lambda_a^{-\varepsilon})$ is invariant for all $a > A_q$ and $0 \le i \le q$, where $A_q > 0$ is some constant. Hence (2.31) is meaningful. Now for any a such that $I_a > q$, (2.28)-(2.30) imply that (2.32) and (2.33) hold. $\qquad \square$

3. Proof of the Main Theorem

In this section, we give a proof of Theorem 1.1 by using the mean index identity of [WHL1], Morse inequality and the index iteration theory developed by Y. Long and his coworkers.

Proof of Theorem 1.1. By the assumption (F) at the beginning of Section 2, we denote by $\{(\tau_j, y_j)\}_{1 \le j \le k}$ all the geometrically distinct closed characteristics on Σ, and by $\gamma_j \equiv \gamma_{y_j}$ the associated symplectic path of (τ_j, y_j) on Σ for $1 \le j \le k$. Then by Lemma 15.2.4 of [Lon4], there exist $P_j \in \mathrm{Sp}(2n)$ and $M_j \in \mathrm{Sp}(2n-2)$ such that

$$\gamma_j(\tau_j) = P_j^{-1}(N_1(1, 1) \diamond M_j) P_j, \quad \forall \, 1 \le j \le k, \tag{3.1}$$

where recall $N_1(1, b) = \begin{pmatrix} 1 & b \\ 0 & 1 \end{pmatrix}$ for $b \in \mathbf{R}$. Thus when $(\tau, y) \in \tilde{\mathcal{J}}(\Sigma)$ is hyperbolic, by (12) of Theorem 8.2.2 of [Lon4] there holds

$$i(y, m) = m(i(y, 1) + 1) - 1, \quad \forall \, m \in \mathbf{N}, \tag{3.2}$$

where $i(y, m)$ for $m \in \mathbf{N}$ denotes the index of the m-th iteration of y defined by C. Conley and E. Zehnder in [CoZ1], Y. Long and E. Zehnder in [LZe1] and Y. Long in [Lon1] (cf. Chap. 5-6 of [Lon4]).

Note that for any closed characteristic (τ, y) on Σ there holds

$$i(y^m) = i(y, m) - n, \qquad \forall m \in \mathbf{N}, \tag{3.3}$$

by Theorem 15.1.1 of [Lon4]. Recall that here $i(y^m)$ is the Ekeland index defined below (2.6) and in (2.22). Hence a hyperbolic closed characteristic (τ, y) on Σ must possess integral mean index

$$\hat{i}(y) = i(y, 1) + 1 \in \mathbf{Z}. \tag{3.4}$$

Note that by Theorem 1.6 of [Lon3] and the main result of [HWZ1], Theorem 1.1 holds for $n = 2$. By Theorem 1.3 of [LoZ1] (cf. Theorem 15.5.2 of [Lon4]), there are at least $\left[\frac{n}{2}\right]$ geometrically distinct closed characteristics on Σ possessing irrational mean indices which are non-hyperbolic by (3.4). Therefore Theorem 1.1 holds for $n \geq 4$. Thus in order to prove Theorem 1.1, it suffices to consider next the case of $n = 3$ only.

Without loss of generality, by Theorem 1.3 of [LoZ1], we may assume that (τ_1, y_1) has irrational mean index. Hence by Theorem 8.3.1 of [Lon4], $M_1 \in \mathrm{Sp}(4)$ in (3.1) can be connected to $R(\theta_1) \diamond N_1$ within $\Omega^0(M_1)$ for some $\frac{\theta_1}{\pi} \notin \mathbf{Q}$ and $N_1 \in \mathrm{Sp}(2)$, where $R(\theta) = \begin{pmatrix} \cos\theta & -\sin\theta \\ \sin\theta & \cos\theta \end{pmatrix}$ for $\theta \in \mathbf{R}$. Here we use notation from Definition 1.8.5 and Theorem 1.8.10 of [Lon4]. By Theorem 2.7, the following identity holds

$$\frac{\hat{\chi}(y_1)}{\hat{i}(y_1)} + \sum_{2 \leq j \leq k} \frac{\hat{\chi}(y_j)}{\hat{i}(y_j)} = \frac{1}{2}. \tag{3.5}$$

Now we have the following four cases according to the classification of basic norm forms (cf. Definition 1.8.9 of [Lon4]).

Case 1. $N_1 = R(\theta_2)$ *with* $\frac{\theta_2}{\pi} \notin \mathbf{Q}$ *or* $N_1 = D(\pm 2) \equiv \begin{pmatrix} \pm 2 & 0 \\ 0 & \pm\frac{1}{2} \end{pmatrix}$.

In this case, by Theorems 8.1.6 and 8.1.7 of [Lon4], we have $\nu(y_1^m) \equiv 1$, i.e., y_1^m is non-degenerate for all $m \in \mathbf{N}$. Hence it follows from (2.26) that $\hat{\chi}(y_1) \neq 0$. Now (3.5) implies that at least one of the y_js for $2 \leq j \leq k$ must have irrational mean index. Then (3.4) implies that our theorem holds in this case.

Case 2. $N_1 = N_1(1, b)$ *with* $b = \pm 1, 0$.

We have two subcases according to the value of $\hat{\chi}(y_1)$.

Subcase 2.1. $\hat{\chi}(y_1) \neq 0$.

Here the same argument in the proof of Case 1 works in this subcase and yields the theorem.

Subcase 2.2. $\hat{\chi}(y_1) = 0$.

Note that by Theorems 8.1.4 and 8.1.7 of [Lon4] and our above Proposition 2.5, we have $K(y_1) = 1$. Since $\nu(y_1) \leq 3$, it follows from Proposition 2.6 and (2.25):

$$0 = \hat{\chi}(y_1) = (-1)^{i(y_1)}(k_0(y_1) - k_1(y_1) + k_2(y_1)). \qquad (3.6)$$

Now (iv) of Proposition 2.6 yields $k_l(y_1) = 0$ for all $l \in \mathbf{Z}$. Hence it follows from Proposition 2.3 and Definition 2.4 that

$$C_{S^1, q}(\Psi_a, S^1 \cdot u_1^p) = 0, \qquad \forall p \in \mathbf{N}, q \in \mathbf{Z}, \qquad (3.7)$$

where we denote by u_1 the critical point of Ψ_a corresponding to (τ_1, y_1). In other words, y_1^m is homologically invisible for all $m \in \mathbf{N}$.

Now to prove the theorem by contradiction we assume that (τ_j, y_j) is hyperbolic and then (3.2)-(3.4) hold for $2 \leq j \leq k$. Let

$$T = \prod_{2 \leq j \leq k} (i(y_j, 1) + 1) \equiv \prod_{2 \leq j \leq k} \hat{i}(y_j). \qquad (3.8)$$

Remark 3.1. Note that here when the hypersurface Σ is given, the k closed characteristics (τ_j, y_j) for $1 \leq j \leq k$ on Σ are uniquely determined. Therefore the above integer T is uniquely determined. Then we can choose the truncation level a in Section 2 to be large enough so that all the critical levels of the functional Ψ_a corresponding to iterations of y_js with Morse indices not greater than $2T$ are below I_a. Therefore our discussion is restricted to the correct range of the Hamiltonian function introduced in Section 2 and thus is meaningful.

Now by (3.2)-(3.4), we have

$$i(u_j^m) = i(y_j^m) = i(y_j, m) - 3 = m\hat{i}(y_j) - 4, \qquad \forall\, 2 \leq j \leq k. \qquad (3.9)$$

By Proposition 2.3, we have

$$C_{S^1, q}(\Psi_a, S^1 \cdot u_j^m) = \delta_{i(u_j^m)}^q \mathbf{Q}, \qquad \text{if } \hat{i}(y_j) \in 2\mathbf{Z}, \qquad (3.10)$$

$$C_{S^1, q}(\Psi_a, S^1 \cdot u_j^m) = \begin{cases} \delta_{i(u_j^m)}^q \mathbf{Q}, & \text{if } m \in 2\mathbf{N} - 1, \\ 0, & \text{otherwise} \end{cases} \qquad \text{if } \hat{i}(y_j) \in 2\mathbf{Z} + 1, \qquad (3.11)$$

for $q < I_a$ and $2 \leq j \leq k$. Here we denote by $\delta_p^q = 1$ if $p = q$ and $\delta_p^q = 0$ if $p \neq q$.

Next we compute the Morse type numbers M_q up to index $q = 2T - 3$. Note that here q measures the Ekeland index $i(y^m) = i(y, m) - 3$.

Note that $\hat{i}(y) \in \mathbf{Z}$ by (3.4) for $2 \leq j \leq k$. From $2T - 4 = i(y_j^{m_j}) = i(y_j, m_j) - 3 = m_j \hat{i}(y_j) - 4$, we obtain

$$m_j = \frac{2T}{\hat{i}(y_j)}.$$

Note that for each $2 \leq j \leq k$, we have $i(y_j^m) > 2T - 2$ when $m \geq m_j + 1$, which follows from (3.9) and the fact that $\hat{i}(y_j) > 2($ cf. the statement below (3.23)).

Hence we claim:

$$M_{2T-3} + M_{2T-5} + \cdots + M_3 + M_1 = \sum_{\substack{\hat{i}(y_j) \in 2\mathbf{Z}+1 \\ 2 \leq j \leq k}} \frac{T}{\hat{i}(y_j)}. \qquad (3.12)$$

In fact, here the left hand side of (3.12) counts the total sum of Morse type numbers M_q with odd $q \leq 2T - 3$ which is the total dimension of all non-trivial critical groups of order q contributed by all possible iterations of those (τ_j, y_j)s for $2 \leq j \leq k$. By (3.9), the Ekeland index $i(u_j^m)$ is always even when $\hat{i}(y_j) \in 2\mathbf{Z}$ for all $m \in \mathbf{N}$ and $2 \leq j \leq k$. Thus by (3.9)-(3.10) all (τ_j, y_j)s with $\hat{i}(y_j) \in 2\mathbf{Z}$ have no contribution to the left hand side of (3.12), and thus should not be counted in the right hand side of (3.12). Then the right hand side of (3.12) counts only the sum of the total dimensions of all non-trivial critical groups $C_{S^1, q}(\Psi_a, S^1 \cdot u_j^m)$ of order $q \leq 2T - 3$ contributed by iterations of each (τ_j, y_j) with $\hat{i}(y_j) \in 2\mathbf{Z} + 1$ for $2 \leq j \leq k$. Note that by (3.11), only odd order iterations of those y_js possess non-trivial critical groups. Hence in the right hand side of (3.12) only half of iterations of such y_js are counted. This proves the claim (3.12).

Similarly when $\hat{i}(y)$ is even, we obtain

$$M_{2T-4} + M_{2T-6} + \cdots + M_2 + M_0 = \sum_{\substack{\hat{i}(y_j) \in 2\mathbf{Z} \\ 2 \leq j \leq k}} \frac{2T}{\hat{i}(y_j)}. \qquad (3.13)$$

Note that here different from the right hand side of (3.12), in the right hand side of (3.13) all the iterations of each y contribute and are counted by (3.10).

Note that by (2.32), we have

$$b_{2T-3} - b_{2T-4} + b_{2T-5} + \cdots + b_3 - b_2 + b_1 - b_0 = -(T - 1). \qquad (3.14)$$

Therefore (3.7), (3.12)-(3.14) and Theorem 2.8 yield

$$\sum_{\substack{\hat{i}(y_j)\in 2\mathbf{Z}+1 \\ 2\leq j\leq k}} \frac{T}{\hat{i}(y_j)} - \sum_{\substack{\hat{i}(y_j)\in 2\mathbf{Z} \\ 2\leq j\leq k}} \frac{2T}{\hat{i}(y_j)} \geq -(T-1)\,. \tag{3.15}$$

On the other hand, by (2.26), (3.7) and (3.9), (3.5) becomes

$$\sum_{\substack{\hat{i}(y_j)\in 2\mathbf{Z} \\ 2\leq j\leq k}} \frac{1}{\hat{i}(y_j)} - \sum_{\substack{\hat{i}(y_j)\in 2\mathbf{Z}+1 \\ 2\leq j\leq k}} \frac{1}{2\hat{i}(y_j)} = \frac{1}{2}\,. \tag{3.16}$$

Now (3.15) and (3.16) yield

$$-T = \sum_{\substack{\hat{i}(y_j)\in 2\mathbf{Z}+1 \\ 2\leq j\leq k}} \frac{T}{\hat{i}(y_j)} - \sum_{\substack{\hat{i}(y_j)\in 2\mathbf{Z} \\ 2\leq j\leq k}} \frac{2T}{\hat{i}(y_j)} \geq -(T-1)\,. \tag{3.17}$$

This contradiction proves the theorem in the subcase 2.2.

Case 3. $N_1 = N_1(-1, 1)$.

In this case, by Theorems 8.1.4, 8.1.5 and 8.1.7 of [Lon4], we have

$$i(y_1, m) = mi(y_1, 1)+2E\left(\frac{m\theta_1}{2\pi}\right)-2, \ \nu(y_1, m) = 1+\frac{1+(-1)^m}{2}, \ \forall m \in \mathbf{N}\,,$$

with $i(y_1, 1) \in 2\mathbf{Z} + 1$. Hence $K(y_1) = 2$ by Proposition 2.5. We have $k_l(y_1) = \delta_0^l$ for all $l \in \mathbf{Z}$ by (2.17), (2.18) and (2.20). By (3.3), we have $i(y_1) = i(y_1, 1) - 3 \in 2\mathbf{Z}$ and $i(y_1^2) - i(y_1) = i(y_1, 2) - i(y_1, 1) \in 2\mathbf{Z} + 1$. Hence $k_0(y_1^2) = 0$ by (v) of Proposition 2.6. Then (2.25) implies

$$\hat{\chi}(y_1) = \frac{1 + k_1(y_1^2)}{2} \neq 0\,. \tag{3.18}$$

Now the proof of Case 1 yields the theorem.

Case 4. $N_1 = N_1(-1, b)$ *with* $b = 0, -1$ *or* $N_1 = R(\theta_2)$ *with* $\frac{\theta_2}{2\pi} = \frac{L}{N} \in \mathbf{Q}$ *with* $(L, N) = 1$.

Note first that if $N_1 = N_1(-1, b)$ with $b = 0, -1$, then Theorems 8.1.5 and 8.1.7 of [Lon4] imply that their index iteration formulae coincide with that of a rotational matrix $R(\theta)$ with $\theta = \pi$. Hence in the following we shall only consider the case $N_1 = R(\theta_2)$ with $\theta_2/\pi \in (0, 2) \cap \mathbf{Q}$. The same argument also shows that the theorem is true for $N_1 = N_1(-1, -1)$.

By Theorems 8.1.4 and 8.1.7 of [Lon4], we have

$$i(y_1, m) = m(i(y_1, 1) - 1) + 2E\left(\frac{m\theta_1}{2\pi}\right) + 2E\left(\frac{m\theta_2}{2\pi}\right) - 3, \qquad \forall m \in \mathbf{N}\,, \tag{3.19}$$

with $i(y_1, 1) \in 2\mathbf{Z} + 1$. It follows from Proposition 2.5 that $K(y_1) = N$. We have $k_l(y_1^m) = \delta_0^l$ for $1 \le m \le N - 1$ by (2.17), (2.18) and (2.20). By (3.3), we have $i(y_1) = i(y_1, 1) - 3 \in 2\mathbf{Z}$. Then (2.25) implies

$$\hat{\chi}(y_1) = \frac{N - 1 + k_0(y_1^N) - k_1(y_1^N) + k_2(y_1^N)}{N}. \tag{3.20}$$

This follows from $\nu(y_1^m) \le 3$ for all $m \in \mathbf{N}$.

We have two subcases according to the value of $\hat{\chi}(y_1)$.

Subcase 4.1. $\hat{\chi}(y_1) \ne 0$.

In this subcase, the proof of Case 1 yields the theorem.

Subcase 4.2. $\hat{\chi}(y_1) = 0$.

In this subcase, it follows from (3.20) and (iv) of Proposition 2.6 that

$$k_1(y_1^N) = N - 1. \tag{3.21}$$

Assuming that (τ_j, y_j) is hyperbolic for $2 \le j \le k$, next we prove the theorem by contradiction.

Using the common index jump theorem (Theorems 4.3 and 4.4 of [LoZ1], Theorems 11.2.1 and 11.2.2 of [Lon4]), we obtain some $(T, m_1, \ldots, m_k) \in \mathbf{N}^{k+1}$ such that $\frac{m_1\theta_2}{\pi} \in \mathbf{Z}$ (cf. (11.2.18) of [Lon4]) and the following hold by (11.2.6), (11.2.7) and (11.2.26) of [Lon4]:

$$i(y_j, 2m_j) \ge 2T - \frac{e(\gamma_j(\tau_j))}{2}, \tag{3.22}$$

$$i(y_j, 2m_j) + \nu(y_j, 2m_j) \le 2T + \frac{e(\gamma_j(\tau_j))}{2} - 1, \tag{3.23}$$

$$i(y_j, 2m_j + 1) \ge 2T + i(y_j, 1). \tag{3.24}$$

Note that $e(\gamma_1(\tau_1)) = 6$ and $e(\gamma_j(\tau_j)) = 2$ for $2 \le j \le k$. Since $\frac{m_1\theta_2}{\pi} \in \mathbf{Z}$, we have $\nu(y_1, 2m_1) = 3$. By the assumption of (τ_j, y_j) being hyperbolic, we have $\nu(y_j, 2m_j) = 1$ for $2 \le j \le k$. By Corollary 15.1.4 of [Lon4], we have $i(y_j, 1) \ge 3$ for $1 \le j \le k$. Hence (3.2), (3.19) and (3.22)-(3.24) yield

$$2T - 3 \le i(y_1, 2m_1) \le 2T + \frac{6}{2} - 1 - \nu(y_1, 2m_1) = 2T - 1, \tag{3.25}$$

$$i(y_j, 2m_j) = 2T - 1, \qquad \forall\, 2 \le j \le k, \tag{3.26}$$

$$i(y_j, 2m_j + m) \ge 2T + 3, \qquad \forall\, m \ge 1, 1 \le j \le k. \tag{3.27}$$

Note that here $i(y_1, 2m_1)$ is odd by (3.19). By Proposition 2.3, we have

$$C_{S^1, q}(\Psi_a, S^1 \cdot u_1^m) = \delta_{i(u_1^m)}^q \mathbf{Q}, \qquad \text{if} \quad m \neq 0 \, mod \, N, \qquad (3.28)$$

$$C_{S^1, q}(\Psi_a, S^1 \cdot u_1^m) = \delta_{i(u_1^m)+1}^q \mathbf{Q}^{k_1(y_1^N)}, \qquad \text{if} \quad m = 0 \, mod \, N, \quad (3.29)$$

for $q < I_a$ where I_a is determined by (2.28).

Note that here we need Remark 3.1 too to determine T, a and I_a respectively so that our discussion is meaningful.

Next we compute the Morse type numbers M_q for $q \leq 2T - 3$.

It follows from (3.9) and (3.26) as in our discussion for the subcase 2.2 that

$$2m_j = \frac{2T}{\hat{i}(y_j)}, \qquad \forall \, 2 \leq j \leq k. \qquad (3.30)$$

Since $\frac{m_1 \theta_2}{\pi} \in \mathbf{Z}$, we have $\frac{2m_1 L}{N} \in \mathbf{Z}$, and then $N | 2m_1$. Hence we claim:

$$M_{2T-3} + M_{2T-5} + \cdots + M_3 + M_1 = \sum_{\substack{\hat{i}(y_j) \in 2\mathbf{Z}+1 \\ 2 \leq j \leq k}} \frac{T}{\hat{i}(y_j)} + \frac{2m_1(N-1)}{N}. \quad (3.31)$$

In fact, here the left hand side of (3.31) counts the total sum of Morse type numbers M_q with odd $q \leq 2T - 3$. As in our discussion of (3.12), by (3.9)-(3.10) all (τ_j, y_j)s with $\hat{i}(y_j) \in 2\mathbf{Z}$ for $2 \leq j \leq k$ have no contribution to the left hand side of (3.31) and thus should not be counted in the right hand side of (3.31). The first summation term in the right hand side of (3.31) counts the sum of the total dimensions of all non-trivial critical groups $C_{S^1, q}(\Psi_a, S^1 \cdot u_j^m)$ of order $q \leq 2T - 3$ contributed by iterations of each (τ_j, y_j) with $\hat{i}(y_j) \in 2\mathbf{Z} + 1$ for $2 \leq j \leq k$ as in our discussion for (3.12). Note that by (3.9), (3.11) and (3.30), in the right hand side of (3.12) only half of iterations of such y_js are counted. The second term in the right hand side of (3.31) counts the sum of all the odd order non-trivial critical groups of iterations of u_1. By (3.19) with $i(y_1, 1) \in 2\mathbf{Z} + 1$ and (3.29), all degenerate iterations of u_1, i.e., those u_1^ms with $m = 0$ mod N, have no contribution to the left hand side of (3.31) and thus should not appear in its right hand side. By (3.19) and (3.28), all non-degenerate iterations of u_1 have odd indices. Here we separate corresponding iterations of u_1 into $2m_1/N$ pieces each of which has precisely N terms among which precisely $N - 1$ of them are non-degenerate. Therefore the claim (3.31) is proved.

Similarly for the total sum of Morse type numbers M_q with even $q \leq 2T - 3$, we obtain

$$M_{2T-4} + M_{2T-6} + \cdots + M_2 + M_0 = \sum_{\substack{\hat{i}(y_j) \in 2\mathbf{Z} \\ 2 \leq j \leq k}} \frac{2T}{\hat{i}(y_j)} + \frac{2m_1 k_1(y_1^N)}{N}. \quad (3.32)$$

Note that here the second term in the right hand side of (3.32) counts the sum of all the even order non-trivial critical groups of iterations of u_1. By (3.29) each of them corresponds to degenerate iteration of u_1 and is of dimension $k_1(y_1^N)$, and there are precisely $2m_1/N$ such terms.

On the other hand, by (2.32), we have

$$b_{2T-3} - b_{2T-4} + b_{2T-5} + \cdots + b_3 - b_2 + b_1 - b_0 = -(T - 1). \quad (3.33)$$

Therefore (3.21), (3.31)-(3.33) and Theorem 2.8 yield

$$\sum_{\substack{\hat{i}(y_j) \in 2\mathbf{Z}+1 \\ 2 \leq k \leq k}} \frac{T}{\hat{i}(y_j)} - \sum_{\substack{\hat{i}(y_j) \in 2\mathbf{Z} \\ 2 \leq k \leq k}} \frac{2T}{\hat{i}(y_j)} \geq -(T - 1). \quad (3.34)$$

By (2.26) and (3.9), (3.5) becomes

$$\sum_{\substack{\hat{i}(y_j) \in 2\mathbf{Z} \\ 2 \leq k \leq k}} \frac{1}{\hat{i}(y_j)} - \sum_{\substack{\hat{i}(y_j) \in 2\mathbf{Z}+1 \\ 2 \leq k \leq k}} \frac{1}{2\hat{i}(y_j)} = \frac{1}{2}. \quad (3.35)$$

Now (3.34) and (3.35) yield a contradiction as in (3.17) which proves our theorem in this subcase.

Combining all the cases together, the proof of our Theorem 1.1 is complete. $\square$

References

[BaL1]   V. Bangert and Y. Long, The existence of two closed geodesics on every Finsler 2-sphere (2005) Preprint.

[Cha1]   K. C. Chang, Infinite Dimensional Morse Theory and Multiple Solution Problems, Birkhäuser, Boston, 1993.

[CoZ1]   C. Conley and E. Zehnder, Morse-type index theory for flows and periodic solutions for Hamiltonian equations, *Comm. Pure. Appl. Math.* 37 (1984) 207–253.

[DDE1]   Dell'Antonio, G., B. D'Onofrio and I. Ekeland, Les systém hamiltoniens convexes et pairs ne sont pas ergodiques en general, *C. R. Acad. Sci. Paris*, Series I, 315 (1992) 1413–1415.

[Eke1] I. Ekeland, Une théorie de Morse pour les systèmes hamiltoniens convexes, *Ann. IHP. Anal. non Linéaire*, 1 (1984) 19–78.

[Eke2] I. Ekeland, An index theory for periodic solutions of convex Hamiltonian systems, *Proc. Symp. in Pure Math.* 45 (1986) 395–423.

[Eke3] I. Ekeland, Convexity Methods in Hamiltonian Mechanics, Springer-Verlag, Berlin, 1990.

[EkH1] I. Ekeland and H. Hofer, Convex Hamiltonian energy surfaces and their closed trajectories, *Comm. Math. Phys.* 113 (1987) 419–467.

[EkL1] I. Ekeland and L. Lassoued, Multiplicité des trajectoires fermées d'un systéme hamiltonien sur une hypersurface d'energie convexe, *Ann. IHP. Anal. non Linéaire*, 4 (1987) 1–29.

[GrM1] D. Gromoll and W. Meyer, On differentiable functions with isolated critical points, *Topology*, 8 (1969) 361–369.

[HWZ1] H. Hofer, K. Wysocki and E. Zehnder, The dynamics on three-dimensional strictly convex energy surfaces, *Ann. of Math.* 148 (1998) 197–289.

[LLZ1] C. Liu, Y. Long and C. Zhu, Multiplicity of closed characteristics on symmetric convex hypersurfaces in $\mathbf{R}^{2n}$, *Math. Ann.* 323 (2002) 201–215.

[Lon1] Y. Long, Maslov-type index, degenerate critical points and asymptotically linear Hamiltonian systems, *Science in China*, Series A, 33 (1990) 1409–1419.

[Lon2] Y. Long, Hyperbolic closed characteristics on compact convex smooth hypersurfaces in $\mathbf{R}^{2n}$, *J. Diff. Equa.* 150 (1998) 227–249.

[Lon3] Y. Long, Precise iteration formulae of the Maslov-type index theory and ellipticity of closed characteristics, *Advances in Math.* 154 (2000) 76–131.

[Lon4] Y. Long, Index Theory for Symplectic Paths with Applications. Progress in Math. 207, Birkhäuser, Basel, 2002.

[Lon5] Y. Long, Index iteration theory for symplectic paths with applications to nonlinear Hamiltonian systems. *Proc. of Inter. Congress of Math. 2002*, Vol. II, 303–313, Higher Edu. Press, Beijing, 2002.

[Lon6] Y. Long, Index iteration theory for symplectic paths and multiple periodic solution orbits, *Frontiers of Math. in China*, 1 (2006) 178–200.

[LZe1] Y. Long and E. Zehnder, Morse theory for forced oscillations of asymptotically linear Hamiltonian systems. In *Stoc. Proc. Phys. and Geom.*, S. Albeverio et al., ed. World Sci. (1990) 528–563.

[LoZ1] Y. Long and C. Zhu, Closed characteristics on compact convex hypersurfaces in $\mathbf{R}^{2n}$, *Ann. of Math.* 155 (2002) 317–368.

[MaW1] J. Mawhin and M. Willem, Critical Point Theory and Hamiltonian Systems, Springer, New York, 1989.

[Rab1] P. H. Rabinowitz, Periodic solutions of Hamiltonian systems, *Comm. Pure Appl. Math.* 31 (1978) 157–184.

[Szu1] A. Szulkin, Morse theory and existence of periodic solutions of convex Hamiltonian systems, *Bull. Soc. Math. France*, 116 (1988) 171–197.

[Vit1] C. Viterbo, Equivariant Morse theory for starshaped Hamiltonian systems, *Trans. Amer. Math. Soc.* 311 (1989) 621–655.

[WHL1] W. Wang, X. Hu and Y. Long, Resonance identity, stability and multiplicity of closed characteristics on compact convex hypersurfaces (2006) Preprint.

[Wei1] A. Weinstein, Periodic orbits for convex Hamiltonian systems, *Ann. of Math.* 108 (1978) 507–518.

(From left to right) Lei FU (NIM), Zaijiu SHANG (Academia Sinica), Chaofeng ZHU (NIM), Chungen LIU (Nankai University), Weiping ZHANG (NIM), S.-S. CHERN, John MATHER (Princeton), Yiming LONG (NIM).

Chapter 13

η-INVARIANT AND FLAT VECTOR BUNDLES II

Xiaonan Ma

Centre de Mathématiques Laurent Schwartz,
UMR 7640 du CNRS, École Polytechnique,
91128 Palaiseau Cedex, France
ma@math.polytechnique.fr

Weiping Zhang

Chern Institute of Mathematics & LPMC,
Nankai University, Tianjin 300071, P. R. China
weiping@nankai.edu.cn

We first apply the method and results in the previous paper to give a new proof of a result (holds in $\mathbf{C}/\mathbf{Z}$) of Gilkey on the variation of η-invariants associated to non self-adjoint Dirac type operators. We then give an explicit local expression of certain η-invariant appearing in recent papers of Braverman-Kappeler on what they call refined analytic torsion, and propose an alternate formulation of their definition of the refined analytic torsion. A refinement in $\mathbf{C}$ of the above variation formula is also proposed.

Keywords: Flat vector bundle, η-invariant, refined analytic torsion.

2000 MR Subject Classification: 58J

1. Introduction

In a previous paper,[14] we have given an alternate formulation of (the mod $\mathbf{Z}$ part of) the η-invariant of Atiyah-Patodi-Singer[1–3] associated to non-unitary flat vector bundles by identifying explicitly its real and imaginary parts.

On the other hand, Gilkey has studied this kind of η-invariants systematically in,[13] and in particular proved a general variation formula for them.

However, it lacks in[13] the identification of the real and imaginary parts of the η-invariants as we did in.[14]

In this article, we first show that our results in[14] lead to a direct derivation of Gilkey's variation formula Theorem 3.7.[13]

The second purpose of this paper is to apply the results in[14] to examine the η-invariants appearing in the recent papers of Braverman-Kappeler[7-9] on refined analytic torsions. We show that the imaginary part of the η-invariant appeared in these articles admits an explicit local expression which suggests an alternate formulation of the definition of the refined analytic torsion there. This reformulation provides an analytic resolution of a problem due to Burghelea[10,11] on the existence of a univalent holomorphic function on the representation space having the Ray-Singer analytic torsion as its absolute value.

Finally, using the extension (to the case of non-self-adjoint operators) given in[18] of the concept of spectral flow,[3] we propose a refinement in **C** of the above variation formula for η-invariants.

Acknowledgements. We would like to thank Maxim Braverman for bringing[13] to our attention, and for helpful discussions. The work of the second author was partially supported by the National Natural Science Foundation of China.

2. η-Invariant and the Variation Formula

Let M be an odd dimensional oriented closed spin manifold carrying a Riemannian metric g^{TM}. Let $S(TM)$ be the associated Hermitian vector bundle of spinors. Let (E, g^E) be a Hermitian vector bundle over M carrying a unitary connection ∇^E. Moreover, let (F, g^F) be a Hermitian vector bundle over M carrying a flat connection ∇^F. We do not assume that ∇^F preserves the Hermitian metric g^F on F.

Let $D^{E \otimes F} : \Gamma(S(TM) \otimes E \otimes F) \longrightarrow \Gamma(S(TM) \otimes E \otimes F)$ denote the corresponding (twisted) Dirac operator.

It is pointed out in Page 93[3] that one can define the reduced η-invariant of $D^{E \otimes F}$, denoted by $\overline{\eta}(D^{E \otimes F})$, by working on (possibly) non-self-adjoint elliptic operators.

In this section, we will first recall the main result in Ref. 14 on $\overline{\eta}(D^{E \otimes F})$ and then show how it leads directly to a proof of the variation formula of Gilkey, Theorem 3.7.[13]

2.1. Chern-Simons classes and flat vector bundles

We fix a square root of $\sqrt{-1}$ and let $\varphi : \Lambda(T^*M) \to \Lambda(T^*M)$ be the homomorphism defined by $\varphi : \omega \in \Lambda^i(T^*M) \to (2\pi\sqrt{-1})^{-i/2}\omega$. The formulas in what follows will not depend on the choice of the square root of $\sqrt{-1}$.

If W is a complex vector bundle over M and ∇_0^W, ∇_1^W are two connections on W. Let W_t, $0 \leq t \leq 1$, be a smooth path of connections on W connecting ∇_0^W and ∇_1^W. We define the Chern-Simons form $CS(\nabla_0^W, \nabla_1^W)$ to be the differential form given by

$$CS\left(\nabla_0^W, \nabla_1^W\right) = -\left(\frac{1}{2\pi\sqrt{-1}}\right)^{\frac{1}{2}} \varphi \int_0^1 \mathrm{Tr}\left[\frac{\partial \nabla_t^W}{\partial t} \exp\left(-\left(\nabla_t^W\right)^2\right)\right] dt. \tag{2.1}$$

Then (cf. Chapter 1[17])

$$dCS\left(\nabla_0^W, \nabla_1^W\right) = \mathrm{ch}\left(W, \nabla_1^W\right) - \mathrm{ch}\left(W, \nabla_0^W\right). \tag{2.2}$$

Moreover, it is well-known that up to exact forms, $CS(\nabla_0^W, \nabla_1^W)$ does not depend on the path of connections on W connecting ∇_0^W and ∇_1^W.

Let (F, ∇^F) be a flat vector bundle carrying the flat connection ∇^F. Let g^F be a Hermitian metric on F. We do not assume that ∇^F preserves g^F. Let $(\nabla^F)^*$ be the adjoint connection of ∇^F with respect to g^F.

From (4.1), (4.2)[6] and §1(g),[5] one has

$$\left(\nabla^F\right)^* = \nabla^F + \omega\left(F, g^F\right) \tag{2.3}$$

with

$$\omega\left(F, g^F\right) = \left(g^F\right)^{-1}\left(\nabla^F g^F\right). \tag{2.4}$$

Then

$$\nabla^{F,e} = \nabla^F + \frac{1}{2}\omega\left(F, g^F\right) \tag{2.5}$$

is a Hermitian connection on (F, g^F) (cf. (4.3)[6]).

Following (2.6)[14] and (2.47),[15] for any $r \in \mathbf{C}$, set

$$\nabla^{F,e,(r)} = \nabla^{F,e} + \frac{\sqrt{-1}r}{2}\omega\left(F, g^F\right). \tag{2.6}$$

Then for any $r \in \mathbf{R}$, $\nabla^{F,e,(r)}$ is a Hermitian connection on (F, g^F).

On the other hand, following (0.2),[5] for any integer $j \geq 0$, let $c_{2j+1}(F, g^F)$ be the Chern form defined by

$$c_{2j+1}\left(F, g^F\right) = \left(2\pi\sqrt{-1}\right)^{-j} 2^{-(2j+1)} \mathrm{Tr}\left[\omega^{2j+1}\left(F, g^F\right)\right]. \tag{2.7}$$

Then $c_{2j+1}(F, g^F)$ is a closed form on M. Let $c_{2j+1}(F)$ be the associated cohomology class in $H^{2j+1}(M, \mathbf{R})$, which does not depend on the choice of g^F.

For any $j \geq 0$ and $r \in \mathbf{R}$, let $a_j(r) \in \mathbf{R}$ be defined as

$$a_j(r) = \int_0^1 \left(1 + u^2 r^2\right)^j du \,. \tag{2.8}$$

With these notation we can now state the following result first proved in Lemma 2.12.[15]

Proposition 1: *The following identity in $H^{\mathrm{odd}}(M, \mathbf{R})$ holds for any $r \in \mathbf{R}$,*

$$CS\left(\nabla^{F,e}, \nabla^{F,e,(r)}\right) = -\frac{r}{2\pi} \sum_{j=0}^{+\infty} \frac{a_j(r)}{j!} c_{2j+1}(F) \,. \tag{2.9}$$

2.2. η-invariant associated to flat vector bundles

Let

$$D^{E \otimes F, e} : \Gamma(S(TM) \otimes E \otimes F) \longrightarrow \Gamma(S(TM) \otimes E \otimes F) \tag{2.10}$$

denote the Dirac operator associated to the connection $\nabla^{F,e}$ on F and ∇^E on E. Then $D^{E \otimes F, e}$ is formally self-adjoint and one can define the associated reduced η-invariant as in.[1]

In view of Proposition 1, one can restate the main result of,[14] which is Theorem 2.2,[14] as follows,

$$\overline{\eta}\left(D^{E \otimes F}\right) \equiv \overline{\eta}\left(D^{E \otimes F, e}\right) + \int_M \widehat{A}(TM)\mathrm{ch}(E) CS\left(\nabla^{F,e}, \nabla^F\right) \quad \mathrm{mod}\ \mathbf{Z}, \tag{2.11}$$

where $\widehat{A}(TM)$ and $\mathrm{ch}(E)$ are the $\widehat{A}$ class of TM and the Chern character of E respectively.[17]

Now let $\widetilde{\nabla}^F$ be another flat connection on F. We use the notation with $\sim$ to denote the objects associated with this flat connection.

Then one has

$$\overline{\eta}\left(\widetilde{D}^{E \otimes F}\right) \equiv \overline{\eta}\left(\widetilde{D}^{E \otimes F, e}\right) + \int_M \widehat{A}(TM)\mathrm{ch}(E) CS\left(\widetilde{\nabla}^{F,e}, \widetilde{\nabla}^F\right) \quad \mathrm{mod}\ \mathbf{Z}. \tag{2.12}$$

By the variation formula for η-invariants associated to self-adjoint Dirac operators,[1,4] one knows that

$$\overline{\eta}\left(\widetilde{D}^{E\otimes F,e}\right) - \overline{\eta}\left(D^{E\otimes F,e}\right) \equiv \int_M \widehat{A}(TM)\mathrm{ch}(E)CS\left(\nabla^{F,e}, \widetilde{\nabla}^{F,e}\right) \quad \mathrm{mod}\ \mathbf{Z}.$$
(2.13)

From (2.11)–(2.13), one deduces that

$$\overline{\eta}\left(\widetilde{D}^{E\otimes F}\right) - \overline{\eta}(D^{E\otimes F}) \equiv \int_M \widehat{A}(TM)\mathrm{ch}(E)CS\left(\nabla^{F,e}, \widetilde{\nabla}^{F,e}\right) \quad (2.14)$$

$$- \int_M \widehat{A}(TM)\mathrm{ch}(E)CS\left(\nabla^{F,e}, \nabla^F\right) + \int_M \widehat{A}(TM)\mathrm{ch}(E)CS\left(\widetilde{\nabla}^{F,e}, \widetilde{\nabla}^F\right)$$

$$= \int_M \widehat{A}(TM)\mathrm{ch}(E)CS\left(\nabla^F, \widetilde{\nabla}^F\right) \quad \mathrm{mod}\ \mathbf{Z},$$

which is exactly the Gilkey formula, Theorem 1.6[13] for the operator $P = D^E$ therein.

Remark 2: As was indicated in Remark 2.4,[14] the main result in[14] holds also for general Hermitian vector bundles equipped with a (possibly) non-Hermitian connection. Indeed, if we do not assume that ∇^F is flat, then at least (2.3)-(2.6) still holds. Thus for any $r \in \mathbf{R}$, we have well-defined (formally self-adjoint) operator $D^{E\otimes F}(r)$ which is associated to the Hermitian connection $\nabla^{F,e,(r)}$ on F. For any $r \in \mathbf{R}$, one then has the variation formula[1,4]

$$\overline{\eta}\left(D^{E\otimes F}(r)\right) - \overline{\eta}(D^{E\otimes F,e}) \equiv \int_M \widehat{A}(TM)\mathrm{ch}(E)CS\left(\nabla^{F,e}, \nabla^{F,e,(r)}\right) \quad \mathrm{mod}\ \mathbf{Z}.$$
(2.15)

By (2.1), one sees easily that the right hand side of (2.15) is a holomorphic function (indeed a polynomial) of r. Thus, by analytic continuity, as in,[14] one gets that for any $r \in \mathbf{C}$, (2.15) still holds. In particular, if we set $r = \sqrt{-1}$, we get

$$\overline{\eta}\left(D^{E\otimes F}\right) \equiv \overline{\eta}\left(D^{E\otimes F,e}\right) + \int_M \widehat{A}(TM)\mathrm{ch}(E)CS\left(\nabla^{F,e}, \nabla^F\right) \quad \mathrm{mod}\ \mathbf{Z},$$
(2.16)

which generalizes (2.11). Then by proceeding as above, we see that (2.14) holds without the assumption of the flatness of connections ∇^F and $\widetilde{\nabla}^F$.

By (2.1) and (2.6),

$$
CS\left(\nabla^{F,e}, \nabla^{F,e,(r)}\right) = \frac{-1}{2\pi} \int_0^1 \mathrm{Tr}\left[\frac{r}{2}\omega\left(F, g^F\right)\right.
$$
$$
\left. \times \exp\left(\frac{-1}{2\pi\sqrt{-1}}\left(\nabla^{F,e,(tr)}\right)^2\right)\right] dt
$$
$$
= \sum_{i=0}^{\dim M} a_i\left(\nabla^F, g^F\right) r^i . \tag{2.17}
$$

By (2.6), one has

$$
\left(\nabla^{F,e,(r)}\right)^2 = \left(\nabla^{F,e}\right)^2 + \frac{\sqrt{-1}r}{2}\left(\nabla^{F,e}\omega\left(F, g^F\right)\right) - \frac{r^2}{4}\left(\omega\left(F, g^F\right)\right)^2 . \tag{2.18}
$$

Note that

$$
\nabla^{F,e}\omega\left(F, g^F\right) = \left[\nabla^{F,e}, \omega\left(F, g^F\right)\right] = 0, \quad \text{if } \nabla^F \text{ is flat.} \tag{2.19}
$$

By taking adjoint of (2.18), we see that when $r \in \mathbf{C}$ is purely imaginary, one has

$$
\left(\frac{1}{2\pi\sqrt{-1}}\left(\nabla^{F,e,(r)}\right)^2\right)^* = \frac{1}{2\pi\sqrt{-1}}\left(\left(\nabla^{F,e}\right)^2 - \frac{r^2}{4}\left(\omega\left(F, g^F\right)\right)^2\right)
$$
$$
- \left(\frac{1}{2\pi\sqrt{-1}}\right)\frac{\sqrt{-1}r}{2}\left(\nabla^{F,e}\omega\left(F, g^F\right)\right) . \tag{2.20}
$$

From (2.1), (2.17) and (2.20), one sees that when $r \in \mathbf{C}$ is purely imaginary, then

$$
\mathrm{Re}\left(CS\left(\nabla^{F,e}, \nabla^{F,e,(r)}\right)\right) = \sum_{i \text{ even}} a_i\left(\nabla^F, g^F\right) r^i ,
$$
$$
\mathrm{Im}\left(CS\left(\nabla^{F,e}, \nabla^{F,e,(r)}\right)\right) = \frac{1}{\sqrt{-1}}\sum_{i \text{ odd}} a_i\left(\nabla^F, g^F\right) r^i . \tag{2.21}
$$

Thus when $r \in \mathbf{C}$ is purely imaginary, from (2.16) and (2.21), we have

$$
\mathrm{Re}(\bar{\eta}(D^{E\otimes F}(r))) \equiv \bar{\eta}(D^{E\otimes F,e}) + \sum_{i \text{ even}} r^i \int_M \widehat{A}(TM)\mathrm{ch}(E)a_i(\nabla^F, g^F) \bmod \mathbf{Z},
$$
$$
\mathrm{Im}\left(\bar{\eta}\left(D^{E\otimes F}(r)\right)\right) = \frac{1}{\sqrt{-1}}\sum_{i \text{ odd}} r^i \int_M \widehat{A}(TM)\mathrm{ch}(E)a_i\left(\nabla^F, g^F\right) . \tag{2.22}
$$

In particular, by setting $r = \sqrt{-1}$, we get

$$\mathrm{Re}\left(\overline{\eta}(D^{E\otimes F})\right) \equiv \overline{\eta}\left(D^{E\otimes F,e}\right)$$

$$+ \sum_{i \text{ even}} (-1)^{\frac{i}{2}} \int_M \widehat{A}(TM)\mathrm{ch}(E)a_i(\nabla^F, g^F) \mod \mathbf{Z},$$

$$\mathrm{Im}\left(\overline{\eta}\left(D^{E\otimes F}\right)\right) = \sum_{i \text{ odd}} (-1)^{\frac{i-1}{2}} \int_M \widehat{A}(TM)\mathrm{ch}(E)a_i\left(\nabla^F, g^F\right). \qquad (2.23)$$

This generalizes the main result in Ref. 14.

3. η-Invariant and the Refined Analytic Torsion of Braverman-Kappeler

Recently, in a series of preprints,[7–9] Braverman and Kappeler introduce what they call refined analytic torsion. The η-invariant associated with flat vector bundles plays a role in their definition. In this section, we first examine the imaginary part of the η-invariant appearing in,[7–9] from the point of view of the previous sections and propose an alternate definition of the refined analytic torsion. We then combine this refined analytic torsion with the η-invariant to construct analytically a univalent holomorphic function on the space of representations of $\pi_1(M)$ having the absolute value equals to the Ray-Singer torsion, thus resolving a problem posed by Burghelea.[11]

3.1. η-*invariant and the refined analytic torsion of Braverman-Kappeler*

Since there needs no spin condition in,[7–9] here we start with a closed oriented smooth odd dimensional manifold M with dim $M = 2n+1$. Let g^{TM} be a Riemannian metric on TM. For any $X \in TM$, let $X^* \in T^*M$ denote its metric dual and $c(X) = X^* \wedge -i_X$ denote the associated Clifford action acting on $\Lambda^*(T^*M)$, where $X^* \wedge$ and i_X are the notation for the exterior and interior multiplications of X respectively.

Let $e_1, \ldots, e_{2n+1}$ be an oriented orthonormal basis of TM. Set

$$\Gamma = \left(\sqrt{-1}\right)^{n+1} c(e_1) \cdots c(e_{2n+1}). \qquad (3.1)$$

Then $\Gamma^2 = \mathrm{Id}$ on $\Lambda^*(T^*M)$.

Let (F, g^F) be a Hermitian vector bundle over M equipped with a flat connection ∇^F which need not preserve the Hermitian metric g^F on F. Then the exterior differential d on $\Omega^*(M) = \Gamma(\Lambda^*(T^*M))$ extends naturally to the twisted exterior differential d^F acting on $\Omega^*(M, F) = \Gamma(\Lambda^*(T^*M) \otimes F)$.

We define the twisted signature operator D_{Sig}^F to be

$$D_{\text{Sig}}^F = \frac{1}{2}\left(\Gamma d^F + d^F\Gamma\right) : \Omega^{\text{even}}(M, F) \to \Omega^{\text{even}}(M, F). \qquad (3.2)$$

It coincides with the odd signature operator $\frac{1}{2}\mathcal{B}_{\text{even}}$ in.[7-9]

Let $\nabla^{\Lambda^{\text{even}}(T^*M) \otimes F}$ (resp. $\nabla^{\Lambda^{\text{even}}(T^*M) \otimes F, e}$) be the tensor product connections on $\Lambda^{\text{even}}(T^*M) \otimes F$ obtained from ∇^F (resp. $\nabla^{F,e}$) and the canonical connection on $\Lambda^{\text{even}}(T^*M)$ induced by the Levi-Civita connection ∇^{TM} of g^{TM}.

From (3.2), it is easy to verify that

$$D_{\text{Sig}}^F = \Gamma\left(\sum_{i=1}^{2n+1} c(e_i)\nabla_{e_i}^{\Lambda^{\text{even}}(T^*M) \otimes F}\right). \qquad (3.3)$$

Set

$$D_{\text{Sig}}^{F,e} = \Gamma\left(\sum_{i=1}^{2n+1} c(e_i)\nabla_{e_i}^{\Lambda^{\text{even}}(T^*M) \otimes F, e}\right). \qquad (3.4)$$

Then $D_{\text{Sig}}^{F,e}$ is formally self-adjoint.

Since locally one has identification $S(TM) \otimes S(TM) = \Lambda^{\text{even}}(T^*M)$, one sees that one can apply the results in the previous section to the case $E = S(TM)$ to the current situation.

In particular, we get

$$\text{Re}\left(\overline{\eta}\left(D_{\text{Sig}}^F\right)\right) \equiv \overline{\eta}\left(D_{\text{Sig}}^{F,e}\right) \mod \mathbf{Z},$$

$$\text{Im}\left(\overline{\eta}\left(D_{\text{Sig}}^F\right)\right) = \frac{1}{\sqrt{-1}}\int_M \text{L}\left(TM, \nabla^{TM}\right) CS\left(\nabla^{F,e}, \nabla^F\right)$$

$$= -\frac{1}{2\pi}\int_M \text{L}(TM)\sum_{j=0}^{+\infty}\frac{2^{2j}j!}{(2j+1)!}c_{2j+1}(F), \qquad (3.5)$$

where $\text{L}(TM, \nabla^T M)$ is the Hirzebruch L-form defined by

$$\text{L}\left(TM, \nabla^T M\right) = \varphi \det^{1/2}\left(\frac{R^{TM}}{\tanh\left(R^{TM}/2\right)}\right), \qquad (3.6)$$

with $R^{TM} = (\nabla^{TM})^2$ the curvature of ∇^{TM}, and $\text{L}(TM)$ is the associated class.

Remark 3: By proceeding as in Section 2, we can get Theorem 3.7[13] easily by using the results in Remark 2.

Proposition 4: *The function*

$$\Psi\left(F, \nabla^F\right) = \mathrm{Im}\left(\bar{\eta}\left(D_{\mathrm{Sig}}^F\right)\right) + \frac{1}{2\pi}\int_M \mathrm{L}(TM)c_1(F) \qquad (3.7)$$

is a locally constant function on the set of flat connections on F. In particular, $\Psi(F, \nabla^F) = 0$ if ∇^F can be connected to a unitary flat connection through a path of flat connections.

Proof. Let ∇_t^F, $0 \le t \le 1$, be a smooth pass of flat connections on F.
From (3.5), we get

$$\sqrt{-1}\mathrm{Im}\left(\bar{\eta}\left(D_{\mathrm{Sig},1}^F\right)\right) - \sqrt{-1}\mathrm{Im}\left(\bar{\eta}\left(D_{\mathrm{Sig},0}^F\right)\right)$$

$$= \int_M \mathrm{L}\left(TM, \nabla^{TM}\right)CS\left(\nabla_1^{F,e}, \nabla_1^F\right) - \int_M \mathrm{L}\left(TM, \nabla^{TM}\right)CS\left(\nabla_0^{F,e}, \nabla_0^F\right)$$

$$= \sqrt{-1}\int_M \mathrm{L}\left(TM, \nabla^{TM}\right)\mathrm{Im}\left(CS\left(\nabla_1^{F,e}, \nabla_0^{F,e}\right) - CS\left(\nabla_1^F, \nabla_0^F\right)\right)$$

$$= \sqrt{-1}\int_M \mathrm{L}\left(TM, \nabla^{TM}\right)\mathrm{Im}\left(CS\left(\nabla_0^F, \nabla_1^F\right)\right). \qquad (3.8)$$

Now consider the path of flat connections ∇_t^F, $0 \le t \le 1$. Since for any $t \in [0,1]$, $(\nabla_t^F)^2 = 0$, from (2.1), (2.5), one gets

$$CS\left(\nabla_0^F, \nabla_1^F\right) = \left(\frac{1}{2\pi\sqrt{-1}}\right)\mathrm{Tr}\left(\nabla_0^F - \nabla_1^F\right) = \left(\frac{1}{2\pi\sqrt{-1}}\right)\mathrm{Tr}\left(\nabla_0^{F,e} - \nabla_1^{F,e}\right)$$

$$- \left(\frac{1}{2\pi\sqrt{-1}}\right)\mathrm{Tr}\left(\frac{1}{2}\omega_0(F, g^F) - \frac{1}{2}\omega_1(F, g^F)\right). \qquad (3.9)$$

Thus, one has

$$\sqrt{-1}\mathrm{Im}\left(CS\left(\nabla_0^F, \nabla_1^F\right)\right) = -\frac{1}{2\pi\sqrt{-1}}\mathrm{Tr}\left(\frac{1}{2}\omega_0(F, g^F) - \frac{1}{2}\omega_1(F, g^F)\right)$$

$$= -\frac{1}{2\pi\sqrt{-1}}\left(c_1\left(F, \nabla_0^F\right) - c_1\left(F, \nabla_1^F\right)\right). \qquad (3.10)$$

From (3.8) and (3.10), we get

$$\mathrm{Im}\left(\bar{\eta}\left(D_{\mathrm{Sig},1}^F\right)\right) + \frac{1}{2\pi}\int_M \mathrm{L}\left(TM, \nabla^{TM}\right)c_1\left(F, \nabla_1^F\right)$$

$$= \mathrm{Im}\left(\bar{\eta}\left(D_{\mathrm{Sig},0}^F\right)\right) + \frac{1}{2\pi}\int_M \mathrm{L}\left(TM, \nabla^{TM}\right)c_1\left(F, \nabla_0^F\right), \qquad (3.11)$$

from which Proposition 4 follows. Q.E.D.

Remark 5: Formula (3.11) is closely related to Theorem 12.3.[7] Moreover, for any representation α of the fundamental group $\pi_1(M)$, let $(F_\alpha, \nabla^{F_\alpha})$ be the associated flat vector bundle. One has

$$\exp\left(\pi\Psi\left(F_\alpha, \nabla^{F_\alpha}\right)\right) = r(\alpha), \qquad (3.12)$$

where $r(\alpha)$ is the function appearing in Lemma 5.5.[9] While from (3.5) and (3.7), one has

$$\Psi\left(F, \nabla^F\right) = -\frac{1}{2\pi}\int_M L(TM)\sum_{j=1}^{+\infty}\frac{2^{2j}j!}{(2j+1)!}c_{2j+1}(F). \qquad (3.13)$$

Combining with (3.12), this gives an explicit local expression of $r(\alpha)$ as well as the locally constant function r_C defined in Definition 5.6.[9]

Remark 6: To conclude this subsection, we recall the recent modification due to Braverman-Kappeler (Braverman mentioned this in a recent Oberwolfach conference) themselves of the original definition of the refined analytic torsion in[7-9] as follows: for any Hermitian vector bundle equipped with a flat connection ∇^F over an oriented closed smooth odd dimensional manifold M equipped with a Riemannian metric g^{TM}, let $\rho(\nabla^F, g^{TM})$ be the element defined in (2.13).[9] Then the modified definition of the refined analytic torsion is given by

$$\rho'_{\mathrm{an}}\left(\nabla^F, g^{TM}\right) = \rho\left(\nabla^F, g^{TM}\right)e^{\pi\sqrt{-1}\mathrm{rk}(F)\overline{\eta}(D_{\mathrm{sig}})}, \qquad (3.14)$$

where $\overline{\eta}(D_{\mathrm{sig}})$ is the reduced η invariant in the sense of Atiyah-Patodi-Singer[1] of the signature operator coupled with the trivial complex line bundle over M (i.e. $D_{\mathrm{sig}} := D_{\mathrm{sig}}^{\mathbf{C}}$). There are two advantages of this reformulation. First, by multiplying the local factor $e^{-\pi\Psi(F,\nabla^F)}$ makes the comparison formula [9, (5.8)] of the refined analytic torsion has closer resemblance in comparing with the formulas of Cheeger-Müller and Bismut-Zhang (cf.[6]). The advantage of this reformulation is that since $\overline{\eta}(D_{\mathrm{sig}})$ various smoothly with respect to the metric g^{TM} (as the dimension of $\ker(D_{\mathrm{sig}})$ does not depend on the metric g^{TM}), the ambiguity of the power of $\sqrt{-1}$ disappears if one uses $e^{\pi\sqrt{-1}\mathrm{rk}(F)\overline{\eta}(D_{\mathrm{sig}})}$ to replace the factor $e^{\frac{\pi\sqrt{-1}\mathrm{rk}(F)}{2}\int_N L(p,g^M)}$ in (2.14).[9]

3.2. Ray-Singer analytic torsion and univalent holomorphic functions on the representation space

Let (F, ∇^F) be a complex flat vector bundle. Let g^F be an Hermitian metric on F. We fix a flat connection $\widetilde{\nabla}^F$ on F (note here that we do not assume that ∇^F and $\widetilde{\nabla}^F$ can be connected by a smooth path of flat connections).

Let g^{TM} be a Riemannian metric on TM and ∇^{TM} be the associated Levi-Civita connection.

Let $\widetilde{\eta}(\nabla^F, \widetilde{\nabla}^F) \in \mathbf{C}$ be defined by

$$\widetilde{\eta}\left(\nabla^F, \widetilde{\nabla}^F\right) = \int_M \mathrm{L}\left(TM, \nabla^{TM}\right) CS\left(\widetilde{\nabla}^{F,e}, \nabla^F\right). \tag{3.15}$$

One verifies easily that $\widetilde{\eta}(\nabla^F, \widetilde{\nabla}^F) \in \mathbf{C}$ does not depend on g^{TM}, and is a holomorphic function of ∇^F. Moreover, by (3.5) one has

$$\mathrm{Im}\left(\widetilde{\eta}\left(\nabla^F, \widetilde{\nabla}^F\right)\right) = \mathrm{Im}\left(\overline{\eta}\left(D_{\mathrm{Sig}}^F\right)\right). \tag{3.16}$$

Recall that the refined analytic torsion of[7-9] has been modified in (3.14).

Set

$$\mathcal{T}_{\mathrm{an}}\left(\nabla^F, g^{TM}\right) = \rho'_{\mathrm{an}}\left(\nabla^F, g^{TM}\right) \exp\left(\sqrt{-1}\pi\widetilde{\eta}\left(\nabla^F, \widetilde{\nabla}^F\right)\right). \tag{3.17}$$

Then $\mathcal{T}_{\mathrm{an}}(\nabla^F, g^{TM})$ is a holomorphic section in the sense of Definition 3.4.[9]

By Theorem 11.3[8] (cf. (5.13)[9]), (3.14), (3.16) and (3.17), one gets the following formula for the Ray-Singer norm of $\mathcal{T}_{\mathrm{an}}(\nabla^F, g^{TM})$,

$$\left\|\mathcal{T}_{\mathrm{an}}\left(\nabla^F, g^{TM}\right)\right\|^{\mathrm{RS}} = 1. \tag{3.18}$$

In particular, when restricted to the space of acyclic representations, $\mathcal{T}_{\mathrm{an}}(\nabla^F, g^{TM})$ becomes a (univalent) holomorphic function such that

$$\left|\mathcal{T}_{\mathrm{an}}\left(\nabla^F, g^{TM}\right)\right| = T^{RS}(\nabla^F), \tag{3.19}$$

the usual Ray-Singer analytic torsion. This provides an analytic resolution of a question of Burghelea.[11]

Remark 7: If one considers $\mathcal{T}_{\mathrm{an}}^2$, then one can further modify it to

$$\mathcal{T}_{\mathrm{an}}^2\left(\nabla^F, g^{TM}\right)' = \mathcal{T}_{\mathrm{an}}^2\left(\nabla^F, g^{TM}\right)$$
$$\times \exp\left(2\pi\sqrt{-1}\left(\overline{\eta}\left(\widetilde{D}_{\mathrm{Sig}}^{F,e}\right) - \mathrm{rk}(F)\overline{\eta}\left(D_{\mathrm{Sig}}\right)\right)\right), \tag{3.20}$$

which does not depend on the choice of $\widetilde{\nabla}^F$, and thus gives an intrinsic definition of a holomorphic section of the square of the determinant line bundle, having the same norm as that of $\mathcal{T}_{\mathrm{an}}^2(\nabla^F, g^{TM})$. The dependence of $\mathcal{T}_{\mathrm{an}}$ on α indicates in part the subtleness of the analytic meaning of the phase of the Turaev torsion (cf.[12,16]).

Next, we show how to modify the Turaev torsion[12,16] to get a holomorphic section with Ray-Singer norm equal to one.

Let ε be an Euler structure on M and $\mathbf{o}$ a cohomological orientation. We use the notation as in[9] to denote the associated Turaev torsion by $\rho_{\varepsilon,\mathbf{o}}$.

Let $c(\varepsilon) \in H_1(M, \mathbf{Z})$ be the canonical class associated to the Euler structure ε (cf.[16] or Section 5.2[12]). Then for any representation α_F corresponding to a flat vector bundle (F, ∇^F), by Theorem 10.2[12] one has

$$\|\rho_{\varepsilon,\mathbf{o}}(\alpha_F)\|^{\mathrm{RS}} = |\det \alpha_F(c(\varepsilon))|^{1/2}. \tag{3.21}$$

Let $\mathrm{L}_{\dim M-1}(TM) \in H^{\dim M-1}(M, \mathbf{Z})$ be the degree $\dim M - 1$ component of the characteristic class $\mathrm{L}(TM)$. Let $\widehat{\mathrm{L}}_1(TM) \in H_1(M, \mathbf{Z})$ denote its Poincaré dual. Then one verifies easily that

$$\left|\det \alpha_F\left(\widehat{\mathrm{L}}_1(TM)\right)\right| = \exp\left(\int_M \mathrm{L}\left(TM, \nabla^{TM}\right) c_1\left(F, \nabla^F\right)\right). \tag{3.22}$$

On the other hand, by Corollary 5.9,[9] $\widehat{\mathrm{L}}_1(TM) + c(\varepsilon) \in H_1(M, \mathbf{Z})$ is divisible by two, and one can define a class $\beta_\varepsilon \in H_1(M, \mathbf{Z})$ such that

$$-2\beta_\varepsilon = \widehat{\mathrm{L}}_1(TM) + c(\varepsilon). \tag{3.23}$$

From Proposition 4, (3.22) and (3.23), one finds

$$|\det \alpha_F(c(\varepsilon))|^{1/2} = |\det \alpha_F(\beta_\varepsilon)|^{-1} \exp\left(-\pi\Phi\left(F, \nabla^F\right) + \pi\mathrm{Im}\left(\overline{\eta}\left(D^F_{\mathrm{Sig}}\right)\right)\right), \tag{3.24}$$

where $\Phi(F, \nabla^F)$ is the locally constant function given by (3.13).

We now define a modified Turaev torsion as follows:

$$\mathcal{T}_{\varepsilon,\mathbf{o}}\left(F, \nabla^F\right) = \rho_{\varepsilon,\mathbf{o}}(\alpha_F) e^{\pi\Phi\left(F, \nabla^F\right) + \sqrt{-1}\pi\tilde{\eta}\left(\nabla^F, \tilde{\nabla}^F\right)} \left(\det \alpha_F(\beta_\varepsilon)\right). \tag{3.25}$$

Clearly, $\mathcal{T}_{\varepsilon,\mathbf{o}}(F, \nabla^F)$ is a holomorphic section in the sense of Definition 3.4.[9] Moreover, by (3.21), (3.24) and (3.25), its Ray-Singer norm equals to one. Thus it provides another resolution of Burghelea's problem mentioned above which should be closely related to what in.[10]

Combining with (3.18) we get

$$\left|\frac{\mathcal{T}_{\mathrm{an}}\left(\nabla^F, g^{TM}\right)}{\mathcal{T}_{\varepsilon,\mathbf{o}}\left(F, \nabla^F\right)}\right| = 1, \tag{3.26}$$

which, in view of (3.12), is equivalent to (5.10).[9]

On the other hand, since now $\mathcal{T}_{\mathrm{an}}(\nabla^F, g^{TM})/\mathcal{T}_{\varepsilon,\mathbf{o}}(F, \nabla^F)$ is a holomorphic function with absolute value identically equals to one, one sees that there is a real locally constant function $\theta_{\varepsilon,\mathbf{o}}(F, \nabla^F)$ such that

$$\frac{\mathcal{T}_{\mathrm{an}}\left(\nabla^F, g^{TM}\right)}{\mathcal{T}_{\varepsilon,\mathbf{o}}\left(F, \nabla^F\right)} = e^{\sqrt{-1}\theta_{\varepsilon,\mathbf{o}}(F,\nabla^F)}, \tag{3.27}$$

which is equivalent to (5.8).[9]

Remark 8: While the univalent holomorphic sections $\mathcal{T}_{\mathrm{an}}$ and $\mathcal{T}_{\varepsilon,\mathbf{o}}$ depend on the choice of an "initial" flat connection $\widetilde{\nabla}^F$, the quotients in the left hand sides of (3.26) and (3.27) do not involve it.

Remark 9: One of the advantages of (3.26) and (3.27) is that they look in closer resemblance to the theorems of Cheeger, Müller and Bismut-Zhang[6] concerning the Ray-Singer and Reidemeister torsions.

Now let ∇_1^F and ∇_2^F be two acyclic unitary flat connections on F. We do not assume that they can be connected by a smooth path of flat connections.

By (14.11)[7] (cf. (6.2)[9]), (3.15), (3.17) and the variation formula for η-invariants,[1,3,4] one finds

$$\frac{\mathcal{T}_{\mathrm{an}}(\nabla_1^F, g^{TM})}{\mathcal{T}_{\mathrm{an}}(\nabla_2^F, g^{TM})} = \frac{T^{\mathrm{RS}}(\nabla_1^F)}{T^{\mathrm{RS}}(\nabla_2^F)} \cdot \frac{\exp(-\sqrt{-1}\pi\overline{\eta}(D_{\mathrm{Sig},1}^F) + \sqrt{-1}\pi\widetilde{\eta}(\nabla_1^F, \widetilde{\nabla}^F))}{\exp(-\sqrt{-1}\pi\overline{\eta}(D_{\mathrm{Sig},2}^F) + \sqrt{-1}\pi\widetilde{\eta}(\nabla_2^F, \widetilde{\nabla}^F))}$$

$$= \frac{T^{\mathrm{RS}}(\nabla_1^F)}{T^{\mathrm{RS}}(\nabla_2^F)} \cdot \frac{\exp(-\sqrt{-1}\pi\overline{\eta}(D_{\mathrm{Sig},1}^F) + \sqrt{-1}\pi\overline{\eta}(D_{\mathrm{Sig},2}^F))}{\exp(-\sqrt{-1}\pi\int_M L(TM, \nabla^{TM})CS(\nabla_2^F, \nabla_1^F))}$$

$$= \frac{T^{\mathrm{RS}}(\nabla_1^F)}{T^{\mathrm{RS}}(\nabla_2^F)} \cdot \exp(\sqrt{-1}\pi \cdot \mathrm{sf}(D_{\mathrm{Sig},1}^F, D_{\mathrm{Sig},2}^F)), \tag{3.28}$$

where $D_{\mathrm{Sig},1}^F$ and $D_{\mathrm{Sig},2}^F$ are the signature operators associated to ∇_1^F and ∇_2^F respectively, while $\mathrm{sf}(D_{\mathrm{Sig},1}^F, D_{\mathrm{Sig},2}^F)$ is the spectral flow of the linear path connecting $D_{\mathrm{Sig},1}^F$ and $D_{\mathrm{Sig},2}^F$, in the sense of Atiyah-Patodi-Singer.[3]

Remark 10: Since we do not assume that ∇_1^F and ∇_2^F can be connected by a path of flat connections, our formula extends the corresponding formula in Proposition 6.2.[9]

Corollary 11: *The ratio $\mathcal{T}_{\mathrm{an}}(\nabla^F, g^{TM})/T^{\mathrm{RS}}(\nabla^F)$ is a locally constant function on the set of acyclic unitary flat connections on F.*

Example 12: Let ∇^F be an acyclic unitary flat connection on F. Let $g \in \Gamma(U(F))$ be a smooth section of unitary automorphisms of F. Then

$g^{-1}\nabla^F g$ is another acyclic unitary flat connection on F. A standard calculation shows that

$$\text{sf}\left(D_{\text{Sig}}^{F,\nabla^F}, D_{\text{Sig}}^{F,g^{-1}\nabla^F g}\right) = \int_M \text{L}(TM)\text{ch}(g), \qquad (3.29)$$

where $\text{ch}(g) \in H^{\text{odd}}(M, \mathbf{R})$ is the odd Chern character associated to g (cf.[17]). From (3.29), one sees that if $\int_M \text{L}(TM)\text{ch}(g)$ is nonzero, then ∇^F and $g^{-1}\nabla^F g$ do not lie in the same connected component in the set of acyclic unitary flat connections on F.

3.3. More on η-invariants, spectral flow and the phase of the refined analytic torsion

We would like to point out that the (reduced) η-invariant for non-self-adjoint operators we used above, when considered as a $\mathbf{C}$-valued function, is the original η invariant appeared in[3] (see also[13]). In this section, we show that the $\mathbf{R}$-valued variation formula for η-invariants (which has been used in (3.28)) admits an extension to a $\mathbf{C}$-valued variation formula valid also for the non-self-adjoint operators discussed in the present paper.

First, the concept of spectral flow can be extended to non-self-adjoint operators, and this has been done in[18] in a general context.

For our specific situation, if $D_{\text{Sig},t}^F$, $0 \leq t \leq 1$, is a smooth path of (possibly) non-self-adjoint signature operators, following,[18] we define the spectral flow of this path to be, tautologically,

$$\begin{aligned}
\text{sf}\left(D_{\text{Sig},0}^F, D_{\text{Sig},1}^F\right) = \\
\# \left\{\text{spec}\left(D_{\text{Sig},0}^F\right) \cap \{\text{Re}(\lambda) \geq 0\} \to \text{spec}\left(D_{\text{Sig},1}^F\right) \cap \{\text{Re}(\mu) < 0\}\right\} \\
- \# \left\{\text{spec}\left(D_{\text{Sig},0}^F\right) \cap \{\text{Re}(\lambda) < 0\} \to \text{spec}\left(D_{\text{Sig},1}^F\right) \cap \{\text{Re}(\mu) \geq 0\}\right\},
\end{aligned}$$
$$(3.30)$$

which simply replaces the number zero in the original definition for self-adjoint operators[3] by the axis of purely imaginary numbers.

Now let ∇_t^F, $0 \leq t \leq 1$, be a smooth path of (not necessary unitary and/or flat) connections on F. Let $D_{\text{Sig},t}^F$, $0 \leq t \leq 1$, be the corresponding path of signature operators. With the definition of spectral flow, one then sees easily that the following variation formula holds in $\mathbf{C}$,

$$\bar{\eta}(D_{\text{Sig},1}^F) - \bar{\eta}(D_{\text{Sig},0}^F) = \text{sf}(D_{\text{Sig},0}^F, D_{\text{Sig},1}^F) + \int_M \text{L}(TM, \nabla^{TM})CS(\nabla_0^F, \nabla_1^F).$$
$$(3.31)$$

Now we observe that in,[7-9] Braverman and Kappeler propose an alternate definition of (reduced) η invariant, which if we denote by η_{BK}, then (cf. Definition 4.3[7] and Definition 5.2[9])

$$\eta_{BK}\left(D_{\mathrm{Sig}}^{F}\right) = \overline{\eta}\left(D_{\mathrm{Sig}}^{F}\right) - m_{-}\left(D_{\mathrm{Sig}}^{F}\right), \tag{3.32}$$

where $m_{-}(D_{\mathrm{Sig}}^{F})$ is the number of purely imaginary eigenvalues of D_{Sig}^{F} of form $\lambda\sqrt{-1}$ with $\lambda < 0$.

Formulas (3.31) and (3.32) together give a variation formula for η_{BK}, which can be used to extend (3.28) to non-unitary acyclic representations.

References

1. M. F. Atiyah, V. K. Patodi and I. M. Singer, Spectral asymmetry and Riemannian geometry I. *Proc. Camb. Philos. Soc.* 77 (1975), 43–69.

2. M. F. Atiyah, V. K. Patodi and I. M. Singer, Spectral asymmetry and Riemannian geometry II. *Proc. Camb. Philos. Soc.* 78 (1975), 405–432.

3. M. F. Atiyah, V. K. Patodi and I. M. Singer, Spectral asymmetry and Riemannian geometry III. *Proc. Camb. Philos. Soc.* 79 (1976), 71–99.

4. J.-M. Bismut and D. S. Freed, The analysis of elliptic families, II. *Commun. Math. Phys.* 107 (1986), 103–163.

5. J.-M. Bismut and J. Lott, Flat vector bundles, direct images and higher real analytic torsion, *J. Amer. Math. Soc.* 8 (1995), 291–363.

6. J.-M. Bismut and W. Zhang, An extension of a theorem by Cheeger and Müller, *Astérisque*, n. 205, Paris, 1992.

7. M. Braverman and T. Kappeler, Refined analytic torsion. *Preprint*, math.DG/0505537.

8. M. Braverman and T. Kappeler, Refined analytic torsion as an element of the determinant line, *Preprint*, math.DG/0510523.

9. M. Braverman and T. Kappeler, Ray-Singer type theorem for the refined analytic torsion, *Preprint*, math.DG/0603638.

10. D. Burghelea and S. Haller, Euler structures, the variety of representations and the Milnor-Turaev torsion, *Preprint*, math.DG/0310154.

11. D. Burghelea and S. Haller, Torsion, as a function on the space of representations, *Preprint*, math.DG/0507587.

12. M. Farber and V. Turaev, Poincaré-Reidemeister metric, Euler structures, and torsion, *J. Reine Angew. Math.* 520 (2000), 195–225.

13. P. B. Gilkey, The eta invariant and secondary characteristic classes of locally flat bundles. *Algebraic and Differential Topology — Global Differential Geometry*, Teubner-Texte Math., vol. 70, Teubner, Leipzig, 1984, pp. 49–87.

14. X. Ma and W. Zhang, η-invariant and flat vector bundles, *Chinese Ann. Math.* 27B (2006), 67–72.

15. X. Ma and W. Zhang, Eta-invariants, torsion forms and flat vector bundles, *Preprint*, math.DG/0405599.

16. V. Turaev, Euler structures, nonsingular vector fields, and Reidemeister-type torsions, *Math. USSR Izvestia* 34 (1990), 627–662.

17. W. Zhang, *Lectures on Chern-Weil Theory and Witten Deformations*, Nankai Tracts in Mathematics, Vol. 4, World Scientific, Singapore, 2001.

18. C. Zhu and Y. Long, Maslov-type index theory for symplectic paths and spectral flow (I), *Chinese Ann. Math.* 20B (1999), 413–424.

Chapter 14

ON PLANAR WEBS WITH INFINITESIMAL AUTOMORPHISMS

D. Marín

Departament de Matemàtiques, Universitat Autònoma de Barcelona,
E-08193 Bellaterra (Barcelona), Spain
davidmp@mat.uab.es

J. V. Pereira

Instituto de Matemática Pura e Aplicada,
Est. D. Castorina, 110, 22460-320, Rio de Janeiro, RJ, Brasil
jvp@impa.br

L. Pirio

IRMAR, Campus de Beaulieu, 35042 Rennes Cedex, France
luc.pirio@univ-rennes1.fr

We investigate the space of abelian relations of planar webs admitting infinitesimal automorphisms. As an application we construct $4k-14$ new algebraic families of global exceptional k-webs on the projective plane, for each $k \geq 5$.

1. Introduction and Statement of the Results

Throughout this paper we will work in the holomorphic category.

1.1. *Planar webs*

A germ of regular k-web $\mathcal{W} = \mathcal{F}_1 \boxtimes \cdots \boxtimes \mathcal{F}_k$ on $(\mathbb{C}^2, 0)$ is a collection of k germs of smooth foliations $\mathcal{F}_i$ subjected to the condition that any two of these foliations have distinct tangent spaces at the origin.

The second author is supported by Cnpq and Instituto Unibanco. The third author was partially supported by the International Cooperation Agreement Brazil-France.

One of the most intriguing invariants of a web is its *space of abelian relations* $\mathcal{A}(\mathcal{W})$. If the foliations $\mathcal{F}_i$ are induced by 1-forms ω_i then by definition

$$\mathcal{A}(\mathcal{W}) = \left\{ (\eta_i)_{i=1}^{k} \in (\Omega^1(\mathbb{C}^2, 0))^k \,\middle|\, \forall i \; d\eta_i = 0, \; \eta_i \wedge \omega_i = 0 \text{ and } \sum_{i=1}^{k} \eta_i = 0 \right\}.$$

The dimension of $\mathcal{A}(\mathcal{W})$ is commonly called the *rank* of $\mathcal{W}$ and noted by $\mathrm{rk}(\mathcal{W})$. It is a theorem of Bol that $\mathcal{A}(\mathcal{W})$ is a finite-dimensional $\mathbb{C}$-vector space and moreover

$$(1) \qquad\qquad \mathrm{rk}(\mathcal{W}) \leq \frac{1}{2}\,(k-1)(k-2)\,.$$

An interesting chapter of the theory of webs concerns the characterization of webs of *maximal rank*, *i.e* webs for which (1) is in fact an equality. It follows from Abel's Addition Theorem that all the webs $\mathcal{W}_C$ obtained from reduced plane curves C by projective duality are of maximal rank (*cf.* §4.1 for details). The webs analytically equivalent to some $\mathcal{W}_C$ are the so called *algebrizable webs*.

It can be traced back to Lie a remarkable result that says that all 4-webs of maximal rank are in fact algebrizable. In the early 1930's Blaschke claimed to have extended Lie's result to 5-webs of maximal rank. Not much latter Bol came up with a counter-example: a 5-web of maximal rank that is not algebrizable.

The non-algebrizable webs of maximal rank are nowadays called *exceptional webs*. For a long time Bol's web remained as the only example of exceptional planar web in the literature. The following quote illustrates quite well this fact.

> (...) *we cannot refrain from mentioning what we consider to be the fundamental problem on the subject, which is to determine the maximum rank non-linearizable webs. The strong conditions must imply that there are not many. It may not be unreasonable to compare the situation with the exceptional simple Lie groups.*

Chern and Griffiths in [8].

A comprehensive account of the current state of the art concerning the exceptional webs is available at [14, Introduction §3.2.1], [17] and [15, §1.4]. Here we will just mention that before this work no exceptional k-web with $k \geq 10$ appeared in the literature.

At first glance, the list of known exceptional webs up today does not reveal common features among them. Although at a second look one sees that many of them (but not all, not even the majority) have one property in common: infinitesimal automorphisms.

1.2. *Infinitesimal automorphisms*

In [4], É. Cartan proves that *a 3-web which admits a 2-dimensional continuous group of transformations is hexagonal*. It is then an exercise to deduce that a k-web ($k > 3$) which admits 2 linearly independent infinitesimal automorphisms is parallelizable and in particular algebrizable.

Cartan's result naturally leads to the following question:

> *What can be said about webs which admit one infinitesimal automorphism?*

In fact, Cartan answers this question for 3-webs. In *loc. cit.* he establishes that such a web is equivalent to those induced by the 1-forms $dx, dy, dy - u(x + y)dx$, where u is a germ of holomorphic function.

It is very surprising that this story stops here... To our knowledge, there is no other study concerning planar webs with infinitesimal automorphisms, although they are particularly interesting.[1] Indeed, on the one hand their study is considerably simplified by the presence of an infinitesimal automorphism, but on the other hand, these webs can be very interesting from an analytico-geometrical point of view: we will show they are connected to the theory of exceptional webs.

1.3. *Variation of the rank*

Let $\mathcal{W}$ be a regular web in $(\mathbb{C}^2, 0)$ which admits an infinitesimal automorphism X, *i.e.* X is a germ of vector field whose local flow preserves the foliations of $\mathcal{W}$. As we will see in §2 the Lie derivative $L_X = i_X d + d i_X$ with respect to X induces a linear operator on $\mathcal{A}(\mathcal{W})$. Most of our results will follow from an analysis of such operator.

In §3.1 we use this operator to give a simple description of the abelian relations of $\mathcal{W}$ and from this we will deduce in §3.2 what we consider our main result:

Theorem 1. *Let $\mathcal{W}$ be a k–web which admits a transverse infinitesimal automorphism X. Then*

$$\mathrm{rk}(\mathcal{W} \boxtimes \mathcal{F}_X) = \mathrm{rk}(\mathcal{W}) + (k - 1).$$

---

[1] However several results have been established concerning higher dimensional webs admitting infinitesimal symmetries (see [1, §7.3]) but never in relation with the notion of abelian relation.

In particular, $\mathcal{W}$ is of maximal rank if and only if $\mathcal{W} \boxtimes \mathcal{F}_X$ is of maximal rank.

We will derive from Theorem 1 the existence of new families of exceptional webs.

1.4. New families of exceptional webs

If we start with a reduced plane curve C invariant under an algebraic $\mathbb{C}^*$-action on $\mathbb{P}^2$ then we obtain a dual algebraic $\mathbb{C}^*$-action on $\check{\mathbb{P}}^2$, letting invariant the algebraic web $\mathcal{W}_C$ (*cf.* §4.1 for details). Combining this construction with Theorem 1 we deduce our second main result:

Theorem 2. *For every $k \geq 5$ there exist a family of dimension at least $\lfloor k/2 \rfloor - 1$ of pairwise non-equivalent exceptional global k-webs on $\mathbb{P}^2$.*

In fact, for each $k \geq 5$, we obtain $4k - 15$ other families of smaller dimension.

Theorem 2 is in sharp contrast with the recent algebrization theorem of Trépreau, generalizing and completing previous works of Bol [3] and Chern-Griffiths [7], which says that a maximal rank 1-codimensional k-web is algebrizable when the ambient space has dimension at least 3 and k is sufficiently large[2].

The classification of the exceptional 5-webs of the type $\mathcal{W} \boxtimes \mathcal{F}_X$ where X is an infinitesimal automorphism of $\mathcal{W}$ follows easily from Theorem 2 (*cf.* Corollary 4.1).

2. Generalities on Webs with Infinitesimal Automorphisms

Let $\mathcal{F}$ be a regular foliation on $(\mathbb{C}^2, 0)$ induced by a (germ of) 1-form ω. We say that a (germ of) vector field X is an infinitesimal automorphism of $\mathcal{F}$ if the foliation $\mathcal{F}$ is preserved by the local flow of X. In algebraic terms: $L_X \omega \wedge \omega = 0$.

When the infinitesimal automorphism X is transverse to $\mathcal{F}$, *i.e.* when $\omega(X) \neq 0$, then a simple computation (*cf.* [12, Corollary 2]) shows that the 1-form

$$\eta = \frac{\omega}{i_X \omega}$$

---

[2]In fact, Trépreau's result is stronger, see [18] for a precise statement

is closed and satisfies $L_X \eta = 0$. By definition, the integral

$$u(z) = \int_0^z \eta$$

is the *canonical first integral* of $\mathcal{F}$ (with respect to X). Clearly, we have $u(0) = 0$ and $L_X(u) = 1$.

Keeping in mind that the local flow of X sends leaves into leaves we can geometrically interpret the first integral $u(z)$ as the time that such local flow takes to transport the leaf through 0 to the leaf through z.

Now let $\mathcal{W}$ be a germ of regular k-web on $(\mathbb{C}^2, 0)$ induced by the (germs of) 1-forms $\omega_1, \ldots, \omega_k$ and let X be an infinitesimal automorphism of $\mathcal{W}$. Here, of course, we mean that X is an infinitesimal automorphism for all the foliations in $\mathcal{W}$.

By hypothesis, we have $L_X \omega_i \wedge \omega_i = 0$ for $i = 1, \ldots, k$. Then because the Lie derivative L_X is linear and commutes with d, it induces a linear map

$$(2) \qquad L_X : \mathcal{A}(\mathcal{W}) \longrightarrow \mathcal{A}(\mathcal{W})$$
$$(\eta_1, \ldots, \eta_k) \longmapsto (L_X \eta_1, \ldots, L_X \eta_k).$$

This map is central in this paper: all our results come from an analysis of the L_X-invariant subspaces of $\mathcal{A}(\mathcal{W})$.

3. Abelian Relations versus Infinitesimal Automorphisms

3.1. *Description of $\mathcal{A}(\mathcal{W})$ in presence of an infinitesimal automorphism*

In this section, $\mathcal{W} = \mathcal{F}_1 \boxtimes \cdots \boxtimes \mathcal{F}_k$ denotes a k-web in $(\mathbb{C}^2, 0)$ which admits an infinitesimal automorphism X, regular and transverse to the foliations $\mathcal{F}_i$ in a neighborhood of the origin.

Let $i \in \{1, \ldots, k\}$ be fixed. We note $\mathcal{A}^i(\mathcal{W})$ the vector subspace of $\Omega^1(\mathbb{C}^2, 0)$ spanned by the ith components α_i of abelian relations $(\alpha_1, \ldots, \alpha_k) \in \mathcal{A}(\mathcal{W})$. If $u_i = \int \eta_i$ denotes the canonical first integral of $\mathcal{F}_i$ with respect to X, then for $\alpha_i \in \mathcal{A}^i(\mathcal{W})$, there exists a holomorphic germ $f_i \in \mathbb{C}\{t\}$ such that $\alpha_i = f_i(u_i) \, du_i$.

Assume now that $\mathcal{A}^i(\mathcal{W})$ is not trivial and let $\{\alpha_i^\nu = f_\nu(u_i) \, du_i \mid \nu = 1, \ldots, n_i\}$ be a basis. Since $L_X : \mathcal{A}^i(\mathcal{W}) \to \mathcal{A}^i(\mathcal{W})$ is a linear map, there exists complex constants $c_{\nu\mu}$ such that, for $\nu = 1, \ldots, n_i$ we have

$$(3) \qquad\qquad L_X(\alpha_i^\nu) = \sum_{\mu=1}^{n_i} c_{\nu\mu}\, \alpha_i^\mu\,.$$

But $L_X(\alpha_i^\nu) = L_X\big(f_\nu(u_i)\, du_i\big) = X\big(f_\nu(u_i)\big) du_i + f_\nu(u_i)\, L_X\big(du_i\big) = f_\nu'(u_i)\, du_i$ for any ν, so relations (3) are equivalent to the scalar ones

$$(4) \qquad\qquad f_\nu' = \sum_{\mu=1}^{n_i} c_{\nu\mu}\, f_\mu\,, \qquad \nu = 1, \ldots, n_i\,.$$

Now let $\lambda_1, \ldots, \lambda_\tau \in \mathbb{C}$ be the eigenvalues of the map L_X acting on $\mathcal{A}(\mathcal{W})$ corresponding to maximal eigenspaces with corresponding dimensions $\sigma_1, \ldots, \sigma_\tau$. The differential equations (4) give us the following description of $\mathcal{A}(\mathcal{W})$:

Proposition 3.1. *The abelian relations of $\mathcal{W}$ are of the form*

$$P_1(u_1)\, e^{\lambda_i u_1}\, du_1 + \cdots + P_k(u_k)\, e^{\lambda_i u_k}\, du_k = 0$$

where $P_1, \ldots, P_k$ are polynomials of degree less or equal to $\sigma_i - 1$.

We will now explain how we can use Proposition 3.1 to effectively determine $\mathcal{A}(\mathcal{W})$. The key point is to determine the possible non-zero eigenvalues of the map (2). Once this is done we can easily determine the abelian relations by simple linear algebra.

We claim that 0 is an eigenvalue of (2) if, and only if, for every germ of vector field Y the Wronskian

$$(5) \qquad\qquad \det \begin{pmatrix} u_1 & \cdots & u_k \\ Y(u_1) & \cdots & Y(u_k) \\ \vdots & \ddots & \vdots \\ Y^{k-1}(u_1) & \cdots & Y^{k-1}(u_k) \end{pmatrix}$$

is identically zero. In fact, if this is the case then we have two possibilities: the functions $u_1, \ldots, u_k$ are $\mathbb{C}$-linearly dependent or all the leaves of Y are cutted out by some element of the linear system generated by $u_1, \ldots, u_k$, cf. [11, theorem 4]. In particular if Y is a vector field of the form $Y = \mu x \frac{\partial}{\partial x} + y \frac{\partial}{\partial y}$, with $\mu \in \mathbb{C} \setminus \mathbb{Q}$, then the leaves of Y accumulate at 0 and are not cutted out by any regular holomorphic function. Therefore the vanishing of (5) implies the existence of an abelian relation of the form

$$\sum c_i u_i = 0\,,$$

where the c_i's are complex constants.

To determine the possible complex numbers λ which are eigenvalues of the map (2) first notice that the corresponding eigenvectors can be readen as a functional equation of the form $c_1 e^{\lambda u_1} + \cdots + c_k e^{\lambda u_k} = 0$, where, as before, the c_i's are complex constants: just take the interior product of the displayed equation in Proposition 3.1 with X. In the same spirit of what we have just made for the zero eigenvalue case consider the holomorphic function given by

$$
(6) \qquad \det \begin{pmatrix} \exp(\lambda u_1) & \cdots & \exp(\lambda u_k) \\ Y(\exp(\lambda u_1)) & \cdots & Y(\exp(\lambda u_k)) \\ \vdots & \ddots & \vdots \\ Y^{k-1}(\exp(\lambda u_1)) & \cdots & Y^{k-1}(\exp(\lambda u_k)) \end{pmatrix}
$$

for an arbitrary germ of vector field Y.

Notice that (6) is of the form $\exp(\lambda(u_1 + \cdots + u_k))\lambda^{k-1} P_Y(\lambda)$, where P_Y is a polynomial in λ, of degree at most $\frac{(k-1)(k-2)}{2}$, with germs of holomorphic functions as coefficients. The common constant roots of these polynomials, when Y varies, are exactly the eigenvalues of the map (2).

Let us now exemplify these ideas. In practice we do not have to consider all the vector fields Y but just a suitably chosen one.

Example 3.1. The k-web $\mathcal{W}$ generated by the functions $f_i(x,y) = y + x^i$, $i = 1, \ldots, k$, has no abelian relations.

Proof. Clearly the vector field $X = \frac{\partial}{\partial y}$ is an infinitesimal automorphism of $\mathcal{W}$ and $X(df_i) = 1$, $i = 1, \ldots, k$. It follows that $u_i = f_i$ are the canonical first integrals of $\mathcal{W}$. On the other hand, if we consider the vector field $Y = \frac{\partial}{\partial x}$ we can easily see that that $P_Y(\lambda)|_{x=y=0} = (-1)^{k-1} \prod_{n=1}^{k-1} n!$. Consequently, the only candidate for a eigenvalue of the map (2) is $\lambda = 0$. But clearly the functions f_i are linearly independent over $\mathbb{C}$. $\qquad \square$

Let us see how to use this approach to recover the abelian relations of one of the exceptional webs found by the third author in [13]

Example 3.2. The 5-web $\mathcal{W}$ induced by the functions $x, y, x+y, x-y, x^2 + y^2$ has rank 6.

Proof. Clearly the radial vector field $R = x\frac{\partial}{\partial x} + y\frac{\partial}{\partial y}$ is an infinitesimal automorphism of $\mathcal{W}$. The canonical first integrals are $u_1 = \log x$, $u_2 = \log y$, $u_3 = \log(x + y)$, $u_4 = \log(x - y)$ and $u_5 = \frac{1}{2}\log(x^2 + y^2)$. If we take $Y = x\frac{\partial}{\partial x} - y\frac{\partial}{\partial y}$ then one can easily check that the polynomial P_Y is a complex multiple of

$$x^7 y^7 \lambda (\lambda - 1)^2 (\lambda - 2)^2 (\lambda - 4)(\lambda - 6).$$

According to Proposition 3.1, we have only to look for abelian relations of the form $\sum_{i=1}^{5} P_{i\lambda}(\log g_i) g_i^{\lambda} \frac{dg_i}{g_i} = 0$, for $\lambda = 0, 1, 2, 4, 6$, where $P_{i\lambda}$ are polynomials and $g_i = \exp(u_i)$. We have thus reduced our search to a simple problem of linear algebra, i.e. to find linear dependences on finite dimensional vector spaces indexed by $\lambda \in \{0, 1, 2, 4, 6\}$. It turns out that for $\lambda = 1$ we obtain two linearly independent abelian relations

$$g_1 + g_2 - g_3 = 0, \qquad g_1 - g_2 - g_4 = 0.$$

For $\lambda = 2$ one finds two more independent abelian relations

$$g_1^2 + g_2^2 - g_5^2 = 0, \qquad 2\, g_1^2 + 2\, g_2^2 - g_3^2 - g_4^2 = 0.$$

Finally, for $\lambda = 4$ and $\lambda = 6$, respectively, there are two more independent abelian relations:

$$5\, g_1^4 + 5\, g_2^4 + g_3^4 + g_4^4 - 6\, g_5^4 = 0, \qquad 8\, g_1^6 + 8\, g_2^6 + g_3^6 + g_4^6 - 10\, g_5^6 = 0.$$

Thus, we have found 6 independent abelian relations. $\qquad\square$

3.2. *Proof of Theorem 1*

With Proposition 3.1 at hand we are able to prove our main result.

Let $\mathcal{W} = \mathcal{F}_1 \boxtimes \cdots \boxtimes \mathcal{F}_k$ and for $i = 1 \ldots k$, set $\eta_i = du_i$ as the differential of the canonical first integral of $\mathcal{F}_i$ relatively to X. We note x a first integral of the foliation $\mathcal{F}_X$, normalized such that $x(0) = 0$.

When j varies from 2 to k, we have

$$i_X(\eta_1 - \eta_j) = 0 \qquad \text{and} \qquad L_X(\eta_1 - \eta_j) = 0.$$

Consequently there exists $g_j \in \mathbb{C}\{x\}$ such that

$$(7) \qquad\qquad du_1 - du_j - g_j(x)\, dx = 0.$$

Clearly these are abelian relations for the web $\mathcal{W} \boxtimes \mathcal{F}_X$. They span a $(k-1)$-dimensional vector subspace $\mathcal{V}$ of the maximal eigenspace of L_X associated to the eigenvalue zero, noted $\mathcal{A}_0(\mathcal{W} \boxtimes \mathcal{F}_X)$.

Observe that $\mathcal{V}$ fits in the following exact sequence (i is the natural inclusion):

$$(8) \qquad\qquad 0 \to \mathcal{V} \xrightarrow{\;i\;} \mathcal{A}_0(\mathcal{W} \boxtimes \mathcal{F}_X) \xrightarrow{\;L_X\;} \mathcal{A}_0(\mathcal{W}).$$

Indeed, the kernel $K := \ker\{L_X : \mathcal{A}_0(\mathcal{W} \boxtimes \mathcal{F}_X) \to \mathcal{A}_0(\mathcal{W})\}$ is generated by abelian relations of the form $\sum_{i=1}^{k} c_i du_i + g(x)\, dx = 0$, where $c_i \in \mathbb{C}$ and $g \in \mathbb{C}\{x\}$. Since $i_X du_i = 1$ for each i, it follows that the constants c_i satisfy $\sum_{i=1}^{k} c_i = 0$. It implies that the abelian relations in the kernel of L_X can be written as linear combinations of abelian relations of the form (7). Therefore

$$(9) \qquad\qquad K = \mathcal{V}$$

and consequently $\ker L_X \subset \operatorname{Im} i$. The exactness of (8) follows easily.

From general principles we deduce that the sequence

$$0 \to \frac{\mathcal{V}}{\mathcal{A}_0(\mathcal{W}) \cap \mathcal{V}} \xrightarrow{\ i\ } \frac{\mathcal{A}_0(\mathcal{W} \boxtimes \mathcal{F}_X)}{\mathcal{A}_0(\mathcal{W})} \xrightarrow{\ L_X\ } \frac{\mathcal{A}_0(\mathcal{W})}{L_X \mathcal{A}_0(\mathcal{W})},$$

is also exact. Thus to prove the Theorem it suffices to verify the following assertions:

(a) $\mathcal{V}$ is isomorphic to

$$\frac{\mathcal{V}}{\mathcal{A}_0(\mathcal{W}) \cap \mathcal{V}} \oplus \frac{\mathcal{A}_0(\mathcal{W})}{L_X \mathcal{A}_0(\mathcal{W})};$$

(b) the morphism $L_X : \mathcal{A}_0(\mathcal{W} \boxtimes \mathcal{F}_X) \to \mathcal{A}_0(\mathcal{W})$ is surjective;

(c) the vector spaces

$$\frac{\mathcal{A}_0(\mathcal{W} \boxtimes \mathcal{F}_X)}{\mathcal{A}_0(\mathcal{W})} \quad \text{and} \quad \frac{\mathcal{A}(\mathcal{W} \boxtimes \mathcal{F}_X)}{\mathcal{A}(\mathcal{W})}$$

are isomorphic.

To verify assertion (a), notice that the nilpotence of L_X on $\mathcal{A}_0(\mathcal{W})$ implies that $\frac{\mathcal{A}_0(\mathcal{W})}{L_X \mathcal{A}_0(\mathcal{W})}$ is isomorphic to $\mathcal{A}_0(\mathcal{W}) \cap K$. Combined with (9), it implies assertion (a).

To prove assertion (b), it suffices to construct a map $\Phi : \mathcal{A}_0(\mathcal{W}) \to \mathcal{A}_0(\mathcal{W} \boxtimes \mathcal{F}_X)$ such that $L_X \circ \Phi = \operatorname{Id}$. Proposition 3.1 implies that $\mathcal{A}_0(\mathcal{W})$ is spanned by abelian relations of the form $\sum_{i=1}^{k} c_i u_i^r du_i = 0$, where c_i are complex numbers and r is a non-negative integer. For such an abelian relation, since

$$\sum_{i=1}^{k} c_i u_i^r du_i = \frac{1}{r+1} L_X \left(\sum_{i=1}^{k} c_i u_i^{r+1} du_i \right) = 0,$$

there exists an unique $g \in \mathbb{C}\{x\}$ satisfying $\sum_{i=1}^{k} c_i u_i^{r+1} du_i + g(x)\, dx = 0$.
If we set

$$\Phi\left(\sum_{i=1}^{k} c_i u_i^r du_i\right) = \frac{1}{r+1}\left(\sum_{i=1}^{k} c_i u_i^{r+1} du_i + g(x)\, dx\right)$$

then $L_X \circ \Phi = \mathrm{Id}$ on $\mathcal{A}_0(\mathcal{W})$ and assertion (b) follows.

To prove assertion (c) we first notice that

$$\mathcal{A}(\mathcal{W} \boxtimes \mathcal{F}_X) = \mathcal{A}_0(\mathcal{W} \boxtimes \mathcal{F}_X) \oplus \mathcal{A}_*(\mathcal{W} \boxtimes \mathcal{F}_X)$$

where $\mathcal{A}_*(\mathcal{W} \boxtimes \mathcal{F}_X)$ denotes the sum of eigenspaces corresponding to non-zero eigenvalues. Of course $\mathcal{A}_*(\mathcal{W} \boxtimes \mathcal{F}_X)$ is invariant and moreover we have the equality

$$L_X\left(\mathcal{A}_*(\mathcal{W} \boxtimes \mathcal{F}_X)\right) = \mathcal{A}_*(\mathcal{W} \boxtimes \mathcal{F}_X).$$

But L_X *kills* the $\mathcal{F}_X$-components of abelian relations. In particular, it implies

$$L_X\left(\mathcal{A}_*(\mathcal{W} \boxtimes \mathcal{F}_X)\right) \subset \mathcal{A}_*(\mathcal{W}).$$

This is sufficient to show that $\mathcal{A}_*(\mathcal{W} \boxtimes \mathcal{F}_X) = \mathcal{A}_*(\mathcal{W})$ and deduce assertion (c) and, consequently that

$$\mathrm{rk}(\mathcal{W} \boxtimes \mathcal{F}_X) = \mathrm{rk}(\mathcal{W}) + (k-1).$$

Because $k(k-1)/2 = (k-1)(k-2)/2 + (k-1)$, the above equality implies immediately the last assertion of Theorem 1. $\qquad\square$

4. New Families of Exceptional Webs

4.1. *Algebrizable webs with infinitesimal automorphisms*

Let $C \subset \mathbb{P}^2$ be a degree k reduced curve. If $U \subset \check{\mathbb{P}}^2$ is a simply-connected open set not intersecting $\check{C}$ and if $\gamma_1, \ldots, \gamma_k : U \to C$ are the holomorphic maps defined by the intersections of lines in U with C then Abel's Theorem implies that

$$\mathrm{Tr}(\omega) = \sum_{i=1}^{k} \gamma_i^* \omega = 0$$

for every $\omega \in H^0(C, \omega_C)$, where ω_C denotes the dualizing sheaf of C.

The maps γ_i define the k-web $\mathcal{W}_C$ on U and the trace formula above associates an abelian relation of $\mathcal{W}_C$ to each $\omega \in H^0(C, \omega_C)$. Since $h^0(C, \omega_C) = (k-1)(k-2)/2$, the web $\mathcal{W}_C$ is of maximal rank.

Suppose now that C is invariant by a $\mathbb{C}^*$-action $\varphi : \mathbb{C}^* \times \mathbb{P}^2 \to \mathbb{P}^2$. Notice that φ induces a dual action $\check{\varphi} : \mathbb{C}^* \times \check{\mathbb{P}}^2 \to \check{\mathbb{P}}^2$ satisfying $\varphi_t \circ \gamma_i = \gamma_i \circ \check{\varphi}_t$ for $i = 1, \ldots, k$. Consequently the web $\mathcal{W}_C$ admits an infinitesimal automorphism.

In a suitable projective coordinate system $[x : y : z]$, a plane curve C invariant by a $\mathbb{C}^*$-action is cut out by an equation of the form

$$(10) \qquad x^{\epsilon_1} \cdot y^{\epsilon_2} \cdot z^{\epsilon_3} \cdot \prod_{i=1}^{k} (x^a + \lambda_i y^b z^{a-b})$$

where $\epsilon_1, \epsilon_2, \epsilon_3 \in \{0, 1\}$, $k, a, b \in \mathbb{N}$ are such that $k \geq 1$, $a \geq 2$, $1 \leq b \leq a/2$, $\gcd(a, b) = 1$ and the λ_i are distinct non zero complex numbers (*cf.* [2, §1] for instance). Notice that here the $\mathbb{C}^*$-action in question is

$$(11) \qquad \begin{aligned} \varphi : \mathbb{C}^* \times \mathbb{P}^2 &\to \mathbb{P}^2 \\ (t, [x : y : z]) &\mapsto [t^{b(a-b)} x : t^{a(a-b)} y : t^{ab} z]. \end{aligned}$$

Moreover once we fix $\epsilon_1, \epsilon_2, \epsilon_3, k, a, b$ we can always choose $\lambda_1 = 1$ and in this case the set of $k - 1$ complex numbers $\{\lambda_2, \ldots, \lambda_k\}$ projectively characterizes the curve C. In particular one promptly sees that there exists a $(d - 1)$-dimensional family of degree $2d$ (or $2d + 1$) reduced plane curves all projectively distinct and invariant by the same $\mathbb{C}^*$-action: for a given $2d + \delta$ with $\delta \in \{0, 1\}$ set $a = 2$, $b = 1$, $\epsilon_1 = \delta$ and $\epsilon_2 = \epsilon_3 = 0$.

A moment of reflection shows that the number of discrete parameters giving distinct families of degree d curves of the form (10) is

$$\underbrace{\left\lfloor \frac{d}{2} \right\rfloor}_{\epsilon_1 = \epsilon_2 = \epsilon_3 = 0} + \underbrace{3 \left\lfloor \frac{d-1}{2} \right\rfloor}_{\epsilon_i = \epsilon_j = 0,\, \epsilon_k = 1} + \underbrace{3 \left\lfloor \frac{d-2}{2} \right\rfloor}_{\epsilon_i = \epsilon_j = 1,\, \epsilon_k = 0} + \underbrace{\left\lfloor \frac{d-3}{2} \right\rfloor}_{\epsilon_1 = \epsilon_2 = \epsilon_3 = 1} - 2 = 4d - 10.$$

Notice that the -2 appears on left hand side because the curves $\{y = 0\}$ and $\{z = 0\}$ are indistinguishable when $a = 2$.

4.2. *Proof of Theorem 2*

If C is a reduced curve of the form (10) then $\mathcal{W}_C$ is invariant by an algebraic $\mathbb{C}^*$-action $\check{\varphi}$. We will note by X the infinitesimal generator of $\check{\varphi}$ and by $\mathcal{F}_X$ the corresponding foliation. From the discussion on the last paragraph, Theorem 2 follows at once from the stronger:

Theorem 4.1. *If* $\deg C \geq 4$ *then* $\mathcal{W}_C \boxtimes \mathcal{F}_X$ *is exceptional. Moreover if* C' *is another curve invariant by* φ *then* $\mathcal{W}_C \boxtimes \mathcal{F}_X$ *is analytically equivalent to* $\mathcal{W}_{C'} \boxtimes \mathcal{F}_X$ *if and only if the curve* C *is projectively equivalent to* C'.

Proof. Since $\mathcal{W}_C$ has maximal rank it follows from Theorem 1 that $\mathcal{W}_C \boxtimes \mathcal{F}_X$ is also of maximal rank. Suppose that its localization at a point $p \in \mathbb{P}^2$ is algebrizable and let $\psi : (\mathbb{P}^2, p) \to (\mathbb{C}^2, 0)$ be a holomorphic algebrization. Since both $\mathcal{W}_C$ and $\psi_*(\mathcal{W}_C)$ are linear webs of maximal rank it follows from a result of Nakai [10] that ψ is the localization of an automorphism of $\mathbb{P}^2$. But the generic leaf of $\mathcal{F}_X$ is not contained in any line of $\mathbb{P}^2$ and consequently $\psi_*(\mathcal{W} \boxtimes \mathcal{F}_X)$ is not linear. This concludes the proof of the theorem. □

Remark 4.1. We do not know if the families above are *irreducible* in the sense that the generic element does not admit a deformation as an exceptional web that is not contained in the family. Due to the presence of automorphism one could imagine that they are indeed degenerations of some other exceptional webs.

4.3. *A characterization result*

Combining Theorem 1 with Lie's Theorem we can easily prove the

Corollary 4.1. *Let $\mathcal{W}$ be a 4-web that admits a transverse infinitesimal automorphism Y. If $\mathcal{W} \boxtimes \mathcal{F}_Y$ is exceptional then it is analytically equivalent to an exceptional 5-web $\mathcal{W}_C \boxtimes \mathcal{F}_X$ described in Theorem 4.1.*

Proof. It follows from Theorem 1 that $\mathcal{W}$ is of maximal rank. Lie's Theorem implies that $\mathcal{W}$ is analytically equivalent to $\mathcal{W}_C$ for some reduced plane quartic C. Since the local flow of Y preserves $\mathcal{W}$ there exists a (germ) of vector field X whose local flow preserves $\mathcal{W}_C$. Using again Nakai's result we deduce that the germs of automorphisms on the local flow of X are indeed projective automorphisms. This is sufficient to conclude that X is a global vector field preserving $\mathcal{W}_C$. □

Remark 4.2. Theorem 4.1 does not give all the exceptional webs admitting an infinitesimal automorphism. As we have seen in the Example 3.2, the web $\mathcal{W}$ induced by the functions $x, y, x + y, x - y, x^2 + y^2$ is exceptional and it admits the radial vector field $R = x \, \partial/\partial_x + y \, \partial/\partial_y$ as a transverse infinitesimal automorphism. Theorem 1 implies that the 6-web $\mathcal{W} \boxtimes \mathcal{F}_R$ is also exceptional. This result was previously obtained by determining an explicit basis of the space of abelian relations, see [14, p. 253].

5. Problems

5.1. *A conjecture about the nature of the abelian relations*

It is clear from Proposition 3.1 that for webs $\mathcal{W}$ admitting infinitesimal automorphisms there exists a Liouvillian extension of the field of definition of $\mathcal{W}$ containing all its abelian relations. We believe that a similar statement should hold for arbitrary webs $\mathcal{W}$.

Conjecture 5.1. *The abelian relations of a web $\mathcal{W}$ are defined on a Liouvillian extension of the definition field of $\mathcal{W}$.*

Our belief is supported by the recent works of Hénaut [9] and Ripoll [16] on abelian relations and of Casale [5] on non-linear differential Galois Theory.

When $\mathcal{W}$ is of maximal rank the main result of [9] shows that there exists a Picard-Vessiot extension of the field of definition of $\mathcal{W}$ containing all the abelian relations. In the general case, one should be able to deduce a similar result from the above mentioned work of Ripoll.

On the other hand, and at least over polydiscs, [5, Theorem 6.4] implies that the foliations with first integrals on Picard-Vessiot extension are transversely projective. Since the first integrals in question are components of abelian relations they are of finite determinacy and hopefully this should imply that they are indeed Liouvillian.

5.2. *Restricted Chern's problem*

With the techniques now available, the classification of all exceptional 5-webs (*"Chern's problem"* see [6, page 27]) seems completely out of reach. So we propose the

Problem 5.1. *Classify exceptional 5-webs admitting infinitesimal automorphisms.*

Notice that this restricted version is not completely hopeless. The linear map L_X can be "integrated" giving birth to a holomorphic action on $\mathbb{P}(\mathcal{A}(\mathcal{W}))$. The Poincaré-Blaschke curves will be orbits of this action and the dual action will induce an automorphism of the associated Blaschke surface. This seems valuable extra data that may lead to a solution of the restricted Chern's problem.

For a definition of the above mentioned concepts see [14, Chapter 8].

References

[1] M. Akivis and V. Goldberg, *Differential geometry of webs*, in "Handbook of differential geometry", Vol. **1**, pp. 1–152, North-Holland, 2000.

[2] P. Aluffi and C. Faber, *Plane curves with small linear orbits II*, Internat. J. Math. **11** (2000), pp. 591–608.

[3] G. Bol, *Flächengewebe im dreidimensionalen Raum*, Abh. Math. Semin. Hamb. Univ. **10** (1934), pp. 119–133.

[4] É. Cartan, *Les sous-groupes des groupes continus de transformations*, Œuvres complètes, Vol. 3, pp. 78–83.

[5] G. Casale, *Le groupoïde de Galois d'un germe de feuilletage de codimension un*, to appear in Ann. Inst. Fourier.

[6] S.-S. Chern, *Wilhelm Blaschke and Web Geometry*, Wilhelm Blaschke Gesammelte Werke, Vol. 5, Thales, Essen (1985), p. 21–23.

[7] S.-S. Chern and P.A. Griffiths, *Abel's theorem and webs*, Jahresber. Deutsch. Math.-Verein. **80** (1978), pp. 13–110.

[8] S.-S. Chern and P.A. Griffiths, *Corrections and addenda to our paper: "Abel's theorem and webs"*, Jahresber. Deutsch. Math.-Verein. **83** (1981), pp. 78–83.

[9] A. Hénaut, *On planar web geometry through abelian relations and connections*, Ann. of Math. (2) **159** (2004), pp. 425–445.

[10] I. Nakai, *Topology of complex webs of codimension one and geometry of projective space curves*, Topology **26** (1987), pp. 475–504.

[11] J.V. Pereira, *Vector Fields, Invariant Varieties and Linear Systems*, Annales de L'Institut Fourier, **51** n. 5 (2001), pp. 1385–1405.

[12] J.V. Pereira and P.F. Sánchez, *Transformation groups of holomorphic foliations*, Comm. Anal. Geom. **10** (2002), pp. 1115–1123.

[13] L. Pirio, *Sur les tissus plans de rang maximal et le problème de Chern*, C. R. Math. Acad. Sci. Paris **339** (2004), pp. 131–136.

[14] L. Pirio, *Équation fonctionnelle abélienne et géométrie des tissus*, Thèse de l'Université Paris VI, defended in 2004.

[15] L. Pirio and J.-M. Trépreau, *Tissus Plans Exceptionnels et Fonctions Thêta*, Ann. Inst. Fourier **55** (2005), pp. 2209–2237.

[16] O. Ripoll, *Détermination du rang des tissus du plan et autres invariants géométriques*, C.R. Acad. Sci. Paris, Ser. I **341** (2005), pp. 247–252.

[17] G. Robert, *Relations fonctionnelles polylogarithmiques et tissus plans*, Prépublication **146**, Université Bordeaux 1 (2002).

[18] J.-M. Trépreau, *Sur l'algébrisation des tissus, le théorème de Bol en toute dimension > 2*, this volume.

Chapter 15

PROJECTIVE LINKING AND BOUNDARIES OF POSITIVE HOLOMORPHIC CHAINS IN PROJECTIVE MANIFOLDS, PART II

F. Reese Harvey

Mathematics Department, Rice University,
Houston, TX 77251, USA

H. Blaine Lawson, Jr.[*]

Mathematics Department, Stony Brook University,
Stony Brook, NY 11794, USA

Dedicated with our deepest esteem to S. S. Chern

Abstract

Part I introduced the notion of the *projective linking number* $\mathrm{Link}_{\mathbf{P}}(\Gamma, Z)$ of a compact oriented real submanifold Γ of dimension $2p - 1$ in complex projective n-space $\mathbf{P}^n$ with an algebraic subvariety $Z \subset \mathbf{P}^n - \Gamma$ of codimension p. It is shown here that a basic conjecture concerning the projective hull of real curves in $\mathbf{P}^2$ implies the following result:

Γ is the boundary of a positive holomorphic p-chain of mass $\leq \Lambda$ in $\mathbf{P}^n$ if and only if the $\widetilde{\mathrm{Link}}_{\mathbf{P}}(\Gamma, Z) \geq -\Lambda$ for all algebraic subvarieties Z of codim-p in $\mathbf{P}^n - \Gamma$

where $\widetilde{\mathrm{Link}}_{\mathbf{P}}(\Gamma, Z) = \mathrm{Link}_{\mathbf{P}}(\Gamma, Z)/p!\deg(Z)$. An analogous result is implied in any projective manifold X.

Table of Contents

---

[*]Partially supported by the N.S.F.

1. Introduction

In Part I we introduced a linking pairing for certain cycles in projective space as follows. Suppose that $\Gamma \subset \mathbf{P}^n$ is a compact oriented submanifold of dimension $2p - 1$, and let $Z \subset \mathbf{P}^n - \Gamma$ be an algebraic subvariety of codimension p. The **projective linking number** of Γ with Z is defined to be

$$\mathrm{Link}_{\mathbf{P}}(\Gamma, Z) \equiv N \bullet Z - \deg(Z) \int_N \omega^p$$

where ω is the standard Kähler form on $\mathbf{P}^n$ and N is any integral 2p-chain with $\partial N = \Gamma$ in $\mathbf{P}^n$. This definition is independent of the choice of N. The associated **reduced linking number** is defined to be

$$\widetilde{\mathrm{Link}}_{\mathbf{P}}(\Gamma, Z) \equiv \frac{1}{p! \deg(Z)} \mathrm{Link}_{\mathbf{P}}(\Gamma, Z).$$

The basic result proved here is the following.

Theorem 1.1. *Let $\Gamma \subset \mathbf{P}^n$ be a compact oriented real analytic submanifold of dimension $2p - 1$ with possible integer multiplicities on each component. If Conjecture B holds (see below), then the following are equivalent:*

(1) *Γ is the boundary of a positive holomorphic p-chain of mass $\leq \Lambda$ in $\mathbf{P}^n$.*
(2) *$\widetilde{\mathrm{Link}}_{\mathbf{P}}(\Gamma, Z) \geq -\Lambda$ for all algebraic subvarieties Z of codimension p in $\mathbf{P}^n - \Gamma$.*

A compact subset $K \subset \mathbf{P}^n$ is called **stable** if the best constant function is bounded on the projective hull $\widehat{K}$ (See [HL$_{3,4}$]).

It is known that for any stable real analytic curve $\gamma \subset \mathbf{P}^n$, the set $\widehat{\gamma} - \gamma$ is a 1-dimensional complex analytic subvariety of $\mathbf{P}^n - \gamma$ [HLW]. Conjecture A from Part I is the statement that any compact real analytic curve $\gamma \subset \mathbf{P}^n$ is stable. Even more likely is the following.

Conjecture B. Let $\gamma \subset \mathbf{P}^2$ be a compact embedded real analytic curve such that for some choice of orientation and positive integer multiplicity on each component, condition (2) above is satisfied. Then γ is stable.

In Part I the conclusion of Theorem 1.1 was established for any stable real analytic curve $\Gamma \subset \mathbf{P}^n$ (with orientation and multiplicity on each component). The main point of Part II is to prove that this result for $p = 1$ implies the result for all $p > 1$, provided one can drop the stability hypothesis in the $p = 1$ case.

Note incidentally that there is no assumption of maximal complexity on the cycle Γ.

Theorem 1.1 represents a projective analogue of a result of H. Alexander and J. Wermer [AW].

If the cycle Γ in Theorem 1.1 bounds a holomorphic p-chain T, then there is a unique such chain T_0 of least mass with $dT_0 = \Gamma$. (All others are obtained by adding positive algebraic p-cycles to T_0.)

Corollary 1.2. *Let Γ be as in Theorem 1.1 and suppose that $\Gamma = dT$ for some positive holomorphic p-chain T. Then T is the chain of least mass with boundary Γ if and only if*

$$\inf_Z \left\{ \frac{T \bullet Z}{\deg Z} \right\} = 0$$

where the infimum is taken over all positive algebraic $(n - p)$-cycles in $\mathbf{P}^n - \Gamma$.

The linking hypothesis (2) in Theorem 1.1 can be replaced by other hypotheses. This is discussed in §3. Another interesting consequence of Theorem 1.1 is the following result, whose proof follows exactly the lines given in Part I for the case $p = 1$.

Theorem 1.3. *Let $M \subset \mathbf{P}^n$ be a compact embedded real analytic submanifold of dimension $2p - 1$ and assume Conjecture B. Then a class $\tau \in H_{2p}(\mathbf{P}^n, M; \mathbf{Z})$ contains a positive holomorphic chain T (with $\operatorname{supp} dT \subseteq M$) if and only if*

$$\tau \bullet u \geq 0$$

for all classes $u \in H_{2n-2p}(\mathbf{P}^n - M; \mathbf{Z})$ which are represented by positive algebraic cycles.

Theorem 1.1 and many of its consequences carry over to general projective manifolds. This is done in §4.

We recall our convention that $d^C = \frac{i}{2\pi}(\overline{\partial} - \partial)$.

2. The Projective Alexander-Wermer Theorem

Let Γ be a compact smooth oriented submanifold of dimension $2p-1$ in $\mathbf{P}^n$. We recall that (even if Γ is only class C^1) any irreducible complex analytic subvariety $V \subset \mathbf{P}^n - \Gamma$ has finite Hausdorff $2p$-measure and defines a current $[V]$ of dimension $2p$ in $\mathbf{P}^n$ by integration on the canonically oriented manifold of regular points. Furthermore, the boundary of this current is of the form $d[V] = \sum_j \epsilon_j \Gamma_j$ where $\Gamma_1, \ldots, \Gamma_\ell$ represent the connected components of Γ and $\epsilon_j = 1, 0,$ or -1. (See [H] for example.) We now allow Γ to carry positive integer multiplicities on each component, so it is of the form $\Gamma = \sum_j m_j \Gamma_j$.

Definition 2.1. By a **positive holomorphic p-chain with boundary** Γ we mean a finite sum $T = \sum_k n_k[V_k]$ where each $n_k \in \mathbf{Z}^+$ and each $V_k \subset \mathbf{P}^n - \Gamma$ is an irreducible subvariety of dimension p, so that

$$dT = \Gamma \qquad \text{(as currents on } \mathbf{P}^n)$$

By the **mass** of such a chain $T = \sum_k n_k[V_k]$ we mean its weighted volume: $\mathbf{M}(T) \equiv \sum_k n_k \mathcal{H}^{2p}(V_k) = T(\Omega_p)$ where $\mathcal{H}^{2p}$ denotes Hausdorff $2p$-measure and

$$\Omega_p \equiv \frac{1}{p!} \omega^p$$

Proposition 2.2. *Suppose T is a positive holomorphic p-chain with boundary Γ as above. Then*

$$\widetilde{\text{Link}}_{\mathbf{P}}(\Gamma, Z) \geq -\mathbf{M}(T)$$

for all positive algebraic cycles Z with support in $\mathbf{P}^n - \Gamma$.

Proof. Note that since $dT = \Gamma$ we have

$$\widetilde{\text{Link}}_{\mathbf{P}}(\Gamma, Z) = \frac{T \bullet Z}{p! \deg Z} - T(\Omega_p) \geq -T(\Omega_p) = -\mathbf{M}(T)$$

since $T \bullet Z \geq 0$ by the positivity of T and Z. $\qquad\qquad \square$

Note that Proposition 2.2 holds for positive holomorphic chains with quite general boundaries Γ. This brings us to the main result.

Theorem 2.3. *Under the assumption of Conjecture B the following are equivalent.*

(1) $\Gamma = dT$ *where T is a positive holomorphic p-chain with $\mathbf{M}(T) \leq \Lambda$ in $\mathbf{P}^n$.*

(2) $\widetilde{\text{Link}}_{\mathbf{P}}(\Gamma, Z) \geq -\Lambda$ *for all $(n - p)$-dimensional algebraic varieties $Z \subset \mathbf{P}^n - \Gamma$.*

Proof. Proposition 2.2 states that (1) $\Rightarrow$ (2). For the converse we shall show that the linking condition persists for hyperplane slices, and then proceed by induction on dimension.

Proposition 2.4. *Suppose that Γ satisfies the Λ-linking condition* (2) *in* $\mathbf{P}^n$. *If $H \cong \mathbf{P}^{n-1}$ is a hyperplane which intersects Γ transversely, then* $\Gamma_H \equiv \Gamma \cap H$ *satisfies the Λ'-linking condition in H where*

$$\Lambda' = p\Lambda + \int_\Gamma d^C u \wedge \Omega_{p-1}$$

and $u = \log(|Z_0|/\|Z\|)$ where Z_0 is the linear function defining H.

Proof. Since bordism and homology agree in $\mathbf{P}^n$ there exists a compact oriented $2p$-manifold N with boundary and a smooth map $f : N \to \mathbf{P}^n$ such that f is an immersion near ∂N and

$$f\big|_{\partial N} \colon \partial N \to \Gamma \qquad \text{is an oriented diffeomorphism.}$$

Since Γ is transversal to H, f is also transversal to H near the boundary. By standard transversality theory we can perturb f, keeping it fixed near the boundary, so that it is everywhere transversal to H. Let $N_H \equiv f^{-1}(H)$ oriented by N and the normal bundle to H, and let $f_H : N_H \to H$ be the restriction of f. Then $(f_H)_*[N_H]$ defines a $(2p-2)$-dimensional current in H with boundary Γ_H. We denote this current simply by $[N_H]$.

Suppose now that $Z \subset H - \Gamma_H$ is an $(n-p)$-dimensional algebraic subvariety. We may assume, again by a small perturbation, that f misses the singular set of Z and is transversal to $\mathrm{Reg}(Z)$. It is then straightforward to check that

$$[N] \bullet Z = [N_H] \bullet_H Z \tag{2.1}$$

where "$\bullet_H$" denotes the intersection pairing in H (defined as in §3 of Part I).

By assumption we have that

$$\widetilde{\mathrm{Link}}_{\mathbf{P}}(\Gamma, Z) = \frac{1}{p!}\left\{\frac{1}{\deg Z}([N] \bullet Z) - \int_N \omega^p\right\} \geq -\Lambda. \tag{2.2}$$

Now the function u above satisfies the Poincaré-Lelong equation

$$dd^C u = H - \omega. \tag{2.3}$$

Substituting (2.1) and (2.3) into (2.2) gives

$$\frac{1}{\deg Z}([N_H] \bullet_H Z) - \int_N (H - dd^C u) \wedge \omega^{p-1}$$

$$= \frac{1}{\deg Z}([N_H] \bullet_H Z) - \int_{N_H} \omega^{p-1} + \int_N dd^C u \wedge \omega^{p-1}$$

$$= \widetilde{\mathrm{Link}}_\mathbf{P}(\Gamma_H, Z)(p-1)! + \int_\Gamma d^C u \wedge \omega^{p-1}$$

$$\geq -\Lambda p!$$

where the first equality is straightforwardly justified using transversality.   □

Corollary 2.5. *Assume that* (2) $\Rightarrow$ (1) *for all manifolds* Γ *of dimension* $2p - 3$ *in projective space. Suppose that* Γ *is a* $(2p - 1)$*-manifold satisfying* (2) *and that* H *is a hyperplane transversal to* Γ. *Then there exists* $\Lambda' > 0$ *so that* $\Gamma_{H'} \equiv \Gamma \cap H'$ *bounds a positive holomorphic* $(p - 1)$*-chain of mass* $\leq \Lambda'$ *for all hyperplanes* H' *in a neighborhood* U *of* H.

Proof. If H is transversal to Γ then so are all hyperplanes H' in a neighborhood of H. Furthermore, the integral $\int_\Gamma d^C(u_{H'}) \wedge \omega^{p-1}$ depends continuously on H' in that neighborhood, where $u_{H'} = \log(|(A_{H'}, Z)|/\|Z\|)$ and $A_{H'}$ is a continuous choice of vectors with $H' = \{[Z] \in \mathbf{P}^n : (A_{H'}, Z) = 0\}$. It follows that the constant Λ' in Proposition 2.4 is uniformly bounded below in a neighbohhood of H. One then applies the inductive hypothesis.   □

Proposition 2.6. *Let* U *be the neighborhood given in Corollary* 2.5. *For each hyperplane* H' *in* U *let* $T_{H'}$ *be the positive holomorphic* $p - 1$ *chain of least mass with* $dT_{H'} = \Gamma_{H'}$. *Then* $T_{H'}$ *is uniquely determined by* H' *and the mapping* $H' \mapsto T_{H'}$ *is continuous on* U.

Proof. We first prove uniqueness. For future reference we formulate this result for Γ instead of $\Gamma_{H'}$.

Lemma 2.7. *Suppose* Γ *bounds a positive holomorphic* p*-chain. Then the positive holomorphic* p*-chain of least mass with boundary* Γ *is unique.*

Proof. Suppose that $T = \sum_i n_i[V_i]$ and $T' = \sum_j n'_j[V'_j]$ are positive holomorphic p-chains of least mass having the same boundary $dT = dT' = \Gamma$ in $\mathbf{P}^n$. By the least mass hypothesis we know that

$$d[V_i] \neq 0 \quad \text{and} \quad d[V'_j] \neq 0 \qquad \text{for all } i, j. \tag{2.4}$$

In fact $d[V_i]$ and $d[V_j']$ each consist of a finite number of oriented connected components of Γ, each with multiplicity one. Let Γ_1 be an oriented connected component of Γ. (If Γ has multiplicity greater than 1 along Γ_1, we ignore that multiplicity for the moment.) Then there must exists a component of T, say V_1, such that Γ_1 forms part of the oriented boundary dV_1. Similarly there is a component, say V_1' of T' such that Γ_1 is part of dV_1'. By boundary regularity [HL$_1$], and local and global uniqueness these two irreducible subvarieties of $\mathbf{P}^n - \Gamma$ must coincide. Hence, $S \equiv T - V_1$ and $S' \equiv T' - V_1'$ are positive holomorphic p chains with $dS = dS'$. By continuing this process one of the two chains will eventually be reduced to zero. However, the other must also be 0 since its boundary is zero and its remaining components satisfy condition (2.4)   □

To prove continuity it will suffice to show that every convergent sequence $H_j \to H$ has a subsequence such that $T_{H_j} \to T_H$. By the local uniform bound on the mass, the fact that $dT_{H_j} \to dT_H = \Gamma_H$, and the compactness of positive holomorphic chains, we know that there is a subsequence which converges to some positive holomorphic chain T with boundary Γ_H. We then apply the uniquness.   □

We recall that Γ is said to be **maximally complex** if

$$\dim_{\mathbf{C}}(T_x\Gamma \cap JT_x\Gamma) = p - 1 \qquad \text{for all } x \in \Gamma$$

where J is the almost complex structure on $\mathbf{P}^n$.

Proposition 2.8. *Assume Conjecture B. If Γ satisfies (2), then Γ is maximally complex.*

Proof. The result is trivial when $\dim\gamma = 1$ so we first consider the case $\Gamma = \Gamma^3 \subset \mathbf{P}^3$. We want to show that $\int_\Gamma \alpha = 0$ for all $(3,0)$-forms α on $\mathbf{P}^n$.

Choose a line $L \cong \mathbf{P}^1 \subset \mathbf{P}^3$ with $\Gamma \cap L = \emptyset$ and a linear projection $\pi : \mathbf{P}^3 - L \to \mathbf{P}^1$. Fix a point $x_\infty \in \mathbf{P}^1$ and choose affine coordinates (z_0, z_1, z_2) on $\pi^{-1}(\mathbf{P}^1 - \{x_\infty\}) \cong \mathbf{C}^3$. We shall show that

$$\int_\Gamma g(z_0)dz_0 \wedge dz_1 \wedge dz_2 = 0 \tag{2.5}$$

for all $g \in C_0^\infty(\mathbf{C})$. Such forms, taken over all possible choices above, are dense in $\mathcal{E}^{3,0}$ on a neighborhood of Γ. Hence, $\int_\Gamma \alpha = 0$ for all $(3,0)$-forms α, which implies that Γ is maximally complex.

To prove (2.5) we choose a 4-chain N with compact support in $\mathbf{P}^3 - L$ such that $dN = \Gamma$. Then (2.5) can be rewritten as

$$\int_N \frac{\partial g}{\partial \bar{z}_0} dz_0 \wedge d\bar{z}_0 \wedge dz_1 \wedge dz_2 = 0$$

for all $g \in C_0^\infty(\mathbf{C})$. For this it will suffice to prove that

$$(N \wedge dz_1 \wedge dz_2, \pi^*\eta) = (\pi_*(N \wedge dz_1 \wedge dz_2), \eta) = 0$$

for any (1,1)-form η with compact support in $\mathbf{C}$. For this it will suffice to consider $\eta = \delta(z_0 - t)dz_0 \wedge d\bar{z}_0$ for $t \in \mathbf{C}$, in other words we want to show that the slice at t:

$$\{\pi_*(N \wedge dz_1 \wedge dz_2)\}_t = \pi_*\{N_t \wedge dz_1 \wedge dz_2\} = \int_{N_t} dz_1 \wedge dz_2 = 0 \quad (2.6)$$

for all $t \in \mathbf{C}$.

Observe now that $dN_t = \Gamma_t$, the slice of Γ by π at t, and by Proposition 2.4 this Γ_t satisfies the projective linking condition (2). Hence by Theorem 6.1 in [HL$_4$] and our hypothesis that Γ_t is stable, we conclude that $\Gamma_t = dT_t$ where T_t is a positive holomorphic 1-chain in $\mathbf{P}^2 = $ the closure of $\pi^{-1}(t)$. Thus our desired condition (2.6) is established by the following.

Lemma 2.9. *Let γ be a curve, or in fact any rectifiable 1-cycle with compact support in $\mathbf{C}^2 \subset \mathbf{P}^2$. Suppose $\gamma = dT$ where T is a positive holomorphic chain in $\mathbf{P}^2$. Then for any $S \in \mathcal{D}'_{2,\mathrm{cpt}}(\mathbf{C}^2)$ with $dS = \gamma$, one has $S(dz_1 \wedge dz_2) = 0$.*

Proof. It suffices to construct one current S with these properties. We can assume that the line at infinity $\mathbf{P}^1_\infty = \mathbf{P}^2 - \mathbf{C}^2$ meets supp T only at regular points and is transversal there. (The general result follows directly.) Choose $x \in$ supp $T \cap \mathbf{P}^1_\infty$ and let L be the tangent line to supp T at x. Then after an affine transformation of the (z_1, z_2)-coordinates, we may assume $L \cong z_1$-axis. This transformation can be chosen with determinant one, so the form $dz_1 \wedge dz_2$ remains unchanged.

Near the point x, the current T is given by a positive multiple of the graph $\Sigma_R \equiv \{(z_1, f(\frac{1}{z_1})) : |z_1| \geq R\}$ where f is holomorphic in the disk of radius $1/R$ and satisfies

$$f(0) = f'(0) = 0.$$

In particular we have that

$$\lim_{z_1 \to 0} z_1 f(1/z_1) = 0. \quad (2.7)$$

We now modify T by replacing (the appropriate multiple of) Σ_R with the current $L_R + U_R$ where

$$L_R \equiv \{(z_1, 0) : |z_1| \leq R\} \quad \text{and} \quad U_R \equiv \{(z_1, tf(1/z_1)) : |z_1| = R \text{ and } 0 \leq t \leq 1\}$$

with orientations chosen so that $d(L_R + U_R) = d\Sigma_R$. Observe that

$$\int_{L_R+U_R} dz_1 \wedge dz_2 = \int_{U_R} dz_1 \wedge dz_2$$

$$= -\int_{dU_R} z_2 dz_1 = \int_0^{2\pi} f(e^{i\theta}/R) \frac{ie^{i\theta}}{R} d\theta \to 0$$

as $R \to \infty$ by (2.7).

Carrying out this procedure at each point of $\operatorname{supp} T \cap \mathbf{P}^1_\infty$ we obtain a current $T(R_1, \ldots, R_\ell)$ with compact support in $\mathbf{C}^2$ and with $dT(R_1, \ldots, R_\ell) = \gamma$. Since $dz_1 \wedge dz_2$ is closed we have

$$S(dz_1 \wedge dz_2) = T(R_1, \ldots, R_\ell)(dz_1 \wedge dz_2) = \sum_{k=1}^\ell \int_{U_{R_k}} dz_1 \wedge dz_2 \to 0$$

as $R_1, \ldots, R_\ell \to \infty$.   □

We have now proved the proposition for 3-folds in $\mathbf{P}^3$. The result for 3-folds in $\mathbf{P}^n$ follows by considering the family of projections $\mathbf{P}^n ---> \mathbf{P}^3$ which are well defined on Γ. The result for general Γ follows by intersecting with hyperplanes and applying Proposition 2.4.   □

We now show that Γ bounds a positive holomorphic chain by applying the main result in [DH]. Let $L \subset \mathbf{P}^n$ be a linear subspace of (complex) codimension $p - 1$ which is transversal to Γ. We assume that L meets every component of Γ. This can be arranged by taking a Veronese embedding $\mathbf{P}^n \subset \mathbf{P}^N$ of sufficiently high degree, and working with linear subspaces there. One checks that the projective linking numbers of Γ are also bounded in $\mathbf{P}^N$. By applying Proposition 2.4 inductively we see that for all linear subspaces L' in a neighborhood of L, the intersections $\Gamma_{L'} = \Gamma \bullet L'$ satisfy the projective linking condition for oriented curves with multiplicities. Therefore by [HL$_4$] and our assumption of Conjecture B, each slice $\Gamma_{L'}$ bounds a positive holomorphic 1-chain. With this property and maximal complexity, It follows directly from [DH] that Γ bounds a holomorphic p-chain. The unique minimal such chain T will be supported in the subvariety $\mathbf{P}^n \subset \mathbf{P}^N$ because Γ is. Furthermore, T must be positive. If not, there would be a negative component, say T_0. Since T is minimal, $dT_0 \neq 0$. Now L must meet T_0 since it meets all components of Γ. It follows that the minimal holomorphic 1-chain with boundary Γ_L is not positive – a contradiction.

Note. This last paragraph could be replaced by an argument based on the results in [HL$_2$].

So we have proved that $\Gamma = dT$ where T is a positive holomorphic p-chain. We may assume that T is the unique such chain of least mass. It remains to prove that $\mathbf{M}(T) \leq \Lambda$.

Suppose not. Then

$$\mathbf{M}(T) = T(\Omega_p) = \Lambda + r \tag{2.8}$$

for $r > 0$, and we see that

$$\frac{T \bullet Z}{p!\, \deg Z} = \frac{T \bullet Z}{p!\, \deg Z} - T(\Omega_p) + T(\Omega_p)$$

$$= \widetilde{\mathrm{Link}}_{\mathbf{P}}(\Gamma, Z) + T(\Omega_p) \geq -\Lambda + \Lambda + r = r$$

for all algebraic subvarieties Z of codimension p in $\mathbf{P}^n - \Gamma$. Hence, it will suffice to prove that

$$\inf_Z \left\{ \frac{T \bullet Z}{\deg Z} \right\} = 0 \tag{2.9}$$

where Z varies over the codimension p subvarieties of $\mathbf{P}^n - \Gamma$.

Lemma 2.10. *Let T be a positive holomorphic p-chain in $\mathbf{P}^n$ with boundary Γ. Then T is the unique such chain of least mass if and only if every irreducible component of $\operatorname{supp} T$ in $\mathbf{P}^n - \Gamma$ has a non-empty boundary (consisting of components of Γ).*

Proof. If T has least mass, it can have no components with boundary zero, since one could remove these components and thereby reduce the mass without changing the boundary. If T is not of least mass, let T_0 be the least mass solution and note that $d(T - T_0) = 0$. It follows from [HS] (actually, elementary arguments involving local uniqueness at the boundary will suffice here) that $T - T_0 = S - S_0$ where S and S_0 are positive algebraic p-cycles, i.e., $dS = dS_0 = 0$. We assume that S and S_0 have no components in common. From the equation $T + S_0 = T_0 + S$ and the uniqueness of the decomposition of analytic subvarieties into irreducible components, we see that the components of S must be components of T and similarly the components of S_0 must be components of T_0. Since T_0 is least mass, we have $S_0 = 0$ and $T = T_0 + S$. Since T is not the least mass solution, $S \neq 0$. $\square$

Our proof now proceeds by induction on p. We assume that condition (2.9) holds for least mass chains of dimension $< p$, and we shall show that it hold in dimension p. The case $p = 1$ has already been established in Part I, Corollary 6.8.

We return to our positive holomorphic p-chain T of least mass. A positive divisor D in $\mathbf{P}^n$ is defined to be **totally transverse** to T if:

(i) D is smooth,

(ii) D meets every component of $dT = \Gamma$, and each of these intersections is transverse,

(iii) D is transversal to every stratum of the singular stratification of $\operatorname{supp} T$.

This condition is open, and it is non-empty for divisors of sufficiently high degree. Let D be such a divisor with degree d. Let $\mathbf{P}^n \subset \mathbf{P}^N$ be the order d Veronese embedding, and let $H_0 \subset \mathbf{P}^N$ be the hyperplane with $D = H_0 \bullet \mathbf{P}^n$. Then H_0 is totally transverse to T in $\mathbf{P}^N$. Assume there exists H in a neighborhood of H_0 which is totally transverse to T and such that $T_H \equiv H \bullet T$ has no irreducible components with boundary zero. Then there exists a sequence $\{Z_j\}_{j \geq 0}$ of subvarieties of codimension $p-1$ in $H - \Gamma$ such that

$$\lim_{j \to \infty} \left\{ \frac{T_H \bullet_H Z_j}{\deg Z_j} \right\} = 0.$$

Note that $T_H \bullet_H Z_j = T \bullet_{\mathbf{P}^N} Z_j$, and so

$$\lim_{j \to \infty} \left\{ \frac{T \bullet_{\mathbf{P}^N} Z_j}{\deg Z_j} \right\} = 0.$$

Now by a small perturbation we may assume that each Z_j is transversal to $\mathbf{P}^n$. Then $W_j \equiv Z_j \cap \mathbf{P}^n$ is a subvariety of codimension p in $\mathbf{P}^n$ with $T \bullet_{\mathbf{P}^n} W_j = T \bullet_{\mathbf{P}^N} Z_j$ and with degree $\deg W_j = d^p \deg Z_j$. It follows that

$$\lim_{j \to \infty} \left\{ \frac{T \bullet_{\mathbf{P}^n} W_j}{\deg W_j} \right\} = 0$$

as desired.

So we are done unless every totally transverse hyperplane section T_H has components with zero boundary. We will show that this cannot happen for H in a neighborhood of H_0. For the sake of clarity, we assume first that each component of $\operatorname{supp} T$ is smooth. Then by transversality each component of $\operatorname{supp} T_H$ is smooth. Suppose there exists a component V of $\operatorname{supp} T_H$ which is without boundary. Then V will be contained in one of the components of $\operatorname{supp} T$, which we denote by W. Note that W is a smooth submanifolds with real analytic boundary.

Now let V_ϵ be a neighborhood of V in W. Since V_ϵ is a subvariety defined in a neighborhood of a hyperplane, it extends to an irreducible **algebraic** subvariety Y of dimension p. It follows that W is a subdomain with real analytic boundary in Y.

Now the generic hyperplane section of an irreducible variety is again irreducible when $p > 1$ (See [Ha, Prop. 18.10]). It follows that $W \cap H = V$. However, this is impossible since H meets every component of Γ and so it must meet the components of Γ which are contained in W.

When $\mathrm{supp}\, T$ is not smooth the argument is similar. By total transversality and our assumption, there exists a component V of $\mathrm{supp}\, T_H$ (for generic H) with no boundary and with the property that V extends to an irreducible p-dimensional subvariety V_ϵ in a neighborhood of H. By [HL₁, Thm. 9.2], V_ϵ extends to an irreducible algebraic subvariety Y of dimension p in $\mathbf{P}^N$. The remainder of the argument is the same.   □

There is an "affine" version of Theorem 2.3 parallel to the "affine" version (Theorem 6.6) of Theorem 6.1 in Part I. However the reader should note that the hypothesis that Γ is contained in some affine chart is not satisfied generically when $\dim(\Gamma) > 1$.

Theorem 2.11. *Let Γ is as in Theorem 2.3 and suppose $\Gamma \subset \mathbf{C}^n$ for some affine chart $\mathbf{C}^n \subset \mathbf{P}^n$. Assuming Conjecture B, the following are equivalent:*

(1) *There exists a constant Λ so that the classical linking number satisfies*

$$\mathrm{Link}_{\mathbf{C}^n}(\Gamma, Z) \geq -\Lambda\, p!\, \deg Z$$

for all $(n - p)$-dimensional algebraic varieties $Z \subset \mathbf{C}^n - \Gamma$.

(2) *$\Gamma = dT$ where T is a positive holomorphic p-chain in $\mathbf{P}^n$ with*

$$\mathbf{M}(T) \leq \Lambda + \frac{1}{p} \int_\Gamma d^C \log \sqrt{1 + \|z\|^2} \wedge \Omega_{p-1}\,.$$

Proof. We recall that in terms of the affine coordinate z on $\mathbf{C}^n$, the Kaehler form is given by $\omega = \frac{1}{2} dd^c \log(1 + \|z\|^2)$. Now let N be a $2p$-chain in $\mathbf{C}^n$ with $dN = \Gamma$. Then $\widetilde{\mathrm{Link}}_{\mathbf{P}}(\Gamma, Z) p! = \frac{1}{\deg Z} N \bullet Z - \int_N \omega^p = \frac{1}{\deg Z} \mathrm{Link}_{\mathbf{C}^n}(\Gamma, Z) - \int_\Gamma d^C \log \sqrt{1 + \|z\|^2} \wedge \omega^{p-1}$. The result now follows directly from Theorem 2.3.   □

3. Theorems for General Projective Manifolds

The results established above generalize from $\mathbf{P}^n$ to any projective manifold. Let X be a compact complex n-manifold with a positive holomorphic line bundle λ. Fix a hermitian metric on λ with curvature form $\omega > 0$, and give X the Kähler metric associated to ω. Let Γ be a $(2p-1)$-cycle on X with properties as in §2 (i.e., an oriented $(2p-1)$-dimensional submanifold with integral weights), and assume $[\Gamma] = 0$ in $H_{2p-1}(X; \mathbf{Z})$.

Definition 3.1. Let Z be a positive algebraic $(n - p)$-cycle on X which has cohomology class $\ell[\omega^p]$ for some $\ell \geq 1$. If Z does not meet Γ, we can define the **linking number** and the **reduced linking number** by

$$\text{Link}_\lambda(\Gamma, Z) \equiv N \bullet Z - \ell \int_N \omega^p \quad \text{and} \quad \widetilde{\text{Link}}_\lambda(\Gamma, Z) \equiv \frac{1}{\ell\, p!} \text{Link}(\Gamma, Z)$$

respectively, where N is any $2p$-chain in X with $dN = \Gamma$ and where the intersection pairing $\bullet$ is defined as in §3 of Part I with $\mathbf{P}^n$ replaced by X.

To see that this is well-defined suppose that N' is another $2p$-chain with $dN' = \Gamma$. Then $(N - N') \bullet Z - \ell \int_{N-N'} \omega^p = (N - N') \bullet (Z - \ell[\omega^p]) = 0$ because $Z - \ell\omega^p$ is cohomologous to zero in X.

Theorem 3.2. *Under the assumption of Conjecture B the following are equivalent*:

(1) $\Gamma = dT$ *where T is a positive holomorphic p-chain on X with mass $\leq \Lambda$.*

(2) $\widetilde{\text{Link}}_\lambda(\Gamma, Z) \geq -\Lambda$ *for all positive algebraic $(n-p)$-cycles $Z \subset X - \Gamma$ of cohomology class $\ell[\omega]^p$ for $\ell \in \mathbf{Z}^+$.*

Proof. That $(1) \Rightarrow (2)$ follows as in the proof of Proposition 2.2. In fact this shows that (2) holds with no restriction on the cohomology class of Z.

For the converse we may assume (by replacing λ with λ^m if necessary) that the full space of sections $H^0(X, \mathcal{O}(\lambda))$ gives an embedding $X \subset \mathbf{P}^N$. An algebraic subvariety $\widetilde{Z} \subset \mathbf{P}^N$ of codimension p is said to be *transversal* to X if each level of the singular stratification of $\widetilde{Z}$ is transversal to X. More generally, a positive algebraic cycle $\widetilde{T} = \sum_\alpha n_\alpha \widetilde{Z}_\alpha$ of codimension p is *transversal* to X if each Z_α is. Such cycles are dense in the Chow variety of all positive algebraic $(N-p)$-cycles in $\mathbf{P}^N$. This follows from the Transversality Theorem for Families applied to the family $\mathrm{GL}_{\mathbf{C}}(N + 1) \cdot \widetilde{Z}$ and the submanifold X in $\mathbf{P}^N$ (cf. [HL$_1$, App. A]).

It is straightforward to check that if $\widetilde{Z}$ is transversal to X, then $Z = \widetilde{Z} \cap X$ is an algebraic subvariety of codimension p in X with cohomology class $\ell[\omega^p]$ where $\ell = \deg\widetilde{Z}$. Let $\widetilde{Z}$ be such a cycle with the property that $\widetilde{Z} \cap \Gamma = \emptyset$. Let N be a $2p$-chain in X with boundary Γ which meets $Z =$

$\widetilde{Z} \cap X$ transversely at regular points. Then local computation of intersection numbers shows that $N \bullet_X Z = N \bullet_{\mathbf{P}^N} \widetilde{Z}$. Consequently, hypothesis (2) implies that

$$\widetilde{\mathrm{Link}}_{\mathbf{P}}(\Gamma, \widetilde{Z}) = \widetilde{\mathrm{Link}}_{\lambda}(\Gamma, Z) \geq -\Lambda.$$

Since this holds for a dense set of subvarieties of $\mathbf{P}^n - \Gamma$ it holds for all such subvarieties. Theorem 2.3 now implies that Γ bounds a holomorphic chain T in $\mathbf{P}^N$. Since Γ is supported in X, so also is T. $\square$

4. Variants of the Main Hypothesis

The linking hypothesis (2) in Theorem 3.2 can be replaced by several quite different conditions thereby yielding several geometrically distinct results. In this section we shall examine these conditions.

Let X, λ, ω and Γ all be as in §3. Suppose $Z \subset X$ is an algebraic subvariety of codimension-p with cohomology class $\ell[\omega^p]$. Then a **spark** associated to Z with curvature $\ell \omega^p$ is a current $\alpha \in \mathcal{D}'^{2p-1}(X)$ which satisfies the *spark equation*

$$d\alpha = Z - \ell\omega^p. \tag{4.1}$$

Such sparks form the basis of (one formulation of) the theory of differential characters (cf. [HLZ]). Two sparks α, α' satisfying (4.1) will be called **commensurate** if $\alpha' = \alpha + d\beta$ for $\beta \in \mathcal{D}'^{2p-2}(X)$.

A. λ-Winding Numbers. Suppose now that $\Gamma \cap Z = \emptyset$. Up to commensurability we may assume α is smooth in a neighborhood of Γ (See [HLZ, Prop. 4.2]). The **reduced λ-winding number** can then be defined as

$$\widetilde{\mathrm{Wind}}_{\lambda}(\Gamma, \alpha) \equiv \frac{1}{\ell\, p!} \int_{\Gamma} \alpha.$$

It follows from the spark equation (4.1) that

$$\widetilde{\mathrm{Wind}}_{\lambda}(\Gamma, \alpha) = \widetilde{\mathrm{Link}}_{\lambda}(\Gamma, Z). \tag{4.2}$$

In particular, this winding number is independent of the choice of α. This can be directly verified using deRham theory and the fact that $[\Gamma] = 0$ in $H_{2p-1}(X; \mathbf{Z})$.

B. Positivity. Recall that a smooth (p, p)-form ψ on X is called **weakly positive** if $\psi(\xi) \geq 0$ for all simple $2p$-vectors ξ representing canonically oriented complex tangent p-planes to X. A current $T \in \mathcal{D}'_{p,p}(X)$ of bidimension p,p is called **positive** if $T(\psi) \geq 0$ for all weakly positive (p, p)-forms ψ on X. In this case we write $T \geq 0$.

C. Algebraic Homology. Set $H_{2k}^+ \equiv \{z \in H_{2k}(X - |\Gamma|; \mathbf{Z}) : (\omega^k, z) \geq 0\}$, where $|\Gamma| \equiv \operatorname{supp}\Gamma$, and define

$$H_{2k,\mathrm{alg}}^+ \subseteq H_{2k}^+$$

to be the subset of classes z which can be represented by a positive holomorphic k-cycle.

Theorem 4.1. *Suppose Conjecture B holds. Then the cycle Γ bounds a positive holomorphic p-chain of mass $\leq \Lambda$ in X if and only if any of the following conditions holds*:

(a) $\widetilde{\operatorname{Link}}_\lambda(\Gamma, Z) \geq -\Lambda$ *for all positive algebraic cycles $Z \subset X - \Gamma$ of codimension-p with cohomology class $\ell[\omega]^p$ for $\ell \in \mathbf{Z}^+$.*

(b) $\widetilde{\operatorname{Wind}}_\lambda(\Gamma, \alpha) \geq -\Lambda$ *for all sparks α satisfying equation (4.1) with Z as above.*

(c) $\frac{1}{p!}\int_\Gamma \beta \geq -\Lambda$ *for all smooth forms $\beta \in \mathcal{E}^{2p-1}(X)$ for which $d^{p,p}\beta + \omega^p \geq 0$ is weakly positive on X.*

(d) *There exists $\tau \in H_{2p}(X, |\Gamma|; \mathbf{Z})$ with $\partial\tau = [\Gamma]$ such that $\tau \bullet [Z] \geq 0$ for all $[Z]$ as above and $\Lambda = (\tau, \frac{1}{p!}[\omega^p])$*

Proof. Condition (a) represents Theorem 3.2 above. Conditions (a) and (b) are obviously equivalent by equation (4.2). To check the necessity of Condition (c) suppose that $\Gamma = dT$ where T is a positive holomorphic p-chain. Then one has $\int_\Gamma \beta = \int_T d\beta = \int_T d^{p,p}\beta = \int_T(d^{p,p}\beta + \omega^p - \omega^p) \geq -\int_T \omega^p = -p!\mathbf{M}(T)$. On the other hand, Condition (c) implies Condition (a) since the subvarieties Z in question arise by intersection with subvarieties $\widetilde{Z}$ in the ambient projective space, where we can mollify to obtain smooth $(p-1, p-1)$-forms α_ϵ with $d\alpha_\epsilon = \omega^p - \widetilde{Z}_\epsilon$ and $\widetilde{Z}_\epsilon \to \widetilde{Z}$.

For Condition (d), suppose there exists T as above. Then the class $\tau = [T] \in H_{2p}(X, |\Gamma|; \mathbf{Z})$ has the stated properties. Conversely, given τ, choose any $2p$-chain $N \in \tau$. Then $dN = \Gamma$ and for any Z as above we have $0 \leq \tau \bullet [Z] = N \bullet Z = N \bullet Z - \ell \int_N \omega^p + \ell \int_N \omega^p = \ell p! \widetilde{\operatorname{Link}}_\lambda(\Gamma, Z) + \ell\tau(\omega^p)$. $\qquad\square$

Corollary 4.2. *If $H_{2(n-p),\mathrm{alg}}^+$ is contained in a proper subcone of $H_{2(n-p)}^+$, that is, if there exists $\tau \in H_{2p}(X, |\Gamma|; \mathbf{Z})$ with $d\tau \neq 0$ and $\tau \bullet u \geq 0$ for $u \in H_{2(n-p),\mathrm{alg}}^+$, then there exists a positive holomorphic p-chain T on X with non-empty boundary supported in $|\Gamma|$.*

This question of holomorphic representability is discussed in detail in [HL$_5$].

References

[AW] H. Alexander and J. Wermer, *Linking numbers and boundaries of varieties*, Ann. of Math. **151** (2000), 125–150.

[DH] P. Dolbeault and G. Henkin, *Chaînes holomorphes de bord donné dans* $\mathbf{CP}^n$, Bull. Soc. Math. de France, **125** (1997), 383–445.

[F] H. Federer, Geometric Measure Theory, Springer–Verlag, New York, 1969.

[Ha] B. Harris, *Differential characters and the Abel-Jacobi map*, pp. 69-86 in "Algebraic K-theory; Connections with Geometry and Topology", Jardine and Snaith (eds.), Kluwer Academic Publishers, 1989.

[H] F.R. Harvey, Holomorphic chains and their boundaries, pp. 309-382 in "Several Complex Variables, Proc. of Symposia in Pure Mathematics XXX Part 1", A.M.S., Providence, RI, 1977.

[HL1] F. R. Harvey and H. B. Lawson, Jr, *On boundaries of complex analytic varieties, I*, Annals of Mathematics **102** (1975), 223–290.

[HL2] F. R. Harvey and H. B. Lawson, Jr, *Boundaries of varieties in projective manifolds*, J. Geom. Analysis, **14**, no. 4 (2005), 673–695. ArXiv:math.CV/0512490.

[HL3] F. R. Harvey and H. B. Lawson, Jr, *Projective hulls and the projective Gelfand transformation*, Asian J. Math. **10**, no. 2 (2006), 279–318. ArXiv:math.CV/0510286.

[HL4] F. R. Harvey and H. B. Lawson, Jr, *Projective linking and boundaries of positive holomorphic chains in projective manifolds, Part I*, ArXiv:math. CV/0512379.

[HL5] F. R. Harvey and H. B. Lawson, Jr, *Relative holomorphic cycles and duality*, Preprint, Stony Brook, 2006.

[HLW] F. R. Harvey, H. B. Lawson, Jr. and J. Wermer, *On the projective hull of certain curves in* $\mathbf{C}^2$, Stony Brook Preprint, 2006.

[HLZ] F. R. Harvey, H. B. Lawson, Jr. and J. Zweck, *A deRham-Federer theory of differential characters and character duality*, Amer. J. of Math. **125** (2003), 791–847.

[HS] F.R. Harvey and B. Shiffman, *A characterization of holomorphic chains*, Ann. of Math., **99** (1974), 553–587.

Chapter 16

ASPECTS OF METRIC GEOMETRY OF FOUR MANIFOLDS

Gang Tian[*]

Department of Mathematics, Princeton University and Beijing University

In this note, we will give a brief tour of some recent progress on metric geometry of 4-manifolds and discuss some open problems.

1. Canonical Structure of 4-Manifolds

Let M be a 4-manifold and g be a Riemannian metric on it. It induces a Hodge operator $\star : \Lambda^2 M \mapsto \Lambda^2 M$ by

$$\varphi \wedge \star\psi = (\varphi, \psi)_g dv_g \,,$$

where $\varphi, \psi \in \Lambda^2 M$, $(\cdot, \cdot)_g$ denotes the inner product on $\Lambda^2 M$ induced by g and dv_g denotes the volume form of g.

If $e_1, \ldots, e_4$ form an orthonormal basis of the cotangent space $T_p^* M$ at any $p \in M$, then $\star^2 = \mathrm{Id}$ and

$$\star(e_1 \wedge e_2) = e_3 \wedge e_4, \star(e_1 \wedge e_3) = e_4 \wedge e_2, \star(e_1 \wedge e_4) = e_2 \wedge e_3 \,.$$

This $\star$ decomposes $\Lambda^2 M$ into self-dual part $\Lambda_+ M$ and anti-self-dual part $\Lambda_- M$, each of which is of dimension 3. Accordingly, we have a decomposition of g's curvature operator $Rm(g)$:

$$Rm = \begin{pmatrix} W_+ + \dfrac{S}{12} & Z \\ Z & W_- + \dfrac{S}{12} \end{pmatrix}$$

where $W = W_+ + W_-$ is the Weyl tensor whose vanishing implies that g is locally conformally flat, Z is the traceless Ricci curvature, that is, $Z = Ric(g) - \frac{S}{4}g$, and S is the scalar curvature.

---

[*]Supported partially by NSF grants and a Simons fund.

There are two types of canonical metrics:

1. **Einstein metrics**: g is Einstein if its curvature $Rm(g)$ is self-dual or equivalently, its Ricci curvature $Ric(g)$ is a constant multiple of g;

2. **Anti-self-dual metrics**: g is anti-self-dual and of constant scalar curvature, that is, $W_+ = 0$, $S = $ constant.

More generally, one can perturb the anti-self-dual equation: Let f be a function on $\mathbb{R}$ taking values in traceless 3×3 matrices, we can have

$$W_+ = f(S), \ S = \text{const.} \tag{1.1}$$

In the case of Kähler geometry, there are extremal metrics, that is, Kähler metrics with constant scalar curvature S, in this case, the half Weyl tensor W_+ can be written as a function $f(S)$ of S, where $f(S) = \text{diag}(S/6, -S/12, -S/12)$.

2. Topological Constraints

The existence of canonical metrics impose topological constraints on underlying 4-manifolds. Here we collect a few known ones which can be derived from the Gauss-Bonnet-Chern formula.

By the Gauss-Bonnet-Chern formula for the Euler number, we have

$$\chi(M) = \frac{1}{12\pi^2} \int_M \left(|W|^2 - |Z|^2 + \frac{S^2}{24} \right) dV, \tag{2.1}$$

where $\chi(M)$ denotes the Euler number.

On the other hand, by the index theorem, we have the following for the signature $\tau(M)$:

$$\tau(M) = \frac{1}{8\pi^2} \int_M (|W_+|^2 - |W_-|^2) dV. \tag{2.2}$$

It follows from these: If M admits an Einstein metric, then we have the Hitchin-Thorpe inequality

$$|\tau(M)| \leq \frac{2}{3}\chi(M).$$

Furthermore, in [Hi], Hitchin proved that the equality holds if and only if M is diffeomorphic to $K3$ surfaces, which include quartic surfaces in $\mathbb{C}P^3$.

As a corollary of the Hitchin-Thorpe inequality, we can easily show that $\mathbb{C}P^2 \sharp k \overline{\mathbb{C}P^2}$ does not have Einstein metrics for $k \geq 9$, while it does admit an Einstein metric for $k \leq 8$ except $k = 2$, see [TY1]. An interesting question is whether or not there is an Einstein metric for $k = 2$. I believe that there is in fact such a metric which is conformal to a Kähler metric.

Assuming that M has non-vanishing Seiberg-Witten invariant, in [Le3], Lebrun proved that $\tau(M) \leq \frac{1}{3}\chi(M)$, which generalizes the Miyaoka-Yau inequality for complex surfaces.

We can state another corollary of the Hitchin-Thorpe inequality: If a simply-connected spin 4-manifold M is a connected sum of smooth 4-manifolds which are homeomorphic to Einstein 4-manifolds, then $8\,b_2(M) \geq -11\,\tau(M)$.

Anti-self-dual metrics also impose constraints on M. First the signature $\tau(M)$ has to be non-positive if M admits an anti-self-dual metric. If $\tau(M) = 0$, then M is locally conformal flat. If M is further simply-connected, M has to be a standard 4-sphere.

3. Curvature Estimates for Einstein Metrics

In dimension 2 or 3, any Einstein metric is of constant sectional curvature, hence, its underlying manifold is simply a quotient of a space form by a discrete group. In dimension 4, there are many more Einstein metrics than metrics of constant sectional curvature. For example, by the Aubin-Calabi-Yau theorem, Calabi-Yau theorem and my theorem ([Au], [Ya], [Ti]), any complex surface with definite first Chern class admits Kähler-Einstein metrics if its Lie algebra of holomorphic vector fields is reductive. On the other hand, as we have seen in last section, the existence of an Einstein metrics does impose strong constraints on the underlying 4-manifold. Hence, it is natural to study the conditions for a closed 4-manifold to admit an Einstein metric. Any approach to this existence problem by geometric analytic methods, e.g. by Ricci flow, will lead to the question of how solutions to the Einstein equation can develop singularities, or equivalently, how Einstein metrics can degenerate. This is a program of Cheeger and mine to give a complete understanding how Einstein metrics develop singularity. The technical parts in this program are the curvature and injectivity radius estimates proved in [CTi].

Consider the normalized Einstein equation

$$\text{Ric}(g) = \lambda g, \tag{3.1}$$

where the Einstein constant λ is $-3, 0$ or 3, and if $\lambda = 0$, we add the additional normalization that the volume is equal to 1.

The first technical result is an ϵ-regularity theorem.

Theorem 3.1. [CTi](ϵ-regularity). *There exists $\epsilon > 0$, c, such that the following holds: If g is a solution to (3.1) and $B_s(p)$ have compact closure* *in a geodesic ball $B_r(p)$ for all $s \leq r$ and some $r \leq 1$, and if*

$$\int_{B_r(p)} |Rm(g)|^2 \leq \epsilon, \qquad (3.2)$$

then

$$\sup_{B_{\frac{1}{2}r}(p)} |Rm(g)| \leq c \cdot r^{-2}, \qquad (3.3)$$

where $Rm(g)$ denotes the sectional curvature of g.

Remark 3.2. *If $\lambda = 0$, then the assumption $r \leq 1$ can be dropped. This can easily seen by scaling. In particular, it implies that a complete Ricci-flat 4-manifold with finite L^2-norm of curvature has quadratic curvature decay.*

One can compare this to well-known ϵ-regularity theorems for Yang-Mills and harmonic maps. They can be proved by Moser iteration and using a bound on the Sobolev constant of the underlying domain. Since in these cases the domain is effectively a standard ball, such a bound is always available. In [An1], [Na], [Ti], the Moser iteration argument was extended to Einstein 4-manifolds to get a version of the above theorem under the assumption that the L^2-norm of the curvature is sufficiently small against the Sobolov constant. The latter can be bounded in terms of a lower bound on the volume of the geodesic ball involved.

The proof of Theorem 3.1 is considerably more difficult than those of the earlier ϵ-regularity theorems and employs entirely different techniques. Also, as a consequence of our ϵ-regularity, we know essentially the topology of the geodesic ball $B_{r/2}(p)$, that is, it is essentially a quotient of an euclidean ball by euclidean isometries. This also tells a difference between our theorem and previous ones for Yang-Mills etc.: We determine the topology as well as analytic property of the geodesic ball considered at the same time.

The next result gives an estimate on the injectivity radius.

Theorem 3.3. [CTi](Lower Bound on Injectivity Radius). *For any $\delta >$ and $v > 0$, there exists $w = w(v, \delta, \chi) > 0$, such that if (M, g) is a complete Einstein 4-manifold with L^2-norm of curvature equal to $12\pi^2\chi$ [1], $\mathrm{vol}(M, g) \geq v$ and $\lambda = \pm 3$, then the set $\mathbf{S}_w$ of $p \in M$ where the injectivity radius at p is less than w has measure less than δ.*

---

[1]If M is compact, χ is just the Euler number of M.

The proof for both theorems above can be proved by using a refined version of Chern's transgression [Che] for the Gauss-Bonnet-Chern formula for the Euler number. Let P_χ be the Chern-Gauss-Bonnet form. On subsets of Riemannian manifolds which are sufficiently collapsed with locally bounded curvature, there is an essentially canonical transgression form $\mathcal{T}P_\chi$, satisfying

$$d\mathcal{T}P_\chi = P_\chi, \quad \text{and} \quad |\mathcal{T}P_\chi| \leq c(4) \cdot (r_{|Rm|}(p))^{-3},$$

where $r_{|Rm|}(p)$ is the supremum of those r such that the curvature Rm is bounded by $1/r^2$ on $B_r(p)$. In fact, this can be done for any dimensions. However, if (M, g) is an Einstein 4-manifold, we have

$$P_\chi = \frac{1}{8\pi^2} \cdot |R|^2 \cdot dV. \tag{3.4}$$

It follows that near those points where (M, g) is sufficiently collapsed with locally bounded curvature, local L^2-norm of curvature for g can be controlled by the curvature on the boundary in a weaker norm. Then the above theorems can be deduced by applying this estimate. We refer the readers to [CTi] for details. In [CTi], we also used a different approach to achieve this same estimate by an equivariant chopping theorem extending [CGr].

4. Compactifying Moduli of Einstein 4-Manifolds

In this section, we describe a compactifcation of moduli of Einstein 4-manifolds following [CTi].

Denote by $\mathcal{M}(\lambda, \chi, v)$ the moduli of all solutions to (3.1) with volume equal to v and such that the underlying manifold is closed and has the Euler number χ. In most cases, this moduli is not compact. It is natural to study how degenerated metrics may arise in its compactification. More specifically, we would like to know the properties of a natural compactification, which is analogous to the Deligne-Mumford compactification in the 2-dimensional case.

Let (M_i, g_i) be a sequence of Einstein 4-manifolds with fixed Euler number and Einstein constant and volume. Let $y_i \in M_i$ be a sequence of base points. After passing to a subsequence if necessary, there is a limit (M_∞, y_∞) of the sequence (M_i, g_i, y_i) in a suitable weak geometric sense, the pointed Gromov-Hausdorff sense. This limit space can be thought of as Einstein manifolds with singularities, although apriori they are length spaces and might not have any smooth points whatsoever. The program of

Cheeger and mine provides understanding of smooth structure of this limit space in general cases.

If $\lambda = 3$, then the diameter of (M, g) in $\mathcal{M}(\lambda, \chi, v)$ is uniformly bounded. It is a non-collapsing case. It has been known since late 80's that the limit M_∞ is an Einstein orbifold with isolated singularities and (M_i, g_i) converges to M_∞ in the Cheeger-Gromov topology[2], see [An1], [Na], and [Ti] in the Kähler case. As a consequence, the moduli can be compactified by adding Einstein 4-dimensional orbifolds with isolated singularities. In this non-collapsing case, the Sobolev constant is uniformly bounded, then the curvature estimate in Theorem 3.1 can be achieved by following Uhlenbeck's arguments in proving curvature estimate for Yang-Mills connections in 4-dimension.

If $\lambda = -3$, the diameter for a sequence of Einstein metrics can diverge to ∞. It is crucial to bound injectivity radius uniformly from below at almost every point in order to have a fine structure for the limit M_∞. Using the curvature and injectivity estimates stated in last section, Cheeger and I proved the following in [CTi]

Theorem 4.1. *Let (M_i, g_i) be a sequence of Einstein 4-manifolds in $\mathcal{M}(-3, \chi, v)$. Then by taking a subsequence if necessary, there is a sequence of N-tuples $(y_{i,1}, \ldots, y_{i,N})$ satisfying:*

1. *N is bounded by a constant depending only on χ;*
2. *$y_{i,\alpha} \in M_i$ and for any distinct α, β, $\lim_{i\to\infty} d_{g_i}(y_{i,\alpha}, y_{i,\beta}) - \infty$;*
3. *For each $1 \leq \alpha \leq N$, in the Gromov-Hausdorff topology, $(M_i, g_i, y_{i,\alpha})$ converges to a complete Einstein orbifold $(M_{\infty,\alpha}, g_{\infty,\alpha}, y_{\infty,\alpha})$ with only finitely many isolated quotient singularities and $\lim_{i\to\infty} y_{i,\alpha} = y_{\infty,\alpha}$;*
4. *$(M_i, g_i, y_{i,1}, \ldots, y_{i,N})$ converges to $(M_\infty, g_\infty, y_{\infty,1}, \ldots, y_{\infty,N})$ in the Cheeger-Gromov topology, where $M_\infty = \bigcup_\alpha M_{\infty,\alpha}$ and $g_\infty|_{M_{\infty,\alpha}} = g_{\infty,\alpha}$. In particular, $\lim_{i\to\infty} \mathrm{vol}(M_i, g_i) = \mathrm{vol}(M_\infty, g_\infty)$.*

If $\lambda = 0$, the sequence (M_i, g_i) can collapse i.e. the injectivity radius can go uniformly to 0. This should be the only case in which collapse can take place. We can scale the metrics g_i such that $(M_i, \mu_i g_i)$ has diameter 1. Then by taking a subsequence, this has a limit M_∞ in the Gromov-Hausdorff topology. Moreover, by Theorem 3.1, the curvature of

---

[2]This means that for any $\epsilon > 0$, there is a compact subset K of the smooth part of M_∞ and diffeomorphisms ϕ_i from a neighborhood of K into M_i such that $M_i \backslash \phi_i(K)$ has measure less than ϵ and $\phi_i^* g_i$ converges to the Einstein orbifold metric of M_∞ in the smooth topology.

g_i is uniformly bounded outside finitely many points, so one has some understanding of the topology of M_i compared to that of M_∞.

The remaining part of my program with J. Cheeger is to understand collapsing parts in the convergence of Einstein 4-manifolds. More precisely, if $x_i \in M_i$ and by taking a subsequence if necessary, (M_i, g_i, x_i) converges to a collapsed limit (Y, d) in the Gromov-Hausdorff topology, where d is a metric inducing the length structure on Y. This includes two cases: 1.The case when $\lambda = 0$ and diameter uniformly bounded as in last paragraph; 2. When $\lambda = -1$, x_i diverges to infinity from any points where local volume is bounded from below. It follows from results stated in last section, outside finitely many points, Y is a limit with bounded curvature, so it has smooth points. In fact, using arguments in [Fu], one may be able to prove that Y is a quotient by a smooth manifold by a group action outside finitely many points. Cheeger and I have shown that a sufficiently collapsed part is covered by open subsets quasi-isometric to open sets in Einstein 4-manifolds with non-trivial Killing fields. Therefore, to understand the structure of Einstein 4-manifolds near their collapsed parts, we are led to classifying all complete Einstein 4-manifolds (M, g) which admit a nilpotent group N action of positive dimension k by isometries. There are 3 cases: $k = 1, 2, 3$. If $k = 3$, it can be easily done. In fact, Cheeger and I verified that if (M, g) is a complete Einstein 4-manifold of scalar curvature -12 and N is a 3-dimensional nilpotent group which acts on M freely by isometries, then (M, g) is an either real or complex hyperbolic space. We have also had results in the case that $k = 1$ or 2.[3]

If M is a Kähler surface, $\mathcal{M}(\lambda, \chi, v)$ contains a component $\mathcal{M}(\Omega, c_2)$ of Kähler-Einstein metrics with Kähler class Ω, where c_1 and c_2 denotes the first and second Chern classes of M. Note that if $c_1 \neq 0$, then $c_1 = \pm 3\,\Omega$.

If $c_1 > 0$ and $\mathcal{M}(c_1/3, c_2)$ is of positive dimension, then M is a Del-Pezzo surface obtained by blowing up $\mathbb{C}P^2$ at m points in general position, where $5 \leq m \leq 8$. In [Ti], it was proved that $\mathcal{M}(c_1/3, c_2)$ can be compactified by adding Kähler-Einstein orbifolds, and furthermore, there are strong constraints on quotient singularities. It was conjectured in [Ti] that *$\mathcal{M}(c_1/3, c_2)$ can be compactified by adding Kähler-Einstein orbifolds with only rational double points.* Indeed, it is true when M is a blow-up of $\mathbb{C}P^2$ at 5 point, see [MaMu].

If $c_1 = 0$, then M is either a complex 2-torus or a K3 surface. It is a collapsing case and is related to problems on large complex limits in the Mirror symmetry.

---

[3]These results will appear in a paper we are preparing.

If $c_1 < 0$ and (M, g_i) be a sequence in $\mathcal{M}(-c_1/3, c_2)$, let (M_∞, g_∞) be one of its limit as in Theorem 4.1, and let $M_{\infty,1}, \ldots, M_{\infty,N}$ be its irreducible components. We know that each $(M_{\infty,\alpha}, g_\infty)$ is a complete Kähler-Einstein orbifold. It should be possible to identify these irreducible components more explicitly. For simplicity, assume that $M_{\infty,\alpha}$ is smooth, then we expect:

$M_{\infty,\alpha}$ *is of the form* $\bar{M} \backslash D$, *where* $\bar{M}$ *is a projective surface and* D *is a divisor with normal crossings, such that* $K_{\bar{M}} + D$ *is positive outside* D *and each component of* D *has either positive genus, or at least two intersection points with components of* D.

Note that the main theorem in [TY3] implies: Given $\bar{M}$ and D as above, there is a Kähler-Einstein metric on $\bar{M} \backslash D$.

5. Complete Ricci-Flat 4-Manifolds

Let (M_i, g_i) be a sequence of Ricci-flat 4-manifolds. In last section, we normalize the diameter of g_i to be 1 and analyze possible collapsing limit. One can also normalize them differently to get non-collapsed limits: complete Ricci-flat manifolds with curvature L^2-bounded. More precisely, if (M_i, g_i) is a sequence of compact Ricci-flat manifolds with diameter one and bounded Euler number, then there are two cases: 1. If they are not collapsing, it has been known that by taking a subsequence, (M_i, g_i) converges to a Ricci-flat orbifold (see [An1], [Na], [Ti]); 2. If they are collapsing, then we can define for any $x \in M_i$,

$$\nu(x) = r(x)^2 \sup_{y \in B_{r(x)}(x)} |Rm(g_i)|_{g_i},$$

where $r(x)$ is the injectivity radius of g_i at x. If $\max_{x \in M_i} \nu(x)$ is uniformly small, then by Cheeger-Gromov, M_i has vanishing Euler number, so g_i is flat and limits can be understood in a very easy way. Otherwise, there are $x_i \in M_i$ such that $\nu(x_i)$ are uniformly bounded from below, then by taking a subsequence if necessary, $(M_i, r(x_i)^{-2} g_i, x_i)$ converge to a complete Ricci-flat orbifold with bounded L^2-norm of curvature. This leads to a natural problem of understanding complete Ricci-flat manifolds with L^2-bounded curvature. Since all known examples of complete Ricci-flat manifolds are Kählerian, we discuss only the Kähler case. Examples of non-flat Ricci-flat Kähler manifolds were first constructed on minimal resolution of $\mathbb{C}^2$ by a finite group in $SU(2)$ by physicists, Hitchin and Calabi explicitly or by using the Twistor theory. First compact examples follow by applying the Calabi-Yau theorem to K3 surfaces, see [Ya]. They were also called Calabi-Yau manifolds. Further complete Calabi-Yau 4-manifolds metrics were constructed by Kronheimer[Kr1], Tian-Yau[TY2], Cherkis-Kapustin [ChKu], Cherkis-Hitchin [ChHi] et al.. A natural question is to see if they or togther with a few more which can be constructed by similar methods, are those complete Calabi-Yau 4-manifolds with L^2-bounded curvature. The work of Cheeger and myself may shed a light on answering this question.

Theorem 5.1. [CTi] *If (M, g) is a complete Ricci-flat manifold with L^2-bounded curvature, then its curvature decays quadratically.*

A related question is the uniqueness of Calabi-Yau metrics on $\mathbb{C}^2$ raised by Calabi long time ago. There are indeed complete non-flat Calabi-Yau metrics, like the Taub-Nut metric, see [Le1]. However, a few years ago, I proved the following result which was not published yet.

Theorem 5.2. *Any complete Calabi-Yau metric on $\mathbb{C}^2$ with maximal volume growth must be flat.*

Proof. Let us outline ideas of its proof here. Let g be such a metric. Define $g_\lambda = \lambda^2 g$ ($\lambda > 0$). Because $\text{vol}(B_r(x, g))/r^4$ is decreasing and converges to a positive constant as $r \to \infty$, by taking a subsequence if necessary, we may assume that $(\mathbb{C}^2, g_\lambda)$ converges to a volume cone as λ goes to 0. It follows from a result of Cheeger and myself, see [Ch], this limiting volume is smooth outside the vortex and the convergence of g_λ to this limit is in smooth topology outside the vortex of the cone. Next we use arguments in [Ba] to prove that curvature of $(\mathbb{C}^2, g)$ decays sufficiently fast. It follows that (M, g) is an ALE space. Then using a result of Kronheimer [Kr2], we conclude that g is flat. $\square$

Remark 5.3. *The arguments actually show that any complete Calabi-Yau 4-manifolds with maximal volume growth must be a deformation of a minimal resolution of the quotient of $\mathbb{C}^2$ by a finite subgroup in $SU(2)$.*

Also we conjecture that the same holds for higher dimensional cases, that is, any complete Calabi-Yau metrics on $\mathbb{C}^n$ with maximal volume growth is flat.

6. Problems on Ricci Flow in 4-Dimension

Consider the Ricci flow introduced by R.Hamilton:

$$\frac{\partial g}{\partial t} = -2\operatorname{Ric}(g), \quad g(0) = g_0. \tag{6.1}$$

In dimension 4, if the initial metric has positive curvature operator, Hamilton proved in [Ha] that after normalizing the volume, the flow has a global solution which converges to a metric of constant sectional curvature. It is for sure that the flow will develop singularities in general cases. A reasonable guess might be that the Ricci flow may develop singularities near hypersurfaces which are circle bundles over Riemann surfaces, more precisely, (6.1) has a solution with surgery $(M_t, g_t)^4$ such that either M_t is empty or consists of finitely many 4-manifolds $M_{t1}, \ldots, M_{tk}$ such that the initial manifold $M = M_0$ is a connected sum of these M_{ti} along hypersurfaces which are circle bundles over Riemann surfaces and some "standard" pieces.

If (M, g_0) is a compact Kähler surface, every $g(t)$ along the Ricci flow is Kähler. It follows from [TZh] that if the Kodaira dimension of M is non-negative[5], then (6.1) has a global solution with surgery, say (M_t, g_t) $(t \geq 0)$, such that each M_t is a Kähler surface obtained by blowing down rational curves of self-intersection -1 successively from previous $M_{t'}$, topologically, M is a connected sum of M_t and finitely many copies of $\mathbb{C}P^2$ with reversed orientation.

If $g(t)$ is a global solution of the normalized Ricci flow:

$$\frac{\partial g}{\partial t} = -2\left(\operatorname{Ric}(g) - \frac{r}{4}g\right), \tag{6.2}$$

where r denotes the integral of scalar curvature. Then the volume of $g(t)$ stays as a constant. We expect that the curvature and injectivity radius estimates in [CTi] hold for $g(t)$ as t goes to ∞. Using Perelman's W-functional, we further expect:

1. If $r > 0$, then $g(t)$ converges to a Ricci orbifold soliton;
2. If $r < 0$, then $g(t)$ converges to a finitely union of complete Einstein orbifolds with negative scalar curvature.
3. If $r = 0$, $g(t)$ either converges to Ricci-flat orbifold with bounded diameter or collapses to a generalized Einstein metric (possibly with singularities) on a lower dimension space.

---

[4]For its definition, see [MoTi].
[5]This is true for almost all Kähler surfaces.

For the case of Kähler surfaces, see [STi].

7. Anti-Self-Dual Metrics and Generalizations

As we have already seen, the anti-self-dual metrics impose strong constraints on underlying 4-manifolds. There have been very effective methods of constructing many examples of anti-self-dual metrics. In [Ta], Taubes proved that given any 4-manifold M, after making connected sum of it with sufficiently many copies of $\overline{CP^2}$, the resulting 4-manifold admits one anti-self-dual metric. We refer the readers to [Ta] for more detail and references on anti-self-dual metrics. There is another powerful method of constructing anti-self-dual metrics, the Twistor space method, see [DoFr], [Le2] and their references.

It has been hoped that the moduli space of anti-self-dual metrics can be used for constructing new differentiable invariants for 4-manifolds. If such an invariant exists, one can compute it and use it to establish existence of anti-self-dual metrics on a 4-manifold. However, there are two major difficulties to be overcome in order to define the invariant: *Compactness and transversality.*

Denote by $\mathcal{AS}(M)$ the moduli space of all anti-self-dual metrics on a 4-manifold M modulo diffeomorphisms and conformal changes of metrics. In general, it is not compact. Therefore, we need to study the following problem: *Given a sequence of anti-self-dual metrics g_i on M, what are possible limits of g_i as i tends to ∞ modulo conformal changes?*

By the signature formula, the L^2-norm of the Weyl tensor is fixed:

$$\int_M ||W(g_i)||^2 dV_{g_i} = 12\pi^2 \tau(M).$$

By the Gauss-Bonnet-Chern formula, the L^2-norm of curvature tensor is bounded in terms of $\tau(M)$, $\chi(M)$ and scalar curvature S, that is,

$$\int_M ||Rm(g_i)||^2 dV_{g_i} = 12\pi^2 \tau(M) - 8\pi^2 \chi(M) + \frac{1}{12}\int_M S(g_i)^2 dV_{g_i}.$$

If the scalar curvature $S(g_i)$ has uniformly bounded L^2-norm, then we have a prior L^2-bound on curvature tensor $Rm(g_i)$. Since the Weyl tensor is a conformal invariant, we can make conformal changes to g_i. Recall the Yamabe constant:

$$Q(M, g_i) = \inf_{u>0} \frac{\int_M \left(|\nabla_{g_i} u|^2 + S(g_i)u^2\right) dV_{g_i}}{\left(\int_M u^4 dV_{g_i}\right)^{\frac{1}{2}}}.$$

Using the Aubin-Schoen solution of the Yamabe conjecture, see [Au2], [Sc], there is a u attaining $Q(M, g_i)$, so we simply take g_i with volume 1 and such that the scalar curvature $S(g_i)$ is $Q(M, g_i)$. Then we have

$$\int_M S(g_i)^2 dV_{g_i} = Q(M, g_i)^2 .$$

Therefore, if g_i form a sequence of anti-self-dual metrics with bounded Yamabe constant, then we may assume that their curvature are uniformly L^2-bounded and they have fixed volume. One can ask two questions:

1. *Given a compact 4-manifold, is there a uniform bound on the Yamabe constant for anti-self-dual metrics?*
2. *What are possible limits of anti-self-dual metrics g_i with uniformly bounded Yamabe constant?*

As a corollary of Theorem 1.3 in [TV], one has the following partial answer to the second question:

Theorem 7.1. *Let g_i be a sequence of anti-self-dual metrics on M with bounded Yamabe constant. We further assume that there is a uniform constant c such that for any function f,*

$$\left(\int_M f^4 dV_{g_i} \right)^{\frac{1}{2}} \le c \int_M (|df|^2_{g_i} + f^2) dV_{g_i} .$$

Then by taking a subsequence if necessary, we have that g_i converges to a multi-fold (M_∞, g_∞) in the Cheeger-Gromov topology[6].

Remark 7.2. *A compactness result similar to the above theorem was claimed in [An2]. However, arguments are not sufficient in the proof which concern the structure of geodesic spheres when the volume estimate is being derived. This is a technical, but crucial, part in the proof.*

I believe that the assumption on the Sobolev constant in the above theorem can be removed by extending effective transgression arguments in [CTi] to anti-self-dual metrics. In order to do it, we also need an estimate on isoperimetric constants. The following conjectured estimate should be true: If g is an anti-self-dual metric with constant scalar curvature or belongs to a certain class of metrics with W^+ controlled by its scalar curvature, for

---

[6]A multi-fold is a connected sum of finitely many orbifolds, see [TV] for definition.

any geodesic ball $B_r(x)$ of radius $r \leq 1$, we have

$$\frac{r\,s(r,g)}{(\mathrm{vol}(B_r(x)))^{\frac{1}{4}}} \geq c\,, \tag{7.1}$$

where c depends on the scalar curvature of g and $\int_{B_r(x)} |Rm(g)|_g^2 dV_g$, $s(r,g)$ denotes the Sobolev constant for functions with compact support in $B_r(x)$. Clearly, this is a non-increasing function on r. It is well-known that the estimate of this sort holds for manifolds with bounded Ricci curvature. Also it may be easier to show this estimate assuming the L^2-norm is sufficiently small.

Now let us discuss the issue of transversality. For this purpose, we need to consider a more general equation which perturbs the anti-self-dual equation.

Let $\mathcal{M}$ be the space of all sufficiently smooth metrics and $\mathcal{D}$ be the group of diffeomorphisms on M. It was shown by Bourguignon long time ago that the quotient $\mathcal{M}/\mathcal{D}$ is a stratified Hausdorff space whose main stratum of all metrics without any non-trivial isometries is a smooth manifold.

Consider the bundle $G(\Lambda^2 M, 3) \mapsto M \times \mathcal{M}/\mathcal{D}$ which consists of all 3-dimensional subspaces in $\Lambda^2 M$. Let $\phi : M \times \mathcal{M}/\mathcal{D} \mapsto G(\Lambda^2 M, 3)$ be a section. It lifts to a section over $M \times \mathcal{M}$, still denoted as ϕ for simplicity. Then for each $x \in M$ and $g \in \mathcal{M}$, $\phi(x,g)$ is a 3-dimensional subspace in $\Lambda_x^2 M$. Since the curvature $Rm(g)$ is an endomorphism of $\Lambda^2 M$, we can consider the following equation: Let π_ϕ be the orthogonal projection from $\Lambda^2 M$ onto ϕ with respect to the metric g, define

$$\phi(Rm) = \pi_\phi \cdot Rm|_\phi : \phi \mapsto \phi\,, \tag{7.2}$$

it is symmetric and let $\phi_0(Rm)$ be its traceless part, then the equation is

$$\phi_0(Rm) = f(S), \quad S = \text{constant}\,, \tag{7.3}$$

where f is a function on $\mathbb{R}$ taking values in trace-free matrices. If $\phi(x,g) = \Lambda_x^+ M$ and $f = 0$, we get just the equation for anti-self-dual metrics.

Next we examine when (7.3) is elliptic. For any $e \in \phi(x) \in \Lambda_x^2 M$, using the metric g, we have an induced endomorphism of $T_x M$, still denoted by e for simplicity.

Lemma 7.3. *The symbol* $\sigma : T_x M \times S_x^2 TM \mapsto S_0^2 \phi(x) \times \mathbb{R}$ *of the linearized operator of (7.3) at* (x,g) *is given as follows: For any* $\xi \in T_x M$ *and* $h \in S_x^2 TM$, $\sigma(\xi, h) = (h_\phi, \mathrm{tr}(h)|\xi|^2)$, *where* $|\cdot|$ *is taken with respect to the inner product* g_x, *where* $h_\phi \in S_0^2 \phi(x)$ *is defined as follows: For* $e, e' \in \phi(x)$, *we have*

$$h_\phi(e, e') = h(e(\xi), e'(\xi)) - \frac{1}{3} \sum_{i=1}^{3} h(e_i(\xi), e_i(\xi)) g_x(e, e'),$$

where $\{e_i\}$ is an orthonormal basis of $\phi(x)$ with respect to g_x.[7]

This lemma can be proved by straightforward computations.

Corollary 7.4. *The equation (7.3) is elliptic modulo diffeomorphisms if and only if for any $\xi \neq 0$, the equations $\operatorname{tr}(h) = 0$, $h_\phi = 0$ and $h(\xi, \cdot) = 0$ ($h \in S^2 TM$) implies that $h = 0$.*

It is easy to check that the above condition holds for the anti-self-dual equation, so the anti-self-dual equation is elliptic.

Let $\mathcal{M}_0$ be the space of Riemannian metrics on M without non-trivial isometries. Then $\mathcal{M}_0/\mathcal{D}$ is smooth. Let $\mathcal{E}$ be the space of smooth sections $\phi : M \times \mathcal{M}_0/\mathcal{D} \mapsto \Lambda^2 M$ such that the resulting equation (7.3) is elliptic. Let $\mathcal{F}$ be the set of smooth functions taking values in matrices with vanishing trace. Put

$$\mathcal{U} = \{([g], \phi, f) \in \mathcal{M}_0/\mathcal{D} \times \mathcal{E} \times \mathcal{F} \mid \phi(Rm(g)) = f(S)\}. \qquad (7.4)$$

There is a natural projection $\pi : \mathcal{U} \mapsto \mathcal{E} \times \mathcal{F}$ which is Fredholm. We further define $\mathcal{U}_0$ to be the set of all $([g], \phi, f)$ such that there is a basis e_1, e_2, e_3 of $\phi(x)$ for some x such that $Rm(g)(e_i)|_x \neq 0$. Then by introducing appropriate norms, one can prove that $\mathcal{U}_0$ is a manifold. This can be proved by proving that the linearized operator of (7.3) in all directions g, ϕ, f is surjective onto its range along $\mathcal{U}_0$. By the Sard-Smale theorem, one can deduce from this

Proposition 7.5. *For a generic (ϕ, f), the moduli of solutions for (7.3) is a smooth manifold of dimension $\frac{1}{2}(15\chi(M) + 29\tau(M))$.*

The dimension is derived by using the index theorem.

It should be possible to extend the compactness theorem in [TV] to this more general equation.

---

[7]We will also denote by g_x the induced inner product on $\Lambda_x^2 M$.

Finally, let me raise a question:

Is there a way of deforming any given metric towards an anti-self-dual metric, like the Ricci flow for the Einstein metric?

This is not possible in a common sense since the linearized equation of the anti-self-dual equation may have non-zero index.

However, one may obtain such a deforming equation by adding auxiliary terms. One possibility can be described as follows: Recall that given a Riemannian 4-manifold, W_+ is a traceless endomorphism of $\Lambda^2_+ M$, so we consider auxiliary fields as traceless endomorphisms $B : \Lambda^2 M \mapsto \Lambda^2 M$, that is, $B \in Hom_0(\Lambda^2 M, \Lambda^2 M)$. With respect to any metric g, we have a decomposition $\Lambda^2 M = \Lambda^2_+ M \oplus \Lambda^2_- M$, correspondingly, there is a decomposition of B. Consider the functional

$$\mathcal{F}(g, B) = \int_M (W_+, B)_g dV_g$$

on the space of metrics with fixed volume form and auxiliary fields. Its gradient flow is given by

$$\frac{\partial B}{\partial t} = W_+ + \frac{1}{2}(\dot{*}B * + * B\dot{*}), \quad \frac{\partial g}{\partial t} = \mathcal{D}_g B, \tag{7.5}$$

where (g, B) is defined on $M \times [0, T)$, $\dot{*}$ denotes the time-derivative of the Hodge star operator as g varies, and $\mathcal{D}_g$ is the adjoint operator of the linearization of the anti-self-dual equation $W_+ = 0$ at g. Note that $\mathcal{D}_g$ maps B into traceless symmetric 2-tensors.

From the second equation, we see that the volume form of g is invariant along the flow since $\mathcal{D}_g B$ is always trace-free. From the first equation, by using the uniqueness theorem of ODE, one can show that if B lies in $Hom_0(\Lambda_+ M, \Lambda_+ M)$ and is symmetric with respect to g at time $t = 0$, so does B for any time $t > 0$ whenever the solution exists.

It may be interesting to see what properties this flow has, such as, local existence, global existence, singularity formation, limiting behavior, etc.

References

[An1]   M. Anderson, Ricci curvature bounds and Einstein metrics on compact manifolds, J. Amer. Math. Soc. 2 (1989), no. 3, 455–490.

[An2]   M. Anderson, Orbifold compactness for spaces of Riemannian metrics and applications, Math. Ann. 331 (2005), no. 4, 739–778.

[Au]   T. Aubin, Équations du type Monge-Ampére sur les varits kähleriennes compactes, C. R. Acad. Sci. Paris Sèr. A-B 283 (1976), no. 3, Aiii, A119–A121.

[Au2]   T. Aubin, Le probléme de Yamabe concernant la courbure scalaire,
          C. R. Acad. Sci. Paris Sèr. A-B 280 (1975), A721–A724.

[Ba]   S. Bando, A. Kasue and H. Nakajima, On a construction of coordinates
          at infinity on manifolds with fast curvature decay and maximal volume
          growth, Invent. Math. 97 (1989), no. 2, 313–349.

[Ch]   J. Cheeger, Degeneration of Einstein metrics and metrics with special
          holonomy, Surveys in differential geometry, Vol. VIII (Boston, MA,
          2002), 29–73, Int. Press, Somerville, MA, 2003.

[Che]   S.S. Chern, A simple intrinsic proof of the Gauss-Bonnet formula for
          closed Riemannian manifolds, Ann. of Math. (2) 45, (1944). 747–752.

[CGr]   J. Cheeger and M. Gromov, Chopping Riemannian manifolds, Differen-
          tial geometry, 85–94, Pitman Monogr. Surveys Pure Appl. Math., 52,
          Longman Sci. Tech., Harlow, 1991.

[CTi]   J. Cheeger and G. Tian, Curvature and injectivity radius estimates for
          Einstein 4-manifolds, J. Amer. Math. Soc. 19 (2006), no. 2, 487–525.

[ChHi]   S. Cherkis and N. Hitchin, Gravitational instantons of type D_k, Comm.
          Math. Phys. 260 (2005), no. 2, 299–317.

[ChKu]   S. Cherkis and A. Kapustin, Singular monopoles and gravitational
          instantons, Comm. Math. Phys. 203 (1999), no. 3, 713–728.

[DoFr]   S. Donaldson and R. Friedman, Connected sums of self-dual manifolds
          and deformations of singular spaces, Nonlinearity 2 (1989), no. 2, 197–
          239.

[Fu]   K. Fukaya, Metric Riemannian geometry, Handbook of differential
          geometry, Vol. II, 189–313, Elsevier/North-Holland, Amsterdam, 2006.

[Ha]   R. Hamilton, Four-manifolds with positive curvature operator, J. Dif-
          ferential Geom. 24 (1986), no. 2, 153–179.

[Hi]   N. Hitchin, Compact four-dimensional Einstein manifolds, J. Differential
          Geometry 9 (1974), 435–441.

[Kr1]   P. Kronheimer, The construction of ALE spaces as hyper-Kähler
          quotients, J. Differential Geom. 29 (1989), no. 3, 665–683.

[Kr2]   P. Kronheimer, A Torelli-type theorem for gravitational instantons,
          J. Differential Geom. 29 (1989), no. 3, 685–697.

[Le1]   C. Lebrun, Complete Ricci-flat Kähler metrics on C^n need not be flat,
          Several complex variables and complex geometry, Proc. Sympos. Pure
          Math., 52, Part 2, Amer. Math. Soc., 1991, 297–304.

[Le2]   C. Lebrun, Anti-self-dual metrics and Kähler geometry, Proceedings of
          the International Congress of Mathematicians, Vol. 1, 2 (Zürich, 1994),
          498–507, Birkhäuser, Basel, 1995.

[Le3]   C. Lebrun, Einstein metrics and Mostow rigidity, Math. Res. Lett. 2
          (1995), no. 1, 1–8.

[MaMu]   T. Mabuchi and S. Mukai, Stability and Einstein-Kähler metric of a
          quartic del Pezzo surface, Einstein metrics and Yang-Mills connections,
          133–160, Lecture Notes in Pure and Appl. Math., 145, Dekker, New
          York, 1993.

[MoTi] J. Morgan and G. Tian, Ricci Flow and the Poincaré Conjecture. A book to be published by the Clay Mathematical Institute through Amer. Math. Soc. (math.DG/0607607).

[Na] H. Nakajima, Hausdorff convergence of Einstein 4-manifolds, J. Fac. Sci. Univ. Tokyo Sect. IA Math. 35 (1988), no. 2, 411–424.

[Sc] R. Schoen, Conformal deformation of a Riemannian metric to constant scalar curvature, J. Differential Geom. 20 (1984), no. 2, 479–495.

[STi] J. Song and G. Tian, The Kähler-Ricci flow on surfaces of positive Kodaira dimension, Preprint 2006 (math.DG/0602150).

[Ta] C. Taubes, The existence of anti-self-dual conformal structures, J. Differential Geom. 36 (1992), no. 1, 163–253.

[Ti] G. Tian, On Calabi's conjecture for complex surfaces with positive first Chern class, Invent. Math. 101 (1990), no. 1, 101–172.

[TV] G. Tian and J. Viaclovsky, Moduli spaces of critical Riemannian metrics in dimension four, Adv. Math. 196 (2005), no. 2, 346–372.

[TY1] G. Tian and S. T. Yau, Kähler-Einstein metrics on complex surfaces with $C_1 > 0$, Comm. Math. Phys. 112 (1987), no. 1, 175–203.

[TY2] G. Tian and S. T. Yau, Complete Kähler manifolds with zero Ricci curvature, I. J. Amer. Math. Soc. 3 (1990), no. 3, 579–609.

[TY3] G. Tian and S. T. Yau, Existence of Kähler-Einstein metrics on complete Kähler manifolds and their applications to algebraic geometry, Mathematical aspects of string theory, 574–628, World Sci. Publishing, Singapore, 1987.

[TZh] G. Tian and Z. Zhang, A note on the Kähler-Ricci flow on projective manifolds of general type. Chinese Annals of Mathematics (2006).

[Ya] S. T. Yau, On the Ricci curvature of a compact Kähler manifold and the complex Monge-Ampére equation, I. Comm. Pure Appl. Math. 31 (1978), no. 3, 339–411.

Chapter 17

ALGÉBRISATION DES TISSUS DE CODIMENSION 1 LA GÉNÉRALISATION D'UN THÉORÈME DE BOL

Jean-Marie Trépreau

Université Pierre et Marie Curie–Paris 6,
UMR 7586, 175 rue du Chevaleret, 75013 Paris, France
trepreau@math.jussieu.fr

Nous montrons que, si $n \geq 3$ et $d \geq 2n$, tout d-tissu de codimension un près d'un point de $\mathbb{C}^n$, qui possède $(2d - 3n + 1)$ relations abéliennes dont les 1-jets sont linéairement indépendants, est isomorphe à un tissu algébrique. C'est en particulier le cas des tissus de rang maximal avec $n \geq 3$ et $d \geq 2n$. Le cas $n = 3$ est dû à Bol [4]. Le cas général résout un problème posé par Chern et Griffiths [6]–[7].

We prove that, if $n \geq 3$ and $d \geq 2n$, a d-web in $\mathbb{C}^n$ is isomorphic to an algebraic d-web, if it has $(2d - 3n + 1)$ abelian relations, the 1-jets of which are linearly independant. The case $n = 3$ is a theorem of Bol [4]. The general case solves a problem which was first considered by Chern and Griffiths [6]–[7].

1. Introduction

1.1. *Tissus; tissus de rang maximal*

Pour fixer les idées et les notations, nous nous plaçons d'emblée dans la catégorie analytique complexe. Notre étude est locale au voisinage de 0 dans $\mathbb{C}^n$, avec $n \geq 2$. On notera $\mathbb{C}^n_0$ un tel voisinage, qu'on pourra réduire autant qu'on veut. On note $(x_0, \ldots, x_{n-1})$ les coordonnées d'un point de $\mathbb{C}^n$. Cette convention inhabituelle sera commode dans certains calculs.

On se donne un entier $d \geq 1$ et un d-tissu $\mathcal{T}$ au voisinage de 0 dans $\mathbb{C}^n$, de codimension[(1)] 1, *i.e.* une famille de d feuilletages de codimension 1, en position générale. On note

$$u_\alpha(x), \qquad \alpha = 1, \ldots, d,$$

[(1)] On ne considère ici que des tissus de codimension 1. Nous aborderons le cas des tissus de codimension supérieure dans un prochain travail.

des fonctions de définition des feuilletages : les feuilles du α-ième feuilletage sont les hypersurfaces de niveau $\{u_\alpha(x) = \text{cste}\}$ de la fonction $u_\alpha(x)$.
L'hypothèse de position générale signifie que toute famille d'au plus n
éléments extraite de la famille de différentielles $du_1(0), \ldots, du_d(0)$ est libre.

Une relation abélienne du tissu $\mathcal{T}$ est un d-uplet $z(x) = (z_1(x), \ldots,$
$z_d(x))$ de fonctions analytiques près de 0 tel que

$$(1) \qquad \sum_{\alpha=1}^{d} z_\alpha(x) \, du_\alpha(x) = 0$$

et que, pour tout α, la fonction $z_\alpha(x)$ soit constante le long des feuilles du
α-ième feuilletage, autrement dit :

$$(2) \qquad dz_\alpha(x) \wedge du_\alpha(x) = 0, \qquad \alpha = 1, \ldots, d.$$

Ces relations forment un espace vectoriel dont la dimension est appelée *le
rang du tissu*. Bol dans le cas $n = 2$, puis Chern dans le cas général, ont
montré que le rang d'un tissu $\mathcal{T}$ est au plus égal au nombre entier

$$(3) \qquad \pi(n, d) = \sum_{q=1}^{+\infty} (d - q(n-1) - 1)^+.$$

(Si $m \in \mathbb{Z}$, on note $(m)^+ = \max(0, m)$.) Nous rappelons une démonstration
de cette inégalité dans le §2.2. *Un tissu de rang $\pi(n, d)$ est dit de rang
maximal.*

1.2. Le problème de l'algébrisation

La géométrie algébrique donne des exemples de tissus, en particulier de
tissus de rang maximal. C'est, avec le problème inverse, l'un des thèmes
majeurs de l'école de Blaschke entre 1927 et 1938, voir [3]. Le problème
inverse est de savoir si tous les tissus de rang maximal viennent de la
géométrie algébrique, à difféomorphisme local près.

Soit C une courbe algébrique de degré d dans l'espace projectif complexe
$\mathbb{P}^n$, non-dégénérée, c'est-à-dire qui n'est pas contenue dans un hyperplan
de $\mathbb{P}^n$. Soit x_0 un point de l'espace $\check{\mathbb{P}}^n$ des hyperplans de $\mathbb{P}^n$. On suppose
que l'hyperplan x_0 coupe la courbe C en d points distincts $p_\alpha(x_0)$, ce qui
est vérifié pour x_0 générique. Tout hyperplan x voisin de x_0 coupe aussi
la courbe C en d points distincts $p_\alpha(x)$. Par dualité projective, les points
$p_\alpha(x)$ définissent d hyperplans de $\check{\mathbb{P}}^n$ passant par x. Parce que la courbe
est non-dégénérée, ces hyperplans sont les feuilles d'un d-tissu au voisinage
de x_0.

Les tissus obtenus de cette manière sont dits algébriques. Un tissu est algébrisable s'il est isomorphe à un tissu algébrique par un difféomorphime analytique local.

À ce point, le théorème d'addition d'Abel est fondamental. Avec les notations précédentes, si ω est une 1-forme holomorphe sur C et p_0 un point donné de C, la somme

$$\sum_{\alpha=1}^{d} \int_{p_0}^{p_\alpha(x)} \omega$$

ne dépend pas de l'hyperplan x voisin de x_0. On peut en fait démontrer que l'espace des 1-formes holomorphes sur la courbe est isomorphe à l'espace des relations abéliennes du tissu associé. En particulier, le rang de ce tissu est égal à la dimension de l'espace des 1-formes holomorphes sur C, autrement dit au genre arithmétique de C.

Le nombre $\pi(n, d)$ défini par (3) est aussi la borne de Castelnuovo pour le genre arithmétique d'une courbe non-dégénérée de degré d dans $\mathbb{P}^n$. Une courbe non-dégénérée de genre $\pi(n, d)$ est dite extrémale. On peut ainsi associer un tissu de rang maximal à toute courbe extrémale.

Il y a une importante différence, qui réapparaîtra, entre le cas $n = 2$ et le cas $n \geq 3$. Le genre arithmétique d'une courbe plane de degré d est toujours égal à $\pi(2, d) = (d-1)(d-2)/2$; le tissu associé est donc de rang maximal. Si $n \geq 3$, la situation est toute différente. Les courbes extrémales sont l'exception.

1.3. *Énoncé du résultat principal*

C'est le suivant :

Théorème 1.1. *Si $n \geq 3$ et $d \geq 2n$, tout d-tissu de rang maximal au voisinage d'un point de $\mathbb{C}^n$ est algébrisable.*

On démontre en fait un résultat plus fort, voir le Théorème 2.4. D'autre part et modulo des variations mineures, la démonstration s'applique aussi aux tissus réels de classe C^∞ au voisinage d'un point de $\mathbb{R}^n$.

Pour $n = 3$, l'énoncé précédent est un théorème de Bol publié en 1933, voir [4]. Le cas $d = 2n$ est classique pour tout n ; on en reparlera. En 1978, Chern et Griffiths ont publié une démonstration du résultat général, mais elle comportait une erreur qu'ils n'ont pas pu corriger, voir [6], [7]. À notre connaissance, notre résultat est donc nouveau quel que soit $n \geq 4$ et $d \geq (2n + 1)$.

Pour moduler cette affirmation, il faut dire que la nouveauté, dans la démonstration, se réduit à peu de choses. La stratégie générale de la démonstration, qu'on appellera la « méthode standard », est celle-là même que Bol a introduite pour $n = 3$ et que Chern et Griffiths ont ensuite adaptée à la dimension quelconque. Dans le cadre de cette méthode, le point crucial consiste à montrer qu'une certaine fonction ϕ de $(n + 1)$ variables, construite à partir des relations abéliennes d'un tissu de rang maximal et à valeur dans un certain espace $\mathbb{P}^m$, *prend ses valeurs sur une surface de* $\mathbb{P}^m$. La démonstration de Bol est indirecte et repose sur une analogie entre les équations que vérifient les fonctions de définition d'un tissu de rang maximal et la « géométrie de Weyl ». Chern et Griffiths poursuivent cette idée et cherchent à montrer que ces équations définissent une « géométrie des chemins ». Notre apport consiste en la remarque qu'il s'agit, dans les deux cas, de calculs à l'ordre deux et que donc, si la méthode standard peut mener au résultat escompté, ce qu'on sait être vrai au moins si $n = 3$, un calcul direct doit permettre de montrer que l'application ϕ est de rang 2 et d'échapper à des considérations géométriques plus subtiles. C'est en effet le cas. Il n'est pas exclu, mais nous ne l'avons pas vérifié, que des calculs analogues permettent de terminer la démonstration inachevée de Chern et Griffiths en en conservant la ligne générale. Quoi qu'il en soit, la démonstration directe est plus courte et plus élémentaire.

1.4. *Le théorème d'Abel inverse*

On appelle tissu linéaire un tissu dont les feuilletages sont des feuilletages en (morceaux d') hyperplans. *Un tissu est linéarisable s'il est localement difféomorphe à un tissu linéaire.* Par construction, les tissus algébriques sont linéaires. Les tissus algébrisables sont donc linéarisables.

Le « théorème d'Abel inverse » est un résultat fondamental de la théorie. Il s'énonce ainsi : tout tissu linéaire qui possède *une* relation abélienne $z(x) = (z_1(x), \ldots, z_d(x))$, dont aucune composante $z_\alpha(x)$ n'est identiquement nulle, est algébrisable. Le cas $n = 2$ est dû à Blaschke et Howe et le cas général fait l'objet de la première partie de l'article [4] de Bol ; voir aussi Griffiths [8] pour des versions encore plus générales de ce théorème, en particulier pour des tissus de codimension plus grande. *Pour démontrer le Théorème 1.1, il suffit donc de montrer que, sous les hypothèses de l'énoncé, le tissu est linéarisable.*

1.5. *Ce qui se passe si $n = 2$ ou si $(n+1) \leq d \leq (2n-1)$*

Rappelons-le brièvement. Quel que soit $n \geq 2$, un d-tissu est linéarisable si $1 \leq d \leq n$; il suffit de prendre ses fonctions de définition comme partie d'un système de coordonnées. Un $(n+1)$-tissu n'est pas linéarisable en général, mais un $(n+1)$-tissu qui possède une relation abélienne non triviale (noter que $\pi(n, n+1) = 1$) est linéarisable. En effet, étant donné une relation (1)–(2), on peut, pour tout α, écrire $z_\alpha(x) = f_\alpha(u_\alpha(x))$ et introduire une primitive g_α de f_α nulle en $u_\alpha(0)$. On a alors la relation :

$$\sum_{\alpha=1}^{n+1} g_\alpha(u_\alpha(x)) = 0 \,.$$

On peut prendre les fonctions $g_\alpha(u_\alpha(x))$ comme fonctions de définition des feuilletages. En changeant de notation, on a donc la relation $\sum_{\alpha=1}^{n+1} u_\alpha(x) = 0$. Il suffit maintenant de prendre $u_1(x), \ldots, u_n(x)$ comme nouvelles coordonnées $x_0, \ldots, x_{n-1}$. Dans ce système de coordonnées, le dernier feuilletage est défini par la fonction linéaire $(x_0 + \cdots + x_{n-1})$.

Du point de vue de l'algébrisation des seuls tissus de rang maximal (on pourrait envisager des hypothèses plus faibles que d'être de rang maximal), le cas $n = 2$ des tissus plans est très particulier. La situation est actuellement la suivante. On a dit que tout 4-tissu plan de rang maximal est algébrisable. Bol a donné le premier exemple d'un d-tissu plan de rang maximal non algébrisable, avec $d = 5$. D'autres exemples ont été découverts récemment, pour des valeurs de d comprises entre 5 et 10 ; voir [9] et sa bibliographie. Mis à part ces résultats partiels, la question de l'algébrisation des tissus plans de rang maximal est ouverte.

On suppose maintenant $n \geq 3$ et $(n+2) \leq d \leq (2n-1)$. Considérons la famille des d-tissus $\mathcal{T}$ de la forme suivante. D'une part, ils contiennent le $(n+1)$-tissu $\mathcal{T}_0$ défini par les fonctions coordonnées $u_1(x) = x_0, \ldots, u_n(x) = x_{n-1}$ et la fonction

$$u_{n+1}(x) = x_0 + \cdots + x_{n-1} \,.$$

D'autre part, leurs autres fonctions de définition sont de la forme

$$u_\alpha(x) = U_{\alpha\,0}(x_0) + \cdots + U_{\alpha\,n-1}(x_{n-1}), \qquad \alpha = n+2, \ldots, d \,,$$

en respectant seulement les conditions sur les $dU_{\alpha\,\mu}(0)$ qui assurent que les feuilletages sont en position générale. Quand on les différentie, les définitions précédentes deviennent $(d - n) = \pi(n, d)$ relations abéliennes indépendantes : les tissus de la famille sont de rang maximal par construction.

Si un difféomorphisme local ϕ linéarise un tissu $\mathcal{T}$ de la famille, il transforme en particulier le $(n+1)$-tissu linéaire $\mathcal{T}_0$ en un tissu linéaire $\phi(\mathcal{T}_0)$. Le tissu $\mathcal{T}_0$ est de rang maximal $= 1$. D'après le théorème d'Abel inverse, le tissu $\phi(\mathcal{T}_0)$ est associé à une courbe algébrique de degré $(n+1)$ dans $\mathbb{P}^n$. On vérifie d'autre part que les transformations qui conservent le tissu $\mathcal{T}_0$ sont affines.

On obtient ainsi que les tissus linéarisables de la famille ne dépendent que d'un nombre fini de paramètres. Un tissu de la famille n'est donc pas linéarisable en général.

2. La borne de Chern et une version précisée du Théorème 1.1

2.1. *Rang d'un système de puissances de formes linéaires*

On a besoin de quelques résultats préliminaires sur le rang du système des puissances d'ordre donné d'une famille de formes linéaires. On suppose $n \geq 2$ et $d \geq 1$. Soit

$$l_1(x), \ldots, l_d(x),$$

des formes linéaires en position générale, *i.e.* telles que toute famille extraite de cardinal $\leq n$ soit libre. Posons :

$$(4) \qquad r_q = \dim\left(\mathbb{C}\, l_1(x)^q + \cdots + \mathbb{C}\, l_d(x)^q\right), \qquad q \geq 0.$$

Lemme 2.1. *Pour tout $n \geq 2$ et pour tout $q \geq 0$, on a les inégalités*

$$(5) \qquad r_{q+1} \geq \min(d, r_q + (n-1)), \qquad r_q \geq \min(d, q(n-1)+1).$$

Démonstration — Si $p \in \mathbb{N}$, notons $L(p)$ l'espace engendré par $l_1(x)^p, \ldots, l_d(x)^p$.

On a bien sûr $r_0 = 1$; on a $r_1 = \min(d, n)$ car les formes sont en position générale. Soit maintenant $q \geq 1$. On fait les hypothèses de récurrence :

$$r_q \geq \min(d, r_{q-1} + (n-1)), \quad r_p \geq \min(d, p(n-1)+1) \quad \text{si} \quad p = 0, \ldots, q.$$

On note $r = r_q$. On peut supposer que $L(q)$ est engendré par $l_1(x)^q, \ldots, l_r(x)^q$. Soit $X = \sum_{k=0}^{n-1} a_k \partial/\partial x_k$ un vecteur non nul tel que :

$$X \cdot l_{r+\alpha} = 0, \qquad \alpha = 1, \ldots, s := \min(d-r, n-1).$$

L'hypothèse de position générale permet de supposer en outre

$$X \cdot l_\alpha \neq 0, \qquad \alpha = 1, \ldots, r.$$

Le vecteur X induit une application linéaire

$$X\colon L(q+1) \to L(q).$$

Elle est surjective car

$$(X \cdot l_\alpha^{q+1})(x) = (q+1)(X \cdot l_\alpha)l_\alpha(x)^q$$

avec $X \cdot l_\alpha \neq 0$ si α est compris entre 1 et r. Ceci montre déjà qu'on a $r_{q+1} \geq r_q$. On peut appliquer ce résultat préliminaire aux formes $l_{r+1}(x), \ldots, l_{r+s}(x)$: l'espace engendré par leurs puissances $(q+1)$-ièmes est de dimension $\geq \min(s, q(n-1)+1) = s$.

Comme le noyau de l'application X contient $l_{r+1}(x)^{q+1}, \ldots, l_{r+s}(x)^{q+1}$, il est de dimension $\geq s$. On obtient

$$r_{q+1} \geq r+s \geq r + \min(d-r, n-1) \geq \min(d, r_q + (n-1))$$

et, par hypothèse de récurrence, $r_{q+1} \geq \min(d, (q+1)(n-1)+1)$. $\qquad\square$

Si $n = 2$, on a ainsi $r_q \geq \min(d, q+1)$. Comme l'espace des polynômes homogènes de degré q en deux variables est de dimension $(q+1)$, on en déduit :

Corollaire 2.2. *Si $n = 2$, on a l'égalité $r_q = \min(d, q+1)$ pour tout $q \geq 0$.*

La discussion est plus subtile si $n \geq 3$. On la complètera dans le §2.5 en rappelant un lemme classique de Castelnuovo.

2.2. *Relations abéliennes de valuation donnée*

Revenons à nos tissus. On suppose $n \geq 2$ et $d \geq (n+1)$. On considère un d-tissu $\mathcal{T}$ près de $0 \in \mathbb{C}^n$, défini par des fonctions $u_\alpha(x)$. On écrit

$$u_\alpha(x) = u_\alpha(0) + l_\alpha(x) + O(2), \qquad \alpha = 1, \ldots, d,$$

où $l_\alpha(x)$ est une forme linéaire. On note encore r_q le rang du système $l_1(x)^q, \ldots, l_d(x)^q$.

Soit $E(q)$ l'espace des relations abéliennes

$$z(x) = (z_1(x), \ldots, z_d(x))$$

dont la valuation[2] est $\geq q$. On définit ainsi une suite décroissante d'espaces ; $E(0)$ est l'espace de toutes les relations abéliennes du tissu.

---

[2] Si f est analytique au voisinage de 0, on note $f = O(q)$ et on dit que f est de valuation $\geq q$ si les coefficients des termes de degré $< q$ dans le développement de f en série entière sont nuls.

Soit $z(x) \in E(q)$. Comme $dz_\alpha(x) \wedge du_\alpha(x) = 0$, on peut écrire

$$z_\alpha(x) = a_\alpha u_\alpha(x)^q + O(q+1), \qquad \alpha = 1, \ldots, d.$$

En intégrant la partie principale de l'équation $\sum_{\alpha=1}^d z_\alpha(x)\, du_\alpha(x) = 0$, on obtient :

$$\sum_{\alpha=1}^d a_\alpha l_\alpha(x)^{q+1} = 0.$$

On peut aussi voir la relation ci-dessus comme une équation linéaire en $(a_1, \ldots, a_d)$. Il est clair que l'espace de ses solutions est de dimension $(d - r_{q+1})$. On en déduit :

$$(6) \qquad\qquad \dim E(q) - \dim E(q+1) \leq d - r_{q+1},$$

et en sommant par rapport à $q \geq 0$:

$$(7) \qquad\qquad \dim E(0) \leq \sum_{q=1}^{+\infty} (d - r_q).$$

D'après le Lemme 2.1, on a $(d - r_q) \leq (d - q(n-1) - 1)^+$. On retrouve ainsi l'inégalité de Chern : le rang ρ d'un tissu est au plus égal au nombre $\pi(n,d)$ défini par (3). Le lemme donne aussi :

$$(8) \qquad \rho = \pi(n,d) \;\Rightarrow\; (\; \forall q \geq 0, \;\; r_q = \min(d, q(n-1) + 1)\;).$$

En général, l'inégalité (7) est stricte.

Si les fonctions de définition $u_\alpha(x)$ sont linéaires (le tissu est alors composé de faisceaux d'hyperplans parallèles), les composantes homogènes d'une relation abélienne du tissu sont des relations abéliennes. On a donc :

  Le rang ρ d'un tissu défini par des formes linéaires $l_1(x), \ldots, l_d(x)$ est donné par la formule $\rho = \sum_{q=1}^{+\infty} (d - r_q)$.

  Le rang d'un tissu général est au plus égal au rang du tissu dont les éléments sont les faisceaux d'hyperplans parallèles aux hyperplans tangents aux feuilles du tissu initial issues d'un point donné.

  Comme on a dit, la borne de Chern $\pi(n,d)$ est atteinte par les tissus associés aux courbes extrémales de $\mathbb{P}^n$. On obtient très simplement des exemples de d-tissu de rang maximal $\pi(n,d)$ de la façon suivante. Posons :

$$l(t,x) = \sum_{\mu=0}^{n-1} t^\mu x_\mu.$$

On remarque que $l(t, x)^q$ est un polynôme en t de degré $\leq q(n-1)$. Il en résulte que, si $\theta_1, \ldots, \theta_d$ sont des nombres complexes deux à deux distincts et que l'on choisit

$$l_1(x) = l(\theta_1, x), \ldots, l_d(x) = l(\theta_d, x),$$

le rang r_q du système $l_1(x)^q, \ldots, l_d(x)^q$ est $\leq (q(n-1)+1)$. Compte tenu du Lemme 2.1, $r_q = \min(d, q(n-1)+1)$. Le tissu défini par les formes linéaires $l_1(x), \ldots, l_d(x)$ est donc de rang $\pi(n, d)$. Le Lemme 2.6 ci-dessous montre que la réciproque est vraie : si $d \geq (2n+1)$, un d-tissu défini par des formes linéaires et de rang maximal est de la forme précédente, à une transformation linéaire de $\mathbb{C}^n$ près.

2.3. Un raffinement du Théorème 1.1

On suppose $d \geq 2n$. Il résulte de la discussion précédente que la dimension des espaces $E(0)/E(1)$ et $E(1)/E(2)$ est majorée par $(d-n)$ et $(d-2(n-1)-1)$, respectivement. La dimension de l'espace $E(0)/E(2)$ est donc majorée par $(2d-3n+1)$. D'après (8), on a l'égalité dans les trois cas si le tissu est de rang maximal.

Définition 2.3. Un d-tissu près d'un point de $\mathbb{C}^n$ est de rang maximal en valuation ≤ 1 s'il possède un système de $(2d-3n+1)$ relations abéliennes dont les 1-jets sont linéairement indépendants.

La condition sur les 1-jets est évidemment ouverte. La discussion qui précède l'énoncé montre aussi qu'*un tissu de rang maximal en valuation ≤ 1 a, en tout point voisin de 0, $(d-n)$ relations abéliennes dont les 0-jets sont linéairement indépendants et $(d-2n+1)$ relations abéliennes de valuation 1, dont les 1-jets sont linéairement indépendants.* On démontrera :

Théorème 2.4. *Si $n \geq 3$ et $d \geq 2n$, un d-tissu de rang maximal en valuation ≤ 1 au voisinage d'un point de $\mathbb{C}^n$ est algébrisable.*

Le Théorème 1.1 est un cas particulier. L'idée de ce raffinement n'est pas nouvelle. Dans le cas $n = 3$, il est mentionné dans les compléments au §35 de [3], avec une formulation différente.

2.4. Courbes rationnelles normales

On se donne un entier $m \geq 2$. Les coordonnées homogènes d'un point de $\mathbb{P}^m$ sont notées $(x_0, \ldots, x_m)$. On appelle *points de base de* $\mathbb{P}^m$ les $(m+1)$ points dont toutes les coordonnées sont nulles, sauf une. Une famille de points de $\mathbb{P}^m$ est en position générale si toute famille extraite de cardinal $\leq (m+1)$ est projectivement libre.

On appelle *courbe rationnelle normale de degré m dans* $\mathbb{P}^m$ toute courbe qui admet une paramétrisation de la forme :

$$x(t) = (P_0(t), \ldots, P_m(t)),$$

où $P_0, \ldots, P_m$ forment une base de l'espace des polynômes de degré $\leq m$ en une variable. Par la méthode de Gauss, on voit qu'une courbe rationnelle normale est projectivement équivalente à la courbe C_m donnée par :

$$x(t) = (1, t, \ldots, t^{m-1}, t^m).$$

Toute famille de points d'une courbe rationnelle normale est en position générale. Il suffit de le vérifier pour une famille de $(m+1)$ points de C_m, autrement dit de vérifier que que si $t_0, \ldots, t_m \in \mathbb{C}$ sont deux à deux distincts, la relation $\sum_{j=0}^{m} a_j x(t_j) = 0$ n'est vérifiée que si tous ses coefficients sont nuls, ce qui est facile. On a en fait la propriété un peu plus forte suivante, dont on laisse la démonstration au lecteur :

Un sous-espace $\mathbb{P}^d$ *de dimension* $d \leq (m-1)$ *coupe une courbe rationnelle normale de* $\mathbb{P}^m$ *en au plus* $(d+1)$ *points, compte tenu des multiplicités.*
On a aussi :

Par $(m+3)$ *points de* $\mathbb{P}^m$ *en position générale, il passe une et une seule courbe rationnelle normale.*
On peut supposer que la famille de $(m+3)$ points est constituée des $(m+1)$ points de base et de deux autres points, représentés par $x' = (x'_0, \ldots, x'_m)$ et $x'' = (x''_0, \ldots, x''_m)$.

Les paramétrages des courbes rationnelles normales qui passent par les points de base en temps fini sont donnés par :

$$x(t) = \left(\prod_{j=0}^{m} (t - \theta_j) \right) \left(\frac{k_0}{t - \theta_0}, \ldots, \frac{k_m}{t - \theta_m} \right).$$

Un automorphisme de $\mathbb{P}^1$ permet encore de supposer que la courbe passe par les points représentés par x' et x'' pour $t = \infty$ et $t = 0$ respectivement et que $\theta_0 \neq 0$ est donné. On obtient les conditions :

$$k_j = k' x'_j, \quad k_j / \theta_j = k'' x''_j, \qquad j = 0, \ldots, m,$$

avec deux nouveaux paramètres k' et k'' non nuls. Par homogénéité, on peut supposer $k' = 1$. Le premier système d'équations détermine alors $k_0, \ldots, k_m$. L'équation $k_0 / \theta_0 = k''$ détermine k'' et les autres déterminent uniquement $\theta_1, \ldots, \theta_m$.

2.5. Le lemme de Castelnuovo

L'importance de ce lemme pour notre sujet a d'abord été reconnue par Chern et Griffiths [6]. Nous en rappelons une démonstration, inspirée de [6]. Nous continuons avec les notations du paragraphe précédent.

Soit $\mathcal{O}_{m+1}(2)$ l'espace vectoriel de dimension $(m+1)(m+2)/2$ des polynômes de la forme :

$$Q(x) = \sum_{0 \leq i \leq j \leq m} a_{ij}\, x_i x_j.$$

À toute famille Γ de points de $\mathbb{P}^m$, on associe le sous-espace $V(\Gamma)$ des polynômes $Q \in \mathcal{O}_{m+1}(2)$ qui s'annulent en tous les points de Γ.

On dit que la famille Γ impose r conditions indépendantes aux quadriques qui la contiennent si $V(\Gamma)$ est de codimension r dans $\mathcal{O}_{m+1}(2)$.

Lemme 2.5. *Si $d \leq (2m+1)$, d points en position générale dans $\mathbb{P}^m$ imposent d conditions indépendantes aux quadriques qui les contiennent.*

Démonstration — Soit $\Gamma = \{p_1, \ldots, p_d\}$ une famille de points en position générale dans $\mathbb{P}^m$. On obtient d conditions linéaires sur les coefficients d'un polynôme $Q \in V(\Gamma)$ en écrivant qu'il s'annule aux points $p_1, \ldots, p_d$. Ces conditions sont linéairement indépendantes. En effet, si $i_0 \in \{1, \ldots, d\}$ et compte tenu de l'hypothèse de position générale, il existe une réunion de deux hyperplans (une quadrique !) qui contient les p_i, $i \neq i_0$, mais pas p_{i_0}. □

Si C est une courbe rationnelle normale, paramétrée par un polynôme $x(t)$ de degré m, $Q(x(t))$ est un polynôme de degré $\leq 2m$ pour tout $Q \in \mathcal{O}_{m+1}(2)$. On en déduit que la codimension de $V(C)$ est $\leq (2m+1)$ donc $= (2m+1)$ d'après le lemme précédent. La réciproque fait l'objet du lemme suivant :

Lemme 2.6 (Castelnuovo). *Soit $d \geq (2m+3)$. Si d points en position générale dans $\mathbb{P}^m$ imposent $(2m+1)$ conditions indépendantes aux quadriques qui les contiennent, ils appartiennent à une même courbe rationnelle normale de degré m.*

Démonstration — On peut supposer que la famille de points considérée Γ contient les points de base. D'autre part, les hypothèses de l'énoncé impliquent que Γ n'est pas contenu dans la réunion de deux hyperplans.

Tout $Q \in V(\Gamma)$ est une combinaison linéaire de monômes de la forme $x_i x_j$ avec $i < j$ et s'écrit, de manière unique,

$$Q(x) = x_m L(x') + R(x'),$$

où $x' = (x_0, \ldots, x_{m-1})$, $L(x')$ est une forme linéaire et $R(x')$ appartient à l'espace de dimension $m(m-1)/2$ engendré par les monômes $x_i x_j$, avec $0 \le i < j < m$. Si $R(x')$ est le polynôme nul, on obtient $Q(x) = x_m L(x')$ et Q doit être nul puisque Γ n'est pas contenu dans la réunion de deux hyperplans. L'application qui associe $R(x')$ à $Q(x) \in V(\Gamma)$ est donc injective. Elle est aussi surjective puisque, par hypothèse, $V(\Gamma)$ est de dimension $(m+1)(m+2)/2 - (2m+1) - m(m-1)/2$.

Il existe donc une base de $V(\Gamma)$ composée d'éléments de la forme:

$$x_m L_{ij}(x') + x_i x_j, \qquad 0 \le i < j \le m-1.$$

On en extrait les éléments suivants:

$$Q_j(x) = x_m L_j(x') + x_0 x_j, \qquad j = 1, \ldots, m-1.$$

Pour tout j et k, $j \ne k$, le polynôme

$$L_k(x') Q_j(x) - L_j(x') Q_k(x) = x_0 (L_k(x') x_j - L_j(x') x_k)$$

s'annule sur Γ. Le polynôme $L_k(x') x_j - L_j(x') x_k \in \mathcal{O}_{m+1}(2)$ s'annule aux points de Γ, sauf peut-être aux m points de base de l'hyperplan d'équation $\{x_0 = 0\}$. Il s'annule aussi aux $(m-2)$ points de base du sous-espace d'équation $\{x_0 = x_j = x_k = 0\}$. Ce polynôme s'annule donc au moins en $(d-2) \ge (2m+1)$ des points de Γ. Compte tenu du lemme précédent et de l'hypothèse, il s'annule sur Γ:

$$L_k(x') x_j - L_j(x') x_k \in V(\Gamma), \qquad j, k = 1, \ldots, m-1.$$

En particulier, le coefficient de x_j dans $L_k(x')$ est nul si $j \ne k$ et $j \ne 0$. Les $L_j(x')$ sont donc de la forme suivante:

$$L_j(x') = a_j x_j + b_j x_0, \qquad j = 1, \ldots, m-1.$$

Soit C l'ensemble algébrique, qui contient Γ, défini par les équations:

$$(a_j x_m + x_0) x_j + b_j x_m x_0 = 0, \qquad j = 1, \ldots, m-1,$$

$$(a_k - a_j) x_j x_k + (b_k x_j - b_j x_k) x_0 = 0, \qquad j, k = 1, \ldots, m-1.$$

Si a_j ou b_j était nul, $a_j x_m x_j + b_j x_m x_0 + x_0 x_j \in V(\Gamma)$ serait un produit de formes linéaires, ce qui est impossible. Si l'on avait $j \ne k$ et $a_j = a_k$, $(b_k x_j - b_j x_k) x_0 \in V(\Gamma)$ serait un produit de formes linéaires, ce qui est impossible. On a donc $a_j \neq 0$ et $b_j \neq 0$ pour tout j et $a_j \neq a_k$ pour tout $j \neq k$. On en déduit en particulier que les points de C tels que $x_0 = 0$ sont des points de base. On paramètre les points de C tels que $x_0 \neq 0$ en posant $x_0 = 1$, $x_m = t$ et en résolvant le premier système d'équations. On obtient:

$$x(t) = \left(1, -\frac{b_1 t}{1 + a_1 t}, \ldots, -\frac{b_{m-1} t}{1 + a_{m-1} t}, t \right).$$

C'est une paramérisation homogène d'une courbe rationnelle normale de degré m qui passe *aussi* par les points de base. On conclut que cette courbe coïncide avec C. Le lemme est démontré. □

3. Rang maximal: conditions différentielles d'ordre un, d'ordre deux

3.1. *Introduction*

Les fonctions de définition d'un tissu de rang maximal en valuation ≤ 1 vérifient un système général d'équations différentielles du premier et du second ordre si $n \geq 3$ et $d \geq (2n + 1)$. Ce n'est pas le cas si $n = 2$. Il peut être utile de préciser que ces conditions, à elles seules, ne semblent pas impliquer la linéarisabilité. Pour démontrer le théorème principal, outre ces conditions, nous utiliserons à nouveau l'existence de relations abéliennes en nombre suffisant. La présentation ci-dessous est légèrement différente de celle de [4] et de celle de [6].

On note $\mathcal{O}_n(2)$ l'espace des polynômes g de la forme

(9)
$$g(\xi) = \sum_{0 \leq \mu \leq \nu \leq n-1} g_{\mu\nu}\, \xi_\mu \xi_\nu\,.$$

Au polynôme (9), on associe les opérateurs[3]

$$g(\nabla v)(x) = \sum_{0 \leq \mu \leq \nu \leq n-1} g_{\mu\nu}\, \partial_{x_\mu} v(x)\, \partial_{x_\nu} v(x)\,,$$

$$g(\partial)v(x) = \sum_{0 \leq \mu \leq \nu \leq n-1} g_{\mu\nu}\, \partial^2_{x_\mu x_\nu} v(x)\,.$$

On suppose $n \geq 2$ et $d \geq (2n + 1)$. On considère un d-tissu près de $0 \in \mathbb{C}^n$, de rang maximal en valuation ≤ 1, défini par des fonctions $u_1(x), \ldots, u_d(x)$.

---

[3] On note ∂_{x_μ} au lieu de $\partial/\partial x_\mu$ et $\partial^2_{x_\mu x_\nu}$ au lieu de $\partial^2/\partial x_\mu \partial x_\nu$.

On fixe un point x voisin de 0. On écrit :

$$u_\alpha(x + y) = u_\alpha(x) + l_\alpha(y) + Q_\alpha(y) + O(3), \qquad \alpha = 1, \ldots, d,$$

où $l_\alpha(y) = \sum_{\mu=0}^{n-1} y_\mu\, \partial_{x_\mu} u_\alpha(x)$ et $Q_\alpha(y)$ est un polynôme homogène de degré 2, dont les coefficients dépendent bien sûr de x. Soit $z = (z_1, \ldots, z_d)$ une relation abélienne du tissu. Comme $dz_\alpha \wedge du_\alpha = 0$, on peut écrire :

$$z_\alpha(x + y) = a_\alpha + b_\alpha l_\alpha(y) + O(2), \qquad \alpha = 1, \ldots, d.$$

Le calcul modulo $O(2)$ de la relation $\sum_{\alpha=1}^{d} z_\alpha\, du_\alpha = 0$ donne

$$(10) \qquad \sum_{\alpha=1}^{d} a_\alpha l_\alpha(y) = 0, \qquad \sum_{\alpha=1}^{d} a_\alpha\, Q_\alpha(y) + \frac{1}{2} \sum_{\alpha=1}^{d} b_\alpha l_\alpha(y)^2 = 0.$$

On considère maintenant (10) comme un système d'équations en $a = (a_1, \ldots, a_d)$ et en $b = (b_1, \ldots, b_d)$.

3.2. *Conditions du premier ordre*

On a :

Lemme 3.1. *Si $n \geq 2$ et $d \geq (2n+1)$ et si le tissu $\mathcal{T}$ est de rang maximal en valuation ≤ 1, il existe une base de 1-formes $\omega_0(x), \ldots, \omega_{n-1}(x)$ telle que les fonctions de définition $u_1(x), \ldots, u_d(x)$ de $\mathcal{T}$ vérifient*

$$(11) \qquad du_\alpha(x) = k_\alpha(x) \sum_{\mu=0}^{n-1} \theta_\alpha(x)^\mu\, \omega_\mu(x)$$

pour des fonctions $k_\alpha(x)$ et $\theta_\alpha(x)$ convenables.

On dira d'une telle base de 1-formes qu'elle est adaptée au tissu $\mathcal{T}$.

Démonstration — L'énoncé est invariant par difféomorphisme local. Pour obtenir l'existence d'une base de 1-formes *régulières*, il est commode de supposer que $u_1(x), \ldots, u_n(x)$ sont les fonctions coordonnées.

Comme on l'a vu dans le §2.2, si le tissu est de rang maximal en valuation ≤ 1, il a $(d - 2n + 1)$ relation abéliennes de valuation 1 au point x, dont les 1-jets sont linéairement indépendants. En particulier, l'espace des solutions de la forme $(0, b)$ de (10) est de dimension $(d - 2n + 1)$. Comme

$$l_\alpha(y)^2 = \sum_{\mu,\nu=0}^{n-1} y_\mu y_\nu\, \partial_{x_\mu} u_\alpha(x)\, \partial_{x_\nu} u_\alpha(x), \qquad \alpha = 1, \ldots, d,$$

cela revient à dire que la matrice de dimension $d \times n(n+1)/2$, dont les entrées sont les nombres $(\partial_{x_\mu} u_\alpha(x) \, \partial_{x_\nu} u_\alpha(x))$, est de rang $(2n-1)$, ou encore que l'espace des polynômes $g \in \mathcal{O}_n(2)$ tels que

$$g(\nabla u_\alpha)(x) = 0, \qquad \alpha = 1, \ldots, d\,,$$

est de codimension $(2n-1)$.

Cette propriété se traduit aussi de la façon suivante: les $d \geq (2(n-1)+3)$ points de $\mathbb{P}^{n-1}$ en position générale repésentés par les vecteurs

$$\nabla u_\alpha(x) = (\partial_{x_0} u_\alpha(x), \ldots, \partial_{x_{n-1}} u_\alpha(x))\,,$$

imposent $(2(n-1)+1)$ conditions indépendantes aux quadriques qui les contiennent. D'après le lemme de Castelnuovo, ils appartiennent à une courbe rationnelle normale de degré $(n-1)$.

La construction qu'on a décrite dans le §1.4 montre que la courbe rationnelle normale qui passe par les points de base de $\mathbb{P}^{n-1}$, qui sont représentés par hypothèse par les vecteurs $\nabla u_1(x), \ldots, \nabla u_n(x)$, et les deux autres points représentés par $\nabla u_{n+1}(x)$ et $\nabla u_{n+2}(x)$, admet une représentation paramétrique homogène de la forme

$$x(t) = (p_0(x,t), \ldots, p_{n-1}(x,t))\,,$$

où les $p_j(x,t)$ sont des polynômes de degré $\leq (n-1)$ en t dont les coefficients dépendent régulièrement de x. On peut aussi écrire:

$$x(t) = \sum_{\mu=0}^{n-1} (a_{\mu\,0}(x), \ldots, a_{\mu\,(n-1)}(x))\, t^\mu\,.$$

La base des 1-formes $\omega_\mu(x) = \sum_{\lambda=0}^{n-1} a_{\mu\lambda}(x)\, dx_\lambda$ est adaptée au tissu $\mathcal{T}$. □

3.3. *Conditions du second ordre*

On continue la discussion. Le tissu $\mathcal{T}$ a aussi $(d-n)$ relations abéliennes dont les 0-jets au point x sont indépendants. Ceci entraîne en particulier que, pour toute solution $a \in \mathbb{C}^d$ de la première équation de (10), il existe une solution $b \in \mathbb{C}^d$ de la deuxième. Donc:

$$\sum_{\alpha=1}^d a_\alpha l_\alpha(y) = 0 \;\Rightarrow\; \sum_{\alpha=1}^d a_\alpha\, Q_\alpha(y) \in (\mathbb{C}\, l_1(y)^2 + \cdots + \mathbb{C}\, l_d(y)^2)$$

pour tout $a \in \mathbb{C}^d$. C'est une condition sur les $Q_\alpha(y)$, non triviale si $l_1(y)^2, \ldots, l_d(y)^2$ n'engendrent pas l'espace des polynômes homogènes de degré 2.

Pour l'expliciter, prenons $l_1(y), \ldots, l_n(y)$ comme nouvelle base de l'espace des formes linéaires. Soit :

$$y_\mu = \sum_{j=1}^n a_{\mu j} l_j(y), \qquad \mu = 0, \ldots, n-1,$$

les formules de passage. Si $\alpha \in \{1, \ldots, d\}$, on a

$$l_\alpha(y) = \sum_{j=1}^n \left(\sum_{\mu=0}^{n-1} a_{\mu j} \partial_{x_\mu} u_\alpha(x) \right) l_j(y).$$

On en déduit :

$$Q_\alpha(y) - \sum_{j=1}^n \left(\sum_{\mu=0}^{n-1} a_{\mu j} \partial_{x_\mu} u_\alpha(x) \right) Q_j(y) \in (\mathbb{C}\, l_1(y)^2 + \cdots + l_d(y)^2).$$

Soit $g \in \mathcal{O}_n(2)$ un polynôme tel que :

$$g(\nabla u_{\alpha'})(x) = 0, \qquad \alpha' = 1, \ldots, d.$$

On remarque que :

$$g(\partial_y) l_{\alpha'}^2(0) = \sum_{0 \leq \mu \leq \nu \leq n-1} g_{\mu\nu} \, (\partial^2_{y_\mu y_\nu} l_{\alpha'}^2)(0) = 2g(\nabla u_{\alpha'})(x) = 0,$$

puisque $l_{\alpha'}(0) = 0$, pour tout $\alpha' \in \{1, \ldots, d\}$. On a donc $g(\partial_y)Q(0) = 0$ pour tout $Q(y) \in (\mathbb{C}\, l_1(y)^2 + \cdots + l_d(y)^2)$. On en déduit :

$$g(\partial_y)Q_\alpha(0) = \sum_{j=1}^n \left(\sum_{\mu=0}^{n-1} a_{\mu j} \partial_{x_\mu} u_\alpha(x) \right) g(\partial_y)Q_j(0).$$

D'autre part :

$$g(\partial_y)Q_\alpha(0) = g(\partial)u_\alpha(x), \qquad \alpha = 1, \ldots, d.$$

En posant $m_\mu = \sum_{j=1}^n a_{\mu j}\, g(\partial_y)Q_j(0)$, on obtient ainsi, au point *fixé* x :

$$g(\partial)u_\alpha(x) = \sum_{\mu=0}^{n-1} m_\mu \partial_{x_\mu} u_\alpha(x) \qquad \alpha = 1, \ldots, d.$$

On traduit maintenant ce résultat dans une base de 1-formes adaptée au tissu. On introduit la notation :

$$(12) \qquad \phi(x) = \sum_{\mu=0}^{n-1} (\phi)_\mu(x)\, \omega_\mu(x)$$

pour la décomposition d'une 1-forme $\phi(x)$ dans la base $\omega_0(x), \ldots, \omega_{n-1}(x)$.

Lemme 3.2. *Soit $n \geq 2$ et $d \geq (2n+1)$. Soit $\mathcal{T}$ un tissu de rang maximal en valuation ≤ 1 et $\omega_0(x), \ldots, \omega_{n-1}(x)$ une base de 1-formes adaptée au tissu. Avec les notations du Lemme 3.1, on a :*
Pour tout $\mu \in \{0, \ldots, n-2\}$, il existe des fonctions $m_{\mu 0}(x), \ldots, m_{\mu(n-1)}(x)$ telles que :

(13)
$$(d(k_\alpha \theta_\alpha))_\mu(x) - (dk_\alpha)_{\mu+1}(x) = k_\alpha(x) \sum_{\lambda=0}^{n-1} m_{\mu\lambda}(x)\theta_\alpha(x)^\lambda, \quad \alpha = 1, \ldots, d.$$

Si de plus $n > 2$, il existe des fonctions $n_{\mu 0}(x), \ldots, n_{\mu n}(x)$ telles que :

$$(14) \quad \theta_\alpha(x)(d\theta_\alpha)_\mu(x) - (d\theta_\alpha)_{\mu+1}(x) = \sum_{\lambda=0}^{n} n_{\mu\lambda}(x)\theta_\alpha(x)^\lambda, \quad \alpha = 1, \ldots, d.$$

Remarque 3.3. On peut montrer que la condition (14) est une condition nécessaire de linéarisabilité, pour un tissu de la forme (11), même si $n = 2$. Elle n'est pas vérifiée en général par un tissu plan de rang maximal, mais c'est une condition nécessaire (d'après ce qu'on vient de dire) et suffisante (compte tenu de la suite de la démonstration) pour qu'un tissu plan de rang maximal soit algébrisable.

Démonstration — On passe de la base canonique à la base adaptée par des formules du type

$$dx_\lambda = \sum_{\lambda'=0}^{n-1} b_{\lambda\lambda'}(x)\,\omega_{\lambda'}(x), \quad \omega_\lambda(x) = \sum_{\lambda'=0}^{n-1} c_{\lambda\lambda'}(x)\,dx_{\lambda'}, \quad \lambda = 0, \ldots, n-1.$$

Pour alléger les écritures, notons $u(x)$ l'une quelconque des fonctions $u_\alpha(x)$ et :

$$du(x) = \sum_{\mu=0}^{n-1} k(x)\theta(x)^\mu\,\omega_\mu(x).$$

On a aussi $du(x) = \sum_{\lambda=0}^{n-1} \partial_{x_\lambda}u(x)\,dx_\lambda$, donc :

$$k(x)\theta(x)^\mu = \sum_{\lambda=0}^{n-1} b_{\lambda\mu}(x)\partial_{x_\lambda}u(x).$$

Soit μ, μ', ν, ν' quatre entiers compris entre 0 et $(n-1)$ tels que $\mu + \nu = \mu' + \nu'$. De toute évidence, $k(x)\theta(x)^\mu\,k(x)\theta(x)^\nu = k(x)\theta(x)^{\mu'}\,k(x)\theta(x)^{\nu'}$. On a donc :

$$\sum_{\lambda,\lambda'=0}^{n-1} \left(b_{\lambda\mu}(x)b_{\lambda'\nu}(x) - b_{\lambda\mu'}(x)b_{\lambda'\nu'}(x)\right)\partial_{x_\lambda}u(x)\,\partial_{x_{\lambda'}}u(x) = 0.$$

Cette relation est vérifiée pour tout $u \in \{u_1, \ldots, u_d\}$. Compte tenu de la discussion qui précède l'énoncé, il existe donc des scalaires $m_\lambda(x)$ (ils dépendent de μ, ν, μ' et ν', mais pas du choix de $u \in \{u_1, \ldots, u_d\}$) tels que:

$$(15) \quad \sum_{\lambda,\lambda'=0}^{n-1} \left(b_{\lambda\mu}(x)b_{\lambda'\nu}(x) - b_{\lambda\mu'}(x)b_{\lambda'\nu'}(x)\right) \partial^2_{x_\lambda x_{\lambda'}} u(x) = \sum_{\lambda=0}^{n-1} m_\lambda(x)\,\partial_{x_\lambda} u(x)\,.$$

Comme d'autre part

$$(d(k\theta^\nu))_\mu(x) = \sum_{\lambda=0}^{n-1} b_{\lambda\mu}(x)\,\partial_{x_\lambda}(k\theta^\nu)(x) = \sum_{\lambda,\lambda'=0}^{n-1} b_{\lambda\mu}(x)\,\partial_{x_\lambda}(b_{\lambda'\nu}\partial_{x_{\lambda'}}u)(x)$$

$$= \sum_{\lambda,\lambda'=0}^{n-1} b_{\lambda\mu}(x)\,\partial_{x_\lambda}b_{\lambda'\nu}(x)\,\partial_{x_{\lambda'}}u(x)$$

$$+ \sum_{\lambda,\lambda'=0}^{n-1} b_{\lambda\mu}(x)\,b_{\lambda'\nu}(x)\,\partial^2_{x_\lambda x_{\lambda'}}u(x)\,,$$

on obtient des identités de la forme:

$$(d(k\theta^\nu))_\mu(x) - (d(k\theta^{\nu'}))_{\mu'}(x) = \sum_{\lambda=0}^{n-1} N_\lambda(x)\partial_{x_\lambda}u(x)\,.$$

En substituant $\partial_{x_\lambda}u(x) = \sum_{\lambda'=0}^{n-1} c_{\lambda'\lambda}k(x)\theta(x)^{\lambda'}$, on obtient finalement des identités de la forme:

$$(16) \quad (d(k\theta^\nu))_\mu(x) - (d(k\theta^{\nu'}))_{\mu'}(x) = \sum_{\lambda=0}^{n-1} M_\lambda(x)k(x)\theta(x)^\lambda\,.$$

On choisit d'abord $\nu = 1$, $\nu' = 0$ et $\mu' = \mu + 1$ dans (16), ce qui donne:

$$(d(k\theta))_\mu(x) - (dk)_{\mu+1}(x) = \sum_{\lambda=0}^{n-1} m_{\mu\lambda}(x)k(x)\theta(x)^\lambda\,.$$

C'est la première partie de l'énoncé. Supposons $n \geq 3$. On peut alors choisir $\nu = 2$, $\nu' = 1$ et $\mu' = \mu + 1$ dans (16), ce qui donne:

$$(d(k\theta^2))_\mu(x) - (d(k\theta))_{\mu+1}(x) = \sum_{\lambda=0}^{n-1} p_{\mu\lambda}(x)k(x)\theta(x)^\lambda\,.$$

On a d'autre part les identités:

$$(d(k\theta^2))_\mu(x) = \theta(x)(d(k\theta))_\mu(x) + k(x)\theta(x)(d\theta)_\mu(x)\,,$$

$$(d(k\theta))_{\mu+1}(x) = \theta(x)(dk)_{\mu+1}(x) + k(x)(d\theta)_{\mu+1}(x)\,.$$

Par soustraction membre à membre et compte tenu des formules obtenues juste auparavant, on obtient:

$$\sum_{\lambda=0}^{n-1} p_{\mu\lambda}(x)\theta(x)^{\lambda} = \theta(x)\sum_{\lambda=0}^{n-1} m_{\mu\lambda}(x)\theta(x)^{\lambda} + (\theta(x)(d\theta)_{\mu}(x) - (d\theta)_{\mu+1}(x)).$$

C'est la deuxième partie de l'énoncé. □

4. Début de la démonstration: la méthode standard

4.1. *Introduction*

La fin de l'article est consacrée à la démonstration du Théorème 2.4. Nous suivons le plan général de Bol [4], dont nous allons rappeler les grandes lignes. On pourra consulter [6] pour une présentation plus détaillée de cette méthode standard et de son arrière-plan géométrique.

La méthode repose sur des idées antérieures de Blaschke. Étant donné un d-tissu de rang maximal $\pi(n,d) := (m+1)$, on en choisit une base de relations abéliennes et on lui associe la famille de d applications p_{α} : $\mathbb{C}_0^n \to \mathbb{P}^m$, définie par les colonnes de la matrice dont les lignes sont les $(m+1)$ éléments de la base. Comme $p_{\alpha}(x)$ ne dépend que de $u_{\alpha}(x)$, les points $p_1(x), \ldots, p_d(x)$ décrivent des courbes dans $\mathbb{P}^m$. *Ces courbes sont des invariants projectifs du tissu*: à homographie près, elles ne dépendent du choix, ni des coordonnées, ni des fonctions de définition du tissu, ni de la base de relations abéliennes. Dans [1], Blaschke utilise cette idée pour prouver qu'un 4-tissu plan de rang maximal est algébrisable. En fait, ce résultat est équivalent à un résultat antérieur de Lie sur les surfaces de double translation, et Blaschke dit prendre pour modèle la démonstration qu'a donnée Poincaré du théorème de Lie dans [10]. Cette première idée suffit pour démontrer que les $(2n)$-tissus de rang maximal sont algébrisables.

Si $d \geq (2n+1)$, la deuxième idée essentielle consiste à montrer que, pour $x \in \mathbb{C}_0^n$ fixé, les points $p_1(x), \ldots, p_d(x)$ appartiennent à une courbe rationnelle normale $C(x)$ d'un sous-espace de $\mathbb{P}^m$, une propriété qui est vérifiée par les tissus qui proviennent d'une courbe algébrique extrémale, et à étudier cette famille de courbes. Cette idée apparaît déjà dans [2], dans une situation plus simple. Il faut alors, pour mener l'analyse à son but, montrer que les courbes $C(x)$ décrivent une surface algébrique de $\mathbb{P}^m$. C'est à ce point que nous nous écartons de [4], [3] et [6]. Nous n'utilisons pas l'analogie remarquée par Bol entre les conditons différentielles du §3 et l'équation des géodésiques dans certaines géométries semi-riemanniennes. Nous faisons une démonstration directe.

4.2. L'application de Poincaré

On suppose $n \geq 2$ et $d \geq 2n$. On considère un d-tissu $\mathcal{T}$ en $0 \in \mathbb{C}^n$, de rang maximal en valuation ≤ 1, défini par des fonctions $u_1(x), \ldots, u_d(x)$. On note:

$$(17) \qquad\qquad l := 2d - 3n + 1, \qquad m := 2d - 3n.$$

L'hypothèse nous permet de choisir l relations abéliennes

$$z_i(x) = (z_{i\,1}(x), \ldots, z_{i\,d}(x)), \qquad i = 1, \ldots l,$$

dont les 1-jets sont linéairement indépendants. Avec un peu d'abus, on note $z'_{i,\alpha}(x)$ les fonctions définies par

$$dz_{i\,\alpha}(x) = z'_{i\,\alpha}(x)\, du_\alpha(x).$$

La méthode de Poincaré consiste d'abord à introduire les vecteurs

$$(18) \qquad Z_\alpha(x) = (z_{1\,\alpha}(x), \ldots, z_{l\,\alpha}(x)) \in \mathbb{C}^l, \qquad \alpha = 1, \ldots, d.$$

Par définition d'une relation abélienne, on a:

$$(19) \qquad\qquad \sum_{\alpha=1}^{d} Z_\alpha(x)\, du_\alpha(x) = 0$$

et:

$$dZ_\alpha(x) = Z'_\alpha(x)\, du_\alpha(x), \qquad \alpha = 1, \ldots, d,$$

où:

$$Z'_\alpha(x) = (z'_{1\,\alpha}(x), \ldots, z'_{l\,\alpha}(x)).$$

Introduisons les matrices:

$$M(x) := \begin{pmatrix} z_{1\,1}(x) & \ldots & z_{1\,d}(x) \\ \ldots & \ldots & \ldots \\ \ldots & \ldots & \ldots \\ z_{l\,1}(x) & \ldots & z_{l\,d}(x) \end{pmatrix}, \quad M'(x) := \begin{pmatrix} z'_{1\,1}(x) & \ldots & z'_{1\,d}(x) \\ \ldots & \ldots & \ldots \\ \ldots & \ldots & \ldots \\ z'_{l\,1}(x) & \ldots & z'_{l\,d}(x) \end{pmatrix}.$$

Les l lignes de $M(x)$ représentent les relations abéliennes de la base qu'on a choisie et ses d colonnes représentent les vecteurs $Z_1(x), \ldots, Z_d(x)$.

Lemme 4.1. *Pour tout x voisin de 0, les propriétés suivantes sont vérifiées:*

1) les vecteurs $Z_1(x), \ldots, Z_d(x)$ engendrent un sous-espace de dimension $(d - n)$ et sont en position générale dans ce sous-espace;

2) *toute famille à* $(d - 2n + 1)$ *éléments extraite du système* $Z'_1(x), \ldots, Z'_d(x)$, *engendre, ensemble avec les vecteurs* $Z_1(x), \ldots, Z_d(x)$, *l'espace* $\mathbb{C}^l$.

Démonstration — On peut supposer $x = 0$ et un automorphisme de $\mathbb{C}^l$ permet de choisir la base de relations abéliennes. On en construit une avec $(d - n)$ relations dont les 0-jets sont linéairement indépendants d'une part, $(d - 2n + 1)$ relations de valuation 1 à l'origine et dont les 1-jets sont linéairement indépendants de l'autre. Notons encore

$$l_\alpha(x) = \sum_{\mu=0}^{n-1} x_\mu\, \partial_{x_\mu} u_\alpha(0)\,.$$

Avec le choix qu'on a fait, les matrices $M(0)$ et $M'(0)$ sont respectivement de la forme:

$$M(0) = \begin{pmatrix} A \\ O \end{pmatrix}, \qquad M'(0) = \begin{pmatrix} \star \\ B \end{pmatrix},$$

où O est une matrice nulle, A une matrice de dimension $(d - n) \times d$ et B une matrice de dimension $(d - 2n + 1) \times d$. Les lignes $(a_{i\,1} \cdots a_{i\,d})$ de A et les lignes $(b_{i\,1} \cdots b_{i\,d})$ de B représentent respectivement des bases de solutions des équations

$$\sum_{\alpha=1}^{d} a_\alpha l_\alpha(x) = 0\,, \qquad \sum_{\alpha=1}^{d} b_\alpha l_\alpha(x)^2 = 0\,.$$

La première assertion du lemme peut s'énoncer ainsi : toutes les matrices de dimension $(d - n) \times (d - n)$ extraites de A sont inversibles. Si ce n'était pas le cas, il existerait une combinaison linéaire non triviale des lignes de la matrice A avec au plus n coefficients non nuls, c'est-à-dire une relation de dépendance entre n parmi les d formes $l_\alpha(x)$. C'est impossible, puisque les $l_\alpha(x)$ sont en position générale.

La deuxième assertion du lemme peut s'énoncer ainsi: toutes les matrices de dimension $(d - 2n + 1) \times (d - 2n + 1)$ extraites de B sont inversibles. Si ce n'était pas le cas, il existerait une combinaison linéaire non triviale des lignes de la matrice B avec au plus $(2n - 1)$ coefficients non nuls, c'est-à-dire une relation de dépendance entre $(2n - 1)$ parmi les carrés $l_\alpha(x)^2$. Le Lemme 2.1 montre que c'est impossible. □

On note $\mathbb{C}^{d-n}(x)$ le sous-espace de $\mathbb{C}^l$ engendré par $Z_1(x), \ldots, Z_d(x)$. Soit

$$\pi\colon \mathbb{C}^l \backslash \{0\} \to \mathbb{P}^m$$

la projection canonique. (Rappelons que $l = m + 1$.) Le Lemme précédent montre que les applications $Z_\alpha \colon \mathbb{C}_0^n \to \mathbb{C}^l(x)$ induisent des applications

$$(20) \qquad\qquad p_\alpha \colon \mathbb{C}_0^n \to \mathbb{P}^m, \qquad \alpha = 1, \dots, d\,,$$

et que les points $p_1(x), \dots, p_d(x)$ engendrent un sous-espace $\mathbb{P}^{d-n-1}(x)$ de dimension $(d - n - 1)$. L'application induite

$$(21) \qquad P \colon \mathbb{C}_0^n \to \mathbb{G}(d - n - 1, m), \qquad P(x) = \mathbb{P}^{d-n-1}(x)\,,$$

à valeurs dans la grassmannienne des $(d - n - 1)$-plans de $\mathbb{P}^m$, est « l'application de Poincaré » associée au tissu.

Lemme 4.2. *L'application de Poincaré (21) est une immersion. Pour tout* $x, x' \in \mathbb{C}_0^n$ *tels que* $x \neq x'$,

$$(22) \qquad\qquad \mathbb{P}^{n-2}(x, x') := \mathbb{P}^{d-n-1}(x) \cap \mathbb{P}^{d-n-1}(x')$$

est un sous-espace de dimension $(n - 2)$ *de l'espace* $\mathbb{P}^m$.

Démonstration — Soit $t \mapsto x(t)$ un arc analytique avec $x(0) = x$, $x'(0) \neq 0$. On a:

$$d_t(Z_\alpha \circ x)(0) = \langle du_\alpha(x), x'(0)\rangle Z'_\alpha(x), \qquad \alpha = 1, \dots, d.$$

Le coefficient $\langle du_\alpha(x), x'(0)\rangle$ est nul pour au plus $(n - 1)$ valeurs de α (position générale). Le Lemme 4.1 montre que les vecteurs $d_t(Z_\alpha \circ x)(0)$ et $Z_\alpha(x)$, $\alpha = 1, \dots, d$, engendrent ensemble $\mathbb{C}^l$. En particulier, l'application $t \mapsto \mathbb{P}^{d-n-1}(x(t))$ n'est pas stationnaire en $t = 0$: c'est la première partie de l'énoncé.

Pour la deuxième partie, on note que, puisque l'espace engendré par les vecteurs $(Z_\alpha \circ x)(0)$ et celui engendré par les vecteurs $d_t(Z_\alpha \circ x)(0)$ sont transverses, les espaces $\mathbb{C}^{d-n}(x(t))$ et $\mathbb{C}^{d-n}(x(t'))$ sont transverses pour tout t, t' petits avec $t \neq t'$. Leur intersection est de dimension $2(d - n) - (2d - 3n + 1) = (n - 1)$. C'est la deuxième partie de l'énoncé. $\square$

4.3. $d = 2n$; le cas de Lie, Poincaré, Blaschke

Rappelons que le fait qu'un 4-tissu plan de rang maximal est algébrisable est une autre formulation d'un théorème difficile de Lie. La démonstration qui suit, reprise de [1], montre la puissance de la méthode standard, sous la forme initiale due à Poincaré, en liaison avec le théorème d'Abel inverse.

Si $d = 2n$, alors $m = n$ et la grassmannienne des $(d-n-1)$-plans de $\mathbb{P}^n$ est l'espace projectif $\check{\mathbb{P}}^n$ des hyperplans de $\mathbb{P}^n$. L'application de Poincaré

$$\mathbb{C}_0^n \to \check{\mathbb{P}}^n, \qquad x \mapsto \mathbb{P}^{n-1}(x)\,,$$

est une immersion, donc un difféomorphisme local. On transporte le tissu $\mathcal{T}$ dans $\check{\mathbb{P}}^n$ grâce à ce difféomorphisme. Dans un système affine de coordonnées de $\check{\mathbb{P}}^n$ au voisinage de $\mathbb{P}^{n-1}(0)$, le tissu est linéaire. En effet, si α est compris entre 1 et d, la feuille du α-ième feuilletage qui passe par un point x_0 est définie par l'équation $\{p_\alpha(x) = p_\alpha(x_0)\}$. Son image sous l'action du difféomorphisme qu'on vient d'introduire est localement l'ensemble des hyperplans de $\mathbb{P}^n$ qui passent par le point $p_\alpha(x_0)$. Par dualité projective, c'est un hyperplan.

4.4. *Les courbes rationnelles normales* **C(x)**

On suppose maintenant $n \geq 2$ et $d \geq (2n+1)$. Suivant le Lemme 3.1, on introduit une base adaptée de 1-formes $\omega_0(x), \ldots, \omega_{n-1}(x)$. On écrit :

$$(23) \qquad du_\alpha(x) = k_\alpha(x) \sum_{\mu=0}^{n-1} \theta_\alpha(x)^\mu \, \omega_\mu(x), \qquad \alpha = 1, \ldots, d.$$

On a donc $\sum_{\alpha=1}^{d} Z_\alpha(x) \, du_\alpha(x) = \sum_{\mu=0}^{n-1} \left(\sum_{\alpha=1}^{d} Z_\alpha(x) k_\alpha(x) \theta_\alpha(x)^\mu \right) \omega_\mu(x)$ et la relation (19) devient :

$$(24) \qquad \sum_{\alpha=1}^{d} Z_\alpha(x) k_\alpha(x) \theta_\alpha(x)^\mu = 0, \qquad \mu = 0, \ldots, n-1.$$

On introduit les polynômes en $t \in \mathbb{C}$:

$$P(x,t) = \prod_{\beta=1}^{d} (t - \theta_\beta(x)), \qquad P_\alpha(x,t) = \prod_{\beta \neq \alpha} (t - \theta_\beta(x)),$$

et la fonction $Z_\star \colon \mathbb{C}_0^n \times \mathbb{C} \to \mathbb{C}^{m+1}$ définie par :

$$(25) \qquad Z_\star(x,t) = \sum_{\alpha=1}^{d} P_\alpha(x,t) k_\alpha(x) Z_\alpha(x).$$

C'est un paramétrage homogène d'une courbe rationnelle qui passe par les points $p_1(x), \ldots, p_d(x)$ aux temps $\theta_1(x), \ldots, \theta_d(x)$, respectivement.

Lemme 4.3. *Les points $p_1(x), \ldots, p_d(x)$ appartiennent à une courbe rationnelle normale de degré $(d-n-1)$ dans l'espace $\mathbb{P}^{d-n-1}(x)$ qu'ils engendrent.*

Démonstration — On n'écrit pas la variable x, qui est fixée. On montre d'abord que le polynôme $Z_\star(t)$, qui est par définition de degré $\leq (d-1)$, est en fait de degré $\leq (d-n-1)$.

Définissons les nombres σ_k et $\sigma_k(\alpha)$ par:

$$\prod_{\beta=1}^{d}(t-\theta_\beta) = \sum_{k=0}^{d}\sigma_k t^k, \quad \text{et} \quad \prod_{\beta\neq\alpha}(t-\theta_\beta) = \sum_{k=0}^{d-1}\sigma_k(\alpha)t^k, \quad \alpha=1,\ldots,d.$$

En écrivant $P(t) = (t-\theta_\alpha)P_\alpha(t)$, on obtient les identités $\sigma_{k+1} = \sigma_k(\alpha) - \theta_\alpha\sigma_{k+1}(\alpha)$, dont on tire:

$$\sigma_k(\alpha) = \sum_{l=0}^{d-k-1}\theta_\alpha^l\,\sigma_{k+l+1}, \qquad k=0,\ldots,d-1.$$

On peut alors calculer:

$$Z_\star(t) = \sum_{k=0}^{d-1}\left(\sum_{\alpha=1}^{d}\sigma_k(\alpha)k_\alpha\,Z_\alpha\right)t^k = \sum_{k=0}^{d-1}\left(\sum_{l=0}^{d-k-1}\sigma_{k+l+1}(\sum_{\alpha=1}^{d}\theta_\alpha^l k_\alpha Z_\alpha)\right)t^k.$$

Dans le membre de droite, le coefficient de t^k est nul d'après (24) si $l \leq (n-1)$ pour tout $l \leq (d-k-1)$. Le polynôme $Z_\star(t)$ est donc bien de degré $\leq (d-n-1)$.

Dans un système de coordonnées de $\mathbb{C}^{d-n}(x)$, on a $Z_\star(t) = (Q_1(t), \ldots, Q_{d-n}(t))$, avec des polynômes $Q_1(t), \ldots, Q_{d-n}(t)$ de degrés $\leq (d-n-1)$. Comme les points $Z_1, \ldots, Z_d$ engendrent $\mathbb{C}^{d-n}(x)$, il est clair que ces polynômes forment une base de l'espace des polynômes de degré $\leq (d-n-1)$. Le lemme est démontré. $\qquad\square$

Pour $x \in \mathbb{C}_0^n$, on note $C(x)$ la courbe définie par le lemme précédent. Si $x \neq x'$, les points de l'intersection $C(x) \cap C(x')$ appartiennent à l'espace $\mathbb{P}^{n-2}(x, x')$ et les rappels du §2.4 montrent qu'on a:

Lemme 4.4. *Pour tout $x, x' \in \mathbb{C}_0^n$ tels que $x \neq x'$, les courbes $C(x)$ et $C(x')$ se rencontrent en au plus $(n-1)$ points, compte tenu des multiplicités.*

On peut affirmer que $C(x)$ et $C(x')$ se coupent en exactement $(n-1)$ points dans le cas suivant: $u_\alpha(x) = u_\alpha(x')$ pour exactement $(n-1)$ valeurs de l'indice α. Les courbes ont alors les points correspondants $p_\alpha(x) = p_\alpha(x')$ en commun. *On montrera plus bas que deux courbes $C(x)$ distinctes se coupent toujours en exactement $(n-1)$ points, compte tenu des multiplicités.*

5. Suite de la démonstration : les courbes $C(x)$ engendrent une surface

5.1. *Énoncé du lemme principal*

C'est le suivant :

Lemme 5.1. *On suppose $n \geq 3$ et $d \geq (2n + 1)$. L'application*

$$(26) \qquad p \colon \mathbb{C}_0^n \times \mathbb{P}^1 \to \mathbb{P}^m,$$

induite par (25) *est de rang 2 en tout* $(x, t) \in \mathbb{C}_0^n \times \mathbb{P}^1$.

L'énoncé est vrai si $n = 2$, à condition de supposer que la deuxième propriété du Lemme 3.2 est vérifiée, ce qui dans ce cas n'est pas impliqué par les autres hypothèses.

L'image de l'espace tangent à $\mathbb{C}_0^n \times \mathbb{C}$ en un point (x, t) par l'application dérivée de p est la projection, par la dérivée de la projection canonique $\pi \colon \mathbb{C}^{m+1} \backslash \{0\} \to \mathbb{P}^m$, du sous-espace de $\mathbb{C}^{m+1}$ engendré par les vecteurs :

$$Z_\star(x, t), \;\; \partial_t Z_\star(x, t), \;\; \partial_{x_0} Z_\star(x, t), \ldots, \partial_{x_{n-1}} Z_\star(x, t).$$

Il s'agit de montrer que cet espace est de dimension 3. Introduisons la 1-forme suivante :

$$(27) \qquad \Omega(x, t) := \sum_{\mu=0}^{n-1} t^\mu \, \omega_\mu(x), \qquad (x, t) \in \mathbb{C}_0^n \times \mathbb{C}.$$

Le résultat cherché est alors une conséquence immédiate du suivant.

Lemme 5.2. *On suppose $n \geq 3$ et $d \geq (2n+1)$. Il existe un vecteur $F(x, t)$ et des 1-formes $\Gamma(x, t)$ et $\Delta(x, t)$ telles que*[4] :

$$(28) \quad dZ_\star(x, t) = \Omega(x, t)\, F(x, t) + \Gamma(x, t)\, Z_\star(x, t) + \Delta(x, t)\, \partial_t Z_\star(x, t).$$

De plus les vecteurs $F(x, t)$, $Z_\star(x, t)$ et $\partial_t Z_\star(x, t)$ sont linéairement indépendants.

Ce lemme montre que si $s \mapsto x(s)$ est un arc tel que $x(0) = x$, l'arc $s \mapsto p(x(s), t)$ est transverse à la courbe $C(x)$ au point $p(x, t)$ si et seulement si $x'(0)$ n'appartient pas au noyau de la forme linéaire $\Omega(x, t)$.

---

[4] La notation $dZ_\star(x, t)$ renvoie à la différentielle de $Z(x, t)$ en x. Il en ira de même dans la suite.

5.2. Début de la démonstration du Lemme 5.2

On n'écrit plus les variables (x, t). Plutôt qu'avec Z_*, on travaille avec la fonction suivante :

$$Z = \sum_{\alpha=1}^{d} \frac{k_\alpha Z_\alpha}{t - \theta_\alpha}.$$

Rappelons que la formule $dZ = \sum_{\mu=0}^{n-1} (dZ)_\mu \, \omega_\mu$ définit les fonctions $(dZ)_\mu$. On a :

$$dZ = \sum_{\alpha=1}^{d} \frac{k_\alpha Z_\alpha'}{t - \theta_\alpha} \, du_\alpha + \sum_{\alpha=1}^{d} Z_\alpha \, d\left(\frac{k_\alpha}{t - \theta_\alpha} \right)$$

Compte tenu de la forme (23) des du_α, on a donc :

$$(29) \quad (dZ)_\nu = \sum_{\alpha=1}^{d} \frac{k_\alpha^2 \theta_\alpha^\nu Z_\alpha'}{t - \theta_\alpha} + \sum_{\alpha=1}^{d} Z_\alpha \left(d\left(\frac{k_\alpha}{t - \theta_\alpha} \right) \right)_\nu, \quad \nu = 0, \ldots, n-1.$$

En différentiant la relation $\sum_{\alpha=1}^{d} Z_\alpha k_\alpha = 0$, voir (24), on obtient la relation

$$\sum_{\alpha=1}^{d} Z_\alpha' k_\alpha \, du_\alpha + \sum_{\alpha=1}^{d} Z_\alpha \, dk_\alpha = 0,$$

qu'on décompose dans la base adaptée :

$$(30) \qquad \sum_{\alpha=1}^{d} Z_\alpha' k_\alpha^2 \theta_\alpha^\mu = - \sum_{\alpha=1}^{d} Z_\alpha \, (dk_\alpha)_\mu, \qquad \mu = 0, \ldots, n-1.$$

On peut s'étonner qu'on utilise une seule des relations (24) : les autres sont en fait couvertes par le Lemme 3.2, de même d'ailleurs que celles qu'on obtiendrait en écrivant que le second membre de (23) est une forme fermée.

5.3. Un calcul

La clé du calcul est le lemme suivant :

Lemme 5.3. Il existe des fonctions $f_\mu(x, t)$ et $g_\mu(x, t)$ telles que :

$$(31) \qquad t(dZ)_\mu - (dZ)_{\mu+1} = f_\mu Z + g_\mu \, \partial_t Z, \qquad \mu = 0, \ldots, n-2.$$

Démonstration — Dans ce qui suit, l'entier $\mu \in \{0, \ldots, n-2\}$ est fixé. Notons :

$$I_\mu := t(dZ)_\mu - (dZ)_{\mu+1}.$$

Les formules (29) donnent :

$$I_\mu = \sum_{\alpha=1}^{d} k_\alpha^2 \theta_\alpha^\mu Z_\alpha' + \sum_{\alpha=1}^{d} Z_\alpha \left(t \left(d \left(\frac{k_\alpha}{t - \theta_\alpha} \right) \right)_\mu - \left(d \left(\frac{k_\alpha}{t - \theta_\alpha} \right) \right)_{\mu+1} \right).$$

Les formules (30) permettent d'éliminer les Z_α'. On obtient ainsi :

$$I_\mu = \sum_{\alpha=1}^{d} Z_\alpha \, K_{\mu\alpha} \,,$$

où (rappelons que les différentielles concernent les x, pas les t) :

$$
\begin{aligned}
K_{\mu\alpha} &= -(dk_\alpha)_\mu + t \left(d \left(\frac{k_\alpha}{t - \theta_\alpha} \right) \right)_\mu - \left(d \left(\frac{k_\alpha}{t - \theta_\alpha} \right) \right)_{\mu+1} \\
&= \left(d \left(\frac{k_\alpha \theta_\alpha}{t - \theta_\alpha} \right) \right)_\mu - \left(d \left(\frac{k_\alpha}{t - \theta_\alpha} \right) \right)_{\mu+1} \\
&= \frac{(d(k_\alpha \theta_\alpha))_\mu - (dk_\alpha)_{\mu+1}}{t - \theta_\alpha} + \frac{k_\alpha(\theta_\alpha(d\theta_\alpha)_\mu - (d\theta_\alpha)_{\mu+1})}{(t - \theta_\alpha)^2} \\
&= L_{\mu\alpha} + M_{\mu\alpha} \,.
\end{aligned}
$$

C'est (enfin) le moment de rappeler les formules suivantes du Lemme 3.2 :

$$(d(k_\alpha \theta_\alpha))_\mu - (dk_\alpha)_{\mu+1} = \sum_{\lambda=0}^{n-1} m_{\mu\lambda} k_\alpha \theta_\alpha^\lambda,$$

$$\theta_\alpha(d\theta_\alpha)_\mu - (d\theta_\alpha)_{\mu+1} = \sum_{\lambda=0}^{n} n_{\mu\lambda} \theta_\alpha^\lambda \,.$$

On a donc :

$$L_{\mu\alpha} = \sum_{\lambda=0}^{n-1} \frac{m_{\mu\lambda} k_\alpha \theta_\alpha^\lambda}{t - \theta_\alpha} = \sum_{\lambda=0}^{n-1} \frac{m_{\mu\lambda} k_\alpha t^\lambda}{t - \theta_\alpha} - \sum_{\lambda=1}^{n-1} \sum_{\lambda'=0}^{\lambda-1} m_{\mu\lambda} k_\alpha t^{\lambda-\lambda'-1} \theta_\alpha^{\lambda'} \,.$$

De la même manière, on écrit :

$$
\begin{aligned}
M_{\mu\alpha} &= \sum_{\lambda=0}^{n} \frac{n_{\mu\lambda} k_\alpha \theta_\alpha^\lambda}{(t - \theta_\alpha)^2} = \sum_{\lambda=0}^{n} \frac{n_{\mu\lambda} k_\alpha t^\lambda}{(t - \theta_\alpha)^2} - \sum_{\lambda=1}^{n} \sum_{\lambda'=0}^{\lambda-1} \frac{n_{\mu\lambda} k_\alpha t^{\lambda-\lambda'-1} \theta_\alpha^{\lambda'}}{t - \theta_\alpha} \\
&= \sum_{\lambda=0}^{n} \frac{n_{\mu\lambda} k_\alpha t^\lambda}{(t - \theta_\alpha)^2} - \sum_{\lambda=1}^{n} \frac{\lambda n_{\mu\lambda} k_\alpha t^{\lambda-1}}{t - \theta_\alpha} + \sum_{\lambda=2}^{n} \sum_{\lambda'=1}^{\lambda-1} \sum_{\lambda''=0}^{\lambda'-1} n_{\mu\lambda} k_\alpha t^{\lambda-\lambda''-2} \theta_\alpha^{\lambda''} \,.
\end{aligned}
$$

On a donc, compte tenu de (24) :

$$\sum_{\alpha=1}^{d} L_{\mu\alpha} Z_{\alpha} = \left(\sum_{\lambda=0}^{n-1} m_{\mu\lambda} t^{\lambda}\right) Z,$$

$$\sum_{\alpha=1}^{d} M_{\mu\alpha} Z_{\alpha} = -\left(\sum_{\lambda=0}^{n} n_{\mu\lambda} t^{\lambda}\right) \partial_t Z - \left(\sum_{\lambda=1}^{n} \lambda n_{\mu\lambda} t^{\lambda-1}\right) Z,$$

ce qui termine la démonstration du lemme.   □

5.4. *Fin de la démonstration du Lemme 5.2*

On a montré :

$$(dZ)_{\mu} = t(dZ)_{\mu-1} - f_{\mu-1} Z - g_{\mu-1} \partial_t Z, \qquad \mu = 1, \ldots, n-1.$$

On en déduit

$$(dZ)_{\mu} = t^{\mu}(dZ)_0 - Z \sum_{\lambda=0}^{\mu-1} t^{\lambda} f_{\mu-\lambda-1} - \partial_t Z \sum_{\lambda=0}^{\mu-1} t^{\lambda} g_{\mu-\lambda-1}, \qquad \mu = 1, \ldots, n-1,$$

et une décomposition de la forme :

$$dZ(x,t) = \Omega(x,t) F'(x,t) + \Gamma'(x,t) Z(x,t) + \Delta'(x,t) \partial_t Z(x,t).$$

Comme $Z_\star = PZ$, on en déduit (28) avec

$$F(x,t) = P(x,t)(dZ)_0(x).$$

Il reste à montrer que $F(x,t)$ n'appartient pas à l'espace engendré par $Z(x,t)$ et $\partial_t Z(x,t)$. On a en fait le résultat plus fort suivant : $F(x,t)$ n'appartient pas à l'espace $\mathbb{C}^{d-n}(x)$ engendré par $Z_1(x), \ldots, Z_d(x)$.

On a :

$$(dZ)_0(x,t) = \sum_{\alpha=1}^{d} \frac{k_\alpha(x)^2 Z'_\alpha(x)}{t - \theta_\alpha(x)} + \sum_{\alpha=1}^{d} Z_\alpha(x) \left(d\left(\frac{k_\alpha(x)}{t - \theta_\alpha(x)}\right)\right)_0.$$

Comme le second terme de la somme appartient à $\mathbb{C}^{d-n}(x)$, il suffit de montrer que

$$X(x,t) = \sum_{\alpha=1}^{d} P_\alpha(x,t) k_\alpha(x)^2 Z'_\alpha(x)$$

n'appartient pas à $\mathbb{C}^{d-n}(x)$. On écrit, avec les notations introduites pour démontrer le Lemme 4.3 :

$$X(x,t) = \sum_{\alpha=1}^{d} \sum_{k=0}^{d-1} \sigma_k(\alpha) t^k k_\alpha(x)^2 Z'_\alpha(x)$$

$$= \sum_{\alpha=1}^{d} \sum_{k=0}^{d-1} \sum_{l=0}^{d-k-1} \theta_\alpha^l \sigma_{k+l+1} t^k k_\alpha(x)^2 Z'_\alpha(x).$$

En différentiant les relations (24), on obtient que, si $\mu \in \{0, \ldots, n-1\}$,

$$\sum_{\alpha=1}^{d} Z'_\alpha(x) k_\alpha(x) \theta_\alpha(x)^\mu \, du_\alpha(x) \in \mathbb{C}^{d-n}(x).$$

En écrivant ceci dans la base adaptée, on obtient la même propriété pour les sommes :

$$\sum_{\alpha=1}^{d} Z'_\alpha(x) k_\alpha(x)^2 \theta_\alpha(x)^\mu \, du_\alpha(x), \qquad \mu = 0, \ldots, 2n-2.$$

On en déduit que $X(x,t)$ est la somme d'un élément de $\mathbb{C}^{d-n}(x)$ et d'un polynôme en t de degré $\leq (d-2n)$, soit :

$$(32) \qquad X(x,t) \equiv \sum_{\mu=0}^{d-2n} X_\mu(x) t^\mu \qquad \text{modulo } \mathbb{C}^{d-n}(x).$$

L'espace engendré par $X(x,t)$ modulo $\mathbb{C}^{d-n}(x)$ quand t décrit $\mathbb{C}$ contient la classe de $X(x, \theta_\alpha(x))$, donc la classe de $Z'_\alpha(x)$, pour tout $\alpha = 1, \ldots, d$. Il est donc de dimension $(d-2n+1)$ d'après le Lemme 4.1. Autrement dit, dans (32), les vecteurs $X_0(x), \ldots, X_{d-2n}(x)$ sont linéairement indépendants modulo $\mathbb{C}^{d-n}(x)$ et la classe de $X(x,t)$ modulo $\mathbb{C}^{d-n}(x)$ ne s'annule pour aucune valeur de t. Le lemme est démontré.

6. Fin de la démonstration

6.1. *Introduction*

Armés du Lemme 5.1, nous nous retrouvons en terrain connu. Les articles déjà cités [4], [3] et [7] fournissent plusieurs manières d'achever la linéarisation du tissu $\mathcal{T}$.

Dans leur livre [3], Blaschke et Bol indiquent une voie que nous allons suivre. Résumons-la brièvement.

On sait maintenant que les courbes $C(x)$ sont portées par une surface $S_0 \subset \mathbb{P}^m$. On déduira d'abord, du fait que S_0 contient beaucoup de courbes rationnelles normales, que S_0 est contenu dans une surface algébrique S de $\mathbb{P}^m$. On montrera ensuite que la famille des courbes $C(x)$ est contenue dans une famille algébrique $\mathcal{C}$, de dimension n, de courbes algébriques portées par S. Compte tenu des propriétés d'intersection des courbes $C(x)$ et par continuité, on obtiendra que, par toute famille générique de n points de S, il passe une et une seule courbe de la famille $\mathcal{C}$. Nous serons alors en mesure d'appliquer un théorème classique d'Enriques et de conclure que le système $\mathcal{C}$ est un système linéaire. Autrement dit, il est paramétré par l'espace projectif $\mathbb{P}^n$, de telle façon que l'ensemble de ses éléments qui passent par un point donné de S est paramétré par un hyperplan de $\mathbb{P}^n$. On se retrouvera ainsi dans une situation où l'argument utilisé dans le cas $d = 2n$ pourra s'appliquer : le difféomorphisme local $\mathbb{C}_0^n \to \mathbb{P}^n$ qui envoie x sur le paramètre de la courbe $C(x)$ linéarise le tissu. Compte tenu du théorème d'Abel inverse, le tissu $\mathcal{T}$ est algébrisable.

Remarque 6.1. La démonstration initiale de Bol [4] n'utilise pas le théorème d'Enriques, mais des projections successives bien choisies, jusqu'à se ramener à la situation qu'on rencontre dans le cas $d = 2n$. L'hypothèse que le tissu est de rang maximal, et pas seulement de rang maximal en valuation ≤ 1, est alors importante, au moins si l'on suit la démonstration de Bol pas à pas. Il n'est pas exclu que cette méthode plus élémentaire puisse être adaptée au cas général.

6.2. *Le point sur la situation*

Il est peut être utile de rappeler où nous en sommes dans la démonstration. On suppose toujours $d \geq (2n+1)$ et $n \geq 3^{(5)}$.

On considère un d-tissu $\mathcal{T}$ au voisinage de $0 \in \mathbb{C}^n$, qui possède $m + 1 = (2d - 3n + 1)$ relations abéliennes dont les 1-jets sont linéairement indépendants.

On a associé au tissu $\mathcal{T}$ une famille d'applications :

$$p_\alpha \colon \mathbb{C}_0^n \to \mathbb{P}^m, \qquad \alpha = 1, \ldots, d.$$

L'image de chacune d'elles est une courbe dans $\mathbb{P}^m$. La feuille $\mathcal{F}_\alpha(x_\star)$ du

---

[5] Le cas $n = 2$ est permis à condition de faire l'hypothèse supplémentaire que la conclusion du Lemme 3.2 est vérifiée. Rappelons aussi que $\mathbb{C}_0^n$ désigne un voisinage de 0 dans $\mathbb{C}^n$, qu'on peut restreindre autant qu'on veut.

α−ième feuilletage qui passe par un point $x_\star$ est donnée par

$$\mathcal{F}_\alpha(x_\star) = \{x \in \mathbb{C}_0^n, \ p_\alpha(x) = p_\alpha(x_\star)\}, \qquad \alpha = 1, \ldots, d.$$

On sait que, pour tout x, les points $p_1(x), \ldots, p_d(x)$ engendrent un sous-espace $\mathbb{P}^{d-n-1}(x)$ de dimension $(d-n-1)$ et qu'ils sont en position générale dans ce sous-espace. De plus, l'application de Poincaré ainsi définie

$$\mathbb{P}^{d-n-1} \colon \ \mathbb{C}_0^n \to \mathbb{G}(d-n-1, m)$$

est une immersion. Pour tout couple de points distincts $x, x' \in \mathbb{C}_0^n$, les sous-espaces $\mathbb{P}^{d-n-1}(x)$ et $\mathbb{P}^{d-n-1}(x')$ se coupent (transversalement) suivant un sous-espace $\mathbb{P}^{n-2}(x, x')$ de dimension $(n-2)$.

On sait que, pour tout x, les points $p_1(x), \ldots, p_d(x)$ appartiennent à une courbe rationnelle normale $C(x)$ de $\mathbb{P}^{d-n-1}(x)$, uniquement déterminée. Enfin, d'après le Lemme 5.1, il existe une application de rang constant 2 :

$$p \colon \ \mathbb{C}_0^n \times \mathbb{P}^1 \to \mathbb{P}^m,$$

telle que, pour tout $x \in \mathbb{C}_0^n$, $t \mapsto p(x, t)$ est un isomorphisme de $\mathbb{P}^1$ sur $C(x)$.

D'après le théorème du rang, l'image de p est une sous-variété lisse de $\mathbb{P}^m$, non fermée. On note cette surface S_0.

6.3. *Propriétés d'intersection des courbes $C(x)$*

On a :

Lemme 6.2. *Pour tout couple de points distincts $x, x' \in \mathbb{C}_0^n$, les courbes $C(x)$ et $C(x')$ se coupent en exactement $(n-1)$ points, compte tenu des multiplicités.*

Démonstration — On a vu que $C(x)$ et $C(x')$ se coupent en au plus $(n-1)$ points. (Rappelons que c'est une conséquence du fait que $\mathbb{P}^{d-n-1}(x)$ et $\mathbb{P}^{d-n-1}(x')$ sont transverses d'intersection $\mathbb{P}^{n-2}(x, x')$ et que toute famille de points d'une courbe rationnelle normale est en position générale dans l'espace qu'elle engendre.) Pour $t \in \mathbb{C}$ voisin de 0, soit $x(t) \in \mathbb{C}_0^n$ le point défini par $u_1(x(t)) = 0, \ldots, u_{n-1}(x(t)) = 0$ et $u_n(x(t)) = t$. La courbe $C(x(t))$ contient les points $p_1(0), \ldots, p_{n-1}(0)$, qui sont en position générale dans $\mathbb{P}^m$, donc distincts. Ainsi, pour tout $t \neq 0$ petit, $C(x(t))$ coupe $C(0)$ en exactement $(n-1)$ points. *Compte tenu du fait que les courbes $C(x)$ sont contenues dans une surface*, le nombre de points d'intersection de deux courbes distinctes $C(x)$ et $C(x')$ est constant quand x' varie un peu. D'où le lemme. $\qquad\square$

6.4. La surface S_0 est contenue dans une surface algébrique S

En effet, donnons-nous un point de S_0 et un voisinage S_1 de ce point dans S_0, contenu dans le domaine d'une carte affine de $\mathbb{P}^m$. On se ramène ainsi au cas où $0 \in S_1 \subset \mathbb{C}^m$. Il suffit de montrer que S_1 est contenu dans une surface algébrique de $\mathbb{C}^m$.

Au voisinage de 0, S_1 est défini par un systèmes d'équations, soit $f_\rho(x) = 0$ avec $\rho = 1, \ldots, r$. Considérons l'espace X des polynômes $x : \mathbb{C} \to \mathbb{C}^m$ de degré $\leq (d - n - 1)$ et nuls en $0 \in \mathbb{C}$. Un élément de X est donné par la famille convenablement ordonnée de ses coefficients, soit $a = (a_1, \ldots, a_N)$. On note x_a le polynôme donné par $a \in \mathbb{C}^N$.

Si $\rho \in \{1, \ldots, r\}$, quand on développe $f_\rho(x_a(t))$ en série entière de t, les coefficients de la série obtenue sont des polynômes en a. L'ensemble X' des $x \in X$ qui envoie un voisinage de $0 \in \mathbb{C}$ dans S_1 est donc un ensemble algébrique. Il contient des paramétrisations convenables des courbes $C(x)$ qui passent par 0. L'image de l'application $(a, t) \mapsto x_a(t)$ est une surface algébrique de $\mathbb{C}^m$, qui contient S_0.

6.5. Le système $\mathcal{C}_0$ des courbes $C(x)$ est contenu dans un système algébrique de dimension n

On peut munir l'ensemble $\mathcal{R}$ des courbes rationnelles normales de degré $(d - n - 1)$ dans $\mathbb{P}^m$ d'une structure naturelle de variété analytique lisse. Comme toutes les courbes $C \in \mathcal{R}$ sont équivalentes, il suffit de définir la structure analytique au voisinage d'une courbe, par exemple la courbe C_0 paramétée par :

$$x(t) = (1, t, \ldots, t^{d-n-1}, 0, \ldots, 0) \,.$$

On peut alors introduire une famille suffisante d'hyperplans H_k transverses à la courbe C_0 et paramétrer les courbes voisines de C_0 par leurs intersections avec ces hyperplans ... Les détails sont laissés au lecteur.

Il résulte du fait que l'application de Poincaré $\mathbb{P}^{d-n-1} : \mathbb{C}_0^n \to \mathbb{G}(d - n - 1, m)$ est une immersion que l'application $x \mapsto C(x)$ définit une immersion :

$$(33) \qquad\qquad C : \mathbb{C}_0^n \to \mathcal{R} \,.$$

On note $\mathcal{C}_0$ son image.

La théorie des variétés de Chow montre que la variété $\mathcal{R}$ est une variété quasi-projective. Autrement dit, il existe une sous-variété (fermée) $\mathcal{R}'$ d'un espace projectif, telle que $\mathcal{R}$ soit isomorphe à un ouvert de Zariski dense de

$\mathcal{R}'$. On identifie provisoirement $\mathcal{R}$ à son image dans $\mathcal{R}'$. Plus précisément, il existe aussi une sous-variété $\mathcal{W}$ de $\mathcal{R}' \times \mathbb{P}^m$, de dimension

$$\dim \mathcal{W} = \dim \mathcal{R} + 1\,,$$

qu'on munit des projections canoniques

$$\varpi \colon \mathcal{W} \to \mathcal{R}'\,, \qquad \pi \colon \mathcal{W} \to \mathbb{P}^m\,,$$

et qui a la propriété suivante. Pour tout $\xi \in \mathcal{R}'$, son image réciproque $\varpi^{-1}(\xi)$ est une courbe de $\mathcal{W}$. De plus, si $\xi \in \mathcal{R}$, la projection π induit un isomorphisme de cette courbe sur son image dans $\mathbb{P}^m$ et cette image est la courbe rationnelle normale représentée par ξ.

Lemme 6.3. *La famille $\mathcal{C}_0$ des courbes $C(x)$, $x \in \mathbb{C}_0^n$, est contenue dans une sous-variété algébrique irréductible $\mathcal{C}$ de $\mathcal{R}'$, de dimension n.*

Démonstration — Suivant [3], on considère l'intersection $\mathcal{C}$ de toutes les sous-variétés algébriques de $\mathcal{R}'$ qui contiennent $\mathcal{C}_0$. C'est une variété irréductible qui contient $\mathcal{C}_0$ et qui est donc de dimension $\geq n$. Il s'agit de montrer qu'elle est de dimension n.

Si $\xi \in \mathcal{R}'$, on note aussi ξ la courbe $\pi(\varpi^{-1}(\xi))$. L'ensemble des $\xi \in \mathcal{C}$ qui sont contenus dans S est une sous-variété de $\mathcal{C}$ qui contient $\mathcal{C}_0$. C'est donc $\mathcal{C}$.

Pour tout $x \in \mathbb{C}_0^n$, l'ensemble des $\xi \in \mathcal{C}$ qui coupent $C(x)$ en $(n-1)$ points est un ouvert de Zariski d'une sous variété qui contient $\mathcal{C}_0$, donc un ouvert de Zariski de $\mathcal{C}$.

De même, si ξ est un élément de $\mathcal{C}$ qui coupe toutes les $C(x)$ en $(n-1)$ points, l'ensemble des $\xi' \in \mathcal{C}$ qui coupent ξ en $(n-1)$ points est un ouvert de Zariski d'une sous-variété de $\mathcal{C}$ qui contient $\mathcal{C}_0$, donc un ouvert de Zariski de $\mathcal{C}$.

Finalement, si la dimension de $\mathcal{C}$ était $\geq (n+1)$, deux éléments génériques de $\mathcal{C}$ se couperaient en au moins n points, ce qui contredit ce qu'on vient de montrer. $\qquad\square$

Au cours de la démonstration, on a presque vérifié que la famille $\mathcal{C}$ a les propriétés suivantes:

Lemme 6.4. *Les éléments de la famille algébrique $\mathcal{C}$ sont des courbes contenues dans la surface S. Deux courbes génériques de la famille se coupent en $(n-1)$ points et, par toute famille générique de n points de S, il passe une et une seule courbe de la famille $\mathcal{C}$.*

Démonstration — Il reste à démontrer la deuxième partie de l'énoncé. Il résulte du fait que la variété C est de dimension n que, par n points génériques de S, il passe au moins une courbe de C. Comme deux courbes génériques se coupent en $(n-1)$ points, il en passe une seule.   □

6.6. *Conclusion*

Pour conclure, il suffit d'appliquer le résultat suivant :

Théorème 6.5 (**Enriques**). *Soit S une surface algébrique et C un système algébrique de courbes sur S, de dimension n. Si la courbe générique de C est irréductible et si, par une famille générique de n points de S, il passe une et une seule courbe de C, le système C est un système linéaire.*

C'est en fait un énoncé de géométrie birationnelle et l'énoncé, tel qu'il est formulé est incorrect. Le système linéaire final peut ne coïncider avec le système initial que modulo un ensemble de dimension $< n$.

Quoi qu'il en soit, l'énoncé signifie qu'on peut choisir $C = \mathbb{P}^n$ comme espace des paramètres et que, pour tout point générique $s \in S$, l'ensemble des $\xi \in C$ tels que s appartient à la courbe ξ est un hyperplan de l'espace projectif C, identifié à $\mathbb{P}^n$.

Comme on a dit dans l'introduction à cette section, en associant à tout $x \in \mathbb{C}_0^n$ l'élément de $\mathbb{P}^n$ qui correspond à la courbe $C(x)$ dans l'identification $C = \mathbb{P}^n$, on obtient un difféomorphisme local $\mathbb{C}_0^n \to \mathbb{P}^n$ qui linéarise le tissu.

Références

[1] W. Blaschke, *Abh. Math. Semin. Hamb. Univ.* **9** (1933), 291–298.

[2] W. Blaschke, Über die Tangenten einer ebenen Kurve fünfter Klasse, *Abh. Math. Semin. Hamb. Univ.* **9** (1933), 313–317.

[3] W. Blaschke, G. Bol, *Geometrie der Gewebe*, Die Grundlehren der Mathematischen Wissenschaften, vol. 49, *J. Springer, Berlin*, 1938 .

[4] G. Bol, Flächengewebe im dreidimensionalen Raum, *Abh. Math. Semin. Hamb. Univ.* **10** (1934), 119–133.

[5] F. Enriques, Una questione sulla linearità dei sistemi di curve appartenenti ad una superficie algebrica, *Rend. Acc. Lincei* (1893), 3–8.

[6] S. S. Chern, P. Griffiths, Abel's theorem and webs, *Jahresber. Deutsch. Math.-Verein.* **80** (1978), *no.* 1-2, 13-110.

[7] S. S. Chern, P. Griffiths, Corrections and addenda to our paper : "Abel's theorem and webs", *Jahresber. Deutsch. Math.-Verein.* **83** (1981), *no.* 2, 78–83.

[8] P. Griffiths, Variations on a theorem of Abel, *Inventiones Math.* **35** (1976), 321–390.

[9] L. Pirio, J.-M. Trépreau, Tissus plans exceptionnels et fonctions thêta, *Ann. Inst. Fourier*, **55** (2005), 2209–2237.

[10] H. Poincaré, Sur les surfaces de translation et les fonctions abéliennes, *Bull. Soc. Math. France,* **29** (1901), 61–86.

Chapter 18

CONFORMAL GEOMETRY AND FULLY NONLINEAR EQUATIONS

Jeff Viaclovsky[*]

Department of Mathematics, MIT, Cambridge, MA 02139, USA
and
Department of Mathematics, University of Wisconsin,
Madison, WI, 53706, USA
jeffv@math.wisc.edu

This article is a survey of results involving conformal deformation of
Riemannian metrics and fully nonlinear equations.

August 31, 2006

To the memory of Professor S. S. Chern

1. The Yamabe Equation

One of the most important problems in conformal geometry is the Yamabe
Problem, which is to determine whether there exists a conformal metric
with constant scalar curvature on any closed Riemannian manifold. In
what follows, let (M, g) be a Riemannian manifold, and let R denote the
scalar curvature of g. Writing a conformal metric as $\tilde{g} = v^{\frac{4}{n-2}} g$, the Yamabe
equation takes the form

$$(1.1) \qquad 4\frac{n-1}{n-2}\Delta v + R \cdot v = \lambda \cdot v^{\frac{n+2}{n-2}} ,$$

where λ is a constant. These are the Euler-Lagrange equations of the
Yamabe functional,

$$(1.2) \qquad \mathcal{Y}(\tilde{g}) = Vol(\tilde{g})^{-\frac{n-2}{n}} \int_M R_{\tilde{g}} dvol_{\tilde{g}} ,$$

---

[*]The research of the author was partially supported by NSF Grant DMS-0503506.

for $\tilde{g} \in [g]$, where $[g]$ denotes the conformal class of g. An important related conformal invariant is the *Yamabe invariant* of the conformal class $[g]$:

(1.3) $$Y([g]) \equiv \inf_{\tilde{g} \in [g]} \mathcal{Y}(\tilde{g}).$$

The Yamabe problem has been completely solved through the results of many mathematicians, over a period of approximately thirty years. Initially, Yamabe claimed to have a proof in [Yam60]. The basic strategy was to prove the existence of a minimizer of the Yamabe functional through a sub-critical regularization technique. Subsequently, an error was found by N. Trudinger, who then gave a solution with a smallness assumption on the Yamabe invariant [Tru68]. Later, Aubin showed that the problem is solvable provided that

(1.4) $$Y([g]) < Y([g_{round}]),$$

where $[g_{round}]$ denotes the conformal class of the round metric on the n-sphere, and verified this inequality for $n \geq 6$ and g not locally conformally flat [Aub76b], [Aub76a], [Aub98]. Schoen solved the remaining cases [Sch84]. It is remarkable that Schoen employed the positive mass conjecture from general relativity to solve these remaining most difficult cases.

An important fact is that $SO(n+1,1)$, the group of conformal transformations of the n-sphere S^n with the round metric, is non-compact. Likewise, the space of solutions to the Yamabe equation in the conformal class of the round sphere is non-compact. However, if (M,g) is compact, and not conformally equivalent to the round sphere, then the group of conformal transformations is compact [LF71], [Oba72]. A natural question is then whether the space of all unit-volume solutions (not just minimizers) to (1.1) is compact on an arbitrary compact manifold, provided (M,g) is not conformally equivalent to S^n with the round metric. Schoen solved this in the locally conformally flat case [Sch91], and produced unpublished lecture notes outlining a solution is certain other cases. Many other partial solutions have appeared, see for example [Dru04], [LZ04], [LZ05], [Mar05], [Sch89]. Schoen has recently announced the complete solution of the compactness problem in joint work with Khuri and Marques, assuming that the positive mass theorem holds in higher dimensions. The positive mass theorem is known to hold in the locally conformally flat case in all dimensions [SY79a], and in the general case in dimensions $n \leq 7$ [SY81], [LP87], and in any dimension if the manifold is spin [Wit81], [PT82], [LP87].

2. A Fully Nonlinear Yamabe Problem

The equation (1.1) is a semi-linear equation, meaning the the non-linearities only appear in lower order terms – second derivatives appear in a linear fashion. One may investigate other types of conformal curvature equations, which brings one into the realm of fully nonlinear equations. We recall the Schouten tensor

$$(2.1) \qquad A_g = \frac{1}{n-2} \left(Ric - \frac{R}{2(n-1)} g \right),$$

where Ric denotes the Ricci tensor. This tensor arises naturally in the decomposition of the full curvature tensor

$$(2.2) \qquad Riem = Weyl + A \odot g,$$

where $\odot$ denotes the Kulkari-Nomizu product [Bes87]. This equation also serves to define the Weyl tensor, which is conformally invariant. Thus the behaviour of the full curvature tensor under a conformal change of metric is entirely determined by the Schouten tensor. Let F denote any symmetric function of the eigenvalues, which is homogeneous of degree one, and consider the equation

$$(2.3) \qquad F(\tilde{g}^{-1} A_{\tilde{g}}) = \text{constant}.$$

Note that the $\tilde{g}^{-1}$ factor is present since only the eigenvalues of an endomorphism are well-defined. If we write a conformal metric as $\tilde{g} = e^{-2u} g$, the Schouten tensor transforms as

$$(2.4) \qquad A_{\tilde{g}} = \nabla^2 u + du \otimes du - \frac{|\nabla u|^2}{2} g + A_g.$$

Therefore, equation (2.3) is equivalent to

$$(2.5) \quad F \left(g^{-1} \left(\nabla^2 u + du \otimes du - \frac{|\nabla u|^2}{2} g + A_g \right) \right) = \text{constant} \cdot e^{-2u}.$$

Let σ_k denote the kth elementary symmetric function of the eigenvalues

$$(2.6) \qquad \sigma_k = \sum_{i_1 < \cdots < i_k} \lambda_{i_1} \cdots \lambda_{i_k}.$$

For the case of $F = \sigma_k^{1/k}$, the equation (2.3) has become known as the σ_k-*Yamabe equation*:

$$(2.7) \qquad \sigma_k^{1/k} (\tilde{g}^{-1} A_{\tilde{g}}) = \text{constant}.$$

In the context of exterior differential systems, they arose in a different form in Bryant and Griffiths' research on conformally invariant Poincaré-Cartan forms [BGG03], and these systems were shown to correspond to the σ_k-Yamabe equation in [Via00a].

For $1 \leq k \leq n$, we define the cone (in $\mathbb{R}^n$)

$$(2.8) \qquad \Gamma_k^+ = \{\sigma_k > 0\} \cap \{\sigma_{k-1} > 0\} \cap \cdots \cap \{\sigma_1 > 0\}.$$

These are well-known as ellipticity cones for the σ_k equation, see [Går59], [Ivo83], [CNS85]. We will say that a metric g is strictly k-*admissible* if the eigenvalues of $g^{-1}A_g$ lie in Γ_k^+ at every point $p \in M$. It is an important fact that if the metric $\tilde{g}$ is k-admissible, then the linearization of (2.7) at $\tilde{g}$ is elliptic. On a compact manifold, if the background metric g is k-admissible, then (2.7) is necessarily elliptic at *any* solution [Via02]. Thus, a k-admissiblity assumption on the background metric is an ellipticity assumption, reminiscent of the k-convexity assumption on domains for the k-Hessian equation [CNS85].

3. Variational Characterization

In [Via00a], it was shown that in several cases, the σ_k-Yamabe equation is variational. Let $\mathcal{M}_1$ denote the set of unit volume metrics in the conformal class $[g_0]$.

Theorem 3.1. (Viaclovsky [Via00a]) *If $k \neq n/2$ and $(N, [g_0])$ is locally conformally flat, a metric $g \in \mathcal{M}_1$ is a critical point of the functional*

$$\mathcal{F}_k : g \mapsto \int_N \sigma_k(g^{-1}A_g)\,dV_g$$

restricted to $\mathcal{M}_1$ if and only if

$$\sigma_k(g^{-1}A_g) = C_k$$

for some constant C_k. If N is not locally conformally flat, the statement is true for $k = 1$ and $k = 2$.

For $k = 1$, this is of course well-known, as $\mathcal{F}_1$ is the Hilbert functional, [Hil72], [Sch89]. For $k = n/2$, in [Via00a] it was shown that the integrand is the non-Weyl part of the Chern-Gauss-Bonnet integrand. Therefore, $\mathcal{F}_{n/2}$ is necessarily constant in the locally conformally flat case (when $n = 4$, this holds in general, this will be discussed in detail in Section 6 below). Nevertheless, in this case the equation is still variational, but with a different functional. Fix a background metric h, write $g = e^{-2u}h$, and let

$$
\mathcal{E}_{n/2}(g) = \int_M \int_0^1 \sigma_{n/2}\Big(-t\nabla_h^2 u + t^2 \nabla_h u \otimes \nabla_h u
$$

(3.1)
$$
-\frac{1}{2}t^2 |\nabla_h u|^2 g_0 + A(h) \Big) u\, dt dV_h ,
$$

then for any differentiable path of smooth conformal metrics g_t,

(3.2)
$$
\frac{d}{dt}\mathcal{E}_{n/2}(g_t) = \int_M \sigma_{n/2}(A_{g_t})\, u\, dvol_{g_t} .
$$

This fact was demonstrated in [BV04], see also [CY03]. This is valid also for $n = 4$, [CY95]. Recently, Sheng-Trudinger-Wang have given conditions on when the more general F-Yamabe equation is variational, see [STW05].

A natural question is: what are the critical metrics of the $\mathcal{F}_k$ functionals, when considering all possible metric variations, not just conformal variations? It is a well-known result that the critical points of $\mathcal{F}_1$ restricted to space of unit volume metrics are exactly the Einstein metrics [Sch89]. But for $k > 1$, the full Euler-Lagrange equations are manifestly fourth order equations in the metric. However, in dimension three we have the following

Theorem 3.2. (Gursky-Viaclovsky [GV01]) *Let M be compact and of dimension three. Then a metric g with $\mathcal{F}_2[g] \geq 0$ is critical for $\mathcal{F}_2$ restricted to the space of unit volume metrics if and only if g has constant sectional curvature.*

A similar theorem was proved by Hu-Li for $n \geq 5$, but with the rather stringent condition that the metric be locally conformally flat [HL04]. We mention that Labbi studied some curvature quantities defined by H. Weyl which are polynomial in the full curvature tensor, and proved some interesting variational formulas [Lab04].

4. Liouville Theorems

We next turn to the uniqueness question. In the negative curvature case, the linearization of (2.3) is invertible, so the uniqueness question is trivial. However, in the positive curvature case the uniqueness question is nontrivial. In [Via00a], [Via00b] the following was proved

Theorem 4.1. (Viaclovsky [Via00a]) *Suppose (N, g_0) is of unit volume and has constant sectional curvature $K > 0$. Then for any $k \in \{1, \ldots, n\}$, g_0 is*

the unique unit volume solution in its conformal class of

(4.1) $$\sigma_k(g^{-1}A_g) = \text{constant},$$

unless N is isometric to S^n with the round metric. In this case we have an $(n + 1)$-parameter family of solutions which are the images of the standard metric under the conformal diffeomorphisms of S^n.

For $k = 1$, the constant scalar curvature case, the theorem holds just assuming N is Einstein. This is the well-known theorem of Obata [Oba72].

This theorem falls under the category of a Liouville-type theorem. We let δ_{ij} be the Kronecker delta symbol, and write the conformal factor as $\tilde{g} = u^{-2}g$. In stereographic coordinates, the equation (4.1) is written

(4.2) $$\sigma_k\left(u \cdot \frac{\partial^2 u}{\partial x^i \partial x^j} - \frac{|\nabla u|^2}{2}\delta_{ij}\right) = \text{constant}.$$

This equation is conformally invariant: if $T : \mathbb{R}^n \to \mathbb{R}^n$ is a conformal transformation (i.e., $T \in SO(n+1, 1)$), and $u(x)$ is a solution of (4.2), then

(4.3) $$v(x) = |J(x)|^{-1/n} u(Tx)$$

is also a solution, where J is the Jacobian of T, see [Via00b].

The uniqueness theorem can then be restated as a Liouville Theorem in $\mathbb{R}^n$:

Theorem 4.2. (Viaclovsky [Via00b]) *Let $u(x) \in C^\infty(\mathbb{R}^n)$ be a positive solution to*

(4.4) $$\sigma_k\left(u \cdot \frac{\partial^2 u}{\partial x^i \partial x^j} - \frac{|\nabla u|^2}{2}\delta_{ij}\right) = \text{constant}$$

for some $k \in \{1, \dots, n\}$. Suppose that $v(y) = |y|^2 \cdot u\left(\frac{y^1}{|y|^2}, \dots, \frac{y^n}{|y|^2}\right)$ is smooth and

$$\lim_{y \to 0} v(y) > 0.$$

Then

$$u = a|x|^2 + b_i x^i + c,$$

where a, b_i, and c are constants.

The proof in [Via00a], [Via00b] requires the stringent growth condition at infinity. For the scalar curvature equation, $k = 1$, this theorem was proved without *any* assumption at infinity in the important paper by Caffarelli-Gidas-Spruck [CGS89], using the moving planes technique.

This analogous theorem for $k \geq 2$ is now known to hold without any condition on the behaviour at infinity. For $k = 2$, important work was done in [CGY02a], and [CGY03b], proving the Liouville Theorem for $n = 4, 5$, and for $n \geq 6$ with a finite volume assumption. Maria del Mar González proved a Liouville Theorem for σ_k, $n > 2(k + 1)$ with a finite volume assumption [Gon04b].

Yanyan Li and Aobing Li proved the following theorem for all k in the important paper [LL03].

Theorem 4.3. (Li-Li [LL03]) *Let $u(x) \in C^\infty(\mathbb{R}^n)$ be a positive solution to*

$$(4.5) \qquad \sigma_k \left(u \cdot \frac{\partial^2 u}{\partial x^i \partial x^j} - \frac{|\nabla u|^2}{2} \delta_{ij} \right) = \text{constant} > 0$$

for some $k \in \{1, \dots, n\}$, satisfying

$$(4.6) \qquad u \cdot \frac{\partial^2 u}{\partial x^i \partial x^j} - \frac{|\nabla u|^2}{2} \delta_{ij} \in \Gamma_k^+.$$

Then

$$u = a|x|^2 + b_i x^i + c,$$

where a, b_i, and c are constants.

Their method is based on the moving planes technique. Subsequently, their results have been generalized to much more general classes of symmetric functions F, see [LL05b], [Li03], [LL05a], ,[Li02].

These types of Liouville theorems can be used in deriving *a priori* estimates for solutions of (4.1), we will discuss this below.

5. Local Estimates

We consider the σ_k-Yamabe equation

$$(5.1) \qquad \sigma_k^{1/k} \left(\nabla^2 u + du \otimes du - \frac{|\nabla u|^2}{2} g + A_g \right) = f(x) e^{-2u},$$

with $f(x) \geq 0$. A remarkable property of (5.1) was discovered in [GW03b]. It turns out that *local* estimates are satisfied, a fact which does not hold in general fully nonlinear equations. We say that $u \in C^2$ is k-admissible if $A_u \in \overline{\Gamma}_k^+$. For equation (5.1), Guan and Wang prove

Theorem 5.1. (Guan-Wang [GW03b]) *Let $u \in C^3(M^n)$ be a k-admissible solution of (5.1) in $B(x_0, \rho)$, where $x_0 \in M^n$ and $\rho > 0$. Then there is a*

constant

$$C_0 = C_0(k, n, \rho, \|g\|_{C^2(B(x_0,\rho))}, \|f\|_{C^1(B(x_0,\rho))}),$$

such that

(5.2) $$|\nabla u|^2(x) \leq C_0\left(1 + e^{-2\inf_{B(x_0,\rho)} u}\right)$$

for all $x \in B(x_0, \rho/2)$.

Let $u \in C^4(M^n)$ *be a* k-*admissible solution of* (5.1) *in* $B(x_0, \rho)$, *where* $x_0 \in M^n$ *and* $\rho > 0$. *Then there is a constant*

$$C_0 = C_0(k, n, \rho, \|g\|_{C^3(B(x_0,\rho))}, \|f\|_{C^2(B(x_0,\rho))}),$$

such that

(5.3) $$|\nabla^2 u|(x) + |\nabla u|^2(x) \leq C_0\left(1 + e^{-2\inf_{B(x_0,\rho)} u}\right)$$

for all $x \in B(x_0, \rho/2)$.

These local estimates for (5.1) generalize the global estimates which were first proved in [Via02]. Subsequently, these results have been extended to much more general classes of symmetric functions F, see [Che05], [GW04], [GLW04b], [LL03],[Li06], [Wan06]. Estimates for solutions of σ_2 in dimension four were proved in [Han04] using integral methods.

Equipped with second derivative estimates, one then uses the work of Evans and Krylov [Eva82], [Kry83] to obtain $C^{2,\alpha}$ estimates, that is, a Hölder estimate on second derivatives. This is crucial – the importance of that work in this theory cannot be overstated.

Consider a symmetric function

(5.4) $$F : \Gamma \subset \mathbb{R}^n \to \mathbb{R}$$

with $F \in C^\infty(\Gamma) \cap C^0(\overline{\Gamma})$, where $\Gamma \subset \mathbb{R}^n$ is an open, symmetric, convex cone, and impose the following conditions:

(i) F is symmetric, concave, and homogenous of degree one.
(ii) $F > 0$ in Γ, and $F = 0$ on $\partial\Gamma$.
(iii) F is *elliptic*: $F_{\lambda_i}(\lambda) > 0$ for each $1 \leq i \leq n$, $\lambda \in \Gamma$.

We mention that Szu-Yu Chen proved local C^2-estimates (5.3) for this general class of symmetric functions F [Che05].

An immediate corollary of these local estimates is an ϵ-regularity result:

Theorem 5.2. (Guan-Wang [GW03b]) *There exist constants $\epsilon_0 > 0$ and $C = C(g, \epsilon_0)$ such that any solution $u \in C^2(B(x_0, \rho))$ of (5.1) with*

$$(5.5) \qquad \int_{B(x_0, \rho)} e^{-nu} dvol_g \le \epsilon_0,$$

satisfies

$$(5.6) \qquad \inf_{B(x_0, \rho/2)} u \ge -C + \log \rho.$$

Consequently, there is a constant

$$C_2 = C_2(k, n, \mu, \epsilon_0, \|g\|_{C^3(B(x_0, \rho))}),$$

such that

$$(5.7) \qquad |\nabla^2 u|(x) + |\nabla u|^2(x) \le C_2 \rho^{-2}$$

for all $x \in B(x_0, \rho/4)$.

This type of estimate is crucial in understanding *bubbling*, a phenomenon which is unavoidable when studying conformally invariant problems. It shows that non-compactness of the space of solutions can arise only through volume concentration.

6. Dimension Four

In dimension four, an important conformal invariant is

$$(6.1) \quad \mathcal{F}_2([g]) \equiv 4 \int_M \sigma_2(A_g) dV_g = \int_M \left(-\frac{1}{2} |Ric_g|^2 + \frac{1}{6} R_g^2 \right) dvol_g.$$

By the Chern-Gauss-Bonnet formula ([Bes87]),

$$(6.2) \qquad 8\pi^2 \chi(M) = \int_M |W_g|^2 dvol_g + \mathcal{F}_2([g]).$$

Thus, the conformal invariance of $\mathcal{F}_2$ follows from the well known (pointwise) conformal invariance of the Weyl tensor W_g (see [Eis97]).

One of the most interesting results in this area is

Theorem 6.1. (Chang-Gursky-Yang [CGY02b]) *Let (M, g) be a closed 4-dimensional Riemannian manifold with positive scalar curvature. If $\mathcal{F}_2([g]) > 0$, then there exists a conformal metric $\tilde{g} = e^{-2u} g$ with $R_{\tilde{g}} > 0$*

and $\sigma_2(\tilde{g}^{-1}A_{\tilde{g}}^1) > 0$ *pointwise. In particular, the Ricci curvature of $\tilde{g}$ satisfies*

$$0 < 2Ric_{\tilde{g}} < R_{\tilde{g}}\tilde{g}\,.$$

By combining this with some work of Margerin [Mar98] on the Ricci flow, the authors obtained the following remarkable integral sphere-pinching theorem:

Theorem 6.2. (Chang-Gursky-Yang [CGY03a]) *Let (M^4, g) be a smooth, closed four-manifold for which*

  (i) *the Yamabe invariant $Y([g]) > 0$, and*
 (ii) *the Weyl curvature satisfies*

(6.3) $$\int_{M^4} |W|^2 dvol < 16\pi^2\chi(M^4)\,.$$

Then M^4 is difeomorphic to either S^4 or $\mathbb{RP}^4$.

The proof of Theorem 6.1 in [CGY02b] involved regularization by a fourth-order equation and relied on some delicate integral estimates. Subsequently, a more direct proof was given in [GV03a]: define the tensor

(6.4) $$A_g^t = \frac{1}{2}\left(Ric_g - \frac{t}{6}R_g g\right)\,.$$

Theorem 6.3. (Gursky-Viaclovsky [GV03a]) *Let (M, g) be a closed 4-dimensional Riemannian manifold with positive scalar curvature. If*

(6.5) $$\mathcal{F}_2([g]) + \frac{1}{6}(1 - t_0)(2 - t_0)(Y([g]))^2 > 0\,,$$

for some $t_0 \leq 1$, then there exists a conformal metric $\tilde{g} = e^{-2u}g$ with $R_{\tilde{g}} > 0$ and $\sigma_2(A_{\tilde{g}}^{t_0}) > 0$ pointwise. This implies the pointwise inequalities

(6.6) $$(t_0 - 1)R_{\tilde{g}}\tilde{g} < 2Ric_{\tilde{g}} < (2 - t_0)R_{\tilde{g}}\tilde{g}\,.$$

The proof involves a deformation of the equation through a path of fully nonlinear equations, and an application of the local estimates of Guan-Wang. A similar technique was applied the σ_k equations in the locally conformally flat case by Guan-Lin-Wang [GLW04a].

As applications of Theorem 6.3, consider two different values of t_0. When $t_0 = 1$, we obtain the aforementioned result in [CGY02b]. The second application is to the spectral properties of a conformally invariant differential operator known as the *Paneitz operator*. Let δ denote the L^2-adjoint of the exterior derivative d; then the Paneitz operator is defined by

(6.7)
$$P_g \phi = \Delta^2 \phi + \delta \left(\frac{2}{3} R_g g - 2 Ric_g \right) d\phi.$$

The Paneitz operator is conformally invariant, in the sense that if $\tilde{g} = e^{-2u}g$, then

(6.8)
$$P_{\tilde{g}} = e^{4u} P_g.$$

Since the volume form of the conformal metric $\tilde{g}$ is $dvol_{\tilde{g}} = e^{-4u} dvol_g$, an immediate consequence of (6.8) is the conformal invariance of the Dirichlet energy

$$\langle P_{\tilde{g}} \phi, \phi \rangle_{L^2(M, \tilde{g})} = \langle P_g \phi, \phi \rangle_{L^2(M, g)}.$$

In particular, positivity of the Paneitz operator is a conformally invariant property, and clearly the kernel is invariant as well.

To appreciate the geometric significance of the Paneitz operator, define the associated *Q-curvature*, introduced by Branson:

(6.9)
$$Q_g = -\frac{1}{12} \Delta R_g + 2\sigma_2(g^{-1} A_g^1).$$

Under a conformal change of metric $\tilde{g} = e^{-2u}g$, the Q-curvature transforms according to the equation

(6.10)
$$-Pu + 2Q_g = 2Q_{\tilde{g}} e^{-4u},$$

see, for example, [BO91]. Note that

(6.11)
$$\int_M Q_g dvol_g = \frac{1}{2} \mathcal{F}_2([g]),$$

so the integral of the Q-curvature is conformally invariant.

An application of Theorem 6.3 with $t_0 = 0$, yields

Theorem 6.4. (Gursky-Viaclovsky [GV03a]) *Let (M, g) be a closed 4-dimensional Riemannian manifold with positive scalar curvature. If*

(6.12)
$$\int Q_g dvol_g + \frac{1}{6}(Y[g])^2 > 0,$$

then the Paneitz operator is nonnegative, and $Ker P = \{constants\}$. Therefore, by the results in [CY95], there exists a conformal metric $\tilde{g} = e^{-2u}g$ with $Q_{\tilde{g}} = constant$.

This yields new examples of manifolds admitting constant Q-curvature metrics, see [GV03a].

Another interesting application of σ_2 in dimension 4 was found in [Wan05], which gives a lower estimate on eigenvalues of the Dirac operator in terms of $\mathcal{F}_2$ on a spin manifold.

7. Parabolic Methods

We recall the *Yamabe flow*,

$$(7.1) \qquad \frac{d}{dt}g = -(R_g - r_g)g \,,$$

where r_g denotes the mean value of the scalar curvature. This flow was introduced by Hamilton, who proved existence of the flow for all time and proved convergence in the case of negative scalar curvature. The case of positive scalar curvature however is highly non-trivial. The locally conformally flat case was studied in [Cho92] and [Ye94]. Schwetlick and Struwe [SS03] proved convergence for $3 \leq n \leq 5$ provided an certain energy bound on the initial metric is satsified. In the beautiful paper [Bre05], Simon Brendle proved convergence for $3 \leq n \leq 5$ for *any* initial data.

In the fully nonlinear case, the following flow was first proposed in [GW03a]:

$$(7.2) \qquad \begin{cases} \dfrac{d}{dt}g = -(\log \sigma_k(g) - \log r_k(g)) \cdot g \,, \\[2mm] g(0) = g_0 \,, \end{cases}$$

where $r_k(g)$ is given by

$$r_k(g) = \exp \left(\frac{1}{vol(g)} \int_M \log \sigma_k(g) \, dvol(g) \right).$$

If $g = e^{-2u} \cdot g_0$, then equation (7.2) can be written as the following fully nonlinear flow

$$(7.3) \qquad \begin{cases} 2\dfrac{du}{dt} = \log \sigma_k \left(\nabla^2 u + du \otimes du - \dfrac{|\nabla u|^2}{2} g_0 + A_{g_0} \right) + 2ku - \log r_k \\[2mm] u(0) = u_0 \,. \end{cases}$$

Guan and Wang settled the locally conformally flat case:

Theorem 7.1. (Guan-Wang [GW03a]) *Suppose (M, g_0) be a compact, connected and locally conformally flat manifold. Assume that $A_{g_0} \in \Gamma_k^+$ and smooth, then flow (7.2) exists for all time $0 < t < \infty$ and $g(t) \in C^\infty(M)$*

for all t. For any positive integer l, there exists a constant C depending only on g_0, k, n (independent of t) such that

$$(7.4) \qquad \|g\|_{C^l(M)} \le C \,,$$

where the norm is taken with respect to the background metric g_0. Furthermore, there are a positive number β and a smooth metric $g_\infty \in \Gamma_k^+$ such that

$$(7.5) \qquad \sigma_k(A_{g_\infty}) = \beta \,,$$

and

$$(7.6) \qquad \lim_{t \to \infty} \|g(t) - g_\infty\|_{C^1(M)} = 0 \,,$$

for all l.

Guan and Wang employed a log flow (rather than a gradient flow) due to a technical reason in obtaining C^2 estimates. This solved the σ_k-Yamabe problem in the locally conformally flat case. Independently, Yanyan Li and Aobing Li solved the locally conformally flat σ_k-Yamabe problem using elliptic methods [LL03].

In a subsequent paper [GW04], Guan-Wang employed the log-flow for the quotient equations to obtain some interesting inequalities. Define the scale invariant functionals

$$(7.7) \qquad \mathcal{F}_k(g) = (Vol(g))^{-\frac{n-2k}{n}} \int_M \sigma_k(g^{-1} A_g) dV_g \,.$$

Theorem 7.2. (Guan-Wang [GW04]) *Suppose that (M, g_0) is a compact, oriented and connected locally conformally flat manifold with $A_{g_0} \in \Gamma_k^+$ smooth and conformal metric g with $A_g \in \Gamma_k^+$. Let $0 \le l < k \le n$.*

(A) Sobolev type inequality: *If $0 \le l < k < \frac{n}{2}$, then there is a positive constant $C_S = C_S([g_0], n, k, l)$ depending only on n, k, l and the conformal class $[g_0]$ such that*

$$(7.8) \qquad (\mathcal{F}_k(g))^{\frac{1}{n-2k}} \ge C_S \, (\mathcal{F}_l(g))^{\frac{1}{n-2l}} \,.$$

(B) Conformal quermassintegral type inequality: *If $n/2 \le k \le n$, $1 \le l < k$, then*

$$(7.9) \qquad (\mathcal{F}_k(g))^{\frac{1}{k}} \le \binom{n}{k}^{\frac{1}{k}} \binom{n}{l}^{-\frac{1}{l}} (\mathcal{F}_l(g))^{\frac{1}{l}} \,.$$

(C) Moser-Trudinger type inequality: *If $k = n/2$, then*

$$(n - 2l)\mathcal{E}_{n/2}(g) \geq C_{MT}\left\{\log \int_M \sigma_l(g^{-1}A_g)dV_g - \log \int_M \sigma_l(g_0^{-1}A_{g_0})dV_{g_0}\right\}.$$

(7.10)

Furthermore, the constants are all explicit and optimal, with a complete characterization of the case of equality see [GW04].

In the non-locally conformally flat case, recall from Theorem 3.1 that the σ_2 equation is always variational. Using this fact, the σ_2-Yamabe equation has recently been studied in the general case using parabolic methods. In [GW05a], convergence was proved in dimensions $n > 8$, by constructing an explicit test function. In [STW05], convergence was proved in dimensions $n > 4$, who avoided having to directly construct a test function by employing the solution of the Yamabe problem. Combining this with the work described in Section 8, it follows that the σ_2-Yamabe problem has been solved in all dimensions. Subsequently, the quotient equation σ_2/σ_1 was studied in [GW06] and existence was proved in dimensions $n > 4$.

8. Positive Ricci Curvature

The ellipticity assumption of k-admissibility has geometric consequences on the Ricci curvature. The following inequality was demonstrated in [GVW03]:

Theorem 8.1. (Guan-Viaclovsky-Wang [GVW03]) *Let (M, g) be a Riemannian manifold and $x \in M$. If $A_g \in \Gamma_k^+$ at x for some $k \geq n/2$, then its Ricci curvature is positive at x. Moreover, if $A_g \in \overline{\Gamma}_k^+$ for some $k > 1$, then*

$$Ric_g \geq \frac{2k - n}{2n(k - 1)}R_g \cdot g.$$

In particular if $k \geq \frac{n}{2}$,

$$Ric_g \geq \frac{(2k - n)(n - 1)}{(k - 1)}\binom{n}{k}^{-\frac{1}{k}}\sigma_k^{\frac{1}{k}}(A_g) \cdot g.$$

Therefore, in the case $k > n/2$, any strictly k-admissible metric necessarily has positive Ricci curvature. This fact led to the following definition in [GV04b]:

Definition 8.2. Let (M^n, g) be a compact n-dimensional Riemannian manifold. For $n/2 \leq k \leq n$ the *k-maximal volume* of $[g]$ is

$$\Lambda_k(M^n, [g]) = \sup\{vol(e^{-2u}g)|e^{-2u}g \in \Gamma_k^+(M^n) \text{ with } \sigma_k^{1/k}(g_u^{-1}A_u) \geq \sigma_k\}.$$

(8.1)

If $[g]$ does not admit a k-admissible metric, set $\Lambda_k(M^n, [g]) = +\infty$.

Consider the σ_k-Yamabe equation

$$(8.2) \qquad \sigma_k^{1/k}\left(\nabla^2 u + du \otimes du - \frac{|\nabla u|^2}{2}g + A_g\right) = C_{n,k}e^{-2u}.$$

Where $C_{n,k}$ is the corresponding $\sigma_k^{1/k}$-curvature of the standard round metric on S^n. Using Bishop's inequality, it follows that the invariant Λ_k is nontrivial when $k > n/2$, and in analogy with the classical Yamabe problem, when the invariant is strictly less than the value obtained by the round metric on the sphere one obtains existence of solutions to (8.2):

Theorem 8.3. (Gursky-Viaclovsky [GV04b]) *If $[g]$ admits a k-admissible metric with $k > n/2$, then there is a constant $C = C(n)$ such that $\Lambda_k(M^n, [g]) < C(n)$. If*

$$(8.3) \qquad\qquad \Lambda_k(M^n, [g]) < vol(S^n),$$

where $vol(S^n)$ denotes the volume of the round sphere, then $[g]$ admits a solution $g_u = e^{-2u}g$ of (8.2). Furthermore, the set of solutions of (8.2) is compact in the C^m-topology for any $m \geq 0$.

The compactness proof uses a bubbling argument, combined with the Liouville Theorem of Li-Li [LL03], and existence is obtained using a degree-theoretic argument.

In dimension three, we have the following estimate

Theorem 8.4. (Gursky-Viaclovsky [GV04b]) *Let (M^3, g) be a closed Riemannian three-manifold, and assume $[g]$ admits a k-admissible metric with $k = 2$ or 3. Let $\pi_1(M^3)$ denote the fundamental group of M^3. Then*

$$(8.4) \qquad\qquad \Lambda_k(M^3, [g]) \leq \frac{vol(S^3)}{\|\pi_1(M^3)\|}.$$

The proof of this used an improvement of Bishop's volume comparison theorem in dimension three, due to Hugh Bray [Bra97]. Therefore, if the three-manifold is not simply-connected, the estimate (8.3) is automatically satisfied.

In dimension four, the optimal estimate of Λ_k follows from the sharp integral estimate for $\sigma_2(A)$ due to Gursky [Gur99]:

Theorem 8.5. (Gursky-Viaclovsky [GV04b]) *Let (M^4, g) be a closed Riemannian four-manifold, and assume $[g]$ admits a k-admissible metric with $2 \leq k \leq 4$. Then*

$$(8.5) \qquad\qquad \Lambda_k(M^4, [g]) \leq vol(S^4).$$

Furthermore, equality holds in (8.5) if and only if (M^4, g) is conformally equivalent to the round sphere.

When $k = 2$ the existence was established previously in [CGY02a]. Combining this work with the four-dimensional solution of the Yamabe problem [Sch84], it follows that the σ_k-Yamabe problem is completely solved in dimension four.

We return to the case of general dimension n. Using similar techniques as in the proof of Theorem 8.3, Guan-Wang studied the minimum eigenvalue of the Ricci, and the p-Weitzenbock operator, and proved various existence theorems, see [GW05b]. Their problem encountered some additional technical difficulties since their symmetric function F is not smooth, and also required an application of some work of Caffarelli [Caf89], and some generalizations of this work [Wan92a], [Wan92b].

In recent work, an existence theorem for a much more general class of F was proved. In addition to the structure conditions (i)–(iii) imposed on F above in Section 5, suppose that we also have (iv) $\Gamma \supset \Gamma_n^+$, and there exists a constant $\delta > 0$ such that any $\lambda = (\lambda_1, \ldots, \lambda_n) \in \Gamma$ satisfies

$$(8.6) \qquad \lambda_i > -\frac{(1 - 2\delta)}{(n - 2)}(\lambda_1 + \cdots + \lambda_n) \quad \forall\, 1 \leq i \leq n.$$

To explain the significance of (8.6), suppose the eigenvalues of the Schouten tensor A_g are in Γ at each point of M^n. Then (M^n, g) has positive Ricci curvature: in fact,

$$(8.7) \qquad\qquad Ric_g - 2\delta\sigma_1(A_g)g \geq 0.$$

Define

$$(8.8) \qquad\qquad A_u \equiv \nabla^2 u + du \otimes du - \frac{|\nabla u|^2}{2}g + A_g.$$

For F satisfying (i)–(iv), consider the equation

(8.9) $$F(g^{-1}A_u) = f(x)e^{-2u},$$

where we assume $g^{-1}A_u \in \Gamma$ (i.e., u is Γ-*admissible*).

Theorem 8.6. (Gursky-Viaclovsky [GV04a]) *Suppose $F : \Gamma \to \mathbb{R}$ satisfies* (i)–(iv). *Let (M^n, g) be closed n-dimensional Riemannian manifold, and assume*

(i) g *is Γ-admissible, and*
(ii) (M^n, g) *is not conformally equivalent to the round n-dimensional sphere.*

Then given any smooth positive function $f \in C^\infty(M^n)$ there exists a solution $u \in C^\infty(M^n)$ of

$$F(g^{-1}A_u) = f(x)e^{-2u},$$

and the set of all such solutions is compact in the C^m-topology for any $m \geq 0$.

This theorem in particular completely solved the σ_k-Yamabe problem whenever $k > n/2$. The proof of this theorem involved an bubbling analysis, and the application of various Holder and integral estimates to determine the growth rate of solutions at isolated singular points. For other works analyzing the possible behaviour of solutions at singularities, see [CHY05], [GV06], [Gon04b],[Gon04a], [Li05a], [Li05b], [TW05]. In addition, a remarkable Harnack inequality for k-admissible metrics, $k > n/2$, was demonstrated in [TW05].

Note that the second assumption in the above theorem is of course necessary, since the set of solutions of (8.9) on the round sphere with $f(x) = $ *constant* is non-compact, while for variable f there are obstructions to existence. In particular, there is a "Pohozaev identity" for solutions of (8.9) in the case of σ_k, which holds in the conformally flat case; see [Via00c].

Theorem 8.7. (Viaclovsky [Via00c]) *Let (M, g) be a closed locally conformally flat n-dimensional manifold. Then for any conformal Killing vector field X, and $1 \leq k < n$, we have*

(8.10) $$\int_M X \cdot \sigma_k(A) \, dvol_M = 0.$$

For $k = 1$, this identity is well-known and holds without the locally con-formally flat assumption [Sch88]. This identity yields non-trivial Kazdan-Warner-type obstructions to existence (see [KW74]) in the case (M^n, g) is conformally equivalent to (S^n, g_{round}). We note that similar Pohozaev-type identities were recently studied in [Han06], [Del06].

It is an interesting problem to characterize the functions $f(x)$ which may arise as σ_k-curvature functions in the conformal class of the round sphere. An announcement of some work by Chang-Han-Yang in this direction for σ_2 in dimension four was made in [Han04]. For $k = 1$, this problem is quite famous and has been studied in great depth. We do not attempt to make a complete list of references for this problem, we mention only [CGY93], [CL01], [ES86], [Li95], [Li96].

9. Admissible Metrics

A natural question is: when does a manifold admit globally a strictly k-admissible metric, that is, a metric g with $A_g \in \Gamma_k^+$? In Section 6, we have already discussed the beautiful result for σ_2 in dimension four by Chang-Gursky-Yang. In the locally conformally flat case, there are various topological restrictions. Guan-Lin-Wang showed

Theorem 9.1. (Guan-Lin-Wang [GLW05]) *Let* (M^n, g) *be a compact, locally conformally flat manifold with* $\sigma_1(A) > 0$.

(i) *If* $A \in \overline{\Gamma}_k^+$ *for some* $2 \leq k < n/2$, *then the* qth *Betti number* $b_q = 0$ *for*

$$(9.1) \qquad \left[\frac{n+1}{2}\right] + 1 - k \leq q \leq n - \left(\left[\frac{n+1}{2}\right] + 1 - k\right).$$

(ii) *Suppose* $A \in \Gamma_2^+$, *then* $b_q = 0$ *for*

$$(9.2) \qquad \left[\frac{n - \sqrt{n}}{2}\right] \leq q \leq \left[\frac{n + \sqrt{n}}{2}\right].$$

If $A \in \overline{\Gamma}_2^+$, $p = \frac{n - \sqrt{n}}{2}$, *and* $b_p \neq 0$, *then* (M, g) *is a quotient of* $S^{n-p} \times H^p$.

(iii) *If* $k \geq \frac{n - \sqrt{n}}{2}$ *and* $A \in \Gamma_k^+$, *then* $b_q = 0$ *for any* $2 \leq q \leq n - 2$. *If* $k = \frac{n - \sqrt{n}}{2}$, $A \in \overline{\Gamma}_k^+$, *and* $b_2 \neq 0$, *then* (M, g) *is a quotient of* $S^{n-2} \times H^2$.

The proof of this theorem involves a careful analysis of the curvature terms in the Weitzenböck formula for the Hodge Laplacian. For σ_2, the following was shown

Theorem 9.2. (Chang-Hang-Yang [CHY05]) *Let* $(M^n, g), (n \geq 5)$ *be a smooth locally conformally flat Riemannian manifold such that* $\sigma_1(A) > 0$, *and* $\sigma_2(A) \geq 0$, *then* $\pi_1(M) = 0$ *for any* $2 \leq i \leq [n/2] + 1$, *and* $H^j(M, \mathbb{R}) = 0$ *for any* $n/2 - 1 \leq j \leq n/2 + 1$.

Briefly, the technique in [CHY05] is to use the positivity condition to estimate the Hausdorf dimension of the singular set, using the developing map, as in [SY88]. Subsequently, in her thesis, María del Mar González generalized this to prove the following.

Theorem 9.3. (González [Gon05]) *Let* (M, g) *be compact, locally conformally flat, and* $A \in \Gamma_k^+$, $k < n/2$. *Then for any* $2 \leq i \leq \left[\frac{n}{2}\right] + k - 1$, *the homotopy group* $\pi_i(M) = \{0\}$, *and the cohomology group* $H^i(M, \mathbb{R}) = \{0\}$ *for* $\frac{n-2k}{2} + 1 \leq i \leq \frac{n+2k}{2} - 1$.

A beautiful gluing theorem was proved in [GLW05]:

Theorem 9.4. (Guan-Lin-Wang [GLW05]) *Let* $2 \leq k < n/2$, *and let* M_1^n *and* M_2^n *be two compact manifolds with* $A_1, A_2 \in \Gamma_k^+$. *Then the connected sum* $M_1 \# M_2$ *also admits a metric* $g_\#$ *with* $A \in \Gamma_k^+$. *If in addition,* M_1 *and* M_2 *are locally conformally flat, then* $g_\#$ *can also be taken to be locally conformally flat.*

This result can be viewed as a generalization of the analogous result for positive scalar curvature [GL80] [SY79b]. This yields many new examples of manifolds admitting metrics with $A \in \Gamma_k^+$, we refer the reader to [GLW05] for more details.

10. The Negative Cone

All of the previous results are concerned with the positive curvature case. The negative curvature case exhibits quite different behaviour. In this case, a serious technical difficulty arises in attempting to derive *a priori* second derivative estimates on solutions [Via02]. Consider instead the following generalization of the Schouten tensor. Let $t \in \mathbb{R}$, and define

$$(10.1) \qquad A^t = \frac{1}{n-2}\left(Ric - \frac{t}{2(n-1)}Rg\right).$$

We let $\Gamma_k^- = -\Gamma_k^+$.

Theorem 10.1. (Gursky-Viaclovsky [GV03b]) *Let (M, g) be a compact Riemannian manifold, assume that $A_g^t \in \Gamma_k^-$ for some $t < 1$, and let $f(x) < 0$ be any smooth function on M^n. Then there exists a unique conformal metric $\tilde{g} = e^{2w}g$ satisfying*

$$(10.2) \qquad\qquad \sigma_k^{1/k}(\tilde{g}^{-1}A_{\tilde{g}}^t) = f(x).$$

As noted above, the second derivative estimate encounters technical difficulties for $t = 1$, but for $t < 1$, this difficulty can be overcome. Also, for $t = 1 + \epsilon$, the equation is not necessarily elliptic, therefore $t = 1$ is critical for more than one reason. Some local counterexamples to the second derivative estimate in the negative case have been given in [STW05], but there are no known global counterexamples. The above theorem was subsequently proved by parabolic methods in [LS05].

The above theorem has the following corollary. Using results of [Bro89], [GY86], and [Loh94], every compact manifold of dimension $n \geq 3$ admits a metric with negative Ricci curvature. Therefore applying the theorem when $t = 0$,

Corollary 10.2. (Gursky-Viaclovsky [GV03b]) *Every smooth compact n-manifold, $n \geq 3$, admits a Riemannian metric with Ric < 0 and*

$$(10.3) \qquad\qquad \det(g^{-1}Ric) = \text{constant}.$$

It turns out the equation is also elliptic for $t \geq n - 1$, and this has some interesting consequences, see [SZ05] for details. We also mention that Mazzeo and Pacard considered the σ_k-Yamabe equation in the context of conformally compact metrics, and showed that the deformation problem is unobstructed [MP03].

References

[Aub76a]   Thierry Aubin, *Équations différentielles non linéaires et problème de Yamabe concernant la courbure scalaire*, J. Math. Pures Appl. (9) **55** (1976), no. 3, 269–296.

[Aub76b]   Thierry Aubin, *Problèmes isopérimétriques et espaces de Sobolev*, J. Differential Geometry **11** (1976), no. 4, 573–598.

[Aub98]   Thierry Aubin, *Some nonlinear problems in Riemannian geometry*, Springer-Verlag, Berlin, 1998.

[Bes87]   Arthur L. Besse, *Einstein manifolds*, Springer-Verlag, Berlin, 1987.

[BGG03]   Robert Bryant, Phillip Griffiths, and Daniel Grossman, *Exterior differential systems and Euler-Lagrange partial differential equations*,

Chicago Lectures in Mathematics, University of Chicago Press, Chicago, IL, 2003.

[BO91] Thomas P. Branson and Bent Orsted, *Explicit functional determinants in four dimensions*, Proc. Amer. Math. Soc. **113** (1991), 669–682.

[Bra97] Hubert L. Bray, *The Penrose inequality in general relativity and volume comparison theorems involving scalar curvature*, Dissertation, Stanford University, 1997.

[Bre05] Simon Brendle, *Convergence of the Yamabe flow for arbitrary initial energy*, J. Differential Geom. **69** (2005), no. 2, 217–278.

[Bro89] Robert Brooks, *A construction of metrics of negative Ricci curvature*, J. Differential Geom. **29** (1989), no. 1, 85–94.

[BV04] Simon Brendle and Jeff A. Viaclovsky, *A variational characterization for $\sigma_{n/2}$*, Calc. Var. Partial Differential Equations **20** (2004), no. 4, 399–402.

[Caf89] Luis A. Caffarelli, *Interior a priori estimates for solutions of fully nonlinear equations*, Ann. of Math. (2) **130** (1989), no. 1, 189–213.

[CGS89] Luis A. Caffarelli, Basilis Gidas, and Joel Spruck, *Asymptotic symmetry and local behavior of semilinear elliptic equations with critical Sobolev growth*, Comm. Pure Appl. Math. **42** (1989), no. 3, 271–297.

[CGY93] Sun-Yung A. Chang, Matthew J. Gursky, and Paul C. Yang, *The scalar curvature equation on 2- and 3-spheres*, Calc. Var. Partial Differential Equations **1** (1993), no. 2, 205–229.

[CGY02a] Sun-Yung A. Chang, Matthew J. Gursky, and Paul C. Yang, *An a priori estimate for a fully nonlinear equation on four-manifolds*, J. Anal. Math. **87** (2002), 151–186, Dedicated to the memory of Thomas H. Wolff.

[CGY02b] Sun-Yung A. Chang, Matthew J. Gursky, and Paul C. Yang, *An equation of Monge-Ampère type in conformal geometry, and four-manifolds of positive Ricci curvature*, Ann. of Math. (2) **155** (2002), no. 3, 709–787.

[CGY03a] Sun-Yung A. Chang, Matthew J. Gursky, and Paul C. Yang, *A conformally invariant sphere theorem in four dimensions*, Publ. Math. Inst. Hautes Études Sci. (2003), no. 98, 105–143.

[CGY03b] Sun-Yung A. Chang, Matthew J. Gursky, and Paul C. Yang, *Entire solutions of a fully nonlinear equation*, Lectures on partial differential equations, New Stud. Adv. Math., vol. 2, Int. Press, Somerville, MA, 2003, pp. 43–60.

[Che05] Szu-yu Sophie Chen, *Local estimates for some fully nonlinear elliptic equations*, Int. Math. Res. Not. (2005), no. 55, 3403–3425.

[Cho92] Bennett Chow, *The Yamabe flow on locally conformally flat manifolds with positive Ricci curvature*, Comm. Pure Appl. Math. **45** (1992), no. 8, 1003–1014.

[CHY05] S.Y. Alice Chang, Zheng-Chao Han, and Paul C. Yang, *Classification of singular radial solutions to the σ_k Yamabe equation on annular domains*, J. Differential Equations **216** (2005), no. 2, 482–501.

[CL01]   Chiun-Chuan Chen and Chang-Shou Lin, *Prescribing scalar curvature on S^N. I. A priori estimates*, J. Differential Geom. **57** (2001), no. 1, 67–171.

[CNS85]   L. Caffarelli, L. Nirenberg, and J. Spruck, *The Dirichlet problem for nonlinear second-order elliptic equations. III. Functions of the eigenvalues of the Hessian*, Acta Math. **155** (1985), no. 3-4, 261–301.

[CY95]   Sun-Yung A. Chang and Paul C. Yang, *Extremal metrics of zeta function determinants on 4-manifolds*, Ann. of Math. (2) **142** (1995), no. 1, 171–212.

[CY03]   Sun-Yung A. Chang and Paul C. Yang, *The inequality of Moser and Trudinger and applications to conformal geometry*, Comm. Pure Appl. Math. **56** (2003), no. 8, 1135–1150, Dedicated to the memory of Jürgen K. Moser.

[Del06]   Philippe Delanoë, *On the local k-Nirenberg problem*, preprint, 2006.

[Dru04]   Olivier Druet, *Compactness for Yamabe metrics in low dimensions*, Int. Math. Res. Not. (2004), no. 23, 1143–1191.

[Eis97]   Luther Pfahler Eisenhart, *Riemannian geometry*, Princeton Landmarks in Mathematics, Princeton University Press, Princeton, NJ, 1997, Eighth printing, Princeton Paperbacks.

[ES86]   José F. Escobar and Richard M. Schoen, *Conformal metrics with prescribed scalar curvature*, Invent. Math. **86** (1986), no. 2, 243–254.

[Eva82]   Lawrence C. Evans, *Classical solutions of fully nonlinear, convex, second-order elliptic equations*, Comm. Pure Appl. Math. **35** (1982), no. 3, 333–363.

[Gàr59]   Lars Gàrding, *An inequality for hyperbolic polynomials*, J. Math. Mech. **8** (1959), 957–965.

[GL80]   Mikhael Gromov and H. Blaine Lawson, Jr., *The classification of simply connected manifolds of positive scalar curvature*, Ann. of Math. (2) **111** (1980), no. 3, 423–434.

[GLW04a] Pengfei Guan, Chang-Shou Lin, and Guofang Wang, *Application of the method of moving planes to conformally invariant equations*, Math. Z. **247** (2004), no. 1, 1–19.

[GLW04b] Pengfei Guan, Chang-Shou Lin, and Guofang Wang, *Local gradient estimates for conformal quotient equations*, preprint, 2004.

[GLW05]   Pengfei Guan, Chang-Shou Lin, and Guofang Wang, *Schouten tensor and some topological properties*, Comm. Anal. Geom. **13** (2005), no. 5, 887–902.

[Gon04a] María del Mar González, *Classification of singularities for a subcritical fully non-linear problem*, preprint, to appear in Pac. J. Math., 2004.

[Gon04b] María del Mar González, *Removability of singularities for a class of fully non-linear elliptic equations*, preprint, to appear in Calc. Var., 2004.

[Gon05]   María del Mar González, *Singular sets of a class of locally conformally flat manifolds*, Duke Math. J. **129** (2005), no. 3, 551–572.

[Gur99]   Matthew J. Gursky, *The principal eigenvalue of a conformally invariant differential operator, with an application to semilinear elliptic*

PDE, Comm. Math. Phys. **207** (1999), no. 1, 131–143.

[GV01] Matthew J. Gursky and Jeff A. Viaclovsky, *A new variational characterization of three-dimensional space forms*, Inventiones Mathematicae **145** (2001), no. 2, 251–278.

[GV03a] Matthew J. Gursky and Jeff A. Viaclovsky, *A fully nonlinear equation on four-manifolds with positive scalar curvature*, J. Differential Geom. **63** (2003), no. 1, 131–154.

[GV03b] Matthew J. Gursky and Jeff A. Viaclovsky, *Fully nonlinear equations on Riemannian manifolds with negative curvature*, Indiana Univ. Math. J. **52** (2003), no. 2, 399–419.

[GV04a] Matthew J. Gursky and Jeff A. Viaclovsky, *Prescribing symmetric functions of the eigenvalues of the Ricci tensor*, preprint, arXiv:math.DG/0409187, to appear in Annals of Mathematics, 2004.

[GV04b] Matthew J. Gursky and Jeff A. Viaclovsky, *Volume comparison and the σ_k-Yamabe problem*, Adv. Math. **187** (2004), 447–487.

[GV06] Matthew J. Gursky and Jeff A. Viaclovsky, *Convexity and singularities of curvature equations in conformal geometry*, Int. Math. Res. Not. (2006), Art. ID 96890, 43.

[GVW03] Pengfei Guan, Jeff Viaclovsky, and Guofang Wang, *Some properties of the Schouten tensor and applications to conformal geometry*, Trans. Amer. Math. Soc. **355** (2003), no. 3, 925–933 (electronic).

[GW03a] Pengfei Guan and Guofang Wang, *A fully nonlinear conformal flow on locally conformally flat manifolds*, J. Reine Angew. Math. **557** (2003), 219–238.

[GW03b] Pengfei Guan and Guofang Wang, *Local estimates for a class of fully nonlinear equations arising from conformal geometry*, Int. Math. Res. Not. (2003), no. 26, 1413–1432.

[GW04] Pengfei Guan and Guofang Wang, *Geometric inequalities on locally conformally flat manifolds*, Duke Math. J. **124** (2004), no. 1, 177–212.

[GW05a] Yuxin Ge and Guofang Wang, *On a fully nonlinear Yamabe problem*, preprint, math.DG/0505257, to appear in Ann. Sci. Ecole Norm. Sup, 2005.

[GW05b] Pengfei Guan and Guofang Wang, *Conformal deformation of the smallest eigenvalue of the Ricci tensor*, preprint, arXiv:math.DG/0505083, 2005.

[GW06] Yuxin Ge and Guofang Wang, *On a conformal quotient equation*, preprint, 2006.

[GY86] L. Zhiyong Gao and Shing Tung Yau, *The existence of negatively Ricci curved metrics on three-manifolds*, Invent. Math. **85** (1986), no. 3, 637–652.

[Han04] Zheng-Chao Han, *Local pointwise estimates for solutions of the σ_2 curvature equation on 4-manifolds*, Int. Math. Res. Not. (2004), no. 79, 4269–4292.

[Han06] Zheng-Chao Han, *A Kazdan-Warner type identity for the σ_k curvature*, C. R. Math. Acad. Sci. Paris **342** (2006), no. 7, 475–478.

[Hil72]   D. Hilbert, *Die grundlagen der physik*, Nach. Ges. Wiss., Göttingen, (1915), 461-472.

[HL04]   Zejun Hu and Haizhong Li, *A new variational characterization of n-dimensional space forms*, Trans. Amer. Math. Soc. **356** (2004), no. 8, 3005–3023 (electronic).

[Ivo83]   N. M. Ivochkina, *Description of cones of stability generated by differential operators of Monge-Ampère type*, Mat. Sb. (N.S.) **122(164)** (1983), no. 2, 265–275.

[Kry83]   N. V. Krylov, *Boundedly inhomogeneous elliptic and parabolic equations in a domain*, Izv. Akad. Nauk SSSR Ser. Mat. **47** (1983), no. 1, 75–108.

[KW74]   Jerry L. Kazdan and F. W. Warner, *Curvature functions for compact 2 manifolds*, Ann. of Math. (2) **99** (1974), 14–47.

[Lab04]   M. L. Labbi, *On a variational formula for the H. Weyl curvature invariants*, preprint, arXiv:math.DG/0406548, 2004.

[LF71]   Jacqueline Lelong-Ferrand, *Transformations conformes et quasi-conformes des variétés riemanniennes compactes (démonstration de la conjecture de A. Lichnerowicz)*, Acad. Roy. Belg. Cl. Sci. Mém. Coll. in–8deg (2) **39** (1971), no. 5, 44.

[Li95]   YanYan Li, *Prescribing scalar curvature on S^n and related problems. I*, J. Differential Equations **120** (1995), no. 2, 319–410.

[Li96]   YanYan Li, *Prescribing scalar curvature on S^n and related problems. II. Existence and compactness*, Comm. Pure Appl. Math. **49** (1996), no. 6, 541–597.

[Li02]   YanYan Li, *On some conformally invariant fully nonlinear equations*, Proceedings of the International Congress of Mathematicians, Vol. III (Beijing, 2002) (Beijing), Higher Ed. Press, 2002, pp. 177–184.

[Li03]   YanYan Li, *Liouville type theorems for some conformally invariant fully nonlinear equations*, Atti Accad. Naz. Lincei Cl. Sci. Fis. Mat. Natur. Rend. Lincei (9) Mat. Appl. **14** (2003), no. 3, 219–225 (2004), Renato Caccioppoli and modern analysis.

[Li05a]   YanYan Li, *Conformally invariant fully nonlinear elliptic equations and isolated singularities*, arXiv:math.AP/0504597, to appear in J. Functional Analysis, 2005.

[Li05b]   YanYan Li, *Degenerate conformally invariant fully nonlinear elliptic equations*, arXiv:math.AP/0504598, 2005.

[Li06]   YanYan Li, *Local gradient estimates of solutions to some conformally invariant fully nonlinear equations*, arXiv:math.AP/0605559, 2006.

[LL03]   Aobing Li and YanYan Li, *On some conformally invariant fully nonlinear equations*, Comm. Pure Appl. Math. **56** (2003), no. 10, 1416–1464.

[LL05a]   Aobing Li and YanYan Li, *A Liouville type theorem for some conformally invariant fully nonlinear equations*, Geometric analysis of PDE and several complex variables, Contemp. Math., vol. 368, Amer. Math. Soc., Providence, RI, 2005, pp. 321–328.

[LL05b]   Aobing Li and YanYan Li, *On some conformally invariant fully non-linear equations, part II: Liouville, Harnack, and Yamabe*, Acta Math. **195** (2005), 117–154.

[Loh94]   Joachim Lohkamp, *Metrics of negative Ricci curvature*, Ann. of Math. (2) **140** (1994), no. 3, 655–683.

[LP87]   John M. Lee and Thomas H. Parker, *The Yamabe problem*, Bull. Amer. Math. Soc. (N.S.) **17** (1987), no. 1, 37–91.

[LS05]   Jiayu Li and Weimin Sheng, *Deforming metrics with negative curvature by a fully nonlinear flow*, Calc. Var. Partial Differential Equations **23** (2005), no. 1, 33–50.

[LZ04]   YanYan Li and Lei Zhang, *Compactness of solutions to the Yamabe problem*, C. R. Math. Acad. Sci. Paris **338** (2004), no. 9, 693–695.

[LZ05]   YanYan Li and Lei Zhang, *Compactness of solutions to the Yamabe problem. II*, Calc. Var. Partial Differential Equations **24** (2005), no. 2, 185–237.

[Mar98]   Christophe Margerin, *A sharp characterization of the smooth 4-sphere in curvature terms*, Comm. Anal. Geom. **6** (1998), no. 1, 21–65.

[Mar05]   Fernando Coda Marques, *A priori estimates for the Yamabe problem in the non-locally conformally flat case*, J. Differential Geom. **71** (2005), no. 2, 315–346.

[MP03]   Rafe Mazzeo and Frank Pacard, *Poincaré-Einstein metrics and the Schouten tensor*, Pacific J. Math. **212** (2003), no. 1, 169–185.

[Oba72]   Morio Obata, *The conjectures on conformal transformations of Riemannian manifolds*, J. Differential Geometry **6** (1971/72), 247–258.

[PT82]   Thomas Parker and Clifford Henry Taubes, *On Witten's proof of the positive energy theorem*, Comm. Math. Phys. **84** (1982), no. 2, 223–238.

[Sch84]   Richard M. Schoen, *Conformal deformation of a Riemannian metric to constant scalar curvature*, J. Differential Geom. **20** (1984), no. 2, 479–495.

[Sch88]   Richard M. Schoen, *The existence of weak solutions with prescribed singular behavior for a conformally invariant scalar equation*, Comm. Pure Appl. Math. **41** (1988), no. 3, 317–392.

[Sch89]   Richard M. Schoen, *Variational theory for the total scalar curvature functional for Riemannian metrics and related topics*, Topics in calculus of variations (Montecatini Terme, 1987), Lecture Notes in Math., vol. 1365, Springer, Berlin, 1989, pp. 120–154.

[Sch91]   Richard M. Schoen, *On the number of constant scalar curvature metrics in a conformal class*, Differential geometry, Longman Sci. Tech., Harlow, 1991, pp. 311–320.

[SS03]   Hartmut Schwetlick and Michael Struwe, *Convergence of the Yamabe flow for "large" energies*, J. Reine Angew. Math. **562** (2003), 59–100.

[STW05]   Weimin Sheng, Neil S. Trudinger, and Xu-Jia Wang, *The Yamabe problem for higher order curvatures*, preprint, arXiv:math.DG/0505463, 2005.

[SY79a]   Richard M. Schoen and Shing Tung Yau, *On the proof of the positive mass conjecture in general relativity*, Comm. Math. Phys. **65** (1979), no. 1, 45–76.

[SY79b]   Richard M. Schoen and Shing Tung Yau, *On the structure of manifolds with positive scalar curvature*, Manuscripta Math. **28** (1979), no. 1-3, 159–183.

[SY81]   Richard M. Schoen and Shing Tung Yau, *Proof of the positive mass theorem. II*, Comm. Math. Phys. **79** (1981), no. 2, 231–260.

[SY88]   Richard M. Schoen and Shing Tung Yau, *Conformally flat manifolds, Kleinian groups and scalar curvature*, Inventiones Mathematicae **92** (1988), no. 1, 47–71.

[SZ05]   Weimin Sheng and Yan Zhang, *A class of fully nonlinear equations arising from conformal geometry*, preprint, 2005.

[Tru68]   Neil S. Trudinger, *Remarks concerning the conformal deformation of Riemannian structures on compact manifolds*, Ann. Scuola Norm. Sup. Pisa (3) **22** (1968), 265–274.

[TW05]   Neil S. Trudinger and Xu-Jia Wang, *On Harnack inequalities and singularities of admissible metrics in the Yamabe problem*, arXiv:math.DG/0509341, 2005.

[Via00a]   Jeff A. Viaclovsky, *Conformal geometry, contact geometry, and the calculus of variations*, Duke Math. J. **101** (2000), no. 2, 283–316.

[Via00b]   Jeff A. Viaclovsky, *Conformally invariant Monge-Ampère equations: global solutions*, Trans. Amer. Math. Soc. **352** (2000), no. 9, 4371–4379.

[Via00c]   Jeff A. Viaclovsky, *Some fully nonlinear equations in conformal geometry*, Differential equations and mathematical physics (Birmingham, AL, 1999), Amer. Math. Soc., Providence, RI, 2000, pp. 425–433.

[Via02]   Jeff A. Viaclovsky, *Estimates and existence results for some fully nonlinear elliptic equations on Riemannian manifolds*, Comm. Anal. Geom. **10** (2002), no. 4, 815–846.

[Wan92a]   Lihe Wang, *On the regularity theory of fully nonlinear parabolic equations. I*, Comm. Pure Appl. Math. **45** (1992), no. 1, 27–76.

[Wan92b]   Lihe Wang, *On the regularity theory of fully nonlinear parabolic equations. II*, Comm. Pure Appl. Math. **45** (1992), no. 2, 141–178.

[Wan05]   Guofang Wang, *A Bär type inequality on higher dimensional manifolds*, preprint, 2005.

[Wan06]   Xu-Jia Wang, *A priori estimates and existence for a class of fully nonlinear elliptic equations in conformal geometry*, Chinese Ann. Math. **27(B)** (2006), 169–178.

[Wit81]   Edward Witten, *A new proof of the positive energy theorem*, Comm. Math. Phys. **80** (1981), no. 3, 381–402.

[Yam60]   Hidehiko Yamabe, *On a deformation of Riemannian structures on compact manifolds*, Osaka Math. J. **12** (1960), 21–37.

[Ye94]   Rugang Ye, *Global existence and convergence of Yamabe flow*, J. Differential Geom. **39** (1994), no. 1, 35–50.

Chapter 19

MEMORY OF MY FIRST RESEARCH TEACHER: THE GREAT GEOMETER CHERN SHIING-SHEN

Wu Wentsun*

Mathematics-Mechanization Key Laboratory,
Academy of Mathematics and System Sciences,
Chinese Academy of Sciences (CAS), Beijing 100080, P. R. China

The present article is for the memory of a great geometer, the late Professor CHERN Shiing-Shen who is my first teacher leading me to ride on the broad road of mathematical researches.

This article will be divided into two parts. Part I describes my acquaintance with Professor CHERN and studies under his guidance. Part II describes my introduction of generalized CHERN Classes and CHERN numbers for irreducible complex algebraic varieties *with arbitrary singularities.*

Part I. Memory as a Student of Professor CHERN

Let me first make a brief introduction of myself. I graduated in Shanghai Communication University as a student of mathematics department in the year 1940. In those university years I became interested in point-set topology and I have read almost all such treatises in the university library, including most of related papers of $\ll Fundamenta\ Mathematica\gg$. I have also read the classic $\ll Topologie\ I\gg$ of Alexandroff-Hopf. I was particularly interested in the Part I on the topic of point-set topology, consulting almost all writings mentioned about. But I met difficulties in reading the part on combinatorial topology, being ignorant of the underlying intuitive background. When I graduated in year 1940 from the university, China was still in the difficult period of Sino-Japanese War which began in 1937. In year 1941 the Japanese attacked Pearl Harbor and the Pacific Ocean Warfare began which ended until 1945. In year 1946 the Chinese

---

*This article is for the memory of my first research teacher: Professor Chern.

government at that time launched a plan of sending students to study abroad. I participated the governmental examination for mathematics in the name of Sino-French cultural exchange of students. However, it was only one year later, i.e. in year 1947, that I know I have passed this examination.

In the summer of 1946, a graduate student Mr. TSIEN Shen-Fa of TsingHua University accompanied me to pay a visit to Professor Chern at his home that time in Shanghai. This is the first time that I saw him which was decisive for my future career in mathematics.

In year 1946 the Chinese government decided to move back the Academia Sinica at Kunming to Shanghai. The government had decided also to establish a Preparatory Committee for a new Institute of Mathematics belonging to the Academia, also situated in Shanghai. Professor CHIANG Li-Fu of Nankai University was asked to be the head of the Committee. As Prof. CHIANG had to be in United States for various affairs, CHIANG asked CHERN to be in charge of the Committee as his representative. CHERN asked mathematics departments of various universities to send each a most promising young man to the Institute. Thus, there comes to the Institute Mr. CHEN Kuo-Tsai from TsingHua University, Mr. TSAO Shih-Hua from ZheJiang University, Mr. LU Chien-Ke from WuHan University, Mr. JOU Yuh-Lin from DaTong University at Shanghai. There were two exceptions. One is CHANG Su-Cheng, who had graduated from ZheJiang University for several years. Being a student of Prof. SU BuChin who was a differential geometer educated in Japan, CHANG had already published in foreign journals quite a number of papers on differential geometry. At that time CHANG was a professor in some college in province SiChuan, but CHERN asked him to leave that college to become a senior member of the Institute. Another exception is myself. I had also graduated from university for a number of years and asked CHERN directly to be a member of the Institute without any recommendation of universities.

In order to pay visit to CHERN I brought with me a survey article on point-set topology together with my personel views, with emphasis on Part I of the classic of Alexandroff-Hopf. CHERN received me in his home. In the next time of my visit he criticized that I was not on the right way of researches. For topology one should not fall into the labyrinth of unimportant concepts with entangled logical interrelations between them. One should emphasize on the combinatorial part of topology which will be decisive for future developments. CHERN rejected my article and returned it back to me. However, he admitted me to be a member of the Institute.

I became thus a member bearing some title same as the other young men from various universities. Actually, we are henceforth all research students of CHERN.

The Committee is situated on the second floor of a two-stage building on the southern border of the former French District of Shanghai. The building was originally the Institute of Natural Sciences of Japan. It was given up to China after Japan's surrender. The biggest room on the second floor is reserved for lectures and seminars. CHERN occupied a single room where he received quite often the young students for discussions of mathematics. CHERN gave us lectures for months on combinatorial topology, begun with the classification of surfaces. Sometimes CHERN gave lectures up to 12 hours a week. I began to understand the geometrical background of the combinatorial structure. With a geometric picture in the mind, the previously difficult combinatorial part of Alexandroff-Hopf's classic became clear and simple for me henceforth.

All the other young members occupied some small rooms of the floor. I was somewhat exceptional. I was asked to stay in the library room and to work as a librarian. I felt extremely satisfied since the library, originally belonging to the Japanese Institute, is rich in books and journals in mathematics.

I indulged in the reading of mathematics tracts the whole day. However, in view of CHERN's instructions, I pay no more attention on texts and papers about point-set topology, but turn my interests to combinatorial topology. I was particularly interested in the papers of J. W. Alexander. I read carefully such classical papers of Alexander on knots, isotopy, and combinatorial invariants, or what I called in future years the *topological* but *non-homotopic* problems.

One day when I was absorbed in reading enjoying like *a fish in a lake*, CHERN suddenly visited me in the library. He said that I had read so much that I had to stop reading. Instead I had to pay the debts to the various authors of the reading materials. Seeing me completely bewildered he explained that for paying debts you had to contribute yourself papers as debts to these authors. Thus I had to stop my reading and try to seek some problems to solve.

Thus, I studied some symmetric product of spaces and CHERN aided me to write in French. As an encouragement he sent the article to *Comptes Rendus* which was published as in [WU1]. Clearly the results are quite trivial ones. It turns out however that the notion of *symmetric product* becomes some important tools in my future research works on topology.

CHERN encouraged all of us to read the works of H. Hopf, besides the classic ≪ *Alexandroff-Hopf I* ≫. Once he even joked: *To study Hopf you need a Kopf.* With CHERN's encouragement I read carefully the papers of Hopf himself and his students particularly the papers of Stiefel. I was thus acquainted with the concrete method of computing the cycles arising from tangent vector fields of a differential manifold.

One day CHERN called me to his bureau and asked me some questions about the imbedding of a torus in 4-dimensional Euclidean space. I solved the problem and CHERN asked me to write down a short paper. CHERN sent this paper to some USA journal for publication. Shortly after we received the referee's rejection report. I could not agree with the report. As the result is of no importance I forgot it soon. However, CHERN did not forget. In year 1949 when CHERN visited University of Chicago and gave a lecture there which was published in a mimeographed form, he gave my result therein.

In that period CHERN was writing a paper about the mod 2 homology intersection ring of a Grassmannian, showing that the ring had a basis with ring structure explicitly given. CHERN not only showed me his paper but also his personal copies of Doklady notes of Pontrjagin about study of characteristic cycles through the homology structure of Grassmannian manifolds. All these were of great importance for me when I studied in France afterwards.

The ≪ *Lectures in Topology* ≫ was available in the 40s of last century. It was the report of the University of Michigan Conference of 1940. See [W-A] in the reference. CHERN asked me to read the paper of S. Eilenberg in [W-A] bearing the title *Extension and Classification of Continuous Mappings.* However, I had not much interest in that paper. The paper in [W-A] which raised my great interest is nevertheless the one of H. Whitney, viz. [WH].

One day I met CHERN in the corridor. I asked him about the concepts of *bundle* and *characteristic cycles* which appeared in Whitney's paper. CHERN seems to be very happy about my questions. He not only explained to me some details about these concepts, but also pointed out that the *product formula* of charateristic cycles (discovered by Whitney) described briefly in the paper is very important. Moreover, the proof of this theorem is extremely difficult thus Whitney even planned to write a whole book devoting to the theorem and its proof. CHERN emphasized that a simple proof of this product formula is much desired.

As I had been acquainted with papers of Stiefel it seems to me that:

The characteristic cycles of Whitney determined from normal vectors to a differential manifold imbedded in a Euclidean space seems to be closely connected with the cycles introduced by Stiefel via tangent vectors to the manifold. As one knows how to actually compute the cycles introduced by Stiefel, this will give a means of computing the cycles of Whitney. This may then give a means of proving Whitney's product formula.

As CHERN was a professor of TsingHua University he had the duty to teach in that University. Thus, in the spring time of 1947 CHERN had to fly to Peiping (now Beijing) to give a course on differential geometry. CHERN brought me with him to stay somewhere in TsingHua University. I tried to work out my suppositions about Whitney's product formula. After some painful trials I ultimately succeeded in getting such a proof. CHERN was very satisfied with my proof of Whitney's product formula and he helped me to write (or more exactly he wrote for me) the paper which he sent to Annals of Mathematics and was published as [WU2].

After our return to Shanghai it was known that I had passed the governmental examination for the Franco-China scholarship-interchange in mathematics. I had to go then to the governmental capital Nanjing to get trainings for going abroad. Knowing that I should go to study in France, CHERN advised me it is better not to go to Paris. Paris is a big city too turbulent to do research work which requires a quiet environment. He advised me to go instead to some other city besides Paris. For this sake he wrote to Professor H. Cartan at that time in University of Strasbourg. CHERN introduced me to CARTAN and asked him to admit me as his student. Cartan answered CHERN quickly with enthusiastic acknowledge of CHERN's important works and accepted me as his student. Unfortunately this letter of CARTAN had been lost during the cultural revolution.

CHERN once told me that his main interest is not in topology but in differential geometry. He tries to develop a new discipline of **Global Differential Geometry** or **Differential Geometry in the Large** which requires the tools from topology. It is for this reason that he was devoted to topology as a preparatory for the first year of the Institute. He also told me about the works of E. Cartan about differential geometry. He said that E. Cartan's work is a *gigantic* structure to be thouroughly studied for whole life.

On the point of my leaving for France, CHERN advised me to stay one more year in the Institute. In that year he would teach me the work and method of E. Cartan. After one year he would send me to study in USA instead of France.

I was moved by CHERN's suggestion. Unfortunately it was a little too late. I was on the point of leaving China with the others and I could not postpone any more. If he talked to me earlier, I would be very glad to follow his advice.

After my return from France to China in year 1951, I tried several times to learn E. Cartan's work by myself. It turns out that E. Cartan's work is so difficult to understand, not to say to grasp the delicate method of computations. I can never understand E. Cartan's work until today.

When I arrived at Strasbourg, it turned out that Cartan had moved to Paris. Being ignorant of how to meet this unexpected situation I had to stay in Strasbourg. I became then a student of Professor Ch. Ehresmann of University of Strasbourg. This is in fact not bad for me. EHRESMANN is one of the few specialists in algebraic topology of France in that time. Moreover, he is the pioneer of studies of homologies of Grassmannian manifolds and is also one of the founders of fiber bundle theory. My studies under CHERN enabled me to follow EHRESMANN quickly. In fact, I obtained my Docteur és Science degree early in the summer of 1949. Afterwards I went to Paris to become a student of Professor H. Cartan in attending his well-known Cartan Seminar.

I returned back to China in the summer of 1951. During my nearly 4 years of stay in France, CHERN wrote to me quite often in giving me advices of studying as well as mode of living. Unfortunately all these letters had been lost during the cultural revolution.

Owing to political situations, I lost connections with the western mathematics world and also with CHERN. It was not until the year 1972 that CHERN visited mainland China and I was able to see him again. Henceforth CHERN visited China quite often and he did immense efforts for the development of Chinese mathematics to arise to a world level. As all these are so well-known to the mathematics circle both in China and abroad I shall leave these to historians of mathematics.

It is well-known that Nobel Prize had none for mathematics. CHERN once expressed his opinion that this is perhaps not a bad thing, at least for China. Under the suggestion of Professor YANG ZhenNing, a Nobel Laureate in physics and perhaps the most intimate friend of CHERN, the Hong Kong billionaire SHAW YiFu was persuaded to establish an Eastern Nobel Prize Foundation. Three prizes, each of one million US dollars, will be offered to scientists of high achievements in three directions, which had been neglected by the proper Nobel prizes. Mathematics is one of these three subjects. A prize committee was organized for each of these 3 subjects. The mathematics committee consisted of 5 members and I was appointed to be its head. After nominations from well-known mathematicians and some authorities of universities, research institutes, etc., the committee entered into careful consideration about the possible well-deserved laureate of Shaw mathematics prize. After nearly one year of vigorous discussions, the committee came ultimately to the conclusion that the prize should be offered to CHERN owing to his revolutionary creations of *global differential geometry* as well as the introduction of **CHERN CLASSES & CHERN Numbers**. These classes and numbers bearing his name had not only influenced the development of mathematics as a whole but also that of theoretical physics.

The SHAW prizes were offered the first time in September of 2004 in Hongkong. For the prize ceremony I gave an address in explaining to the audience the achievements of CHERN. For this sake Professor Yang and I wrote an address to be read before the audience. The address ended as follows:

CHERN's contribution to mathematics is both deep and broad. The SHAW Foundation is honored to award its first PRIZE in Mathematical Sciences to Professor CHERN.

All the members of the committee came to Hong Kong for the exciting date of ceremory. When the committee met together in Hongkong Professor P. Griffiths, one of the five members of the committee, pointed out that the mere introduction of mathematics achievements is insufficient to see the immense contributions of CHERN to the mathematical society. It should be included his immense help to the youngsters as a great mentor. As I was unable to express these ideas clearly owing to my poor English, I asked Professor Griffiths to write down these ideas to be added to the original address. GRIFFITHS added some lines to the original address which I reproduced below:

As a teacher and mentor of young mathematicians, and as one who took special pleasure in bringing the work of his colleagues to the attention of the mathematical community, CHERN's influence extended far beyond his own scientific work.

Among the young mathematicians benefitted from CHERN as both a teacher and a mentor, I am clearly one of them.

Someday in November 2005 I paid a visit to CHERN's Nankai Institute of Mathematics at Tianjin. Several members of the Institute and I met CHERN in his building. CHERN seems to be very energetic. We discussed on diverse aspects of mathematical researches and the future of Chinese mathematics. CHERN was particularly enthusiastic in explaining his intending method of proving the non-existence of complex structures on a 6-sphere, a long-standing unsolved problem in mathematics. Remark that CHERN was already 93 years old. His enthusiasm shows clearly his attitude toward mathematics: *He will exert all his efforts and contribute all his best to the country until the last minute of his life.*

A few days after my return to Beijing, I learned suddenly the unexpected death of CHERN. In remembrance of my last meeting with CHERN, I myself, and I hope also Chinese mathematicians of younger generations, should follow the maxim of CHERN:

Exert all his efforts and contribute all his best to the country until the last minute of his life.

Part II. Generalized CHERN Classes and CHERN Numbers of Irreducible Complex Algebraic Varieties with Arbitrary Singularities

1. Introduction

In one of CHERN's classical papers, viz. [CH1], CHERN introduced the monumental *CHERN Classes* and *CHERN Numbers* for arbitrary complex vector space bundles with unitary group structure. In particular, the tangent bundle of a complex manifold is such a one. It follows that for a complex *nonsingular* algebraic variety there are well-defined Chern Classes which had been proved by CHERN himself that they enjoy an algebraico-geometric character, i.e. its dual homology classes include cycles verified by algebraic subvarieties, see [CH2].

Suppose now a complex algebraic variety possesses some singularities so that tangent bundle does not exist and the introduction of CHERN Classes and CHERN Numbers will be in doubt. To overcome the difficulties, R. D. MacPherson and others applied the process of *blowing up* to remove the singularities and then introduced the *generalized CHERN Classes* and *generalized CHERN Numbers* in a very ingenious way (see e.g. [MacPh]). However, as the blowing-up procedure is only implicitly carried out, it seems that it is difficult to determine the generalized CHERN Classes and CHERN Numbers in any concrete case.

On the contrary the present author, based on the notions of *generic point* of an irreducible complex algebraic variety, non-singular or not, and the method of *specialization*, both due to Van der Waerden (see [VdW]), had introduced *Generalized CHERN Classes* and *Generalized CHERN Numbers* which are easily *computed* in any concrete case.

Below we shall successively consider extended point, specialization, generic point, homology of Grassmannian and composite Grassmannians according to Ehresmann, and *Generalized CHERN Classes* and *Generalized CHERN Numbers* for complex irreducible algebraic varieties with arbitrary singularities. We shall indicate how to compute these generalized CHERN Classes and CHERN Numbers in a simple way. In particular, the well-known Miyaoka-Yau inequality (cf. [Miy] & [YAU]) will be shown to be proved in a very simple way and even greatly extended. Finally we shall give some comments about the involved concepts and their possible future applications.

2. Extended Point and Specialization

Specialization of extended points and *generic point* of an irreducible algebraic variety, nonsingular or not, seems both introduced by Van der Waerden, see [VdW], Chap. 4. Unfortunately these powerful concepts and tools had disappeared in the literature in later years, for which the present author could not understand and could not accept. We shall however apply these tools of Van der Waerden for the introduction of *generalized CHERN classes* and *genaralized CHERN numbers* of complex irreducible algebraic varieties *with arbitrary singularities* in a quite simple way.

In what follows we shall follow mostly the original text of Van der Waerden [VdW], quite often direct translation from the Germann statements.

Let $\mathbf{K}$ be a fixed field of characteristic 0, which may be $\mathbf{Q}, \mathbf{R}, \mathbf{C}$, viz. the field of rational numbers, of real numbers, and of complex numbers, or else. Let E^n (E_n in [VdW]) be the n-dimensional vector space over $\mathbf{K}$. Instead of points with coordinates from $\mathbf{K}$, let us extend the point concept, with coordinates either indeterminates or algebraic functions of indeterminates or even more general elements of any extension field of $\mathbf{K}$. Such a "point in extended sense" (or *extended point*) of the vector space E^n is thus a system of n elements $y_1, \ldots, y_n$ of some arbitrary extension field of $\mathbf{K}$. Correspondingly we may define a point in extended sense (or an *extended point*) of an n-dimensional projective space S^n (S_n in [VdW]). We may also extend the concept of linear spaces, the hypersurfaces, etc. if we admit points in extended sense to be points of the linear spaces, or we admit the coefficients of the equations of the hypersurface to be any elements of an extension field of $\mathbf{K}$.

Let S^n be the n-dimensional projective space with homogeneous coordinates $(x_0 : x_1 : \cdots : x_n)$. By a *generic point* of S^n it is meant such a point whose coordinate ratios $\frac{x_1}{x_0}, \ldots, \frac{x_n}{x_0}$ are algebraically independent with respect to the groundfield $\mathbf{K}$. It follows that no algebraic equation $f(\frac{x_1}{x_0}, \ldots, \frac{x_n}{x_0}) = 0$ will hold true, or what is the same, no homogeneous algebraic relation $F(x_0, x_1, \ldots, x_n) = 0$ with coefficients in $\mathbf{K}$ will be true, so far the polynomial f or F is not identically vanishing. For example, one may get a generic point if we take all coordinates $x_0, \ldots, x_n$ to be indeterminates, or in setting $x_0 = 1$ while taking $x_1, \ldots, x_n$ to be indeterminates.

In similar way we may define *generic* hyperplanes in S^n, a *generic* hypersurface of degree m, a *generic* m-dimensional subspace S^m of S^n, etc.

As a complement to Van der Waerden's concept of *extended points* let us consider some concrete representations of such points as described below:

Let us restrict ourselves to the case of points in E^n, the case of S^n is similar.

An extended point $\Xi = (\xi_1, \ldots, \xi_n)$ in E^n may be canonically represented as follows. Let us set $\Xi_i = (\xi_1, \ldots, \xi_i)$ for $i = 1, \ldots, n$, while $\Xi_0 := \emptyset$ for $i = 0$.

Suppose that for $0 < i < c_1$, each ξ_i is transcendental over $\mathbf{K}(\Xi_{i-1})$, while ξ_{c_1} is algebraic over $\mathbf{K}(\Xi_{c_1-1})$ with defining equation

$$F_1 \equiv \xi_{c_1}^{d_1} + f_{11} * \xi_{c_1}^{d_1-1} + \cdots + f_{1d_1} = 0, \quad f_{1j} \in \mathbf{K}(\Xi_{c_1-1}).$$

Suppose for $c_1 < i < c_2$, each ξ_i is transcendental over $\mathbf{K}(\Xi_{c_1-1})$, while Ξ_{c_2} is algebraic over $\mathbf{K}(\Xi_{c_2-1})$ with defining equation

$$F_2 \equiv \xi_{c_2}^{d_2} + f_{21} * \xi_{c_2}^{d_2-1} + \cdots + f_{2d_2} = 0, \quad f_{2j} \in \mathbf{K}(\Xi_{c_2-1}).$$

This may be continued as far as possible. So the extended point $\Xi = (\xi_1, \ldots, \xi_n)$ in E^n may be represented by a set of polynomials $F_i, i = 1, \ldots, r \leq n$ given below:

$$F_i \equiv x_{c_i}^{d_i} + f_{i1} * x_{c_i}^{d_i-1} + \cdots + f_{id_i} \in \mathbf{K}(\Xi_{c_i-1}), \quad (2.1)$$

The above polynomials F_i will be called the *defining polynomials* and the set of F_i the *defining polynomial set* of the extended point $\Xi = (\xi_1, \ldots, \xi_n)$. The corresponding set of equations $F_i = 0$ is called the *defining equations* of Ξ and ξ_{c_i} the *bounded coordinates* while ξ_j with $j \neq$ any c_i the *free coordinates* of Ξ. Remark that if Ξ is an ordinary point in A^n, then the corresponding defining polynomial set is void and $\mathbf{K}(\Xi) = \mathbf{K}$.

It is easy to verify that for an extended point $\Xi = (\xi_1, \ldots, \xi_n)$ with defining polynomial set (2.1) the dimension over $\mathbf{K}$ of the extended point

Ξ is equal to $n - r$. In the extreme cases we have:

$r = n : Dim_{\mathbf{K}}(\Xi) = 0$,

$r = n$ and all $d_i = 1$: $\Xi =$ an ordinary point in E^n,

$r = 0 : Dim_{\mathbf{K}}(\Xi) = n$.

We give some particular examples below:

Ex. 2.1. Let $\mathbf{K} = \mathbf{R}$ and $\Xi = (-i, +i), \Xi' = (+i, -i)$, where $i = \sqrt{(-1)}$, then Ξ, Ξ' are *conjugate* extended points in $\mathbf{R}^2$ with the same defining polynomials

$$x_1^2 + 1, x_2 + x_1 \, ,$$

so that $n = r = 2$ and $Dim_{\mathbf{R}}\Xi = Dim_{\mathbf{R}}\Xi' = 0$.

Ex. 2.2. Let t be an indeterminate and

$$\Xi = \left(\frac{2 * t}{1 + t^2}, \frac{1 - t^2}{1 + t^2} \right), \Xi' = \left(\frac{1 - t^2}{1 + t^2}, \frac{2 * t}{1 + t^2} \right).$$

Then both ξ and ξ' are transcendental over $\mathbf{K}$ and have the same single defining polynomial $x_2^2 + x_1^2$ so that $n = 2, r = 1$. Thus $Dim_{\mathbf{K}}(\Xi) = Dim_{\mathbf{K}}(\Xi') = 1$.

After these preliminaries let us come now to the critical concept of *specialization*. For this sake let us reproduce the original definition of **relation-preserving specialization** or simply **specialization** in S^n of Van der Waerden ([VdW], p. 106) in German translated into English as follows:

A point in extended sense η is called a **relation-preserving specialization** *of the given point ξ also in extended sense if:*

When all homogeneous algebraic equation $F(\xi_0, \dots, \xi_n) = 0$ with coefficients in the ground field $\mathbf{K}$ which hold for the point ξ, will hold also for the point η. In other words, from $F(\xi) = 0$ will always follow $F(\eta) = 0$. As example every point of the space is a relation-preserving specialization of a generic point of this space. Another example is: Let $\xi_0, \xi_1, \dots, \xi_n$ be rational functions of indeterminate parameter t, and $\eta_0, \dots, \eta_n$ be the values of these rational functions for a definite value of t. Note that the above definition of specialization may be transferred to case of A^n.

In what follows let us give some further examples besides the above trivial ones given in [VdW]. We shall consider however points in A^n, the case of points in S^n is similar. These examples are taken from [WU6], Chap. 3, Sect. 3.1.

Ex. 2.3. In E^3, the point $(0, 0, 1)$ is a specialization of the extended point $\Xi = (\xi_1, \xi_2, \frac{\xi_2}{\xi_1})$, where ξ_1, ξ_2 are independent indeterminates over $\mathbf{K}$.

Proof. Let $P(x, y, z)$ be an arbitrary polynomial for which

$$P\left(\xi_1, \xi_2, \frac{\xi_2}{\xi_1}\right) \equiv 0 \tag{2.2}$$

for indeterminate ξ_1, ξ_2. We have to prove that $P(0, 0, 1) = 0$.

Let us divide $P(x_1, x_2, x_3)$ by $x_1 * x_3 - x_2$ with respect to the variable x_3. We will get for some polynomials $Q \in \mathbf{K}[x_1, x_2]$ and $R \in \mathbf{K}[x_1, x_2]$ such that

$$x_1^d * P(x_1, x_2, x_3) \equiv (x_1 * x_3 - x_2) * Q(x_1, x_2) + R(x_1, x_2).$$

Setting $x_3 = \frac{x_2}{x_1}$ we get then from (2.2)

$$R \equiv x_1^d * P\left(x_1, x_2, \frac{x_2}{x_1}\right) \equiv 0$$

and we get therefore

$$x_1^d * P(x_1, x_2, x_3) \equiv (x_1 * x_3 - x_2) * Q(x_1, x_2).$$

It follows that x_1^d divides Q and $x_1 * x_3 - x_2$ divides P so that for some polynomial $P'(x_1, x_2, x_3) \in \mathbf{K}[x_1, x_2, x_3]$ we have

$$P(x_1, x_2, x_3) \equiv (x_1 * x_3 - x_2) * P'(x_1, x_2, x_3).$$

By direct substitution we have then $P(0, 0, 1) = 0$. As P is arbitrary verifying (2.2) $(0, 0, 1)$ is a specialization of $\Xi = (\xi_1, \xi_2, \frac{\xi_2}{\xi_1})$ as to be proved.

Ex. 2.4. In E^2 the points $(0, 1), (0, -1), (1, 0), (-1, 0)$ are all specializations of the extended point $\xi = (\frac{2*t}{1+t^2}, \frac{1-t^2}{1+t^2})$, in which t is an indeterminate.

That $(0, 1), (1, 0), (-1, 0)$ are all specializations of ξ is clear in setting $t = 0, +1, -1$ in ξ. To see that $(0, -1)$ is also a specialization of ξ we need a proof. As the proof is similar to that of Ex.2.3 we shall omit it and refer to the one given in [WU6].

For extended points Ξ, Ξ' in E^n with Ξ' a specialization of Ξ over $\mathbf{K}$, we shall write

$$\Xi \longrightarrow_{\mathbf{K}} \Xi'.$$

The set of all specializations over $\mathbf{K}$ of an extended point Ξ in $\mathbf{K}^n$ will then be denoted by $Spec_{\mathbf{K}}(\Xi)$. Thus, for *Ex.2.4* we may write

$$\xi \longrightarrow_{\mathbf{K}} (1, 0), (-1, 0), (0, 1), (0, -1), \; or$$

$$(1, 0), (-1, 0), (0, 1), (0, -1) \; all \; \in Spec_{\mathbf{K}} \xi.$$

For two extended points in E^n (or S^n), the problem of deciding whether one is the specialization of the other will be solved in the next section.

For specializations in *projective* space S^n we have now the following useful important theorem due to Van der Waerden ([VdW] Sect. 28, p. 107):

Specialization Extension Theorem. Each specialization $\xi \longrightarrow_K \xi'$ may be extended to a specialization $(\xi, \eta) \longrightarrow_K (\xi', \eta')$.

As pointed out by Van der Waerden, the above theorem is not true if ξ, ξ', η are extended points in an *affine* space.

3. Irreducible Algebraic Variety and Generic Point

According to Van der Waerden ([VdW], Sect. 28), an *algebraic variety* (or *Mannigfaltigkeit* in [VdW]) in the n-dimensional projective space S^n is the collection of all (extended) points whose coordinates $\eta_0, \eta_1, \ldots, \eta_n$ satisfies a system of a finite or infinite number of algebraic equations

$$f_i(\eta_0, \ldots, \eta_n) = 0 , \qquad (3.1)$$

with coefficients in a constant-field **K** (of characteristic 0). If there are no such points, then the variety is called an *empty* one. Such case will be out of consideration.

Owing to Hilbert's basis-theorem any system of infinite number of equations can be replaced by an equivalent one of finite number of equations.

An algebraic variety M in S^n is said to be *decomposable* or *reducible*, if it is the union of two true, i.e. different from M itself subvarieties. A non-decomposable variety is then said to be *irreducible*.

After some preparatory arguments Van der Waerden arrived at the following fundamental theorems for which we refer to his original text [VdW], pp. 108–110.

Decomposition Theorem. Each algebraic variety is either irreducible or a union of a finite number of irreducible ones:

$$M = M_1 + M_2 + \cdots + M_r .$$

Uniqueness Theorem. The representation of a variety M as uncontractible union (contractible means a union for which one member of the union is contained in the union of the other members so that it can be deleted) of irreducible ones, is unique irrespective of the ordering of the sequence.

We leave the details of proofs to the original text and come to the fundamental notion of **generic point**, see [VdW] Sect. 29:

A point ξ is called a **generic point** of a variety M, if ξ belongs to M, and if for all homogeneous algebraic equation with coefficients in $\mathbf{K}$ which holds for the point ξ, holds also for all points of M. In other words, ξ belongs to M and all points of M are *specializations* of the point ξ. In the notation of Sect. 2, we have for a generic point ξ of a variety M:

$$M = Spec_{\mathbf{K}}(\xi) \,.$$

Van der Waerden proved successively the following theorems.

(Second) Irreducibility Criterion. If a variety has a generic point ξ, then it is irreducible.

Existence Theorem. Every non-empty irreducible variety M has a generic point ξ in some convenient extension field of $\mathbf{K}$.

A point will be said *normalized*, if the first non-zero coordinate is equal to 1. Each point can be normalized. By renumbering of coordinates we can assume that $\xi_0 \neq 0$ so that we can assume $\xi_0 = 1$. $\xi_1, \ldots, \xi_n$ are then called the inhomogeneous coordinates of ξ.

Uniqueness Theorem. Any two normalized generic points ξ, η of a variety M can be transformed into each other by a field isomorphism $\mathbf{K}(\xi) \cong \mathbf{K}(\eta)$ which leaves elements of $\mathbf{K}$ fixed. The algebraic properties of ξ and η are thus coincident to each other.

Converse Theorem. To each point ξ (whose coordinates belongs to any extension field of $\mathbf{K}$, e.g. algebraic functions with indeterminate parameters), belongs to an (irreducible) algebraic variety M, with ξ as generic point.

Dimension Theorem. If M and M' are irreducible and $M' \subset M$, then the dimension of M' is less than that of M.

Consequence. Each extended point ξ' of M has a dimension $d' \leq d$, where d is the dimension of the irreducible variety M. If $d' = d$, then ξ' is a generic point of M.

For varieties in particular dimensions Van der Waerden shows that:

A zero-dimensional irreducible variety in S^n is a system of points conjugate to each other with respect to the ground field $\mathbf{K}$.

Any pure $(n-1)$-dimensional manifold M in S^n can be given by a single homogeneous equation $h(\eta) = 0$, and every form which has all points of M as zeros, is divisible by $h(\eta)$.

Each hypersurface $f(\eta) = 0$ is a purely $(n-1)$-dimensional variety, and conversely.

In next sections 30, 31 of [VdW] concrete representations of varieties and effective decompositions of a variety in irreducible ones by means of elimination theory are given. We shall however no more follow Van der Waerden and in what follows we shall show how to determine the irreducibility of a variety, how to decompose a variety into its irreducible components, and to express a generic point in explicit form for an irreducible variety. Moreover, all these will be accomplished in an *algorithmic* manner, so that we may achieve all these by means of a computer.

For simplicity let us consider algebraic varieties in a vector space or more exactly an affine space E^n, the case of varieties in projective spaces is similar. Thus, let M be an algebraic variety represented by a system of polynomial equations $P_i(x_1, \ldots, x_n) = 0, i = 1, \ldots, m$, with $P_i \in \mathbf{K}[X], X = \{x_1, \ldots, x_n\}$. The variety is consisting of all extended points which are zeros of the polynomial set P_i so that we may write

$$M \equiv Zero_{\mathbf{K}}(PS), \; or \; simply \; Zero(PS), \; PS = \{P_i \mid i = 1, \ldots, m\}. \quad (3.2)$$

Our task is to give an algorithmic way of determining the totality of such zeros, in particular the generic zero, of the variety M from PS, in case it is irreducible.

For this sake let us first take some ordering of the variables x_i, usually the natural ordering

$$x_1 \prec x_2 \prec \cdots \prec x_n. \quad (3.3)$$

With respect to such an ordering any non-constant polynomial $P \in \mathbf{K}[X]$ may be expressed in the following *canonical form*:

$$P \equiv I_0 * x_c^d + I_1 * x_c^{d-1} + \cdots + I_d, \quad (3.4)$$

in which $c > 0$, $d > 0, I_j \in \mathbf{K}[x_1, \ldots, x_{c-1}]$, and $I_0 \neq 0$. The integers c, d are called respectively the *class* and *degree* of the polynomial P, to be denoted as $cls(P), deg(P)$. The polynomial I_0 of P is called the *initial* of P, to be denoted by *Init(P)*. In case P is a non-zero constant polynomial, we shall leave class, degree, and initial undefined. We may now introduce a partial ordering among all non-constant polynomials according first to their class, and then to their degree. We shall enlarge such a partial ordering in admitting all the non-zero constant polynomials to be in the lowest ordering.

Consider now a sequence of non-constant polynomials $A_1, A_2, \ldots, A_r$ in $\mathbf{K}[X]$ with classes $c_1, c_2, \ldots, c_r$ which are steadily increasing:

$$0 < c_1 < c_2 < \cdots < c_r, \quad (3.5)$$

so that in canonical forms we have:

$$A_j = I_{j0} * x_{c_1}^{d_j} + I_{j1} * x_{c_1}^{d_j-1} + \cdots + I_{jd_j}, \quad (3.6)$$

in which all I_{jk} are in lower ordering with respect to $A_1, \ldots, A_{j-1}$ and $I_{j0} \neq 0$. Such a sequence of polynomials will be called below a (non-trivial) *ascending set* (abbr. *asc-set*). As usual, we may define the irreducibility of such an asc-set. The importance of asc-sets lies in the fact that, in considering $x_i, i \neq$ *any* c_j as independent indeterminates, we may solve successively $x_{c_1}, x_{c_2}, \ldots, x_{c_r}$ as definite algebraic functions of these indeterminates and previous $x's$, so that $Zero(AS)$ may be considered as completely determined.

For any asc-set AS with successive classes as in (3.5) let $I_j, j = c_1, \ldots, c_r$ be the successive initials of polynomials in AS. Then for any polynomial $P \in \mathbf{K}[X]$ we have the following *Remainder Formula* of Ritt (cf. [RITT1,2]):

There are some polynomials $Q_j, j = 1, \ldots, r$, and $R \in \mathbf{K}[X]$ as well as some non-negative integers $h_j, j = 1, \ldots, r$ such that

$$I_1^{h_1} * \cdots * I_r^{h_r} * P = \Sigma_{j=1,\ldots,r} \ Q_j * A_j + R, \tag{3.7}$$

in which R, if non-zero, is of lower ordering than A_1. The polynomial R is then called the *Remainder* of P with respect to the asc-set AS.

With respect to the ordering (3.3) we may introduce some partial ordering among the system of ascending sets which we shall not enter. By considering the possible subsets of non-constant polynomials contained in given polynomial systems we may further introduce some partial ordering among the sets of all polynomial systems which we shall not enter too.

In Yuan Dynasty (1271-1368 A.D.) some scholar ZHU Shijie wrote a mathematics treatise (see [ZHU]) which, among others, describes a general method, based on well-designed successive eliminations, of solving arbitrary system of polynomial equations in arbitrary number of variables. Though the examples given are restricted to integer coefficients with 2, 3, or at most 4 unknowns and the solutions sought for are also restricted to real ones, the line of thought is clear and is valid for general case of arbitrary number of variables and equations. In borrowing some terminolgy and techniques from Ritt (cf. [RITT1,2]) we may express the final result dated back to scholar ZHU in the following form:

There is an algorithm which permits to give for any finite polynomial system PS an asc-set AS such that

$$Zero(PS) = Zero(AS/IP) \bigcup Zero(PS \cup \{IP\}), \ or \tag{3.8}$$

$$Zero(PS) = Zero(AS/IP) \bigcup \bigsqcup_{j=1,\ldots,r} Zero(PS \cup \{I_j\}). \tag{3.8}'$$

In (3.8), (3.8)$'$ IP means the product of all initials $I_j, j = 1, \ldots, r$ of polynomials in the asc-set AS and $Zero(GS/HS) = Zero(GS) \setminus Zero(HS)$, for any two polynomial systems GS, HS. The asc-set AS determined from the given PS will be called a *characteristic set* (abbr. *char-set*) of the given polynomial system PS.

By applying the above algorithm further and further on each of the sets in the $\bigcup$ of (3.8)$'$ we get finally the following *Decomposition Theorem*:

There is an algorithm which will give for any finite polynomial system PS a finite number of asc-sets $AS_k, k = 1, \ldots, s$ with initial products IP_k such that

$$Zero(PS) \equiv \bigcup_{k=1,\ldots s} Zero(AS_k/IP_k) . \tag{3.9}$$

As we have already remarked, each of the sets $Zero(AS_k)$, and hence also each of the sets $Zero(AS_k/IP_k)$ may be considered as well-determined, so the formula (3.9) gives actually a complete determination of $Zero(PS)$ for an arbitrary finite polynomial system. We remark further that a powerful algorithm of determing AS_k from an arbitrary PS had been given by WANG Dingkang, see [WDK]. Moreover, by some factorization procedure which may also be carried out algorithmetically, we may urge that each AS_k appearing in (3.9) to be *irreducible* ones.

We come now to our object of studying specialization and generic point of an algebraic variety considered as the zero-set of a polynomial set. An irreducible affine variety in the affine space E^m will then be defined by an irreducible asc-set $AS = \{A_1, \ldots, A_r\}$ with A_j of the form (3.6) and vice versa. For $\mathbf{K} = $ the complex field $\mathbf{C}$, the ordinary points of the variety will be consisting of the point set of ordinary zeros in $Zero_{\mathbf{C}}(AS)$ for which all $I_j \neq 0$ and their limit points.

Now any extended point Ξ in the affine space will be determined by an irreducible asc-set AS of the form (3.6) and vice versa. By the Ritt's Remainder formula it is readily seen that a polynomial will vanish at this extended point if and only if its remainder with respect to AS is zero. It follows easily that, if Ξ' is another extended point defined by some irreducible asc-set AS', then Ξ' will be a specialization over $\mathbf{K}$ of Ξ if and only if the remainder of any polynomial in AS' with respect to AS is 0. It follows also that the affine variety determined by the polynomial system leading to the irreducible charset AS is irreducible with Ξ as its generic point. A second irreducible affine variety with corresponding irreducible char-set AS' is contained in the one with irreducible charset AS if and only if the generic point determined by AS' is a specialization of that one determined by AS which may be verified by the remainder determination. All these can be achieved by algorithmic computations.

Now given any affine variety as a zero-set of some finite polynomial system we can decompose it into a union of irreducible ones as in (3.9). We may verify whether any one of the components is contained in some one else by remainder computation so that it may be removed from the union. It follows that the above will give an effective way of decomposing any affine variety into a no more contractible union of irreducible subvarieties.

4. Homology of Composite Grassmannians

The homology of Grassmannian manifolds in the complex case was determined already early in 1934 by Ehresmann, see [EH]. Consider thus a complex projective space $P_{\mathbf{C}}^n$ of complex dimension n. The set of all linear subspaces of complex dimension d in $P_{\mathbf{C}}^n$ form then the complex Grassmannian manifold of complex dimension $n * d$, to be denoted as $GR(n, d)$. A homology base for cycles of this manifold had been given by Ehresmann via so-called *Schubert cycles*. Moreover, the variety being homogeneous, its intersection homology ring is well-defined and had been completely determined by Hodge, see [H-P], Chap. 14. In order not to make the present paper too lengthy we shall leave it aside and go directly to the homology of *composite Grassmannian* $GR(n; 0, d)$ consisting of all pair-elements of the form (P_0, L_d) for which L_d is a linear subspace of dimension d in $P_{\mathbf{C}}^n$, while P_0 is an arbitrary point in L_d.

The variety $GR(n; 0, d)$ is in fact a compact complex manifold. According to Ehresmann the total homology group on complex coefficients of $GR(n; 0, d)$ may be described as follows. In $P_{\mathbf{C}}^n$ let us consider an arbitrary sequence $\mathbf{S}$ of linear subspaces L_i of dimension $i = 0, 1, \ldots, n$ such that

$$\mathbf{S} : L_0 \subset L_1 \subset \cdots \subset L_{n-1} \subset L_n = P_{\mathbf{C}}^n. \qquad (4.1)$$

Consider a sequence of d integers $b_i, i = 0, 1, \ldots, d$ such that

$$0 \le b_0 < b_1 < \cdots < b_{d-1} < b_d \le n. \qquad (4.2)$$

Then the subvariety of $GR(n; 0, d)$ of composite elements (L_0', L_d') with L_0' a point in some $L_i, i \in \{0, 1, \ldots, n\}$ and L_d' a d-dimensional linear subspace through L_0' with

$$Dim_{\mathbf{C}}(L_d' \cap L_{b_j}) \ge j, \qquad j = 0, 1, \ldots, d \qquad (4.3)$$

will be called a *Schubert cycle* of **Ehresmann symbol**

$$E = [b_i | b_0 b_1 \dots b_d] \,. \tag{4.4}$$

According to Ehresmann such cycles will belong to the same homology class C_E of $GR(n; 0, d)$ irrespective of the chosen linear subspaces L'_i and is of dimension

$$Dim_{\mathbf{C}} \, C_E = \Sigma_j (b_j - j) + b_i \,. \tag{4.5}$$

We shall call accordingly such a homology class C_E of the composite Grassmannian an *Ehresmann class* of *Ehresmann symbol* E. In particular, the composite Grassmannian itself is a cycle of Ehresmann symbol

$$GR(n; 0, d) = [n | (n - d, n)] \,, \tag{4.6}$$

in which (i, j) for $i \le j$ will stand for the sequence of successive integers $i, i + 1, \dots, j$. The dimension of $GR(n; 0, d)$ is by (4.5)

$$Dim_{\mathbf{C}} GR(n; 0, d) = (n - d) * (d + 1) + d = D, \; say \,. \tag{4.6$'$}$$

Now the composite Grassmannian $GR(n; 0, d)$ is a compact complex differential manifold so that both duality and intersections of homology classes are well-defined. Let us set for cycle E in (4.4) its dual

$$\delta E = [n - b_i | n - b_d \dots n - b_1 \, n - b_0] \,,$$

then the induced duality operator δ_* of homology classes will be given by

$$\delta_* C_E = C_{\delta E} \,. \tag{4.7}$$

The intersections in the composite Grassmannian are rather quite complicated and we shall restrict ourselves to the case of codimensions $\le d$ or dimensions $\ge D - d$. In these dimensions there are two kinds of cycles $\delta P, \delta Q_h, 0 \le h \le d$ of particular importance for which

$$P = [1 | (0, d)], \quad Q_h = [0 | (0, d - 1), d + h] \,. \tag{4.8}$$

In fact, the intersection ring of $GR(n; 0, d)$ in above dimensions is generated by the above cycles and we have for example ($\sim$ means homologous to each other of cycles)

$$\delta[i | (0, i), b_{i+1}, \dots, b_d] \sim \delta[0 | (0, i), b_{i+1}, \dots, b_d] * (\delta P)^i \,, \tag{4.9}$$

$$\delta[0 | a_0, \dots, a_d] * \delta[0 | b_0, \dots, b_d] \sim \Sigma_c \delta[0 | c_0, \dots, c_d] \,, \tag{4.10}$$

in which Σ_c is to be extended over $c = (c_0, \ldots, c_d)$ as in the ordinary Grassmann variety $GR(n; d)$ given in [H-P], v. 2, Chap. 14, Sect. 6 which can be deduced from the well-known formula of Pieri and Giambelli and we shall omit its explicit expression.

Among the Ehresmann classes we have the *Gamkrelidze-Todd classes* GT_{st} for $0 \leq t \leq s \leq d$:

$$GT_{st} = [s - t | (0, d - t), (d - t + 2, d + 1)] \in H_s(GR(n; 0, d)). \quad (4.11)$$

We then have the *CHERN Classes* or *CHERN Numbers* $CH_s, (0 \leq s \leq d)$:

$$CH_s = \Sigma_{0 \leq t \leq s}(-1)^t * \binom{d - t + 1}{d - s + 1} * GT_{st} \in H_s(GR(n; 0, d)). \quad (4.12)$$

For a partition in positive integers $\pi = (p_1, \ldots, p_s), 0 < p_1 \leq \cdots \leq p_s, p = p_1 + \cdots + p_s$ we define *CHERN Classes of partition* π to be

$$CH_\pi = \delta_*(\delta_* CH_{p_1} * \cdots \delta_* CH_{p_s}) \in H_p(GR(n; 0, d)). \quad (4.13)$$

From the above we deduce easily

$$\delta_* CH_{11} = \binom{d + 1}{2}^2 * (\delta_* P)^2 - 2 * (d + 1) * \delta_* P * (\delta_* Q_1) + (\delta_* Q_1)^2, \quad (4.14)$$

$$\delta_* CH_2 = \binom{d + 1}{2} * (\delta_* P)^2 - d * \delta_* P * \delta_* Q_1 + (\delta_* Q_1)^2 - \delta_* Q_2. \quad (4.15)$$

In the case of a hypersurface $V^2 \subset P_{\mathbf{C}}^3$ with $d = 2, n = 3$ we have clearly further

$$([0|014])(V^2) = 0. \quad (4.16)$$

5. Generalized CHERN Classes and CHERN Numbers of Complex Irreducible Varities with Arbitrary Singularities

Let us consider an irreducible complex algebraic variety V_d of complex dimension d in a complex projective space $P_{\mathbf{C}}^n$ of complex dimension n. Let us take an arbitrary *generic point* Ξ of V_d. Then Ξ is known to be a *simple point* with tangent space T_Ξ of complex dimension d well-defined. The pair $(\Xi, T(\Xi))$ is then a well-determined extended point in the composite Grassmannian $GR(n; 0, d)$. This point will determine a well-defined subvariety $\tilde{V}_d \subset GR(n; 0, d)$ for which it is a generic point.

Now for any point (P, L_P) consisting of a point P and a d-dimensional linear space L_P through P in the complex projective space $P_{\mathbf{C}}^n$ the restriction of the map $\iota_* : (P, L_P) \longrightarrow P$ to $\tilde{V}$ will induce a homomorphism of homology groups $\iota_* : H_*(\tilde{V}, \mathbf{C}) \longrightarrow H_*(V, \mathbf{C})$. Consider now in $GR(n; 0, d)$ an arbitrary cycle C_E of codimension $\geq D - d$ and $\leq D$ of Ehresmann symbol E which may be taken to be in general position with $\tilde{V}$. The intersection of C_E with $\tilde{V}$ will give then a cycle of dimension ≥ 0 and $\leq d$ in $\tilde{V}$. The image under ι_* of the homology class of such cycles is then well-defined and will be called an **Ehresmann-Chern class** or a **generalized Chern class** of Ehresmann symbol E. In the case of dimension 0 these will give the respective **Ehresmann-Chern numbers** or **generalized Chern numbers**.

Now in the case of algebraic varieties without singularities it is well-known that among the various Chern numbers there are inequalities to be held. For example, for some kind of hypersurfaces in a 3-dimensional complex projective space there is a celebrated Miyaoka-Yau inequality (see [MIY] and [YAU]), viz.

$$(c_1)^2 \leq 3 * c_2.$$

In applying our method it may however be easily proved by simple computations that this inequality holds for any hypersurface with arbitrary singularities in 3-dimensional complex projective space without any restriction. In fact, let $d = 2, n = 3$ so that we consider an irreducible complex algebraic surface V^2 in $P_{\mathbf{C}}^3$ with arbitrary singularities. By (4.14), (4.15), (4.16) we readily find

$$3 * CH_2(V^2) - CH_{11}(V^2) = 2 * \delta_*(\delta_* Q_1)^2 \geq 0.$$

This is just the Miyaoka-Yau inequality whatever is the hypersurface V^2 in question.

Other kinds of inequalities also exist between such (generalized) Chern numbers in higher dimensional case. In fact, SHI He had discovered and proved a lot of equalities and inequalities among such generalized CHERN numbers for which we refer to the original papers of SHI ([SHI1,2]) and the present author ([WU5,6]).

6. Some Comments

Let us first remark that the present author had introduced generalized CHERN classes and numbers as elements of CHOW ring associated to an algebraic veriety, cf. [CHOW1,2], [CH-VdW], and [WU3,4]. However, in later years they had been shifted to ordinary homology classes as exhibited in previous sections which are more convenient to handle.

We have seen how powerful are the concepts of *specialization* and *generic point* in the introduction of CHERN CLASSES and NUMBERS for complex algebraic varieties with arbitrary singularities. In fact, the powerfulness of these concepts are far beyond the above application. Thus, the fifth chapter of [VdW] is devoted to algebraic correspondence together with its applications. In Sect. 33 of [VdW] Van der Waerden begins by saying: *An irreducible correspondence is (as every irreducible variety), determined by its generic pointpair* $(\xi, \eta), \ldots :$ *all pointpairs of the correspondence take place by specialization of this generic pointpair* (ξ, η). *If one wants defining an irreducible correspondence, one should start from an arbitrarily defined generic pointpair; the totality of pairs* (x, y) *gotten from this generic point pair by specialization is then always an irreducible correspondence.*

Van der Waerden further introduces a new concept *multiplicity* and deduces then the *Principle of Constant Enumeration* with applications including e.g. the interesting problem of 27 lines on a cubic surface. The multiplicity concept is further clarified in a whole chapter (Chap. 8 of [VdW]) together with the *Principle of Number Conservation* and its numerous applications.

In fact, all these had been established in some earlier papers of Van der Waerden ([VdW1-3]). It seems that he is aimed at establishing a rigorous foundation of Schubert's enumerative geometry, see [SCH].

In Prob. 15 of Hilbert's well-known 23 problems (see [HILB]) Hilbert says:

To establish rigorously and with an exact determination of the limits of their validity those geometrical numbers which Schubert especially has determined on the basis of the so-called principle of special positions or conservation of number, by means of the enumerative calculus developed by him.

The topological method already introduced by Van der Waerden (see [VdW3]) and developed further in later generations may be considered to have given a satisfactory positive answer to Hilbert's Problem 15.

However, there remain the delicate problem of *actual determination* of the number of solutions in each concrete case, as in Schubert's original book. This was pointed out by e.g. Kleiman in his paper [KL2] on Hilbert's 15th problem. He says that *in both the statement and the explanation of the problem Hilbert makes clear his interest in the* **efficient production** *of accurate geometrical numbers*. To achieve this it requires a thorough study of homologies in ordinary and composite Grassmannians as done by Ehresmann, Hodge, and others. It requires also a very skillful choice of appropriate procedure of *specialization*, as done by Schubert in his book. Kleiman even pointed out that Chern classes of a variety may play a role in such actual determinations. In present author's opinion, the role of *generalized Chern classes and Chern numbers* of algebraic varieties with arbitrary singularities will also be inavoidable to be applied. In any way, there remains a lot of works waiting to be done by mathematicians in later generations.

Besides the above we remark that the author's mechanical geometry theorem-proving may be considered as some kind of applications of the method of specialialization and generic point. In fact, in taking convenient coordinate system we may suppose that for a given theorem the hypothesis is represented by a system of equations $HYP = 0$ with $HYP = \{H_1, \ldots, H_m\}, H_i \in \mathbf{K}[X], X = \{x_1, \ldots, x_n\}, char(\mathbf{K}) = 0$ and the conclusion is represented by some equation $CONC = 0, CONC \in \mathbf{K}[X]$. For simplicity let us suppose that the char-set of HYP is irreducible with initial-product IP and $CONC$ is also irreducible. Let the generic points of CS and $CONC$ be respectively G_{HYP} and G_{CONC}. Then the theorem in question will be true if G_{CONC} is a specialization of G_{HYP} which may be verified by Ritt's remainder formation. This amounts to say that if the remainder of $CONC$ with respect to CS is 0, then the theorem is true so far the subsidiary condition $IP \neq 0$ is observed. The general case may be reduced to the above one by splitting HYP into irreducible ones as shown in Sect. 4.

Based on the above principle and method hundreds of delicate geometry theorems had been proved and even discovered. See the excellent book [CHOU] and related papers of CHOU Shang-Ching.

Furthermore the present author had applied the method of generic point to the study of problems of surface-fitting in CAGD (Computer-Aided Geometry Design), see the paper [WU7] as well as related passages in the book [WU6]. It seems that many more interesting and important applications of the method of generic point and specialization may be found in sciences and technology, besides mathematics itself. It is for this reason that the present author strongly proposes that the notions of generic point and specialization, both due to Van der Waerden, should be recovered from their obscurity to become important tools of future-day mathematicians.

References

[CH1] CHERN Shiing-Shen, Characteristic Classes of Hermitian Manifolds, Annals of Math., 47 (1946) 88–121.

[CH2] CHERN Shiing-Shen, On the Characteristic Classes of Complex Sphere Bundles and Algebraic Varieties, J. Amer. Math. Soc., 75 (1953) 565–577.

[CHOU] CHOU Shang-Ching, Mechanical Geometry Theorem-Proving, Reidel (1988).

[CHOW1] CHOW Wei-Liang, Algebraic Varieties with Rational Dissections, Proc. Nat. Acad. Sci., 42 (1956) 116–119.

[CHOW2] CHOW Wei-Liang, On the Equivalence Classes of Cycles in an Algebraic Variety, Annals of Math., 64 (1956) 450–479.

[CH-VdW] CHOW Wei-Liang and Van der Waerden, Ueber Zugeordnete Formen und Algebraische Systeme von Mannigfaltigkeiten, Math. Annalen, 113 (1937) 692–704.

[EH] Ch. Ehresmann, Sur la Topologie de Certaines Espaces Homogénes, Annals of Math., 35 (1934) 398–443.

[Gam] P. B. Gamkrelidze, CHERN Cycles of Complex Algebraic Manifolds, Izv. Acad. Scis., CCCP, Math. Series, 20 (1956) 685–706 (in Russian).

[HILB] D. Hilbert, Mathematical Problems (translated), Bull. Amer. Math. Soc., 8 (1902) 437–479.

[H-P] W. V. D. Hodge and D. Pedoe, Methods of Algebraic Geometry, v. 2, Camb. Univ. Press (1952).

[KL1] S. L. Kleiman, An Introduction to the Reprint Edition, in [SCH].

[KL2] S. L. Kleiman, Problem 15, Rigorous Foundation of Schubert's Enumerative Calculus, in [Symp-HILB], v. 2, 445–482.

[MacPh] R. D. MacPherson, CHERN Cycles for Singular Algebraic Varieties, Annals of Math., 42 (1974) 423–432.

[MIY] Y. Miyaoka, On the CHERN Numbers of Surfaces of General Type, Invent. Math., 42 (1977) 225–237.

[Ritt1] J. F. Ritt, Differential Equations from the Algebraic Standpoint, Amer. Math. Soc. (1932).

[Ritt2] J. F. Ritt, Differential Algebra, Amer. Math. Soc. (1950).

[SCH] H. Schubert, Kalkuel der Abzaehlenden Geometrie (Reprint), Springer (1979).

[SHI1] SHI He, On CHERN Characters of Algebraic Hypersurfaces with Arbitrary Singularities, Acta Math. Sinica, New Series, 4 (1988) 289–360.

[SHI2] SHI He, CHERN Classes of Algebraic Varieties with Singularity, *Singularity Theory*, World Scientific, Singapore (1995) 705–730.

[Symp-HILB] Mathematical Developments arising from Hilbert Problems, Proc. Symp. in Pure Mathematics v. 28 (Edit. F. E. Browder) (1974).

[Todd] J. A. Todd, The Geometrical Invariants of Algebraic Loci, Proc. London Math. Soc., Ser. 2, 43 (1937) 127–138.

[VdW] B. L. Van der Waerden, Einfuehrung in die Algebraische Geometrie, Dover Publications (1945).

[VdW1] B. L. Van der Waerden, Der Multiplizitaetsbegriff der Algebraischen Geometrie, Math. Annalen, 97 (1927) 756–774.

[VdW2] B. L. Van der Waerden, Eine Verallgemeinerung des Bézoutschen Theorems, Math. Annalen, 99 (1928) 497–541.

[VdW3] B. L. Van der Waerden, Topologische Begruendung des Kalkuels der abzaehlenden Geometrie, Math. Annalen, 102 (1930) 337–362.

[W-A] R. L. Wilder and W. L. Ayres (Editors), Lectures in Topology, Univ. of Michigan Press (1941).

[WDK] WANG DingKang, Polynomial Equations Solving and Mechanical Geometric Theorem Proving, PhD. Thesis, Inst. Sys. Sci. (1987).

[WH] H. Whitney, On the Topology of Differential Manifolds, in [W-A], 101–141.

[WU1] WU Wentsun, Note sur les Produits Essentiels Symmètriques des Espaces Topologiques, C. R. Paris, 224 (1947) 1139–1141.

[WU2] WU Wentsun, On the Product of Sphere Bundles and the Duality Theorem Modulo Two, Annals of Math., 49 (1948) 641–653.

[WU3] WU Wentsun, On CHERN Characteristic Systems of an Algebraic Variety, (in Chinese), Shuxue Jinzhan, 8 (1965) 395–401.

[WU4] WU Wentsun, On Algebraic Varieties with Dual Rational Dissections (in Chinese), Shuxue Jinzhan, 8 (1965) 402–409.

[WU5] WU Wentsun, On CHERN Numbers of Algebraic Varieties with Arbitrary Singularities, Acta Math. Sinica, New Series, 3 (1987) 227–236.

[WU6] WU Wentsun, MATHEMATICS MECHANIZATION: Mechanical Geometry Theorem-Proving, Mechanical Geometry Problem-Solving and Polynomial Equations-Solving, Science Press/Kluwer Academic Publishers (2000) Chap. 5, Sect. 1–3.

[WU7] WU Wentsun, On Surface Fitting Problem in CAGD, MM-Res. Preprints, No. 10 (1993) 1–10.

[YAU] YAU Shing Tung, Calabi's Conjecture and Some New Results in Algebraic Geometry, Proc. Nat. Acad. Sci., 74 (1977) 1798–1799.

[ZHU] ZHU Shijie, Jade Mirror of Four Elements (in Chinese) (1303).

Chapter 20

SOME OPEN GROMOV-WITTEN INVARIANTS
OF THE RESOLVED CONIFOLD

Jian Zhou

Department of Mathematical Sciences,
Tsinghua University, Beijing, 100084, China
jzhou@math.tsinghua.edu.cn

Dedicated to the memory of Professor Shiing-Shen Chern

We use localization to calculate open Gromov-Witten invariants of the resolved conifold relative to the fixed locus of an antiholomorphic involution. Our calculation generalizes an earlier calculation made by Katz and Liu, and it uses Mariño-Vafa formula and free field realization of Feynman rules arising from localization.

1. Introduction

One of the many famous applications of Chern-Simons theory in theoretical physics is the paper by Witten [25] on link invariants. This work relates several branches of mathematics: link invariants and 3-manifold invariants, representations of quantum groups and Kac-Moody algebras, etc. As a consequence, it is possible to find explicit expression of the colored HOMFLY polynomials of the unknot and the Hopf link in terms of specializations of the Schur functions. (For a mathematical treatment, see e.g. [20].)

In a series of papers by physicists, a surprising connection between Chern-Simons theory and Gromov-Witten theory of some open Calabi-Yau 3-folds has been predicted. Witten [26] suggested that when $N \to \infty$, the $SU(N)$ Chern-Simons theory is related to open string theory on T^*S^3. Gopakumar-Vafa [3], Ooguri-Vafa [24] and Mariño-Vafa [17] used the conifold transition from T^*S^3 to $\mathcal{O}(-1) \oplus \mathcal{O}(-1) \to \mathbb{P}^1$ to relate Chern-Simons theory to string theory on the resolved conifold $\mathcal{O}(-1) \oplus \mathcal{O}(-1)$. This has been further developed into a duality between Chern-Simons theory and Gromov-Witten theory of open Calabi-Yau 3-folds with toric symmetries [2, 1].

Many of these physical predictions can be mathematically proved using the method of localizations. This method has its roots in Chern-Weil theory and Chern's proof of the Gauss-Bonnet formula. In [7], Katz and Liu described a possible mathematical formulation of open Gromov-Witten invariants based on the moduli spaces of bordered stable maps. The proposal in [7] was further elaborated by Liu in [11]. Katz and Liu also carried out a calculation for some open Gromov-Witten invariants of the resolved conifold $(\mathcal{O}(-1) \oplus \mathcal{O}(-1) \to \mathbb{P}^1)$ using localization techniques. This leads to Hodge integrals on the Deligne-Mumford moduli spaces of algebraic curves. By comparing with the expressions for the Chern-Simons link invariants of the unknot, Mariño and Vafa [17] conjectured a formula on such Hodge integrals. This conjecture has been proved by localization techniques [14, 23]. A generalization which involves the Chern-Simons link invariants of the Hopf link was made in [29], and proved in [15], again by localization techniques. Furthermore, a mathematical theory of the topological vertex has been developed [12] by similar ideas. The applications of such Hodge integral formulas to the computations of Gromov-Witten invariants of open Calabi-Yau 3-folds with toric symmetries is again via localizations (see e.g. [28, 30, 12]).

The purpose of this paper is to extend the calculations by Katz and Liu [7] to more general open Gromov-Witten invariants. More precisely, while only stable maps with images on half of the sphere $\mathbb{P}^1$ were considered in [7], we will consider stable maps with images possibly wrapping the whole of $\mathbb{P}^1$. Our main result is presented in Theorem 4.1.

Our calculations will need a combination of techniques: Localization calculations [8, 4, 7], Mariño-Vafa formula for Hodge integrals [17, 14, 23], free field realizations of Feynman rules [28, 30], and operator manipulations of [21, 22].

The rest of the paper is arranged as follows. In Section 2 we explain the localization calculation on the moduli spaces of bordered stable maps to the resolved conifold. In Section 3 we use the Mariño-Vafa formula to reformulate the Feymnan rules obtained by localization. In Section 4 we use free field realization of the Feynman to calculate the partition functions. In Section 5 we use different operator manipulations to get some explicit expressions.

This research is partially supported by research grants from NSFC and Tsinghua University.

2. Localizations on Moduli Spaces of Bordered Stable Maps to $\mathbb{P}^1$

2.1. *A family of anti-holomorphic involution on the resolved conifold*

Slightly different from [7], we consider for each $A > 0$ the following anti-holomorphic involution on $\mathbb{C}^2$:

$$(1) \qquad (z_1, z_2) \mapsto (\bar{z}_2, A\bar{z}_1).$$

This induces the following anti-holomorphic involution σ on $\mathbb{P}^1$:

$$(2) \qquad z \mapsto \frac{A}{\bar{z}}, \qquad z' \mapsto \frac{1}{A\bar{z}'},$$

where $z = z_2/z_1$ and $z' = z_1/z_2$ are affine coordinates related by

$$(3) \qquad z' = \frac{1}{z}.$$

The following anti-holomorphic involution $\tilde{\sigma}$ on $\mathcal{O}_{\mathbb{P}^1}(-1) \oplus \mathcal{O}_{\mathbb{P}^1}(-1)$ covers σ:

$$(z, u, v) \mapsto \left(\frac{A}{\bar{z}}, \frac{1}{A^{1/2}} \bar{z}\bar{v}, \frac{1}{A^{1/2}} \bar{z}\bar{u} \right), \quad (z', u', v') \mapsto \left(\frac{1}{A\bar{z}'}, A^{1/2}\bar{z}'\bar{v}', A^{1/2}\bar{z}'\bar{u}' \right),$$

(4)

where (u, v) and (u', z') are related by:

$$(5) \qquad u' = zu, \quad v' = zv.$$

The fixed points of σ consists of points $z \in \mathbb{C}$ such that $|z| = A$, it is a circle that divides $\mathbb{P}^1$ into two discs:

$$(6) \qquad \mathbb{P}^1 = D_+ \cup_{S^1} D_-.$$

It follows that

$$(7) \qquad H_2(\mathbb{P}^1, S^1; \mathbb{Z}) = \mathbb{Z}\beta_+ \oplus \mathbb{Z}\beta_-,$$

where $\beta_\pm = [D_\pm, S^1] \in H_2(\mathbb{P}^1, S^1; \mathbb{Z})$.

The fixed point set of $\tilde{\sigma}$ is a real 3-manifold:

$$(8) \qquad L := \{(z, u, v) \in \mathbb{C}^3 \mid z = Ae^{i\theta}, u = A^{1/2}e^{-i\theta}\bar{v}\}.$$

2.2. Torus actions

As in [7], we consider the following $U(1)$-action on $\mathbb{P}^1$:

$$(9) \qquad (e^{i\theta}, z) \mapsto e^{i\theta} z, \qquad (e^{i\theta}, z') \mapsto e^{-i\theta} z'.$$

This action commutes with the antiholomorphic involution $\tilde{\sigma}$ and so it preserves $S^1 = \{z \in \mathbb{C} \mid |z| = A\}$. Furthermore it has two fixed points: $z = 0$ and $z = \infty$.

For $a \in \mathbb{Z}$, the above $U(1)$-action has the following lifting to $\mathcal{O}(-1) \oplus \mathcal{O}(-1)$:

$$(10) \qquad \begin{aligned} (e^{i\theta}, (z, u, v)) &\mapsto (e^{i\theta} z, e^{i(a-1)\theta} u, e^{-ia\theta} v), \\ (e^{i\theta}, (z', u', v')) &\mapsto (e^{-i\theta} z', e^{ia\theta} u', e^{i(1-a)\theta} v'). \end{aligned}$$

This action commutes with the antiholomorphic involution $\tilde{\sigma}$ and so it preserves the submanifold L. When $a \neq 0, 1$, this action has two fixed points $(z, u, v) = (0, 0, 0)$ and $(z', u', v') = (0, 0, 0)$; When $a = 0$ or 1, this action has two 2-dimensional fixed point set components: For $a = 0$, $z = u = 0$, $u \in \mathbb{C}$ and $z' = v' = 0$, $u' \in \mathbb{C}$; for $a = 1$, $z = v = 0$, $u \in \mathbb{C}$ and $z' = u' = 0$, $v' \in \mathbb{C}$.

The integer a will be referred to as the *framing*.

2.3. Moduli spaces of bordered stable maps to $(\mathbb{P}^1, S^1)$

Let $\mu^\pm$ be two partitions, $|\mu^\pm| = d^\pm$. As usual, $l(\mu^\pm)$ denotes the length of $\mu^\pm$. For $d \geq 0$, we will consider the moduli space

$$\overline{\mathcal{M}}_{g,l(\mu^+)+l(\mu^-)}(\mathbb{P}^1, S^1 | (d^+ + d)\beta_+ + (d^- + d)\beta_-; \mu^+, -\mu^-).$$

See [7] for notations. For simplicity, we will write it as $\overline{\mathcal{M}}_g(d; \mu^+, -\mu^-)$.

A point in this moduli space is the equivalence class of a bordered stable map

$$f : (\Sigma, \partial \Sigma) \to (\mathbb{P}^1, S^1),$$

where Σ is a connected bordered prestable Riemann surface of type $(g, l(\mu^+) + l(\mu^-))$, such that the degrees of $f|_{\partial \Sigma} \to S^1$ are given by the partitions μ^+ and μ^-, and

$$f_*([\Sigma, \partial \Sigma]) = (d^+ + d)\beta_+ + (d^- + d)\beta_-,$$

In [7] only cases with $d^- = d = 0$ are considered.

2.4. *The obstruction bundle*

One can define an orbifold vector bundle on the moduli space as follows [7]. The fiber at the point represented by $f : (\Sigma, \partial\Sigma) \to (\mathbb{P}^1, S^1)$ is

$$H^1(\Sigma, \partial\Sigma; f^*(\mathcal{O}(-1) \oplus \mathcal{O}(-1)), f^*(\mathcal{O}(-1) \oplus \mathcal{O}(-1))_{\mathbb{R}}).$$

Alternatively, let $f_{\mathbb{C}} : \Sigma_{\mathbb{C}} \to \mathbb{P}^1$ be the complex double of f. Then the fiber is given by

$$H^1(\Sigma_{\mathbb{C}}, f_{\mathbb{C}}^*(\mathcal{O}(-1) \oplus \mathcal{O}(-1)))^{\tilde{\sigma}}.$$

We will denote the obstruction bundle by $V_g(d; \mu^+, -\mu^-)$.

2.5. *Localizations and labelled graphs*

The $U(1)$-actions on $\mathbb{P}^1$ and $\mathcal{O}(-1)\oplus\mathcal{O}(-1)$ mentioned above induce natural actions on the moduli space $\overline{\mathcal{M}}_g(d; \mu^+, -\mu^-)$ and the obstruction bundle $V_g(d; \mu^+, -\mu^-)$. As in [11], one can consider the equivariant fundamental chain of $\overline{\mathcal{M}}_g(d; \mu^+, -\mu^-)$ and try to calculate

$$(11) \qquad \int_{[\overline{\mathcal{M}}_g(d;\mu^+,-\mu^-)]_{U(1)}^{vir}} e_{U(1)} V_g(d; \mu^+, -\mu^-)$$

by localization. We will not address the complicated issues involved in this, but as in [7], carry out a formal calculation and leave the justifications for future investigations. Define

$$F_{\mu^+,\mu^-}^a(\lambda, t_+, t_-) = \sum_{g=0}^{\infty} \sum_d \lambda^{2g-2}(t_+ t_-)^d t_+^{|\mu^+|} t_-^{|\mu^-|}$$

$$\cdot \int_{[\overline{\mathcal{M}}_g(d;\mu^+,-\mu^-)]_{U(1)}^{vir}} e_{U(1)} V_g(d; \mu^+, -\mu^-).$$

Here for $(\mu^+, \mu^-) \neq ((0), (0))$, the summation over d is taken from 0 to ∞; for $(\mu^+, \mu^-) = ((0), (0))$, the summation over d is taken from 1 to ∞. Clearly $F_{(0),(0)}^a(\lambda, t_+, t_-)$ is the free energy of closed Gromov-Witten invariants of the resolved conifold. It will be denoted by $F_{closed}^a(\lambda, t_+, t_-)$. Define the open part of the free energy by

$$F_{open}^a(\lambda, t_+, t_-; x^+, x^-) = \sum_{(\mu^+, \mu^-) \neq ((0),(0))} F_{\mu^+,\mu^-}^a(\lambda, t_+, t_-) \frac{p_{\mu^+}(x^+)}{z_{\mu^+}} \frac{p_{\mu^-}(x^-)}{z_{\mu^-}},$$

where $p_{\mu^\pm}(x^\pm)$ are Newton symmetric functions in $x^\pm = (x_1^\pm, \ldots, x_n^\pm, \ldots)$. The total free energy is defined by

$$F(\lambda, t_+, t_-; x^+, x^-) = F_{closed}(\lambda, t_+, t_-) + F_{open}(\lambda, t_+, t_-; x^+, x^-).$$

We also define the partition function by:

$$Z^a(\lambda, t_+, t_-; x^+, x^-) = \exp F(\lambda, t_+, t_-; x^+, x^-).$$

We write

$$(12) \quad Z^a(\lambda, t_+, t_-; x^+, x^-) = \sum_{\mu^\pm} Z^a_{\mu^+, \mu^-}(\lambda, t_+, t_-) \frac{p_{\mu^+}(x^+)}{z_{\mu^+}} \frac{p_{\mu^-}(x^-)}{z_{\mu^-}}.$$

We first describe the fixed point components of $\overline{\mathcal{M}}_g(d; \mu^+, -\mu^-)$. As usual, we will use labelled graphs to index these components. A fixed point is represented by a bordered stable map $f : (\Sigma, \partial\Sigma) \to (\mathbb{P}^1, S^1)$, where Σ has $l(\mu^+) + l(\mu^-)$ components which are isomorphic to D^2. For the graph of f, we associate a dotted half edge e to each of the disc components. Here by a half edge we mean an edge with only one vertex. We will refer to the vertices of the half edges as the *special vertices*, and denote the disc components by C_e. On these components, f is either given by

$$w \mapsto z = A w^{\mu_i^+}$$

or by

$$w \mapsto z' = \frac{1}{A} w^{\mu_i^-}.$$

In the first case, the special vertex v of e is labelled by a + sign, and mark the edge e by the degree μ_i^+; in the second case, the vertex v is labelled by a − sign, and mark the edge e by the degree μ_i^-.

Other components of Σ on which f has positive degrees d_j are copies of $\mathbb{P}^1$ on which f is given by

$$[w_1 : w_2] \mapsto [w_1^{d_j} : w_2^{d_j}].$$

We associated an edge e to each of these positive degree sphere components, and denote the corresponding component by C_e. We labelled the two vertices of each edge by the + and − sign respectively, and mark the edge by the degree d_j. The edges and the half edges are joined at their vertices in the following fashion. By removing the disc components and the positive degree components, what remains of Σ are possibly a disjoint union of stable marked curves which are contracted to the fixed points in $\mathbb{P}^1$ by f. The graph of f is obtained from Σ by contracting each of these stable curves to a point, and then replace the disc components by half edges, and the positive degree sphere components by edges. For each of the vertices, one puts the extra marking by the genus of the contracted stable curve.

To summarize, the fixed point component of $\overline{\mathcal{M}}_g(d; \mu^+, -\mu^-)$ are indexed by connected labelled graphs Γ of the following kind: It has $l(\mu^\pm)$ half edges incident at $l(\mu^\pm)$ special vertices which are marked by the $\pm$ signs, the half edges are marked by $\mu_i^\pm$. All other vertices are also marked by the $\pm$ signs, two vertices of the same edge has different markings, each vertex v is marked with an integer $g(v) \geq 0$, each edge e is marked by an integer $d_e > 0$. Denote by $E(\Gamma)$ the set of edges of Γ, $E^d(\Gamma)$ the set of dotted half edges of Γ, and $V(\Gamma)$ the set of vertices of Γ. The genus of graph is given by

$$(13) \qquad g(\Gamma) = 1 - |V(\Gamma)| + |E(\Gamma)| .$$

The graph satisfies the following conditions:

$$(14) \qquad \sum_{e \in E(\Gamma)} d_e = d ,$$

$$(15) \qquad \sum_{v \in V(\Gamma)} g(v) + g(\Gamma) = g .$$

It is straightforward to define the automorphisms of Γ. We denote by $\mathrm{Aut}(\Gamma)$ the group of automorphism of Γ.

For $v \in V(\Gamma)$, the valence $val(v)$ is the number of edges and dotted half edges incident at v. A stable vertex is a vertex with

$$(16) \qquad 2g(v) - 2 + \mathrm{val}(v) > 0 .$$

For a stable vertex v, there is a contracted component C_v in Σ, such that C_v together with the nodal points lies in the Deligne-Mumford moduli space $\overline{\mathcal{M}}_{g(v),\mathrm{val}(v)}$. If v is unstable, there are two possibilities: Either $(g(v), \mathrm{val}(v)) = (0, 1)$, or $(g(v), \mathrm{val}(v)) = (0, 2)$. In the first case, there is only one edge or dotted half edge e incident at v, we denote by C_v the corresponding point on C_e. In the second case, there are two edges or dotted edges e_1 and e_2 incident at v, we denote by C_v the corresponding nodal point. Putting all the pieces together, we get the following description of the bordered Riemann surface Σ as follows:

$$(17) \qquad \Sigma = \bigcup_{v \in V(\Gamma)} C_v \cup \bigcup_{e \in E(\Gamma)} C_e \cup \bigcup_{e \in E^d(\Gamma)} C_e .$$

It is then easy to see that the fixed point component of $\overline{\mathcal{M}}_g(d; \mu^+, -\mu^-)$ corresponding to the graph Γ is isomorphic to

$$(18) \qquad \overline{\mathcal{M}}_\Gamma := \prod_{v \in V(\Gamma)} \overline{\mathcal{M}}_{g(v),\mathrm{val}(v)} / A_\Gamma ,$$

where A_Γ fits in an exact sequence:

$$(19) \quad 0 \to \prod_{e \in E(\Gamma)} \mathbb{Z}/d_e \times \prod_{i=1}^{l(\mu^+)} \mathbb{Z}/\mu_i^+ \times \prod_{i=1}^{l(\mu^-)} \mathbb{Z}/\mu_i^- \to A_\Gamma \to \text{Aut}(\Gamma) \to 1 \,.$$

Here we understand that $\overline{\mathcal{M}}_{0,1}$ and $\overline{\mathcal{M}}_{0,2}$ as points.

2.6. *Feynman rule from localization*

Denote by N_Γ the virtual normal bundle of $\overline{\mathcal{M}}_\Gamma$ and by V_Γ the restriction of $V_g(d; \mu^+, -\mu^-)$ to $\overline{\mathcal{M}}_\Gamma$. For $v \in V(\Gamma)$, let $i(v) = 0$ if v is marked with a $+$ sign, let $i(v) = 1$ if v is marked with a $-$ sign. For $e \in E^d(\Gamma)$, let $i(e) = 0$ if its vertex is marked with a $+$ sign, let $i(e) = 1$ if its vertex is marked with a $-$ sign. If e is an edge or a dotted half edge incident at v, the pair (v, e) is referred to as a flag of Γ. Denote by $F(\Gamma)$ the set of flags in Γ, and by $F(v)$ the flags at v. For the contracted curve C_v, it has $\text{val}(v)$ marked points indexed by the flags (v, e), $e \in E(\Gamma)$. These marked points define the standard psi classes on $\overline{\mathcal{M}}_{g(v), \text{val}(v)}$. Denote them by $\psi_{(v,e)}$, $(v, e) \in F(\Gamma)$.

By a straightforward modification of the calculations in [8, 4, 7], one can get

$$e_{U(1)}(N_\Gamma) = \prod_{(v,e) \in F(\Gamma)} \left(\frac{(-1)^{i(v)} t}{d_e} - \psi_{v,e} \right) \cdot \prod_{e \in E(\Gamma)} \prod_{j=1}^{d_e} \left(\frac{jt}{d_e} \cdot \frac{-jt}{d_e} \right)$$

$$\cdot \prod_{e \in E^d(\Gamma)} \prod_{j=1}^{d_e} \frac{j(-1)^{i(e)} t}{d_e}$$

$$(20) \qquad \cdot \prod_{v \in V(\Gamma)} \frac{(-1)^{i(v)} t}{\Lambda_{g(v)}^\vee ((-1)^{i(v)} t) \cdot \prod_{(v,e) \in F(v)} (-1)^{i(v)} t} \,,$$

and

$$e_{U(1)}(V_\Gamma) = \prod_{(v,e) \in F(\Gamma)} ((-1)^{i(v)} (a-1) t \cdot (-1)^{i(v)-1} at)$$

$$\cdot \prod_{e \in E(\Gamma)} \prod_{j=1}^{d_e-1} \left(\left(\frac{j}{d_e} - a \right) (-1)^{i(v_1)} t \cdot \left(\frac{j}{d_e} - a \right) (-1)^{i(v_2)} t \right)$$

$$\cdot \prod_{e \in E^d(\Gamma)} \prod_{j=1}^{d_e-1} \left(\left(\frac{j}{d_e} - a \right) (-1)^{i(e)} t \right)$$

$$(21) \qquad \cdot \prod_{v \in V(\Gamma)} \frac{\Lambda_{g(v)}^\vee ((-1)^{i(v)} (a-1) t) \Lambda_{g(v)}^\vee ((-1)^{i(v)-1} at)}{(-1)^{i(v)} (a-1) t \cdot (-1)^{i(v)-1} at} \,,$$

where v_1 and v_2 denote the two vertices of v. These can be derived from the exact sequences related to the normalization of Σ. For localization calculations in open string theory, see also [5, 18]. It is useful to reformulate the above result as follows:

$$
\frac{e_{U(1)}V_\Gamma}{e_{U(1)}(N_\Gamma)} = \prod_{v \in V(\Gamma)} \Bigg([(-1)^{i(v)}t \cdot (a(1-a)t^2]^{\mathrm{val}(v)-1}
$$

$$
\frac{\Lambda^\vee_{g(v)}((-1)^{i(v)}t)\Lambda^\vee_{g(v)}((-1)^{i(v)}(a-1)t)\Lambda^\vee_{g(v)}((-1)^{i(v)-1}at)}{\prod_{(v,e) \in F(v)} \frac{(-1)^{i(v)}t}{d_e}(\frac{(-1)^{i(v)}t}{d_e} - \psi_{(v,e)})}
$$

$$
(22) \qquad \prod_{(v,e) \in F(v)} \frac{\prod_{j=1}^{d_e-1}(j - d_e a)}{d_e!} \Bigg)
$$

and so

$$
\int_{[\overline{\mathcal{M}}_g(d;\mu^+,-\mu^-)]^{vir}_{U(1)}} e_{U(1)}V_g(d;\mu^+,-\mu^-)
$$

$$
= \sum_{\Gamma \in G_g(d;\mu^+,-\mu^-)} \frac{1}{|A_\Gamma|} \prod_{v \in V(\Gamma)} \Bigg((a(1-a))^{\mathrm{val}(v)-1} \prod_{(v,e) \in F(v)} \frac{\prod_{j=1}^{d_e-1}(j - d_e a)}{d_e!}
$$

$$
(23) \qquad \int_{\overline{\mathcal{M}}_{g(v),\mathrm{val}(v)}} \frac{\Lambda^\vee_{g(v)}(1)\Lambda^\vee_{g(v)}(a-1)\Lambda^\vee_{g(v)}(-a)}{\prod_{(v,e) \in F(v)} \frac{1}{d_e}(\frac{1}{d_e} - \psi_{(v,e)})} \Bigg).
$$

By taking the generating series one then gets

$$
(24) \qquad F^a_{\mu^+,\mu^-} = \sum_{g \geq 0}\sum_{d > 0}\sum_{\Gamma \in G_g(d;\mu^+,-\mu^-)} \frac{1}{|A_\Gamma|}w_\Gamma,
$$

where

$$
(25) \qquad w_\Gamma = \prod_{v \in V(\Gamma)} w_v \cdot \prod_{e \in E(\Gamma) \cup E^d(\Gamma)} w_e
$$

is determined by the following Feynman rule:

$$
w_v = (a(1-a))^{\mathrm{val}(v)-1} \prod_{(v,e) \in F(v)} \frac{d_e \prod_{j=1}^{d_e-1}(j - d_e a)}{(d_e - 1)!}
$$

$$
(26) \qquad \sum_{g(v)=0}^{\infty} \lambda^{2g(v)+\mathrm{val}(v)-2} \int_{\overline{\mathcal{M}}_{g(v),\mathrm{val}(v)}} \frac{\Lambda^\vee_{g(v)}(1)\Lambda^\vee_{g(v)}(a-1)\Lambda^\vee_{g(v)}(-a)}{\prod_{(v,e) \in F(v)}(1 - d_e\psi_{(v,e)})},
$$

and

$$(27) \qquad w_e = \begin{cases} (t_+ t_-)^{d_e}, & e \in E(\Gamma) \,, \\ t_\pm^{d_e}, & e \in E^d(\Gamma), (-1)^{i(e)} = \pm 1 \,. \end{cases}$$

Note that

$$\int_{\overline{\mathcal{M}}_{0,l(\mu)}} \frac{\Lambda_0^\vee(1)\Lambda_0^\vee(-\tau-1)\Lambda_0^\vee(\tau)}{\prod_{i=1}^{l(\mu)}(1-\mu_i\psi_i)} = \int_{\overline{\mathcal{M}}_{0,l(\mu)}} \frac{1}{\prod_{i=1}^{l(\mu)}(1-\mu_i\psi_i)} = |\mu|^{l(\mu)-3}$$

for $l(\mu) \geq 3$, and we use this expression to extend the definition to the case $l(\mu) < 3$. In the above we have not treated the case when C_v is unstable. We leave it for the interested reader to check that with our convention here, the Feynman rule works for all cases.

3. Reformulation of the Feynman Rule by the Mariño-Vafa Formula

In this section we use the Mariño-Vafa formula to reformulate the Feynman rule obtained in the preceding section.

3.1. *Mariño-Vafa formula*

For a partition μ given by

$$\mu_1 \geq \mu_2 \geq \cdots \geq \mu_{l(\mu)} > 0 \,,$$

let $|\mu| = \sum_{i=1}^{l(\mu)} \mu_i$. Denote by χ_μ denotes the character of the irreducible representation of S_d indexed by μ, where $d = |\mu|$. Denote by $C(\nu)$ the conjugacy class of S_d corresponding to the partition ν, and by $\chi_\mu(C(\nu))$ the value of the character χ_μ on $C(\nu)$. The number κ_μ is defined by

$$\kappa_\mu = |\mu| + \sum_i (\mu_i^2 - 2i\mu_i) \,.$$

For each positive integer i,

$$m_i(\mu) = |\{j : \mu_j = i\}| \,.$$

Also define

$$z_\mu = \prod_j m_j(\mu)! j^{m_j(\mu)} \,.$$

Define

$$
C_{g,\mu}(\tau) = -\frac{\sqrt{-1}^{l(\mu)}}{\prod_j m_j(\mu)!} [\tau(\tau+1)]^{l(\mu)-1} \prod_{i=1}^{l(\mu)} \frac{\prod_{a=1}^{\mu_i-1}(\mu_i\tau+a)}{(\mu_i-1)!}
$$

$$
\cdot \int_{\overline{\mathcal{M}}_{g,l(\mu)}} \frac{\Lambda_g^\vee(1)\Lambda_g^\vee(-\tau-1)\Lambda_g^\vee(\tau)}{\prod_{i=1}^{l(\mu)}(1-\mu_i\psi_i)},
$$

$$
C_\mu(\lambda;\tau) = \sum_{g\geq 0} \lambda^{2g-2+l(\mu)} C_{g,\mu}(\tau).
$$

The Mariño-Vafa formula [17, 27, 14, 23] states:

$$
C_\mu(\lambda;\tau) = \sum_{n\geq 1} \frac{(-1)^{n-1}}{n} \sum_{\cup_{i=1}^n \mu^i=\mu} \prod_{i=1}^n \sum_{|\nu^i|=|\mu^i|} \frac{\chi_{\nu^i}(C(\mu^i))}{z_{\mu^i}} e^{\sqrt{-1}\tau\kappa_{\nu^i}\lambda/2} \mathcal{W}_{\nu^i}(q),
$$

(28)

where

(29)
$$
\mathcal{W}_\nu(q) = s_\nu(q^\rho),
$$

where $q^\rho = (q^{-1/2}, q^{-3/2}, \dots)$, $q = e^{\sqrt{-1}\lambda}$. It is clear that

(30)
$$
C_\mu(\lambda; -\tau-1) = (-1)^{|\mu|-l(\mu)} C_\mu(\lambda;\tau).
$$

Remark 3.1. In the above formula, $\mathcal{W}_v$ is related to the invariants of the unknot arising in Chern-Simons link theory (see e.g. [17]).

3.2. *Applications of the Marinño-Vafa formula*

Now we apply the Mariño-Vafa formula to rewrite the Feynman rule (26). The collection $\{d_e \mid (v,e) \in F(v)\}$ defines a partition which we denote by $\mu(v)$. Then we can take

(31)
$$
w_v = \frac{z_{\mu(v)}}{\sqrt{-1}^{l(\mu(v))}} C_{\mu(v)}(\lambda; -a).
$$

One then uses (28) to get the following expressions of w_v:

$$
w_v = \frac{z_{\mu(v)}}{\sqrt{-1}^{l(\mu(v))}} \sum_{n\geq 1} \frac{(-1)^{n-1}}{n}
$$

(32)
$$
\cdot \sum_{\cup_{i=1}^n \mu^i=\mu(v)} \prod_{i=1}^n \sum_{|\nu^i|=|\mu^i|} \frac{\chi_{\nu^i}(C(\mu^i))}{z_{\nu^i}} q^{-a\kappa_{\nu^i}/2} \mathcal{W}_{\nu^i}(q).
$$

4. Free Field Realization of Feynman Rules

In [28, 30] we have used a system of free bosons to realize the localization graphs and the Feynman rule arising from localization in the context of closed string theory. In this section we extend the idea to open string theory localizations considered above.

4.1. *Chemistry of two-colored labelled graphs*

Let $\{\beta_n\}_{n \in \mathbb{Z}}$ be a system of free bosons. We regard them as operators on the space Λ of symmetric functions. They satisfy the Heisenberg commutation relations:

$$(33) \qquad [\beta_m, \beta_n] = m\delta_{m,-n} \,.$$

For a partition μ, define

$$(34) \qquad \beta_\mu = \prod_{i=1}^{l(\mu)} \beta_{\mu_i}, \qquad \beta_{-\mu} = \prod_{i=1}^{l(\mu)} \beta_{-\mu_i} \,.$$

When μ is the empty partition $\emptyset$,

$$\beta_\emptyset = 1 \,.$$

For two collections of elements $\{w_{\pm\mu}\}$ in a suitable coefficient ring, indexed by partitions, define the sum of "atoms" by:

$$(35) \qquad Y_\pm(\beta) = \sum_{|\mu|>0} w_{\pm\mu} \frac{\beta_{\pm\mu}}{z_\mu} t_\pm^{|\mu|} \lambda^{l(\mu)-2}$$

and the generalized vertex operator:

$$(36) \qquad X_\pm = e^{Y_\pm(\beta)} \,.$$

Then we have the following identity for the vacuum expectation value

$$(37) \qquad \langle X_+(\beta) X_-(\beta) \rangle = \sum_{d\geq 0} \sum_{g\geq 0} \sum_{\Gamma \in G_g(d)^\bullet} \frac{1}{|A_\Gamma|} \prod_{v \in V(\Gamma)} w_v \prod_{e \in E(\Gamma)} w_e \,,$$

where for a vertex v of type μ with $i(v) = \pm$, we have

$$(38) \qquad w_v = w_{\pm\mu} \,,$$

and for an edge e, we have

$$(39) \qquad w_e = (t_+ t_-)^{d_e} \,.$$

This is a straightforward consequence of Wick's theorem. Now we consider

$$(40) \qquad \langle X_+(\beta) \cdot \lambda^{-l(\mu^+)} \frac{\beta_{-\mu^+}}{z_{\mu^+}} \cdot \lambda^{-l(\mu^-)} \frac{\beta_{\mu^-}}{z_{\mu^-}} \cdot X_-(\beta) \rangle .$$

To evaluate this, we first use the commutation relation (33) to move $\beta_{-\mu_i^+}$ to the left and move $\beta_{\mu_j^-}$ to the right, then apply the Wick theorem:

$$\langle X_+(\beta) \cdot \lambda^{-l(\mu^+)} \frac{\beta_{-\mu^+}}{z_{\mu^+}} \cdot \lambda^{-l(\mu^-)} \frac{\beta_{\mu^-}}{z_{\mu^-}} \cdot X_-(\beta) \rangle$$

$$(41) \qquad = \sum_{d \geq 0} \sum_{g \geq 0} \sum_{\Gamma \in G_g(d;\mu^+, -\mu^-)^\bullet} \frac{1}{|A_\Gamma|} \prod_{v \in V(\Gamma)} w_v \prod_{e \in E(\Gamma)} w_e .$$

Here the superscript $\bullet$ means we are taking sums over possibly disconnected graphs. We use

$$(42) \qquad \langle X_+(\beta) \cdot \lambda^{-l(\mu^+)} \frac{\beta_{-\mu^+}}{z_{\mu^+}} \cdot \lambda^{-l(\mu^-)} \frac{\beta_{\mu^-}}{z_{\mu^-}} \cdot X_-(\beta) \rangle^\circ$$

to denote the contribution to the sum over connected graphs. It is straightforward to see how to extract the connected contributions. For example,

$$\langle X_+(\beta) \cdot \lambda^{-1} \frac{\beta_{-m}}{m} \cdot X_-(\beta) \rangle = \langle X_+(\beta) \cdot \lambda^{-1} \frac{\beta_{-m}}{m} \cdot X_-(\beta) \rangle^\circ \cdot \langle X_+(\beta) X_-(\beta) \rangle ,$$

(43)

and

$$\langle X_+(\beta) \cdot \lambda^{-2} \frac{\beta_{-m} \beta_{-n}}{mn(1 + \delta_{mn})} \cdot X_-(\beta) \rangle$$

$$= \langle X_+(\beta) \cdot \lambda^{-2} \frac{\beta_{-m} \beta_{-n}}{mn(1 + \delta_{mn})} \cdot X_-(\beta) \rangle^\circ \cdot \langle X_+(\beta) X_-(\beta) \rangle$$

$$+ \langle X_+(\beta) \cdot \lambda^{-1} \frac{\beta_{-m}}{m} \cdot X_-(\beta) \rangle^\circ \langle X_+(\beta) \cdot \lambda^{-1} \frac{\beta_{-n}}{n} \cdot X_-(\beta) \rangle^\circ$$

$$(44) \qquad \cdot \langle X_+(\beta) X_-(\beta) \rangle ,$$

etc. In general, define generating series

$$F(\lambda, t_+, t_-, x^+, x^-) = \sum_{(\mu^+, \mu^-) \neq ((0),(0))} \langle X_+(\beta) \cdot \lambda^{-l(\mu^+)} \beta_{-\mu^+} \cdot \lambda^{-l(\mu^-)} \beta_{\mu^-} X_-(\beta) \rangle^\circ$$

$$\cdot p_{\mu^+}(x^+) p_{\mu^-}(x^-) ,$$

$$Z(\lambda, t_+, t_-; x^+, x^-) = \sum_{\mu^+, \mu^-} \langle X_+(\beta) \cdot \lambda^{-l(\mu^+)} \beta_{-\mu^+} \cdot \lambda^{-l(\mu^-)} \beta_{\mu^-} X_-(\beta) \rangle$$

$$\cdot p_{\mu^+}(x^+) p_{\mu^-}(x^-) .$$

When $x^\pm = 0 = (0, \ldots, 0, \ldots)$,

$$Z(\lambda, t_+, t_-; 0, 0) = \langle X_+(\beta) X_-(\beta) \rangle .$$

It is not hard to see that we have

$$(45) \qquad F(\lambda, t_+, t_-; x^+, x^-) = \log \frac{Z(\lambda, t_+, t_-; x^+, x^-)}{Z(\lambda, t_+, t_-; 0, 0)} .$$

4.2. *An extension of the AMVI Feynman rule*

For closed string partition function of the resolved conifold, Aganagic-Mariño-Vafa [2] have derived the following Feynman rule from duality with Chern-Simons theory (see also Iqbal [6]):

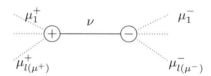

$$(46) \qquad Z = Z_{\emptyset, \emptyset} = \sum_\nu \mathcal{W}_\nu(q) \mathcal{W}_{\nu^t}(q) (-t_+ t_-)^{|\nu|} .$$

We will refer to this as the AMVI rule. This has been proved using localization calculations in [28]. Now we are concerned with diagrams of the following form:

We will find an extension of (46) for $F^a_{\mu^+, \mu^-}$.

By (32) we have

$$Y^a_\pm(\beta) = \sum_{|\mu|>0} \frac{\beta_{\pm\mu}}{z_\mu} t_\pm^{|\mu|} \cdot \frac{z_\mu}{\sqrt{-1}^{l(\mu)}} \sum_{n \geq 1} \frac{(-1)^{n-1}}{n}$$

$$\cdot \sum_{\cup_{i=1}^n \mu^i = \mu} \prod_{i=1}^n \sum_{|\nu^i|=|\mu^i|} \frac{\chi_{\nu^i}(C(\mu^i))}{z_{\mu^i}} e^{-\sqrt{-1} a \kappa_{\nu^i} \lambda/2} \mathcal{W}_{\nu^i}(q)$$

$$= \log \sum_{|\mu| \geq 0} \beta_{\pm\mu} t_\pm^{|\mu|} \cdot \frac{1}{\sqrt{-1}^{l(\mu)}} \sum_{|\nu|=|\mu|} \frac{\chi_\nu(C(\mu))}{z_\mu} e^{-\sqrt{-1} a \kappa_\nu \lambda/2} \mathcal{W}_\nu(q) ,$$

and so

$$X^a_\pm(\beta) = e^{Y^a_\pm(\beta)} = \sum_{|\mu| \geq 0} \beta_{\pm\mu} t^{|\mu|}_\pm \cdot \frac{1}{\sqrt{-1}^{l(\mu)}} \sum_{|\nu|=|\mu|} \frac{\chi_\nu(C(\mu))}{z_\mu} q^{-a\kappa_\nu/2} \mathcal{W}_\nu(q) \,.$$

(47)

For later use, we will make some changes of X_+ as follows. We have shown that

$$(48) \qquad \mathcal{W}_{\nu^t}(q) = q^{-\kappa_\nu/2} \mathcal{W}_\nu(q) = (-1)^{|\nu|} \mathcal{W}_\nu(q^{-1}) \,,$$

and it is well-known that

$$(49) \qquad \kappa_{\nu^t} = -\kappa_\nu, \qquad \chi_{\nu^t}(C(\mu)) = (-1)^{|\mu|-l(\mu)} \chi_\nu(C(\mu)) \,.$$

Now we change ν in the expression X_+ to ν^t and use these identities to get

$$(50) \qquad X^a_+(\beta) = \sum_\mu \beta_\mu t^{|\mu|}_+ \cdot \sqrt{-1}^{l(\mu)} \sum_{|\nu|=|\mu|} \frac{\chi_\nu(C(\mu))}{z_\mu} q^{a\kappa_\nu/2} \mathcal{W}_\nu(q^{-1}) \,.$$

Hence we have

$$Z^a_{\mu^+,\mu^-}(\lambda, t_+, t_-)$$

$$= \langle X^a_+(\beta) \cdot \lambda^{-l(\mu^+)-l(\mu^-)} \frac{\beta_{-\mu^+}}{z_{\mu^+}} \frac{\beta_{\mu^-}}{z_{\mu^-}} \cdot X^a_-(\beta) \rangle$$

$$= \Bigg\langle \sum_{\eta^+} \beta_{\eta^+} t^{|\eta^+|}_+ \cdot \sqrt{-1}^{l(\eta^+)} \sum_{|\nu^+|=|\eta^+|} \frac{\chi_{\nu^+}(C(\eta^+))}{z_{\eta^+}} q^{a\kappa_{\nu^+}/2} s_{\nu^+}(q^{-\rho})$$

$$\cdot \lambda^{-l(\mu^+)-l(\mu^-)} \frac{\beta_{-\mu^+}}{z_{\mu^+}} \frac{\beta_{\mu^-}}{z_{\mu^-}}$$

$$\cdot \sum_{\eta^-} \beta_{-\eta^-} t^{|\eta^-|}_- \cdot \frac{1}{\sqrt{-1}^{l(\eta^-)}} \sum_{|\nu^-|=|\mu^-|} \frac{\chi_\nu(C(\eta^-))}{z_{\eta^-}} q^{-a\kappa_{\nu^-}/2} s_{\nu^-}(q^\rho) \Bigg\rangle \,.$$

Note only terms with $\eta^+ \cup \mu^- = \eta^- \cup \mu^+$ contribute. When this is the case we have

$$(51) \qquad |\eta^+| + |\mu^-| = |\eta^-| + |\mu^+|, \qquad l(\eta^+) + l(\mu^-) = l(\eta^-) + l(\mu^+) \,.$$

Therefore we have

$$Z^a_{\mu^+,\mu^-}(\lambda, t_+, t_-)$$

$$= \sqrt{-1}^{-l(\mu^+)-l(\mu^-)} \cdot \left\langle \sum_{\eta^+} t_+^{|\eta^+|}\beta_{\eta^+} \cdot \sum_{|\nu^+|=|\eta^+|} \frac{\chi_{\nu^+}(C(\eta^+))}{z_{\eta^+}} q^{a\kappa_{\nu^+}/2} s_{\nu^+}(q^{-\rho}) \right.$$

$$\cdot \lambda^{-l(\mu^+)-l(\mu^-)} \frac{\beta_{-\mu^+}}{z_{\mu^+}} \frac{\beta_{\mu^-}}{z_{\mu^-}} \cdot \sum_{\eta^-} t_-^{|\eta^-|}\beta_{-\eta^-}$$

$$\cdot \sum_{|\nu^-|=|\eta^-|} \frac{\chi_{\nu^-}(C(\eta^-))}{z_{\eta^-}} q^{-a\kappa_{\nu^-}/2} s_{\nu^-}(q^\rho) \right\rangle$$

$$= \sqrt{-1}^{-l(\mu^+)-l(\mu^-)} \lambda^{-l(\mu^+)-l(\mu^-)} \cdot \left\langle \sum_{\nu^+} s_{\nu^+}(t_+\beta_+) q^{a\kappa_{\nu^+}/2} s_{\nu^+}(q^{-\rho}) \right.$$

$$\cdot \frac{\beta_{-\mu^+}}{z_{\mu^+}} \frac{\beta_{\mu^-}}{z_{\mu^-}} \cdot \sum_{\nu^-} s_{\nu^-}(t_-\beta_-) q^{-a\kappa_{\nu^-}/2} s_{\nu^-}(q^\rho) \right\rangle,$$

where $s_{\nu^\pm}(\beta_\pm)$ are Schur functions $s_{\mu^\pm}$ for which the Newton polynomials are given by

$$p_{\nu^\pm} = \beta_{\pm\nu^\pm}.$$

here we have used Frobenius character formula:

$$(52) \qquad s_\nu = \sum_\mu \frac{\chi_\nu(C(\mu))}{z_\mu} p_\mu.$$

Introduce the following modified generating series:

$$\tilde{Z}^a(\lambda, t_+, t_-; x^+, x^-) = \sum_{\mu^+,\mu^-} Z^a_{\mu^+,\mu^-}(\lambda, t_+, t_-)$$

$$(53) \qquad \cdot \sqrt{-1}^{-l(\mu^-)-l(\mu^+)} \lambda^{l(\mu^+)+l(\mu^-)} p_{\mu^+}(t_-x^+) p_{\mu^-}(t_+x^-).$$

One can obtain from $\tilde{Z}^a(\lambda, t_+, t_-; x^+, x^-)$ by changing $p_n(x^\pm)$ to $t_\mp^{-n}\lambda^{-1}\sqrt{-1}^{\pm1} p_n(x^\pm)$.

The following is our main result:

Theorem 4.1. *We have*

$$\tilde{Z}^a(\lambda, t_+, t_-; x^+, x^-) = \sum_{\nu^\pm} t_+^{|\nu^+|} q^{a\kappa_{\nu^+}/2} s_{\nu^+}(q^{-\rho}) t_-^{|\nu^-|} q^{-a\kappa_{\nu^-}/2} s_{\nu^-}(q^\rho)$$

$$(54) \qquad \cdot \sum_\eta s_{\nu^-/\eta}(t_+x_-) s_{\nu^+/\eta}(t_-x^+).$$

Proof. Recall the following identity for Schur functions and Newton functions:

$$
(55) \quad \exp \sum_{n=1}^{\infty} \frac{1}{n} p_n(x) p_n(y) = \sum_{\mu} \frac{1}{z_\mu} p_\mu(x) p_\mu(y) = \sum_{\mu} s_\mu(x) s_\mu(y) \,.
$$

It is also well-known that from (33) one can get

$$
\exp \left(\sum_{n=1}^{\infty} \frac{a_n}{n} \beta_n \right) \exp \left(\sum_{n=1}^{\infty} \frac{a_{-n}}{n} \beta_{-n} \right)
$$

$$
(56) \qquad = \exp \left(\sum_{n=1}^{\infty} \frac{a_n a_{-n}}{n} \right) \cdot \exp \left(\sum_{n=1}^{\infty} \frac{a_{-n}}{n} \beta_{-n} \right) \exp \left(\sum_{n=1}^{\infty} \frac{a_n}{n} \beta_n \right) \,.
$$

We use these identities to carry out the following calculations:

$$
\tilde{Z}^a(\lambda, t_+, t_-; x^+, x^-)
$$

$$
= \left\langle \sum_{\nu^+} s_{\nu^+}(t_+ \beta_+) q^{a\kappa_{\nu^+}/2} s_{\nu^+}(q^{-\rho}) \cdot \sum_{\mu^+} \frac{1}{z_{\mu^+}} \beta_{-\mu^+} p_{\mu^+}(t_- x^+) \right.
$$

$$
\left. \cdot \sum_{\mu^-} \frac{1}{z_{\mu^-}} \beta_{\mu^-} p_{\mu^-}(t_+ x^-) \cdot \sum_{\nu^-} s_{\nu^-}(t_- \beta_-) q^{-a\kappa_{\nu^-}/2} s_{\nu^-}(q^\rho) \right\rangle
$$

$$
= \left\langle \sum_{\nu^+} s_{\nu^+}(t_+ \beta_+) q^{a\kappa_{\nu^+}/2} s_{\nu^+}(q^{-\rho}) \cdot \exp \left(\sum_{n=1}^{\infty} \frac{t_-^n}{n} \beta_{-n} p_n(x^+) \right) \right.
$$

$$
\left. \cdot \exp \left(\sum_{n=1}^{\infty} \frac{t_+^n}{n} \beta_n p_n(x^-) \right) \cdot \sum_{\nu^-} s_{\nu^-}(t_- \beta_-) q^{-a\kappa_{\nu^-}/2} s_{\nu^-}(q^\rho) \right\rangle
$$

$$
= \exp \left(\sum_{n=1}^{\infty} -\frac{(t_+ t_-)^n}{n} p_n(x^+) p_n(x^-) \right)
$$

$$
\cdot \left\langle \sum_{\nu^+} s_{\nu^+}(t_+ \beta_+) q^{a\kappa_{\nu^+}/2} s_{\nu^+}(q^{-\rho}) \cdot \exp \left(\sum_{n=1}^{\infty} \frac{t_+^n}{n} \beta_n p_n(x^-) \right) \right.
$$

$$
\left. \cdot \exp \left(\sum_{n=1}^{\infty} \frac{t_-^n}{n} \beta_{-n} p_n(x^+) \right) \cdot \sum_{\nu^-} s_{\nu^-}(t_- \beta_-) q^{-a\kappa_{\nu^-}/2} s_{\nu^-}(q^\rho) \right\rangle
$$

$$= \exp\left(\sum_{n=1}^{\infty} -\frac{(t_+ t_-)^n}{n} p_n(x^+) p_n(x^-)\right)$$

$$\cdot \left\langle \sum_{\nu^+} s_{\nu^+}(t_+ \beta_+) q^{a\kappa_{\nu^+}/2} s_{\nu^+}(q^{-\rho}) \cdot \sum_{\mu^+} s_{\mu^+}(t_+ \beta_+) s_{\mu^+}(x^-) \right.$$

$$\left. \cdot \sum_{\mu^-} s_{\mu^-}(t_- \beta_-) s_{\mu^-}(x^+) \cdot \sum_{\nu^-} s_{\nu^-}(t_- \beta_-) q^{-a\kappa_{\nu^-}/2} s_{\nu^-}(q^\rho) \right\rangle .$$

Now we recall the Littlewood-Richardson coefficients are defined by:

$$(57) \qquad s_\mu(x) s_\nu(x) = \sum_\eta c_{\mu\nu}^\eta s_\eta(s) .$$

Using these coefficients one can define the skew Schur functions:

$$(58) \qquad s_{\eta/\nu}(x) = \sum_\mu c_{\mu\nu}^\eta s_\mu(x) .$$

The skew Schur functions satisfy the following summation formula:

$$(59) \sum_\eta s_{\eta/\mu}(x) s_{\eta/\nu}(y) = \exp(\sum_{n=1}^{\infty} \frac{1}{n} p_n(x) p_n(y)) \cdot \sum_\eta s_{\nu/\eta}(x) s_{\mu/\eta}(y) .$$

Now we continue our calculations of $Z(x^+, x^-)$ using these identities:

$$\tilde{Z}^a(\lambda, t_+, t_-; x^+, x^-)$$

$$= \exp\left(\sum_{n=1}^{\infty} -\frac{(t_+ t_-)^n}{n} p_n(x^+) p_n(x^-)\right)$$

$$\cdot \left\langle \sum_{\mu^+, \nu^+, \eta^+} c_{\mu^+ \nu^+}^{\eta^+} s_{\eta^+}(t_+ \beta_+) q^{a\kappa_{\nu^+}/2} s_{\nu^+}(q^{-\rho}) \cdot s_{\mu^+}(x^-) \right.$$

$$\left. \cdot \sum_{\mu^-, \nu^-, \eta^-} c_{\mu^- \nu^-}^{\eta^-} s_{\eta^-}(t_- \beta_-) s_{\mu^-}(x^+) q^{-a\kappa_{\nu^-}/2} s_{\nu^-}(q^\rho) \right\rangle$$

$$= \exp\left(\sum_{n=1}^{\infty} -\frac{(t_+ t_-)^n}{n} p_n(x^+) p_n(x^-)\right)$$

$$\cdot \sum_{\mu^\pm, \nu^\pm, \eta} c_{\mu^+ \nu^+}^{\eta} t_+^{|\mu^+| + |\nu^+|} q^{a\kappa_{\nu^+}/2} s_{\nu^+}(q^{-\rho}) \cdot s_{\mu^+}(x^-)$$

$$\cdot c_{\mu^- \nu^-}^{\eta} t_-^{|\mu^-| + |\nu^-|} s_{\mu^-}(x^+) q^{-a\kappa_{\nu^-}/2} s_{\nu^-}(q^\rho)$$

$$= \exp \left(\sum_{n=1}^{\infty} -\frac{(t_+ t_-)^n}{n} p_n(x^+) p_n(x^-) \right)$$

$$\cdot \sum_{\nu^{\pm}} t_+^{|\nu^+|} q^{a\kappa_{\nu^+}/2} s_{\nu^+}(q^{-\rho}) t_-^{|\nu^-|} q^{-a\kappa_{\nu^-}/2} s_{\nu^-}(q^\rho)$$

$$\cdot \sum_{\eta} s_{\eta/\nu^+}(t_+ x_-) s_{\eta/\nu^-}(t_- x^+)$$

$$= \sum_{\nu^{\pm}} t_+^{|\nu^+|} q^{a\kappa_{\nu^+}/2} s_{\nu^+}(q^{-\rho}) t_-^{|\nu^-|} q^{-a\kappa_{\nu^-}/2} s_{\nu^-}(q^\rho)$$

$$\cdot \sum_{\eta} s_{\nu^-/\eta}(t_+ x_-) s_{\nu^+/\eta}(t_- x^+).$$

The proof of (54) is completed. □

From the proof one can also get

Corollary 4.1.

$$\tilde{Z}^0(\lambda, t_+, t_-; x^+, x^-)$$

$$= \exp \sum_{n=1}^{\infty} \frac{(t_+ t_-)^n}{n} (p_n(q^{-\rho}) p_n(x^+) + p_n(q^\rho) p_n(x^-) + p_n(q^\rho) p_n(q^{-\rho})).$$

(60)

Note by (60) one can get

$$F^0(\lambda, t_+, t_-; x^+, x^-) = \ln \frac{Z^0(\lambda, t_+, t_-; x^+, x^-)}{Z^0(\lambda, t_+, t_-; 0, 0)}$$

$$= \sum_{n=1}^{\infty} \frac{(t_+ t_-)^n}{n} (p_n(q^{-\rho}) \sqrt{-1} \lambda^{-1} t_-^{-n} p_n(x^+)$$

$$- p_n(q^\rho) \sqrt{-1} \lambda^{-1} t_+^{-n} p_n(x^-))$$

$$= \sum_{n=1}^{\infty} \left(\frac{t_+^n}{2n \sin(n\lambda/2)} p_n(x^+) + \frac{t_-^n}{2n \sin(n\lambda/2)} p_n(x^-) \right).$$

This satisfies the integrality predicted by Ooguri and Vafa [24]. We expect that the framed free energy $\ln Z^a(x^+, x^-)$ satisfy the framed version of integrality predicted by Labastida, Mariño and Vafa [10, 17]. Note also F^0 has the following very strong vanishing property:

(61) $$F^0_{\mu^+, \mu^-}(\lambda, t_+, t_-) = 0$$

unless $l(\mu^+) + l(\mu^-) < 2$; furthermore, $F^0_{(n),(0)}(\lambda, t_+, t_-)$ receives contributions only from $\overline{\mathcal{M}}_g(0; (n), -(0))$. This is very similar to the situation noted in [24, 7, 13].

5. Other Operator Manipulations

In this section we present another way to evaluate Z_{μ^+, μ^-} following the operator manipulations as in [21, 22].

5.1. *Some operators and commutations relations*

We first recall some operators and commutation relations from *loc. cit.* Define

$$(62) \qquad \mathcal{E}_r(z) = \sum_{k \in \mathbb{Z} + \frac{1}{2}} e^{z(k - \frac{r}{2})} E_{k-r,k} + \frac{\delta_{r,0}}{\varsigma(z)},$$

where the function $\varsigma(z)$ is defined by

$$(63) \qquad \varsigma(z) = e^{z/2} - e^{-z/2}.$$

The operators $\mathcal{E}_r$ satisfy

$$(64) \qquad \mathcal{E}_r(z)^* = \mathcal{E}_{-r}(z)$$

and

$$(65) \qquad [\mathcal{E}_a(z), \mathcal{E}_b(w)] = \varsigma(aw - bz)\mathcal{E}_{a+b}(z + w).$$

These operators act on the Fock space of two fermions, and by the boson-fermion correspondence [19], they also act on the space Λ of symmetric functions.

For $k \neq 0$,

$$(66) \qquad \beta_k = \mathcal{E}_k(0).$$

The commutation relation (65) specializes to

$$(67) \qquad [\beta_k, \mathcal{E}_l(z)] = \varsigma(kz)\mathcal{E}_{k+l}(z).$$

We will also need an operator $\mathcal{F}_2$ defined by

$$(68) \qquad \mathcal{F}_2 = \frac{1}{2} \sum_{k \in \mathbb{Z} + \frac{1}{2}} k^2 E_{kk}.$$

Its action on the Schur functions are given by:

$$(69) \quad \mathcal{F}_2 s_\mu(x) = \frac{1}{2} \sum_{i=1}^{l(\mu)} \left[(\mu_i - i + \frac{1}{2})^2 - (-i + \frac{1}{2})^2 \right] \cdot s_\mu(x) = \frac{1}{2} \kappa_\mu s_\mu(x).$$

By [22, (2.14)],

$$(70) \qquad\qquad e^{u\mathcal{F}_2} \beta_{-m} e^{-u\mathcal{F}_2} = \mathcal{E}_{-m}(um).$$

5.2. *Some commutation formulas*

Starting from (67), one can easily prove by induction:

$$(71) \qquad\qquad \beta_k^m \mathcal{E}_{-n}(z) = \sum_{j=0}^{m} \binom{m}{j} \varsigma(kz)^j \mathcal{E}_{-n+jk}(z) \beta_k^{m-j},$$

$$(72) \qquad\qquad \mathcal{E}_{-n}(z) \beta_{-k}^m = \sum_{j=0}^{m} \binom{m}{j} \varsigma(kz)^j \beta_{-k}^{m-j} \mathcal{E}_{-n-jk}(z).$$

It then follows that

$$\exp\left(\frac{t_+^k a_k}{k} \beta_k \right) \cdot \mathcal{E}_{-n}(z) = \sum_{j=0}^{\infty} \frac{(t_+^k a_k)^j}{j! k^j} \varsigma(kz)^j \mathcal{E}_{-n+kj}(z) \cdot \exp\left(\frac{t_+^k a_k}{k} \beta_k \right),$$

$$(73)$$

$$\mathcal{E}_{-n}(z) \cdot \exp\left(\frac{t_-^k a_{-k}}{k} \beta_{-k} \right) = \exp\left(\frac{t_-^k a_{-k}}{k} \beta_{-k} \right)$$

$$(74) \qquad\qquad \cdot \sum_{j=0}^{\infty} \frac{(t_-^k a_{-k})^j}{j!} \varsigma(kz)^j \mathcal{E}_{-n-kj}(z).$$

Repeating these formulas for all k,

$$\exp\left(\sum_{k=1}^{\infty} \frac{t_+^k a_k}{k} \beta_k \right) \cdot \mathcal{E}_{-n}(z) = \sum_{\mu^+} \frac{\tilde{a}_{\mu^+}(z)}{z_{\mu^+}} \mathcal{E}_{-n+|\mu^+|}(z) \cdot \exp\left(\sum_{k=1}^{\infty} \frac{t_+^k a_k}{k} \beta_k \right),$$

$$(75)$$

$$\mathcal{E}_{-n}(z) \cdot \exp\left(\sum_{k=1}^{\infty} \frac{a_{-k}}{k} \beta_{-k} \right) = \exp\left(\sum_{k=1}^{\infty} \frac{t_-^k a_{-k}}{k} \beta_{-k} \right)$$

$$(76) \qquad\qquad \cdot \sum_{\mu^-} \frac{\tilde{a}_{-\mu^-}(z)}{z_{\mu^-}} \mathcal{E}_{-n-|\mu^-|}(z),$$

where

$$\text{(77)} \qquad \tilde{a}_{\pm\mu^\pm} = \prod_k (t_\pm^k a_{\pm k}\varsigma(kz))^{m_k(\mu^\pm)} \,.$$

Using these identities, one can reduce the computation of

$$\text{(78)} \qquad \left\langle \exp\left(\sum_{k=1}^\infty \frac{t_+^k a_k}{k} \beta_k \right) \cdot \mathcal{E}_{a_1}(z_1) \cdots \mathcal{E}_{a_n}(z_n) \cdot \exp\left(\sum_{k=1}^\infty \frac{t_-^k a_{-k}}{k} \beta_{-k} \right) \right\rangle$$

to the computation of

$$\text{(79)} \qquad G^\bullet \begin{pmatrix} a_1 & \cdots & a_n \\ z_1 & \cdots & z_n \end{pmatrix} = \langle \mathcal{E}_{a_1}(z_1) \cdots \mathcal{E}_{a_n}(z_n) \rangle.$$

The procedure to compute the latter has been described in [21]: When $a_1 < 0$,

$$\text{(80)} \qquad G^\bullet \begin{pmatrix} a_1 & \cdots & a_n \\ z_1 & \cdots & z_n \end{pmatrix} = 0 \,;$$

when $a_1 = 0$,

$$\text{(81)} \qquad G^\bullet \begin{pmatrix} a_1 & \cdots & a_n \\ z_1 & \cdots & z_n \end{pmatrix} = \frac{1}{\varsigma(z_1)} G^\bullet \begin{pmatrix} a_2 & \cdots & a_n \\ z_2 & \cdots & z_n \end{pmatrix} \,;$$

when $a_1 > 0$,

$$\text{(82)} \ G^\bullet \begin{pmatrix} a_1 & \cdots & a_n \\ z_1 & \cdots & z_n \end{pmatrix} = \sum_{i=2}^n \varsigma(a_1 z_i - a_i z_1) G^\bullet \begin{pmatrix} a_2 & \cdots & a_i + a_1 & \cdots & a_n \\ z_2 & \cdots & z_i + z_1 & \cdots & z_n \end{pmatrix} \,.$$

For example,

$$\text{(83)} \qquad G^\bullet \begin{pmatrix} 0 \\ z \end{pmatrix} = \frac{1}{\varsigma(z)},$$

$$\text{(84)} \qquad G^\bullet \begin{pmatrix} a & -a \\ z_1 & z_2 \end{pmatrix} = \frac{\varsigma(a(z_1 + z_2))}{\varsigma(z_1 + z_2)}, \qquad a > 0 \,,$$

$$G^\bullet \begin{pmatrix} a & b & -a-b \\ z_1 & z_2 & z_3 \end{pmatrix} = \varsigma((a+b)z_2 + bz_3) \frac{\varsigma(a(z_1 + z_2 + z_3))}{\varsigma(z_1 + z_2 + z_3)}, \qquad a, b > 0 \,.$$

$$\text{(85)}$$

As an example, let us compute the $n = 1$ case of (78).

$$\left\langle \exp\left(\sum_{k=1}^{\infty} \frac{t_+^k a_k}{k} \beta_k\right) \cdot \mathcal{E}_{-n}(z) \cdot \exp\left(\sum_{k=1}^{\infty} \frac{t_-^k a_{-k}}{k} \beta_{-k}\right) \right\rangle$$

$$= \left\langle \sum_{\mu^+} \frac{\tilde{a}_{\mu^+}(z)}{z_{\mu^+}} \mathcal{E}_{-n+|\mu^+|}(z) \cdot \exp\left(\sum_{k=1}^{\infty} \frac{t_+^k a_k}{k} \beta_k\right) \cdot \exp\left(\sum_{k=1}^{\infty} \frac{t_-^k a_{-k}}{k} \beta_{-k}\right) \right\rangle$$

$$= \exp\left(\sum_{k=1}^{\infty} \frac{a_k a_{-k}}{k} (t_+ t_-)^k\right)$$

$$\cdot \left\langle \sum_{\mu^+} \frac{\tilde{a}_{\mu^+}(z)}{z_{\mu^+}} \mathcal{E}_{-n+|\mu^+|}(z) \cdot \exp\left(\sum_{k=1}^{\infty} \frac{t_-^k a_{-k}}{k} \beta_{-k}\right) \right\rangle$$

$$= \exp\left(\sum_{k=1}^{\infty} \frac{a_k a_{-k}}{k} (t_+ t_-)^k\right) \sum_{\mu^+} \frac{\tilde{a}_{\mu^+}(z)}{z_{\mu^+}} \sum_{\mu^-} \frac{\tilde{a}_{-\mu^-}(z)}{z_{\mu^-}} \langle \mathcal{E}_{-n+|\mu|^+ - |\mu^-|}(z)\rangle$$

$$= \frac{1}{\varsigma(z)} \exp\left(\sum_{k=1}^{\infty} \frac{a_k a_{-k}}{k} (t_+ t_-)^k\right) \sum_{|\mu^+|=|\mu^-|+n} \frac{\tilde{a}_{\mu^+}(z)}{z_{\mu^+}} \frac{\tilde{a}_{-\mu^-}(z)}{z_{\mu^-}} \cdot$$

Recall the complete symmetric functions h_n in $x_1, \ldots, x_k, \ldots$ are defined by

$$(86) \qquad \sum_{n=0}^{\infty} h_n(x) t^n = \prod_i (1 - t x_i)^{-1}.$$

By a standard series manipulation we get

$$\sum_{n=0}^{\infty} h_n(x) t^n = \exp\sum_{n=1}^{\infty} \frac{t^n}{n} p_n(x) = \sum_{\mu} \frac{t^{|\mu|}}{z_\mu} p_\mu(x).$$

Hence by comparing the coefficients of t^n we get

$$(87) \qquad \sum_{|\mu|=n} \frac{p_\mu(x)}{z_\mu} = h_n(x).$$

This formula express h_n as a polynomial of p_n's. Hence we have

$$\left\langle \exp\left(\sum_{k=1}^{\infty} \frac{t_+^k a_k}{k} \beta_k\right) \cdot \mathcal{E}_{-n}(z) \cdot \exp\left(\sum_{k=1}^{\infty} \frac{t_-^k a_{-k}}{k} \beta_{-k}\right) \right\rangle$$

$$= \exp\left(\sum_{k=1}^{\infty} \frac{a_k a_{-k}}{k} (t_+ t_-)^k\right) \sum_{k=0}^{\infty} \tilde{h}_{n+k}(z) \cdot \tilde{h}_{-k}(z) \frac{1}{\varsigma(z)},$$

where $\tilde{h}_{\pm k}$ is the specialization of h_k with

$$p_n = a_{\pm n} t_{\pm}^n \varsigma(nz).$$

5.3. Applications to Z_{μ^+,μ^-}^a

In the above we have shown that

$$Z_{\mu^+,\mu^-}^a = \sqrt{-1}^{l(\mu^+)-l(\mu^-)} \lambda^{-l(\mu^+)-l(\mu^-)}$$

$$\cdot \left\langle \sum_{\nu^+} s_{\nu^+}(t_+\beta_+) q^{a\kappa_{\nu^+}/2} s_{\nu^+}(q^{-\rho}) \cdot \frac{\beta_{-\mu^+}}{z_{\mu^+}} \frac{\beta_{\mu^-}}{z_{\mu^-}} \right.$$

$$\left. \cdot \sum_{\nu^-} s_{\nu^-}(t_-\beta_-) q^{-a\kappa_{\nu^-}/2} s_{\nu^-}(q^\rho) \right\rangle.$$

We rewrite the vev as

$$\left\langle \exp\left(\sum_{k=1}^\infty \frac{p_k(t_+ q^{-\rho})}{k}\beta_k\right) q^{a\mathcal{F}_2} \cdot \frac{\beta_{-\mu^+}}{z_{\mu^+}} \frac{\beta_{\mu^-}}{z_{\mu^-}} \cdot q^{-a\mathcal{F}_2} \exp\left(\sum_{k=1}^\infty \frac{p_k(t_- q^\rho)}{k}\beta_{-k}\right) \right\rangle$$

(88)

and for simplicity of notations, write it as

(89) $\langle\langle \mu^+; -\mu^- \rangle\rangle.$

Now we have

$$q^{a\mathcal{F}_2} \cdot \beta_{-\mu^+} \beta_{\mu^-} \cdot q^{-a\mathcal{F}_2} = \prod_{i=1}^{l(\mu^+)} (q^{a\mathcal{F}_2}\beta_{-\mu_i^+} q^{-a\mathcal{F}_2}) \cdot \prod_{i=1}^{l(\mu^-)} (q^{a\mathcal{F}_2}\beta_{\mu_i^-} q^{-a\mathcal{F}_2})$$

$$= \prod_{i=1}^{l(\mu^+)} \mathcal{E}_{-\mu_i^+}(a\mu_i^+) \cdot \prod_{i=1}^{l(\mu^-)} \mathcal{E}_{\mu_i^-}(-a\mu_i^-).$$

Hence the computation of Z_{μ^+,μ^-}^a can be reduced to (78).

For example, let us compute $\langle\langle (n); \emptyset \rangle\rangle$:

$$\langle\langle (n); \emptyset \rangle\rangle$$

$$= \frac{1}{n} \left\langle \exp\left(\sum_{k=1}^\infty \frac{p_k(t_+ q^{-\rho})}{k}\beta_k\right) \cdot \mathcal{E}_{-n}(nai\lambda) \cdot \exp\left(\sum_{k=1}^\infty \frac{p_k(t_- q^\rho)}{k}\beta_{-k}\right) \right\rangle$$

$$= \frac{1}{n\varsigma(nai\lambda)} \exp\left(\sum_{k=1}^\infty \frac{p_k(q^{-\rho})p_k(q^\rho)}{k}(t_+ t_-)^k\right) \sum_{k=0}^\infty \tilde{h}_{n+k}(nai\lambda) \cdot \tilde{h}_{-k}(nai\lambda),$$

where $\tilde{h}_{\pm k}(nai\lambda)$ is the specialization of h_k with

$$p_k = p_k(q^{-\pm\rho})t_{\pm}^k\varsigma(knai\lambda) = \frac{(q^{\pm 1/2}t_{\pm}q^{na/2})^k - (q^{\pm 1/2}t_{\pm}q^{-na/2})^k}{1 - q^{\pm k}}.$$

Note when one has the specialization

$$(90) \qquad\qquad p_k = \frac{a^k - b^k}{1 - q^k},$$

one has [16, p.27]:

$$(91) \qquad\qquad h_k = \prod_{j=1}^{k} \frac{a - bq^{j-1}}{1 - q^j}.$$

Therefore,

$$(92) \qquad \tilde{h}_{\pm k}(nai\lambda) = t_{\pm}^k \prod_{j=1}^{k} \frac{q^{\pm 1/2 + na/2} - q^{\pm(j-1/2) - na/2}}{1 - q^{\pm j}}.$$

6. Concluding Remarks

Even though there are still many issues to resolve in the mathematical theory of open Gromov-Witten invariants for an S^1-equivariant pair, the localization techniques together with Hodge integral identities (e.g. Mariño-Vafa formula) and operator manipulations provide an effective way to compute all genus open Gromov-Witten invariants of the resolved conifold with respect to the fixed locus of an antiholomorphic involution. This method clearly can be generalized to local Calabi-Yau geometries coming from toric Fano surfaces with antiholomorphic involutions. We hope to report on this in a subsequent paper. The results of such calculations can be expressed in terms of link invariants arising from Chern-Simons theory.

We have witnessed many deep applications of Professor Chern's work in mathematical physics, and many more will appear in the future. I dedicate this small piece to his memory.

References

[1] M. Aganagic, A. Klemm, M. Marino, C. Vafa, *The topological vertex*, Comm. Math. Phys. 254, No. 2, 425–478, hep-th/0305132.
[2] M. Aganagic, M. Marino, C. Vafa, *All loop topological string amplitudes from Chern-Simons theory*, Commun. Math. Phys. **247** (2004) 467–512, hep-th/0206164.

[3] R. Gopakumar, C. Vafa, *M-Theory and Topological Strings–I*, hep-th/9809187.

[4] T. Graber, R. Pandharipande, *Localization of virtual classes*, Invent. Math. **135** (1999) no. 2, 487–518, alg-geom/9708001.

[5] T. Graber, E. Zaslow, *Open-string Gromov-Witten invariants: Calculations and a mirror "Theorem"*, hep-th/0109075.

[6] A. Iqbal, *All genus topological amplitudes and 5-brane Webs as Feynman diagrams*, preprint, hep-th/0207114.

[7] S. Katz, C.-C. Liu, *Enumerative geometry of stable maps with Lagrangian boundary conditions and multiple covers of the disc*, Adv. Theor. Math. Phys. **5** (2001) 1–49, math.AG/0103074.

[8] M. Kontsevich, *Enumeration of rational curves via torus actions*, in The moduli space of curves (Texel Island, 1994), 335–368, Progr. Math. **129**, Birkhäuser Boston, Boston, MA, 1995, hep-th/9405035.

[9] J. M. F. Labstida, M. Marino, *Polynomial invariants for torus knots and topological strings*, Comm. Math. Phys. 217 (2001) 423–449, hep-th/0004196.

[10] J. M. F. Labstida, M. Marino, C. Vafa, *Knots, links and branes at large N*, JHEP 0011 (2001) 007, hep-th/0010102.

[11] C.-C. Liu, *Moduli of J-holomorphic curves with Lagrangian boundary conditions and open Gromov-Witten invariants for an S^1-equivariant pair*, math.SG/0210257.

[12] J. Li, C.-C. Liu, K. Liu, J. Zhou, *A mathematical theory of the topological vertex*, math.AG/0408426.

[13] J. Li, Y. S. Song, *Open string instantons and relative morphisms*, Adv. Theor. Math. Phys. 5 (2001) no. 1, 67–91, hep-th/0103100.

[14] C.-C. Liu, K. Liu, J. Zhou, *A proof of a conjecture of Mariño-Vafa on Hodge integrals*, J. Differential Geom. **65** (2003), no. 2, 289–340, math.AG/0306434.

[15] C.-C. Liu, K. Liu, J. Zhou, *A formula on two-partition Hodge integrals*, to appear in Journ. AMS, math.AG/0310272.

[16] I. G. MacDonald, Symmetric functions and Hall polynomials, 2nd edition, Clarendon Press, 1995.

[17] M. Marino, C. Vafa, *Framed knots at large N*, hep-th/0108064.

[18] P. Mayr, *Summing up open string instantons and $N = 1$ string amplitudes*, hep-th/0203237.

[19] T. Miwa, M. Jimbo, E. Date, *Solitons. Differential equations, symmetries and infinite-dimensional algebras*, Cambridge Tracts in Mathematics, 135, Cambridge University Press, 2000.

[20] H. R. Morton, S. G. Lukac, *The HOMFLY polynomial of the decorated Hopf link*, Journal of Knot Theory Ramif. 12:3 (2003) 395–416, math.GT/0108011.

[21] A. Okounkov, R. Pandharipande, *Gromov-Witten theory, Hurwitz theory, and completed cycles*, Ann. Math. **163** (2006) 517–560, math.AG/0204305.

[22] A. Okounkov, R. Pandharipande, *The equivariant Gromov-Witten theory of P^1*, Ann. Math. **163** (2006) 561–605, math.AG/0207233.

[23] A. Okounkov, R. Pandharipande, *Hodge integrals and invariants of the un-knots*, Geom. Topol. **8** (2004) 675–699.

[24] H. Ooguri, C. Vafa, *Knot invariants and topological strings*, Nucl. Phys. B, 577 (2000) 419–438, hep-th/9912123.

[25] E. Witten, *Quantum field theory and the Jones polynomial*, Comm. Math. Phys. **121** (1989) no. 3, 351–399.

[26] E. Witten, Chern-Simons gauge theory as a string theory, in *The Floer Memorial Volume*, 637–678, Progr. Math. **133**, Birkhäuser, Basel, 1995.

[27] J. Zhou, *Hodge integrals, Hurwitz numbers, and symmetric groups*, math.AG/0308024.

[28] J. Zhou, *Proof and interpretation of a string duality*, Math. Res. Lett. **11** (2004) 213–229.

[29] J. Zhou, *A conjecture on Hodge integrals*, math.AG/0310282.

[30] J. Zhou, *Localizations on moduli spaces and free field realizations of Feynman rules*, math.AG/0310283.

江泽民主席、李岚清副总理接见中外数学家

ICM2002 北京 大北照相